Electromagnetism

Theory & Problems

Electrodynamics and Plasma Physics

Electromagnetism

Theory & Problems

Electrodynamics and Plasma Physics

DR. S.L. KAKANI
M.SC. (PHYSICS), PH.D.
Principal (Retd.)
Government Post-Graduate College
Shahpura, Bhilwara-311 404 (Rajasthan)
(Formerly Convener BOS and Member Research Board, M.D.S. University, Ajmer)

DR. C. HEMRAJANI
M.SC. (PHYSICS), PH.D.
Formerly Head, Physics Department
MLV Government P.G. College
Bhilwara-311 001 (Rajasthan)

CBS

CBS Publishers & Distributors Pvt. Ltd.

New Delhi • Bengaluru • Chennai • Kochi • Kolkata • Mumbai
Hyderabad • Uttarakhand • Nagpur • Patna • Pune • Jharkhand

ISBN: 81-239-0871-7

First Edition: 2002
Reprint: 2011, 2014, 2018

Published by **Satish Kumar Jain** and produced by **Varun Jain** for
CBS Publishers & Distributors Pvt. Ltd.,
4819/XI Prahlad Street, 24 Ansari Road, Daryaganj, New Delhi - 110002
delhi@cbspd.com, cbspubs@airtelmail.in • www.cbspd.com
Ph.: 23289259, 23266861, 23266867 • Fax: 011-23243014

Corporate Office: 204 FIE, Industrial Area, Patparganj, Delhi - 110 092
Ph: 49344934 • Fax: 011-49344935
E-mail: publishing@cbspd.com • publicity@cbspd.com

Branches:
- ***Bengaluru:*** 2975, 17th Cross, K.R. Road, Bansankari 2nd Stage, Bengaluru - 70 • Ph: +91-80-26771678/79 • Fax: +91-80-26771680 E-mail: cbsbng@gmail.com, bangalore@cbspd.com
- ***Chennai:*** No. 7, Subbaraya Street, Shenoy Nagar, Chennai - 600030 Ph: +91-44-26681266, 26680620 • Fax: +91-44-42032115 E-mail: chennai@cbspd.com
- ***Kochi:*** Ashana House, 39/1904, A.M. Thomas Road, Valanjambalam, Ernakulum, Kochi • Ph: +91-484-4059061-65 Fax: +91-484-4059065 • E-mail: cochin@cbspd.com
- ***Kolkata:*** 6-B, Ground Floor, Rameshwar Shaw Road, Kolkata - 700014 Ph: +91-33-22891126/7/8 • E-mail: kolkata@cbspd.com
- ***Mumbai:*** 83-C, Dr. E. Moses Road, Worli, Mumbai - 400018 Ph: +91-9833017933, 022-24902340/41 • E-mail: mumbai@cbspd.com

Representatives:
- Hyderabad: 0-9885175004
- Patna: 0-9334159340
- Jharkhand: 0-9811541605
- Nagpur: 0-9021734563
- Pune: 0-9623451994
- Uttarakhand: 0-9716462459

Printed at:
J.S. Offset Printers, Delhi (India)

Preface

This textbook covers the courses on electricity, magnetism, electromagnetic waves and fields, special theory of relativity, relativistic electrodynamics and plasma physics for the B.Sc. Physics (Pass as well as Honours), M.Sc. Physics, B.E. (Electrical) and A.M.I.E. students of Indian and other foreign universities. The subject is treated in such a way as to serve a wider readership for the students of science and engineering.

The topics have been written in a clear and concise manner with explanatory diagrams and tables, wherever necessary. The mathematical details have been given so that the students will have no difficulty in following the logical development of the subject. Many problems have been worked out to train the students in the application of the formalism to specific situations. To make the book useful for UGC–CSIR: NET, SLET and other entrance examinations a large number of short-question answers and objective type questions with answers are provided at the end of each chapter. A variety of problems for the practice and review questions are also given at the end of chapters. SI system of units have been used in the book in keeping with the modern practice adopted internationally.

Authors shall deem their labour rewarded if the book wins the appreciation of students and teachers. Suggestions for further improvement of the book are most welcome.

Bhilwara — S.L. KAKANI

15 May, 2002 — C. HEMRAJANI

Contents

Chapter 1

Vector Analysis

1.1 INTRODUCTION

The use of Vector algebra and Vector Calculus makes the language of electromagnatics most precise and concise. Physical quantities are divided into two main classes (i) Scalars and (ii) Vectors.

(i) Scalar quantities

Scalar quantities are those which can be completely characterized by their magnitude and algebraic sign. These quantities do not involve direction. Mass, length, time, pressure, work, temperature, charge, potential etc. are few examples of scalar quantities.

(ii) Vector quantities

A vector is a quantity that is completely characterized by its magnitude and direction. Displacement, velocity, momentum force, current density, electric and magnetic field intensities, magnetic induction, etc. are few examples of vector quantities.

Figure 1.1. Representation of a vector.

A vector is represented on a diagram by an arrow line whose length is proportional to the magnitude and the arrow indicating the direction of the vector (Fig. 1.1). Vectors are further classified as *localized* and *free* vectors. For the complete specification of a localized vector, one has to specify the point at which the vector acts, where as there is no such restriction for free vectors.

We must note that a directed quantity is a vector provided that it obeys the (commutative) law of vector addition, e.g. a large angular displacement about a given axis is not a vector because its addition to another angular displacement about a different axis is not a commutative.

To distinguish vectors from scalars, *bold-face* type letters e.g., **A**, **B**, **a**, **b** may be used for vectors. The magnitude of a vector **A** is denoted by |**A**| or A. The other suitable or usually used notations for vectors are $\vec{A}$, $\vec{B}$, $\vec{a}$, $\vec{b}$, or **A**, **B**, **a**, **b**, etc. In electromagnetics, one usually come across scalar and vector fields. These are discussed in chapter 2.

1.2 VECTOR ALGEBRA

(i) Addition of Vectors

Let us consider two vectors **A** and **B** representing, for example, two successive displacements of a point. One can obtain their sum **C** = **A** + **B** by placing the tail of **B** at the head of **A**, then the sum **C** = **A** + **B**, is the vector from the tail of **A** to the head of **B** (Fig. 1.2). (This rule generalizes the obvious procedure for combining two displacements). Vector addition *commulative*, i.e.

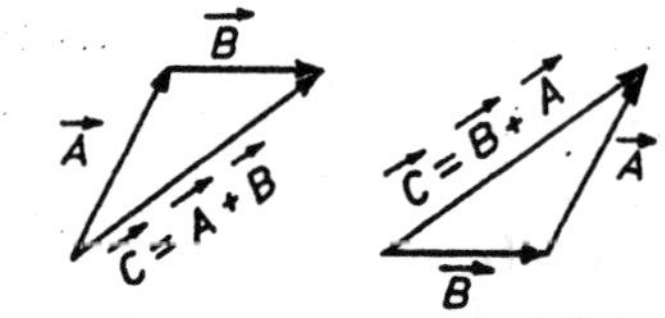

Figure 1.2. Addition of two vectors.

$$\mathbf{A}+\mathbf{B}=\mathbf{B}+\mathbf{A} \tag{1.1}$$

vector sum is independent of the order in which the vectors are taken.

The vector sum is *associative*, i. e. the sum of number of vectors **A**, **B** and **C** is same where we add **A** to the sum of **B** and **C** or we add **B** to the sum of **A** and **C** or we add **C** to the sum of **A** and **B**, i.e.

$$\mathbf{D}=\mathbf{A}+(\mathbf{B}+\mathbf{C})=\mathbf{B}+(\mathbf{A}+\mathbf{C})=\mathbf{C}+(\mathbf{A}+\mathbf{B}) \tag{1.2}$$

The same is evident from Fig. 1.3.

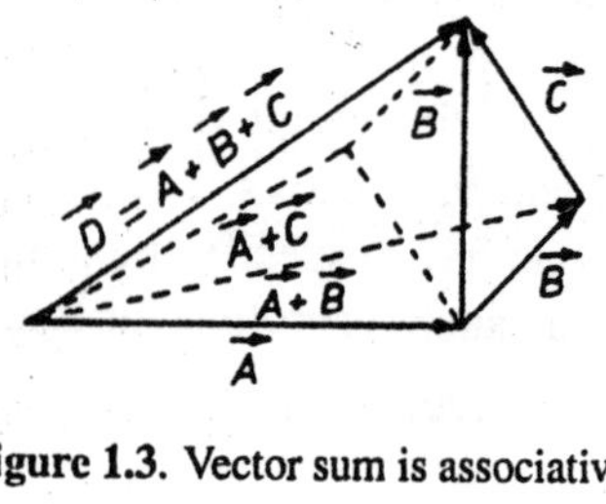

Figure 1.3. Vector sum is associative.

(ii) Subtraction of vectors

Let there are two vectors **A** and **B** and we have to subtract **B** from **A**. Subtraction is just the same thing as adding **−B** to **A**, i.e.

$$\mathbf{C} = \mathbf{A} - \mathbf{B} = \mathbf{A} + (-\mathbf{B}) \tag{1.3}$$

The negative of a vector means a vector of equal magnitude but opposite direction. Obviously, to subtract **B** from **A**, we draw **−B** from the end point of **A** and draw **A** from the initial point of **A** to the end point of **−B** (Fig. 1.4).

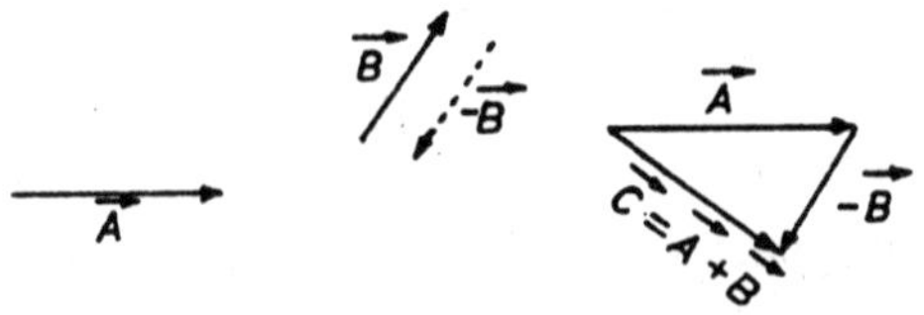

Figure 1.4.

If **A** = **B**, then **A** − **B** = **0**. Thus **0** is called *null vector.*

One can also subtract **B** from **A** as shown in Fig. 1.5. In this process, one has to draw both the given vectors **A** and **B** from the same point, and then draw **C** from the end point of **B** to the end point of **A**. Thus

$$\mathbf{C} = \mathbf{A} + (-\mathbf{B}).$$

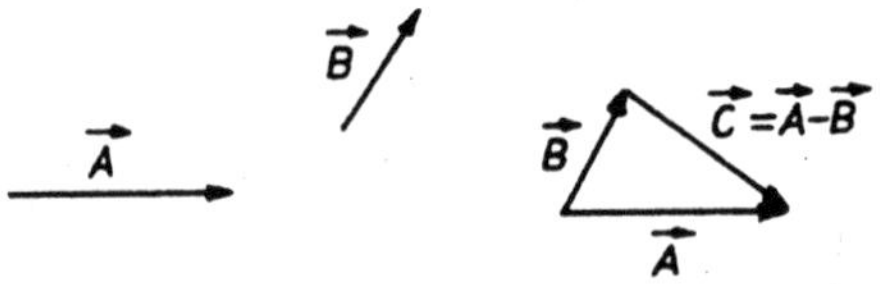

Figure 1.5.

(iii) Multiplication of a vector by a scalar

Multiplication of a vector **A** by a positive scalar a multiplies the *magnitude* but leaves the direction unchanged (Fig. 1.6). If a is negative, the direction is reversed. we may write

$$\mathbf{C} = a\mathbf{A} \tag{1.4}$$

Scalar multiplication is *distributive*, i.e.

$$a(\mathbf{A} + \mathbf{B}) = a\mathbf{A} + a\mathbf{B} \tag{1.5}$$

To divide a vector by a scalar, the vector is multiplied by the reciprocal of the scalar.

When a null vector is multiplied by a finite scalar number, then one obtains the null vector, i.e.

$$n\mathbf{0} = \mathbf{0} \tag{1.6}$$

When a vector is multiplied by 0, one obtains null vector

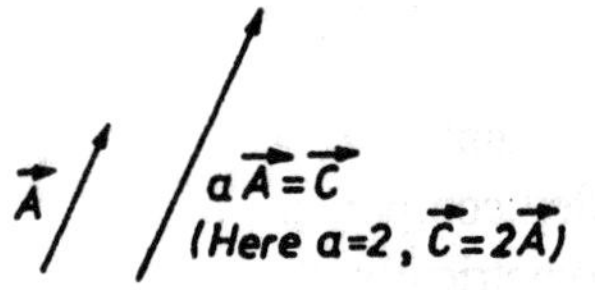

Figure 1.6. Multiplication by a scalar.

$$0\mathbf{A} = \mathbf{0} \tag{1.7}$$

1.3 UNIT VECTOR

A vector **A** may be expressed as $A\hat{a}$, where A is the magnitude of vector **A** and $\hat{a}$ is a vector of unit magnitude and direction that of **A**. The vector $\hat{a}$ is known as an *unit vector*. Obviously,

$$\hat{a} = \frac{\mathbf{A}}{A} \tag{1.8}$$

Thus any vector in the direction of unit vector may be represented by the product of the unit vector and the scalar magnitude of the vector. In cartesian coordinates x, y, z system, usually $\hat{\imath}$, $\hat{\jmath}$ and $\hat{k}$ are taken as unit vectors parallel to x, y and z axes respectively. These are called the *orthogonal unit vectors*. In some texts unit vectors parallel to x, y and z axes in cartesian coordinate system are expressed by $\hat{a}_x$, $\hat{a}_y$ and $\hat{a}_z$ respectively. Unit vectors $\hat{\imath}$, $\hat{\jmath}$ and $\hat{k}$ are also termed as *basis vectors*.

1.4 RESOLUTION OF A VECTOR

If $\mathbf{C} = \mathbf{A} + \mathbf{B}$, then **A** and **B** are said to be the components of **C**. In other words, **C** may be resolved into its component vectors **A** and **B**. Obviously, a given vector can be resolved into components in a variety of ways. The components of a vector **A** are any number of vectors whose algebric sum is **A**. The components most frequently used, are those parallel to x, y and z axes of the right handed cartesian coordinate system and they are called *rectangular* components of a vector.

Let the origin O be one extremity of vector **A**. Now draw a rectangular parallelopiped with the three edges which meet at O lying along the three axes x, y and z and such that **A** is the diagonal through O (Fig.

1.7). Let $\hat{i}$, $\hat{j}$, and $\hat{k}$ unit vectors parallel to x, y and z axes respectively. The vector **A** can be expanded in terms of these basis vectors

$$\mathbf{A} = A_x\hat{i} + A_y\hat{j} + A_z\hat{k} \tag{1.9}$$

The number A_x, and A_y and A_z are called components of **A**, geometrically, they are the projections of **A** along the three coordinate axes.

By geometry of the figure, we have

$$\mathbf{A}_x = A_x\hat{i},\ \mathbf{A}_y\hat{j} \quad \text{and} \quad \mathbf{A}_z = A_z\hat{k}$$

Thus

$$A^2 = A_x^2 + A_y^2 + A_z^2$$

or $$A = \sqrt{A_x^2 + A_y^2 + A_z^2} \tag{1.10}$$

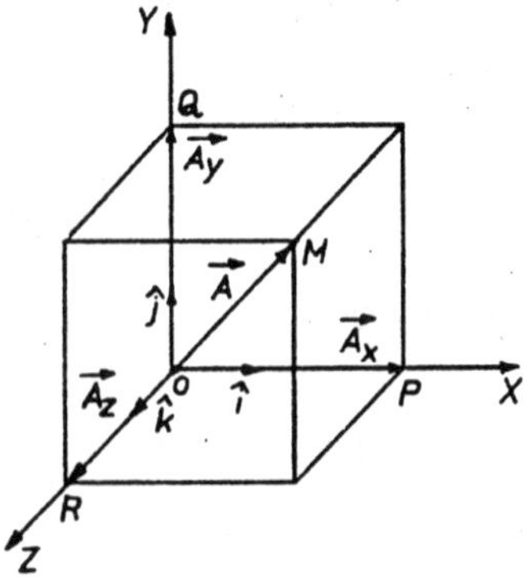

Figure 1.7.

If α, β, γ the angles between the vector **A** and the positive directions of x, y and z axes respectively, then cosine of those angles are called *the direction cosines*. From Fig. 1.7, one can write

$$\begin{aligned} \cos\alpha &= \frac{A_x}{|\mathbf{A}|} = \frac{A_x}{A} \\ \cos\beta &= \frac{A_y}{|\mathbf{A}|} = \frac{A_y}{A} \\ \text{and} \quad \cos\gamma &= \frac{A_z}{|\mathbf{A}|} = \frac{A_y}{A} \end{aligned} \tag{1.11}$$

Obviously,

$$A_x = A\cos\alpha, A_y = A\cos\beta \quad \text{and} \quad A_z = A\cos\gamma \tag{1.11a}$$

and

$$\mathbf{A} = A[\hat{i}\cos\alpha + \hat{j}\cos\beta + \hat{k}\cos\gamma] \tag{1.12}$$

The unit vector $\hat{a}$ corresponding to vector **A** is given by

$$\hat{a} = \frac{\mathbf{A}}{A} = \hat{i}\cos\alpha + \hat{j}\cos\beta + \hat{k}\cos\gamma \tag{1.13}$$

Using 1.11(a), one obtains from (1.10)

$$\cos^2\alpha + \cos^2\beta + \cos^2\gamma = 1 \tag{1.14}$$

Obviously, *the sum of the square of three direction cosines is equal to unity*. Further any three numbers proportional to the direction cosines of a line having the same proportional constant are called *direction ratios*. It a, b and c be the direction ratios of a line and $\cos\alpha$, $\cos\beta$ and $\cos\gamma$ its direction cosines, then

$$\begin{aligned}\frac{\cos\alpha}{a} &= \frac{\cos\beta}{b} = \frac{\cos\gamma}{c} \\ &= \frac{\sqrt{\cos^2\alpha+\cos^2\beta+\cos^2\gamma}}{\sqrt{a^2+b^2+c^2}} = \frac{1}{\sqrt{a^2+b^2+c^2}}\end{aligned} \tag{1.15}$$

This means

$$\begin{aligned}\cos\alpha &= \frac{a}{\sqrt{a^2+b^2+c^2}} \\ \cos\beta &= \frac{b}{\sqrt{a^2+b^2+c^2}} \\ \text{and} \quad \cos\gamma &= \frac{c}{\sqrt{a^2+b^2+c^2}}\end{aligned} \tag{1.16}$$

1.5 PRODUCTS OF VECTORS

The product of two vectors is defined in two different ways:

(i) Scalar or Dot product

The scalar or dot product of two vectors **A** and **B** (written as $\mathbf{A}\cdot\mathbf{B}$ and read as **A** dot **B**) is defined as the magnitude of product of the magnitude of **A** and **B** and the cosine of the smaller angle between them. Thus

$$\mathbf{A}\cdot\mathbf{B} = AB\cos\theta \tag{1.17}$$

The scalar product $AB\cos\theta$ of **A** and **B** may be considered as either the multiplication of the magnitude of $|\mathbf{A}|$ and the projection of **B** along **A** [$AB\cos\theta = A(B\cos\theta)$ as shown in Fig. 1.8(a)] or the multiplication of the magnitude of **B** and the projection of **A** along **B** [$AB\cos\theta = B(A\cos\theta)$ as shown in Fig. 1.8(b)].

It is clear that the scalar multiplication of two vectors gives a scalar. Obviously, the commutative law of multiplication holds, i.e.

$$\mathbf{A}\cdot\mathbf{B} = \mathbf{B}\cdot\mathbf{A} = AB\cos\theta \tag{1.18}$$

If **A** and **B** are parallel to each other and pointing in the same direction, then $\theta = 0$. Thus

$$\mathbf{A}\cdot\mathbf{A} = A^2 \tag{1.19}$$

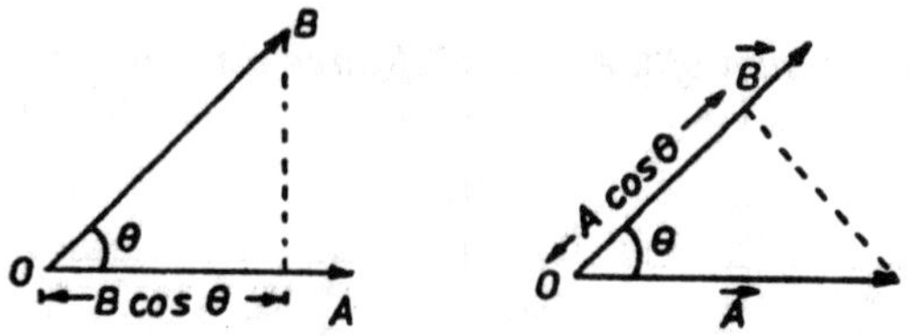

Figure 1.8.

If this property is applied to unit vectors $\hat{i}$, $\hat{j}$, $\hat{k}$, which are mutually perpendicular to each other, then

$$\hat{i}\cdot\hat{i} = \hat{j}\cdot\hat{j} = \hat{k}\cdot\hat{k} = 1 \tag{1.20}$$

In case, if two vectors are perpendicular to each other, then $\cos\theta = \cos\frac{\pi}{2} = 0$. This means

$$\mathbf{A}\cdot\mathbf{B} = AB\cos\frac{\pi}{2} = 0 \tag{1.21}$$

This is the condition for two orthogonal vectors. If we apply this property to unit vectors, $\hat{i}$, $\hat{j}$, $\hat{k}$, then

$$\hat{i}\cdot\hat{j} = \hat{j}\cdot\hat{k} = \hat{k}\cdot\hat{i} = 0 \tag{1.22}$$

Obviously,

$$\begin{aligned}\mathbf{A}\cdot\mathbf{B} &= (A_x\hat{i}+A_y\hat{j}+A_z\hat{k})\cdot(B_x\hat{i}+B_y\hat{j}+B_z\hat{k})\\ &= A_xB_x + A_yB_y + A_zB_z \end{aligned}\tag{1.23}$$

Since the scalar product of two vectors is a scalar, the physical scalar quantity like work, gravitational potential energy, electric potential, electric power etc. can be described as scalar products of two vectors. Eq. 1.23 reveals that the scalar product of two vectors is equal to the sum of the products of their corresponding x, y and z components.

The scalar magnitude of **A** and **B** are $\sqrt{A_x^2+A_y^2+A_z^2}$ and $\sqrt{B_x^2+B_y^2+B_z^2}$. Therefore the angle between **A** and **B** is given by

$$\cos(\mathbf{A},\mathbf{B}) = \frac{\mathbf{A}\cdot\mathbf{B}}{AB} = \frac{A_xB_x+A_yB_y+A_zB_z}{\sqrt{A_x^2+A_y^2+A_z^2}\sqrt{B_x^2+B_y^2+B_z^2}} \tag{1.24}$$

(ii) Vector Product or Cross Product

The vector product of two vectors (**A** and **B**) written with a cross between two vectors as $\mathbf{A}\times\mathbf{B}$ and read "**A** cross **B**", is a vector having

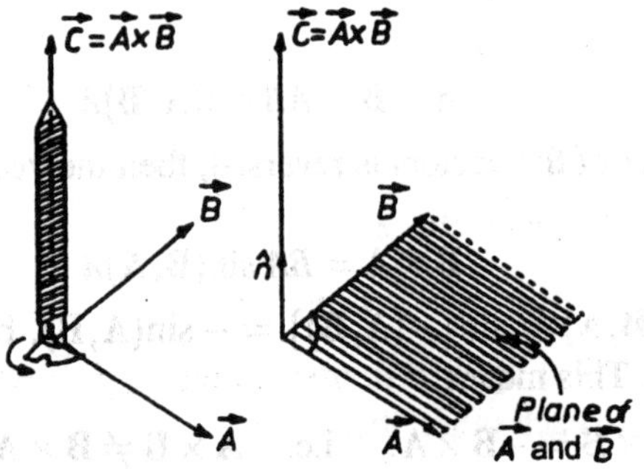

Figure 1.9. Vector product of two vectors

a magnitude equal to the product of the magnitudes of the two vectors and the sine of the angle between them, and a direction perpendicular to the plane containing the tow vectors. Thus the vector product of two vectors **A** and **B** is defined as

$$\mathbf{C} = \mathbf{A} \times \mathbf{B} = AB \sin(\mathbf{A}, \mathbf{B}) \hat{n} \tag{1.25}$$

where A and B are the magnitudes of **A** and **B**; (**A**, **B**) is the angle between them, and $\hat{n}$ is a unit vector perpendicular to the plane of **A** and **B**.

Since the sine of an angle between 0 and π is not negative, the vector product cannot be negative. We must remember that scalar product can be negative.

The direction of $\mathbf{C}(= \mathbf{A} \times \mathbf{B})$ is perpendicular to the plane containing **A** and **B** and in the sense of advance of a right hand screw rotated from **A** (first vector) to **B** (second vector) through the smaller angle between them (Fig. 1.9). Obviously, if a right handed screw whose axis is perpendicular to the plane formed by **A** and **B** is rotated from **A** to **B** through the *smaller* angle between them, then the direction of advance of the screw gives the direction of **C**.

The vector product of two vectors is a vector. Torque, angular momentum, flow of electromagnetic energy, force on a moving charge in a magnetic field etc. are few examples of vector products of two vectors.

Properties of vector products

(a) The vector product is not commutative i.e.,

$$\mathbf{A} \times \mathbf{B} \neq \mathbf{B} \times \mathbf{A} \tag{1.26}$$

we have

$$\mathbf{A}\times\mathbf{B}=AB\sin(\mathbf{A},\mathbf{B})\hat{n}$$

If the order of the vectors is reversed, then the vector product will be

$$\mathbf{B}\times\mathbf{A}=BA\sin(\mathbf{B},\mathbf{A})\hat{n}$$

Now $\sin(\mathbf{B},\mathbf{A})=\sin\{-(\mathbf{A},\mathbf{B})\}=-\sin(\mathbf{A},\mathbf{B})$, because $\sin(-\theta) = -\sin\theta$. This means

$$\mathbf{A}\times\mathbf{B}=-\mathbf{B}\times\mathbf{A},\quad \text{i.e.}\quad \mathbf{A}\times\mathbf{B}\neq\mathbf{B}\times\mathbf{A} \tag{1.26a}$$

Obviously, the vector product of two vectors is not commutative and the order of the vectors must be strictly maintained.

(b) *The vector productive is distributive*

$$\mathbf{A}\times(\mathbf{B}+\mathbf{C})=\mathbf{A}\times\mathbf{B}+\mathbf{A}\times\mathbf{C} \tag{1.27}$$

However, the order of the vectors must remain unchanged.

Let us express **A**, **B**, and **C** in terms of their x, y and z components.

$$\begin{aligned}\mathbf{A}\times(\mathbf{B}+\mathbf{C}) &= (A_x\hat{i}+A_y\hat{j}+A_z\hat{k})\times[(B_x\hat{i}+B_y\hat{j}+B_z\hat{k})\\ &\qquad +(C_x\hat{i}+C_y\hat{j}+C_z\hat{k})]\\ &= (A_x\hat{i}+A_y\hat{j}+A_z\hat{k})[(B_x+C_x)\hat{i}+(B_y+C_y)\hat{j}\\ &\qquad +(B_z+C_z)\hat{k}]\\ &= \begin{vmatrix}\hat{i} & \hat{j} & \hat{k}\\ A_x & A_y & A_z\\ (B_x+C_x) & (B_y+C_y) & (B_z+C_z)\end{vmatrix}\\ &= \begin{vmatrix}\hat{i} & \hat{j} & \hat{k}\\ A_x & A_y & A_z\\ B_x & B_y & B_z\end{vmatrix}+\begin{vmatrix}\hat{i} & \hat{j} & \hat{k}\\ A_x & A_y & A_z\\ C_x & C_y & C_z\end{vmatrix}\\ &= \mathbf{A}\times\mathbf{B}+\mathbf{A}\times\mathbf{C}\end{aligned}$$

one can easily see that the above is true for any number of vectors.

(c) If either vector is multiplied by a scalar (say p) the vector product is also multiplied by a scalar

$$(p\mathbf{A})\times\mathbf{B}=(\mathbf{A}\times p\mathbf{B})=pAB\sin(\mathbf{A},\mathbf{B})\hat{n}$$

(d) The magnitude of the vector product of two vectors mutually at right angles is equal to the product of the magnitude of the vectors. Thus

$$\mathbf{A}\times\mathbf{B}=AB\sin(\mathbf{A},\mathbf{B})\hat{n}=AB\sin(90^\circ)\hat{n}=AB\hat{n}.$$

Obviously the two vectors **A** and **B** and their vector product **A** × **B** are mutually at right angles.

(e) A vector product of two parallel vectors is a null vector (or zero). Thus when two vectors **A** and **B** are parallel to each other, then the angle between them is 0° or 180°. We have

$$\mathbf{A} \times \mathbf{B} = AB\sin(\mathbf{A},\mathbf{B})\hat{n} = AB(\sin 0°)\hat{n} = \mathbf{0} \text{ (null vector)}$$

∴ if **A** × **B** = **0** and **A** and **B** are not null vectors, then **A** and **B** parallel.

(f) The vector product of a vector by itself is a null vector, i.e.

$$\mathbf{A} \times \mathbf{A} = 0 \tag{1.28}$$

we have

$$\mathbf{A} \times \mathbf{A} = AA\sin(\mathbf{A},\mathbf{A})\hat{n} = AA\sin(0°)\hat{n} = \mathbf{0}$$

(g) The vector product of unit orthogonal vectors $\hat{i}$, $\hat{j}$, $\hat{k}$ have the following relations in the right handed coordinate system

$$\hat{i}\times\hat{i} = \hat{j}\times\hat{j} = \hat{k}\times\hat{k} = 0 \tag{1.29}$$

and

$$\begin{aligned} \hat{i}\times\hat{j} &= -\hat{j}\times\hat{i} = \hat{k} \\ \hat{j}\times\hat{k} &= -\hat{k}\times\hat{j} = \hat{i} \\ \hat{k}\times\hat{i} &= -\hat{i}\times\hat{k} = \hat{j} \end{aligned} \tag{1.30}$$

The magnitude of each of unit vector $\hat{i}$, $\hat{j}$ and $\hat{k}$ is 1, and they are along the positive directions of the right-handed cartesian axes x, y and z respectively (Fig. 1.10). Obviously, the angle between the two of them is 90°. Therefore, one obtains

$$\hat{i}\times\hat{j} = (1)(1)\sin 90°\hat{n} = \hat{n}$$

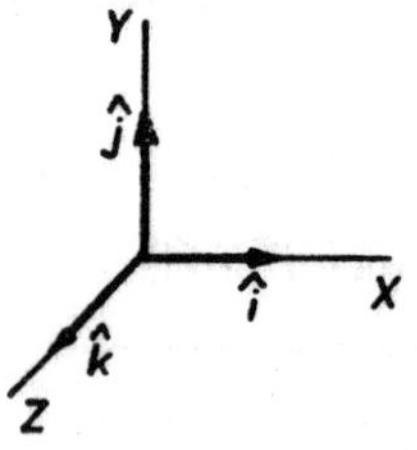

Figure 1.10.

where $\hat{n}$ is a unit vector perpendicular to the plane of $\hat{\imath}$ and $\hat{\jmath}$, i.e. it is just the third unit vector $\hat{k}$, Thus

$$\hat{\imath}\times\hat{\jmath}=\hat{k}$$

Now

$$\begin{aligned} \hat{\imath}\times\hat{\jmath} &= -\hat{\jmath}\times\hat{\imath} \\ \therefore \quad \hat{\imath}\times\hat{\jmath} &= -\hat{\jmath}\times\hat{\imath}=\hat{k} \end{aligned}$$

Similarly, one finds $\hat{\jmath}\times\hat{k}=-\hat{k}\times\hat{\jmath}=\hat{\imath}$ and $\hat{k}\times\hat{\imath}=-\hat{\imath}\times\hat{k}=\hat{\jmath}$. Further, $\hat{\imath}\times\hat{\imath}=(1)(1)\sin 0°\hat{n}=\mathbf{0}=0$. Similarly $\hat{\jmath}\times\hat{\jmath}=0$ and $\hat{k}\times\hat{k}=0$.

(h) The vector product of two vectors in term of their x, y and z components can be expresses as a determinant we have

$$\begin{aligned} \mathbf{A} &= A_x\hat{\imath}+A_y\hat{\jmath}+A_z\hat{k} \\ \text{and} \quad \mathbf{B} &= B_x\hat{\imath}+B_y\hat{\jmath}+B_z\hat{k} \\ \therefore \quad \mathbf{A}\times\mathbf{B} &= (A_x\hat{\imath}+A_y\hat{\jmath}+A_z\hat{k}) \\ &= (A_yB_z-A_zB_y)\hat{\imath}+(A_z+B_x-A_xB_z)\hat{\jmath} \\ &\quad +(A_xB_y-A_yB_x)\hat{k} \end{aligned} \tag{1.31}$$

This cumbersome expression can be written more neatly as a determinant, i.e.

$$\mathbf{A}\times\mathbf{B} = \begin{vmatrix} \hat{\imath} & \hat{\jmath} & \hat{k} \\ A_x & A_y & A_z \\ B_x & B_y & B_z \end{vmatrix} \tag{1.32}$$

$$= \hat{\imath}\begin{vmatrix} A_y & A_z \\ B_y & B_z \end{vmatrix} - \hat{\jmath}\begin{vmatrix} A_x & A_z \\ B_x & B_z \end{vmatrix} + \hat{k}\begin{vmatrix} A_x & A_y \\ B_x & B_y \end{vmatrix} \tag{1.33}$$

1.6 TRIPLE PRODUCTS

Since the cross product of two vectors is itself a vector, it can be dotted or crossed with a third vector to form a triple product.

(i) Scalar triple product

$$\mathbf{A}\cdot(\mathbf{B}\times\mathbf{C}).$$

Geometrically, $|\mathbf{A}\cdot(\mathbf{B}\times\mathbf{C})|$ is the volume of the parallelopiped generated by **A**, **B** and **C**, Since $|\mathbf{B}\times\mathbf{C}|$ is the area of the base, and $|A\cos\theta|$ is the altitude (Fig. 1.11), evidently,

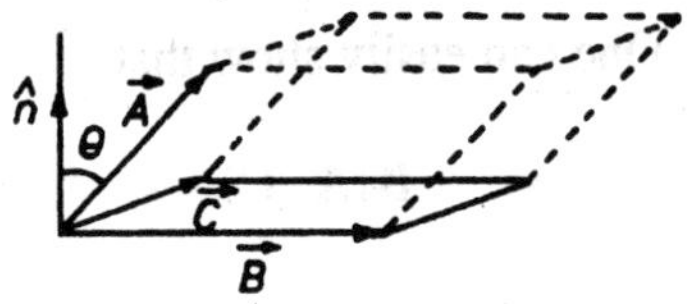

Figure 1.11.

$$\begin{aligned}\mathbf{A}\cdot(\mathbf{B}\times\mathbf{C}) &= \mathbf{B}\cdot(\mathbf{C}\times\mathbf{A}) \\ &= \mathbf{C}\cdot(\mathbf{A}\times\mathbf{B})\end{aligned} \quad (1.34)$$

for they all correspond to the same figure. Note that "alphabetical" order is preserved-in view of Eq. 1.26(a), the "nonalphabetical" triple products

$$\mathbf{A}\cdot(\mathbf{C}\times\mathbf{B}) = \mathbf{B}\cdot(\mathbf{A}\times\mathbf{C}) = \mathbf{C}\cdot(\mathbf{B}\times\mathbf{A})$$

have the opposite sign. In component form

$$\mathbf{A}\cdot(\mathbf{B}\times\mathbf{C}) = \begin{vmatrix} A_x & A_y & A_z \\ B_x & B_y & B_z \\ C_x & C_y & C_z \end{vmatrix} \quad (1.35)$$

We must note that the dot and cross products can be interchanged

$$\mathbf{A}\cdot(\mathbf{B}\times\mathbf{C}) = (\mathbf{A}\times\mathbf{B})\cdot\mathbf{C} \quad (1.36)$$

Like a vector **A**, the product $\mathbf{A}\times\mathbf{C}$ is also a vector. Obviously, their scalar product $\mathbf{A}\cdot(\mathbf{B}\times\mathbf{C})$ gives a scalar. It is equal in magnitude to the volume of the parallelopiped formed by the three vectors **A**, **B** and **C**

We must note that the placement of the parentheses is critical, i.e., $(\mathbf{A}\cdot\mathbf{B})\times\mathbf{C}$ is a nonsensical expression. One cannot make a cross product from a scalar and a vector.

(ii) Vector triple product

By the property of the vector product $\mathbf{A}\times(\mathbf{B}\times\mathbf{C})$ is a vector perpendicular to **A** and also to $(\mathbf{B}\times\mathbf{C})$. Since $(\mathbf{B}\times\mathbf{C})$ is perpendicular to the plane of **B** and **C** and hence $\mathbf{A}\times(\mathbf{B}\times\mathbf{C})$ must lie in the plane of **B** and **C** and perpendicular to **A**. We must note that

$$\mathbf{A}\times(\mathbf{B}\times\mathbf{C}) \neq (\mathbf{A}\times\mathbf{B})\times\mathbf{C}$$

i.e. the associative law does not apply to vector triple products. Obviously, the parenthesis in vector triple product is very important and one cannot avoid parenthesis in vector triple product as we can do in scalar triple product. One can easily show that

$$\mathbf{A}\times(\mathbf{B}\times\mathbf{C})=\mathbf{B}(\mathbf{A}\cdot\mathbf{C})-\mathbf{C}(\mathbf{A}\cdot\mathbf{B}) \tag{1.37}$$

$$\begin{aligned}
\mathbf{A}\times\mathbf{B}\times\mathbf{C} &= (A_x\hat{i}+A_y\hat{j}+A_z\hat{k})\times\begin{vmatrix}\hat{i} & \hat{j} & \hat{k}\\ B_x & B_y & B_z\\ C_x & C_y & C_z\end{vmatrix}\\
&= (A_x\hat{i}+A_y\hat{j}+A_z\hat{k})[\hat{i}(B_yC_z-B_zC_y)\\
&\quad -\hat{j}(B_xC_z-B_zC_x)+\hat{k}(B_xC_y-B_yC_x)]\\
&= \begin{vmatrix}\hat{i} & \hat{j} & \hat{k}\\ A_x & A_y & A_z\\ B_yC_z-B_zC_y & B_zC_x-B_xC_z & B_xC_y-B_yC_x\end{vmatrix}\\
&= \hat{i}(A_yB_xC_y-A_yB_yC_x-A_zB_zC_x+A_zB_xC_z)\\
&\quad -\hat{j}(A_xB_xC_y-A_xB_yC_x-A_zB_yC_z+A_zB_zC_y)\\
&\quad +\hat{k}(A_xB_zC_x-A_xB_xC_z-A_yB_yC_z+A_yB_zC_y)\\
&= \hat{i}B_x(A_yC_y+A_zC_z+A_xC_x)-\hat{i}C_x(A_yB_y+A_zB_z+A_xB_x)\\
&\quad +\hat{j}B_y(A_xC_x+A_zC_z+A_xC_x)-\hat{j}C_y(A_xB_x+A_zB_z+A_yB_y)\\
&\quad +\hat{k}B_z(A_xC_x+A_yC_y+A_zC_z)-\hat{k}C_z(A_xB_x+A_yB_y+A_zB_z)\\
&= (\hat{i}B_x+\hat{j}B_y+\hat{k}B_z)(A_xC_x+A_yC_y+A_zC_z)\\
&\quad -(\hat{i}C_x+\hat{j}C_y+\hat{k}C_z)(A_xB_x+A_yB_y+A_zC_z)\\
&= \mathbf{B}(\mathbf{A}\cdot\mathbf{C})-\mathbf{C}(\mathbf{A}-\mathbf{B})
\end{aligned}$$

One can put the vector triple product in the determinant form as follows:

$$\mathbf{A}\times(\mathbf{B}\times\mathbf{C})=\begin{vmatrix}\mathbf{B} & \mathbf{C}\\ \mathbf{A}\cdot\mathbf{B} & \mathbf{A}\cdot\mathbf{C}\end{vmatrix} \tag{1.38}$$

one can easily verify that the following cyclic relation in vector triple product holds good

$$\mathbf{A}\times(\mathbf{B}\times\mathbf{C})+\mathbf{B}\times(\mathbf{C}\times\mathbf{A})+\mathbf{C}\times(\mathbf{A}\times\mathbf{B})=0 \tag{1.39}$$

1.7 VECTOR DERIVATIVES

(i) *Differentiation of a vector to a scalar*

Let us consider a particle moving along a curve (Fig. 1.12) so that at any instant and it is at the point P-whose position vector is $\mathbf{r}$ relative to an origin O. Obviously, $\mathbf{r}$ is a function of scalar variable t. As t changes to $t+\delta t$, $\mathbf{r}$ changes to $\mathbf{r}+\delta\mathbf{r}$. Clearly, $\frac{\delta r}{\delta t}$ is the average rate of change of $\mathbf{r}$ with t, which is a vector in the direction of $\delta\mathbf{r}$. One can define the differential coefficient, or the derivative of the position vector $\mathbf{r}$ w.r.t. to time t as

$$\frac{d\mathbf{r}}{dt} = \lim_{\delta t\to 0}\frac{\delta\mathbf{r}}{\delta t} \qquad (1.40)$$

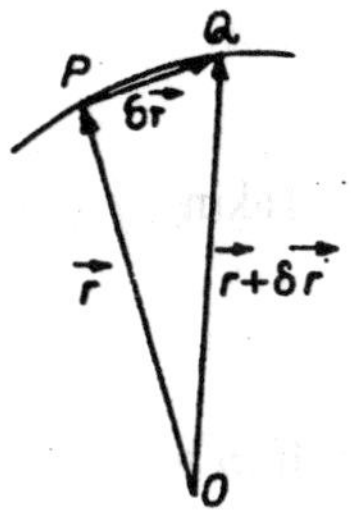

Figure 1.12.

Obviously, the derivative of a vector with respect to a scalar variable is itself a vector. Here $\frac{d\mathbf{r}}{dt}$ is the velocity $\mathbf{V}$ of the particle and $\mathbf{V}$ is a vector. The derivative of $\mathbf{V}$ with respect to time i.e., $\frac{d\mathbf{V}}{dt}$ is the acceleration $\mathbf{a}$ which is also a vector.

If we express $\mathbf{r}$, in rectangular coordinates, then

$$\mathbf{r} = x\hat{i}+y\hat{j}+z\hat{k}$$

$$\therefore \quad \frac{d\mathbf{r}}{dt} = \mathbf{V} = \frac{dx}{dt}\hat{i}+\frac{dy}{dt}\hat{j}+\frac{dz}{dt}\hat{k}$$

and

$$\frac{d^2\mathbf{r}}{dt^2} = \mathbf{a} = \frac{d^2x}{dt^2}\hat{i}+\frac{d^2y}{dt^2}\hat{j}+\frac{d^2z}{dt^2}\hat{k}$$

(ii) *Differentiation of Sum of vectors*

We can see that the general rules of ordinary differential calculus hold good for differentiation of sums and products of vector quantities also. If $\mathbf{A}$ and $\mathbf{B}$ are functions of scalar variable t, then we have

(a) $$\frac{d}{dt}(A+B) = \frac{d\mathbf{A}}{dt}+\frac{d\mathbf{B}}{dt} \qquad (1.41)$$

one can easily prove it. Let

$$\mathbf{V} = \mathbf{A}+\mathbf{B}$$

when t changes to $t+\delta t$, then $\mathbf{A}$ changes to $\mathbf{A}+\delta\mathbf{A}$ and $\mathbf{B}$ to

B + δ**B**. We have

$$\begin{aligned} \mathbf{V}+\delta\mathbf{V} &= (\mathbf{A}+\delta\mathbf{A})+(\mathbf{B}+\delta\mathbf{B}) \\ \therefore \qquad \delta\mathbf{V} &= \delta\mathbf{A}+\delta\mathbf{B} \end{aligned}$$

Dividing by δt, one obtains

$$\frac{\delta\mathbf{V}}{\delta t}=\frac{\delta\mathbf{A}}{\delta t}+\frac{\delta\mathbf{B}}{\delta t}$$

Taking the limit $\delta t \to 0$, one obtains

$$\frac{d\mathbf{v}}{dt}=\frac{d\mathbf{A}}{dt}+\frac{d\mathbf{B}}{dt}=\frac{d}{dt}(\mathbf{A}+\mathbf{B})$$

(b) If m be a scalar function of t, then

$$\frac{d}{dt}(m\mathbf{A})=m\frac{d\mathbf{A}}{dt}+\frac{dm}{dt}\mathbf{A} \qquad (1.42)$$

when t changes to $t+\delta t$, then m changes to $m+\delta m$ and **A** to **A** + δ**A**. Obviously,

$$\begin{aligned} \mathbf{V}+\delta\mathbf{V} &= (m+\delta m)(\mathbf{A}+\delta\mathbf{A}) \\ &= m\mathbf{A}+m\delta\mathbf{A}+\delta m\mathbf{A}+\delta m\delta\mathbf{A} \\ &= \mathbf{V}+m\delta\mathbf{A}+\delta m\mathbf{A} \end{aligned}$$

Here we have ignored the smallar term $\delta m \delta\mathbf{A}$. Thus we have

$$\delta\mathbf{V}=m\delta\mathbf{A}+\delta m\mathbf{A}$$

Dividing by δt, one obtains

$$\frac{\delta\mathbf{V}}{\delta t}=m\frac{\delta\mathbf{A}}{\delta t}+\frac{\delta m}{\delta t}\mathbf{A}$$

Taking the limit $\delta t \to 0$, one obtains

$$\frac{d\mathbf{V}}{dt}=\frac{d}{dt}(m\mathbf{A})=m\frac{d\mathbf{A}}{dt}+\frac{dm}{dt}\mathbf{A}$$

(c)

$$\frac{d}{dt}(\mathbf{A}\cdot\mathbf{B})=\mathbf{A}\cdot\frac{d\mathbf{B}}{dt}+\frac{d\mathbf{A}}{dt}\cdot\mathbf{B} \qquad (1.43)$$

Let $\mathbf{C}=\mathbf{A}\cdot\mathbf{B}$. When t changes to $t+\delta t$, then **A** changes to **A** + δ**A** and **B** changes to **B** + δ**B**. Obviously,

$$\begin{aligned} \mathbf{C}+\delta\mathbf{C} &= (\mathbf{A}+\delta\mathbf{A})\cdot(\mathbf{B}+\delta\mathbf{B}) \\ &= \mathbf{A}\cdot\mathbf{B}+\mathbf{A}\cdot\delta\mathbf{B}+\delta\mathbf{A}\cdot\mathbf{B}+\delta\mathbf{A}\cdot\delta\mathbf{B}. \end{aligned}$$

Here we have ignored the smaller term $\delta\mathbf{A}\cdot\delta\mathbf{B}$. Thus

$$\delta\mathbf{C} = \mathbf{A}\cdot\delta\mathbf{B}+\delta\mathbf{A}\cdot\delta\mathbf{B}$$

or $$\frac{\delta\mathbf{C}}{\delta t} = \mathbf{A}\cdot\frac{\delta\mathbf{B}}{\delta t}+\frac{\delta\mathbf{A}}{\delta t}\cdot\mathbf{B}$$

In the limit $\delta t \to 0$, one obtains

$$\frac{d\mathbf{C}}{dt} = \frac{d}{dt}(\mathbf{A}\cdot\mathbf{B}) = \mathbf{A}\cdot\frac{d\mathbf{B}}{dt}+\frac{d\mathbf{A}}{dt}\cdot\mathbf{B}$$

we may note that the order of factors in any of the above terms is immaterial.

(d) $$\frac{d}{dt}(\mathbf{A}\times\mathbf{B}) = \mathbf{A}\times\frac{d\mathbf{B}}{dt}+\frac{d\mathbf{A}}{dt}\times\mathbf{B} \qquad (1.44)$$

Let $\mathbf{C} = \mathbf{A}\times\mathbf{B}$. When t changes to $t+\delta t$, then $\mathbf{A}$ changes to $\mathbf{A}+\delta\mathbf{A}$ and $\mathbf{B}$ to $\mathbf{B}+\delta\mathbf{B}$. We have

$$\begin{aligned}\mathbf{C}+\delta\mathbf{C} &= (\mathbf{A}+\delta\mathbf{A})\times(\mathbf{B}+\delta\mathbf{B})\\ &= \mathbf{A}\times\mathbf{B}+\mathbf{A}\times\delta\mathbf{B}+\delta\mathbf{A}\times\mathbf{B}+\delta\mathbf{A}\times\delta\mathbf{B}\\ &= \mathbf{C}+\mathbf{A}\times\delta\mathbf{B}+\delta\mathbf{A}\times\mathbf{B}\end{aligned}$$

we have neglected the smaller term $\delta\mathbf{A}\times\delta\mathbf{B}$. Thus

$$\delta\mathbf{C} = \mathbf{A}\times\delta\mathbf{B}+\delta\mathbf{A}\times\mathbf{B}$$

or $$\frac{\delta\mathbf{C}}{\delta t} = \mathbf{A}\times\frac{\delta\mathbf{B}}{\delta t}+\frac{\delta\mathbf{A}}{\delta t}\times\mathbf{B}$$

In the limit $t \to 0$, one obtains

$$\frac{d\mathbf{C}}{dt} = \frac{d}{dt}\mathbf{A}\times\mathbf{B} = \mathbf{A}\times\frac{d\mathbf{B}}{dt}+\frac{d\mathbf{A}}{dt}\times\mathbf{B}$$

We must note that the order of factors in each term should remain unchanged.

1.8 PARTIAL DIFFERENTIATION OF VECTORS

Consider that $\mathbf{D}$ is a vector which depends on more than one scalar variable such as cartesian coordinates x, y, z of a point in space. Thus $\frac{\partial\mathbf{A}}{\partial x}$, $\frac{\partial\mathbf{A}}{\partial y}$ and $\frac{\partial\mathbf{A}}{\partial z}$ are the partial derivatives of vector $\mathbf{A}$ with respect x (when y and z remain constant), y (when x and z remain constant) and z (when x and y remain constant) respectively. Let us consider that if x, y and z change simultaneously by small increments dx, dy and dz, then the total change or total derivative of **A** will be

$$\begin{aligned} d\mathbf{A} &= \frac{\partial \mathbf{A}}{\partial x}dx+\frac{\partial \mathbf{A}}{\partial y}dy+\frac{\partial \mathbf{A}}{\partial z}dz \\ &= \left(\frac{\partial}{\partial x}dx+\frac{\partial}{\partial y}dy+\frac{\partial}{\partial z}dz\right)\mathbf{A} \\ &= \left(\frac{\partial}{\partial x}\hat{i}+\frac{\partial}{\partial y}\hat{j}+\frac{\partial}{\partial z}\hat{k}\right)\cdot(dx\hat{i}+dy\hat{j}+dz\hat{k})\mathbf{A} \\ &= (\nabla\cdot d\mathbf{r})\mathbf{A} \end{aligned}$$

Here $\nabla = \frac{\partial}{\partial x}\hat{i}+\frac{\partial}{\partial y}\hat{j}+\frac{\partial}{\partial z}\hat{k}$ is called '**del**' operator and $\mathbf{r} = x\hat{i}+y\hat{j}+z\hat{k}$ is the position vector of the point (x,y,x) with respect to origin. We must note that ∇ is a vector quantity possessing all the properties of a ordinary vector. $\nabla\cdot\nabla = \nabla^2 = \frac{\partial^2}{\partial x^2}+\frac{\partial^2}{\partial y^2}+\frac{\partial^2}{\partial z^2}$ is however a scalar quantity and is known as '*Laplacian*' *operator.*

EXAMPLE 1.1: Two vectors **A** and **B** are such that $\mathbf{A}+\mathbf{B} = |\mathbf{A}-\mathbf{B}|$. Find the angle between them.

Solution: Let θ be angle between **A** and **B**. Then angle between **A** and $-\mathbf{B}$ is $\pi-\theta$. Thus

$$\begin{aligned} |\mathbf{A}+\mathbf{B}| &= \sqrt{A^2+B^2+2AB\cos\theta} \\ |\mathbf{A}-\mathbf{B}| &= |\mathbf{A}+(-\mathbf{B})| = \sqrt{A^2+B^2+2AB\cos(\pi-\theta)} \\ &= \sqrt{A^2+B^2-2AB\cos\theta} \end{aligned}$$

Since $|\mathbf{A}+\mathbf{B}| = |\mathbf{A}-\mathbf{B}|$, we have $A^2+B^2+2AB\cos\theta = A^2+B^2-2AB\cos\theta$ or $4AB\cos\theta = 0$

$$\text{But} \quad A \neq 0 \text{ and } B \neq 0 \qquad \therefore \cos\theta = 0 \text{ or } \theta = 90^\circ$$

EXAMPLE 1.2: Two vectors **A** and **B** are such that $\mathbf{A}-\mathbf{B} = \mathbf{C}$ and $A-B = C$. Find the angle between them.

Solution: We have

$$\mathbf{C} = \mathbf{A}+(-\mathbf{B})$$

This yields
$$C^2 = A^2+B^2-2AB\cos\theta \qquad (1)$$

$\because \quad C = A - B$, we have

$$C^2 = A^2 + B^2 - 2AB\cos\theta \qquad (2)$$

Comparing (1) and (2), one obtains

$$\cos\theta = 1 \qquad \therefore\ \theta = 0^\circ$$

EXAMPLE 1.3: If $\mathbf{A} = 4\hat{i} - \hat{j}$, $\mathbf{B} = -3\hat{i} + 2\hat{j}$ and $\mathbf{C} = -3\hat{j}$, find the vector sum, its scalar magnitude and direction. Also find a unit vector parallel to $\mathbf{A} + \mathbf{B} + \mathbf{C}$.

Solution: Let

$$\begin{aligned} \mathbf{r} &= \mathbf{A} + \mathbf{B} + \mathbf{C} \\ &= (4\hat{i} - \hat{j}) + (-3\hat{i} + 2\hat{j}) + (-3\hat{j}) \\ &= \hat{i} - 2\hat{j} \end{aligned}$$

Now

$$|\mathbf{r}| = |\mathbf{A} + \mathbf{B} + \mathbf{C}| = |\hat{i} - 2\hat{j}| = \sqrt{(1)^2 + (-2)^2} = \sqrt{5}$$

The angle which the vector $\mathbf{r}$ makes with the positive x-axis, measured counter clockwise is

$$\theta = \tan^{-1}\frac{r_y}{r_x} = \tan^{-1}\left(-\frac{2}{1}\right) = \tan^{-1}(-2) = -63^\circ.$$

As θ is negative, i.e. this is the angle which $\mathbf{r}$ makes with the x-axis measured clockwise. This means the angle which it makes with the x-axis measured counterclockwise is $360^\circ - \theta = 297^\circ$ A unit vector parallel to $\mathbf{r}$,

$$\hat{r} = \frac{\mathbf{r}}{|\mathbf{r}|} = \frac{\hat{i} - 2\hat{j}}{\sqrt{5}} = \frac{i}{\sqrt{5}} - \frac{2}{\sqrt{5}}\hat{j}.$$

EXAMPLE 1.4: If $\mathbf{A} = 3\hat{i} + 2\hat{j} + 4\hat{k}$ and $\mathbf{B} = 4\hat{i} + 2\hat{j} + \hat{k}$, find the magnitude and direction cosines of $\mathbf{A} + \mathbf{B}$ and $\mathbf{A} - \mathbf{B}$.

Solution: We have

$$\mathbf{A} + \mathbf{B} = 7\hat{i} + 4\hat{j} + 5\hat{k} \qquad (1)$$

and

$$\mathbf{A} - \mathbf{B} = -\hat{i} + 3\hat{k} \qquad (2)$$

$$|\mathbf{A}+\mathbf{B}| = \sqrt{(7)^2+(4)^2+(5)^2} = \sqrt{(90)}$$

and $$|\mathbf{A}-\mathbf{B}| = \sqrt{(-1)^2+(3)^2} = \sqrt{10}$$

$$\therefore \quad \cos\alpha = \frac{|\mathbf{A}+\mathbf{B}|_x}{|\mathbf{A}+\mathbf{B}|} = \frac{7}{\sqrt{90}}$$

$$\cos\beta = \frac{|\mathbf{A}+\mathbf{B}|_y}{|\mathbf{A}+\mathbf{B}|} = \frac{4}{\sqrt{90}}$$

$$\cos\gamma = \frac{|\mathbf{A}+\mathbf{B}|_z}{|\mathbf{A}+\mathbf{B}|} = \frac{5}{\sqrt{90}}.$$

Similarly, one obtains the direction cosines of $\mathbf{A}-\mathbf{B}$ as $-\frac{1}{\sqrt{10}}$, 0, $\frac{3}{\sqrt{10}}$.

EXAMPLE 1.5: If l_1, m_1, n_1, and l_2, m_2, n_2 are the direction-cosines of two vectors, show that the angle θ between them is given by

$$\cos\theta = l_1 l_2 + m_1 m_2 + n_1 n_2.$$

Solution: Let **A** and **B** be two vectors and angle between them is θ. We have

$$\mathbf{A}\cdot\mathbf{B} = AB\cos\theta$$

$$\therefore \quad \cos\theta = \frac{\mathbf{A}\cdot\mathbf{B}}{AB} = \frac{A_xB_x + A_yB_y + A_zB_z}{AB}$$

$$= \frac{A_xB_x}{AB} + \frac{A_yB_y}{AB} + \frac{A_zB_z}{AB}$$

Now, $\frac{A_x}{A}, \frac{A_y}{A}$ and $\frac{A_z}{A}$ are direction cosines of **A**, where as $\frac{B_x}{B}, \frac{B_y}{B}$ and $\frac{B_z}{B}$ are direction cosines of **B**. Thus

$$\cos\theta = l_1 l_2 + m_1 m_2 + n_1 n_2$$

EXAMPLE 1.6: If the non parallel vectors **A** and **B** are equal in magnitude, show that vector $(\mathbf{A}+\mathbf{B})$ is perpendicular to vector $(\mathbf{A}-\mathbf{B})$.

Solution:

$$
\begin{aligned}
(\mathbf{A}+\mathbf{B})\cdot(\mathbf{A}-\mathbf{B}) &= \mathbf{A}\cdot\mathbf{A}-\mathbf{A}\cdot\mathbf{B}+\mathbf{B}\cdot\mathbf{A}-\mathbf{B}\cdot\mathbf{B} \\
&= \mathbf{A}\cdot\mathbf{A}-\mathbf{B}\cdot\mathbf{B} \qquad (\because \quad \mathbf{A}\cdot\mathbf{B}=\mathbf{B}\cdot\mathbf{A}) \\
&= A^2-B^2 \\
&= 0 \qquad (\because \quad A=B)
\end{aligned}
$$

Since $\mathbf{A}$ is not parallel to $\mathbf{B}$, the vector $\mathbf{A}-\mathbf{B}$ cannot be zero. This means $(\mathbf{A}+\mathbf{B})$ is perpendicular to $(\mathbf{A}-\mathbf{B})$.

EXAMPLE 1.7: Can $\mathbf{A}\times\mathbf{B}=\mathbf{A}\cdot\mathbf{B}$?

Solution: This is possible only when $\mathbf{A}=\mathbf{B}=0$. If either $\mathbf{A}$ or $\mathbf{B}$ is zero and the other is non-zero then $\mathbf{A}\times\mathbf{B}$ will be a null vector, while $\mathbf{A}\cdot\mathbf{B}$ will be 0, i.e. null scalar.

EXAMPLE 1.8: Given two vector: $\mathbf{A}=4\hat{i}+3\hat{j}$ and $\mathbf{B}=-2\hat{i}+6\hat{j}$. Determine $\mathbf{A}\times\mathbf{B}$ and a unit vector in the direction of $\mathbf{B}$.

Solution: $\mathbf{A}\times\mathbf{B}=\begin{vmatrix} \hat{i} & \hat{j} & \hat{k} \\ 4 & 3 & 0 \\ -2 & 6 & 0 \end{vmatrix}=30\hat{k}$

A unit vector in the direction of $\mathbf{B}$ is given by

$$\hat{b}=\frac{\mathbf{B}}{B}=\frac{-2\hat{i}+6\hat{j}}{\sqrt{[(-2)^2+(6)^2]}}=\frac{1}{\sqrt{40}}(-2\hat{i}+6\hat{j})$$

EXAMPLE 1.9: Find the constant p such that the vector $2\hat{i}-\hat{j}+\hat{k}$, $\hat{i}+2\hat{j}-3\hat{k}$ and $3\hat{i}+p\hat{j}+5\hat{k}$ are coplanar.

Solution: The scalar triple product of vectors is

$$=\begin{vmatrix} 2 & -1 & 1 \\ 1 & 2 & -3 \\ 3 & p & 5 \end{vmatrix}=2(10+3p)+1(5+9)+1(9-6)=28+7p$$

Since the vectors are coplanar, and if their scalar triple product is zero, i.e. if

$$
\begin{aligned}
28+7p &= 0 \\
\therefore \qquad p &= -4
\end{aligned}
$$

EXAMPLE 1.10: If $\hat{r}$ is a unit vector, show that $\frac{d\hat{r}}{dt}$ is perpendicular to $\hat{r}$.

Solution: Since $\hat{r}$ is a unit vector, its scalar product by itself is 1, i.e.

$$\hat{r}\cdot\hat{r} = 1$$

$$\therefore \quad \frac{d}{dt}(\hat{r}\cdot\hat{r}) = 0$$

$$\text{or} \quad \hat{r}\cdot\frac{d\hat{r}}{dt}+\frac{d\hat{r}}{dt}\cdot\hat{r} = 0$$

$$\text{or} \quad 2\hat{r}\cdot\frac{d\hat{r}}{dt} = 0$$

$$\therefore \quad \hat{r}\cdot\frac{d\hat{r}}{dt} = 0$$

Obviously $\hat{r}$ and $\frac{d\hat{r}}{dt}$ are perpendicular to each other.

QUESTIONS

1. What is a unit vector? Show that a unit vector in three dimension can be express as

$$\hat{a} = \hat{i}\cos(\mathbf{a},x)+\hat{j}\cos(\mathbf{a},y)+\hat{k}\cos(\mathbf{a},z)$$

2. If **A** makes angles α, β and γ respectively with the x, y and z axes of a cartesian coordinate system, prove that

$$\cos^2\alpha+\cos^2\beta+\cos^2\gamma = 1$$

3. Define scalar and vector products of two vectors. Give physical examples.

4. Prove that $\mathbf{A}\times\mathbf{B} = \begin{vmatrix} \hat{i} & \hat{j} & \hat{k} \\ A_x & A_y & A_z \\ B_x & B_y & B_z \end{vmatrix}$

5. If $\hat{i}$, $\hat{j}$ and $\hat{k}$ are unit vectors along x, y and z axes respectively, show that

$$\hat{i}\cdot\hat{i} = 1,\ \hat{i}\cdot\hat{j} = 0,\ \hat{i}\times\hat{i} = 0 \qquad \text{and} \qquad \hat{i}\times\hat{j} = \hat{k}$$

6. Show that the component of **A** along **B** may be expressed as $(\mathbf{A}\cdot\mathbf{B})\mathbf{B}/B^2$.

7. Suppose that **A, B** and **C** are any three non-coplanar vectors. They are not necessarily at right angles. Show that

$$\mathbf{A}\cdot(\mathbf{B}\times\mathbf{C})=\mathbf{B}\cdot(\mathbf{C}\times\mathbf{A})$$

8. Show that

$$\mathbf{A}\times(\mathbf{B}\times\mathbf{C})=\mathbf{B}(\mathbf{A}\cdot\mathbf{C})-\mathbf{C}(\mathbf{A}\cdot\mathbf{B})$$

PROBLEMS

1. If $\mathbf{A}=2\hat{i}-4\hat{j}-3\hat{k}$ and $\mathbf{B}=\hat{i}+2\hat{j}+\hat{k}$, find $|\mathbf{A}+\mathbf{B}|$ and $|\mathbf{A}-\mathbf{B}|$ and a unit vector parallel to $(\mathbf{A}+\mathbf{B})$.

   $\left[\textbf{Ans.}\sqrt{17};\sqrt{53};\frac{1}{\sqrt{17}}(3\hat{i}-2\hat{j}+\hat{k})\right]$

2. If $\mathbf{A}=\hat{i}+\hat{j}-2\hat{k}$, $\mathbf{B}=2\hat{i}+\hat{j}+\hat{k}$ and $\mathbf{C}=\hat{i}-2\hat{j}+2\hat{k}$, find the magnitude and direction of vectors $(\mathbf{A}+\mathbf{B}+\mathbf{C})$ and $(\mathbf{A}-\mathbf{B}+\mathbf{C})$

   $\left[\textbf{Ans.}\sqrt{17};\sqrt{5};\left(\frac{4}{\sqrt{17}},0,\frac{1}{\sqrt{17}}\right),\left(0,-\frac{2}{\sqrt{5}},-\frac{1}{\sqrt{5}}\right)\right]$

3. If α, β be the direction cosines of a vector with respect to x-and y-axes and if C is its Z-component, find the magnitude of the vector.

   $\left[\textbf{Ans.}C/\sqrt{[1-\alpha^2-\beta^2]}\right]$

4. Show that the vectors $5\hat{i}+7\hat{j}-3\hat{k}$ and $2\hat{i}-b\hat{j}+c\hat{k}$ will be parallel to each other, if $b=-\frac{14}{5}$ and $c=-\frac{6}{5}$.

5. For what value of n the vectors $4\hat{i}+3\hat{j}-7\hat{k}$ and $5\hat{i}-2\hat{j}-n\hat{k}$ are orthogonal. [**Ans.** $n=-2$]

6. Determine unit vectors which are normal to x-axis and lie in the plane $3x-2y+3z=15$. [**Ans.** $\pm(3\hat{j}+2\hat{k})/\sqrt{13}$]

7. Show that the cross product of each two of the following vectors is parallel to the third

$$(\hat{i}+\hat{j}+\hat{k}),(\hat{i}-2\hat{j}+\hat{k}),(\hat{i}-\hat{k}).$$

8. Prove the following

   (i) $\mathbf{A}\times(\mathbf{B}\times\mathbf{C})+\mathbf{B}\times(\mathbf{C}\times\mathbf{A})+\mathbf{C}\times(\mathbf{A}\times\mathbf{B})=0$

   (ii) $(\mathbf{A}-\mathbf{B})\times(\mathbf{A}+\mathbf{B})=2(\mathbf{A}\times\mathbf{B})$

9. Show that if **A**, **B** and **C** are coplanar vectors,

$$(\mathbf{A} \times \mathbf{B}) \times (\mathbf{C} \times \mathbf{D}) = 0$$

10. Write down the expression for the time derivative or $r\hat{r}$.

$$\left[\textbf{Ans.}\, r\frac{d\hat{r}}{dt} + \hat{r}\frac{dr}{dt}\right]$$

OBJECTIVE QUESTIONS

1. If three vectors **A**, **B** and **C** are 12, 5 and 13 in magnitude such that $\mathbf{C} = \mathbf{A} + \mathbf{B}$, then the angle between **A** and **B** is
   (a) 30°; (b) 60°; (c) 90°; (d) 120° **[c]**
2. Two non zero vectors **A** and **B** are such that $|\mathbf{A} + \mathbf{B}| = |\mathbf{A} - \mathbf{B}|$. The angle between them is
   (a) 180°; (b) 90°; (c) 60°; (d) 0° **[b]**
3. Two vectors **A** and **B** are such that $\mathbf{A} + \mathbf{B} = \mathbf{C}$ and $A^2 + B^2 = C^2$. The angle between them is
   (a) 180°; (b) 120°; (c) 90°; (d) 0° **[c]**
4. If $\mathbf{A} = 4\hat{i} + 3\hat{j} - 2\hat{k}$ and $\mathbf{B} = 8\hat{i} + 6\hat{j} - 4\hat{k}$, the angle between **A** and **B** is
   (a) 90°; (b) 60°; (c) 45°; (d) 0° **[d]**
5. Two vectors **A** and **B** lie in a plane. Another vector **C** lies outside this plane. Then $\mathbf{A} + \mathbf{B} + \mathbf{C}$
   (a) can be zero
   (b) cannot be zero
   (c) lies in the plane containing $|\mathbf{A} + \mathbf{B}|$
   (d) lies in the plane containing $|\mathbf{A} - \mathbf{B}|$. **[b]**
6. The scalar product of vectors $\mathbf{A} = 2\hat{i} + 5\hat{k}$ and $\mathbf{B} = 3\hat{j} + 4\hat{k}$ is
   (a) 20; (b) $5\sqrt{33}$; (c) 26; (d) 23 **[a]**
7. A vector **A** and is along the positive x-axis, If **B** is another vector such that $\mathbf{A} \times \mathbf{B}$ is zero then **B** could be
   (a) $4\hat{j}$; (b) $-4\hat{i}$; (c) $-(\hat{i} + \hat{j})$; (d) $(\hat{j} + \hat{k})$ **[b]**
8. If $\mathbf{A} = 5\hat{i} + 7\hat{j} - 3\hat{k}$ and $\mathbf{B} = 2\hat{i} + 2\hat{j} - ak$ are perpendicular vectors, the value of a is
   (a) 8; (b) −8; (c) −7; (d) −2 **[b]**

9. Three vectors **A**, **B** and **C** satisfy the relation $\mathbf{A}\cdot\mathbf{B}=0$ and $\mathbf{A}\cdot\mathbf{C}=0$. The vector of **A** is parallel to
   (a) $\mathbf{B}\cdot\mathbf{C}$; (b) $\mathbf{B}\times\mathbf{C}$; (c) **B**; (d) **C** [b]
10. A particle moves from position $\mathbf{r}_1=3\hat{\imath}+2\hat{\jmath}-6\hat{k}$ to position $\mathbf{r}_2=14\hat{\imath}+13\hat{\jmath}+9\hat{k}$ under the action of a force $\mathbf{F}=4\hat{\imath}+\hat{\jmath}+3\hat{k}$. The work done by the force is
    (a) 200 units; (b) 100 units; (c) 75 units; (d) 50 units [b]

SHORT ANSWER QUESTIONS

1. If a quantity having a magnitude and direction necessarily a vector?
   [**Ans.** No, A directed quantity is a vector provided it obeys the commutative law of vector addition]
2. If a vector has zero magnitude is it meaning to call it a vector?
   [**Ans.** Yes, it is a null vector]
3. A zero vector is multiplied by a finite number. A finite vector is multiplied by zero Is the result in two cases same.
   [**Ans.** Yes, both are equal to null or zero vector]
4. Give one example each in physics where 'dot' and 'cross' product of two vectors occur as physical quantities.
   [**Ans.** (i) dot product: work (ii) cross product: torque]
5. The two vectors **A** and **B** are at right angles to each other. What will be their scalar product. [**Ans.** Zero]
6. When the two vectors **A** and **B** are parallel to each other, what will be their scalar product? [**Ans.** $\mathbf{A}\cdot\mathbf{B}=AB\cos 0^\circ=AB$]
7. Find $\mathbf{A}\times\mathbf{A}$. [**Ans.** null vector]
8. What is the necessary and sufficient condition for **A**, **B** and **C** to be coplanar?
   [**Ans.** Their scalar triple product $\mathbf{A}\cdot\mathbf{B}\times\mathbf{C}$ must be zero]
9. Can the dot and cross product in a scalar triple product $\mathbf{A}\cdot\mathbf{B}\times\mathbf{C}$ be interchanged? [**Ans.** Yes]

Chapter 2

Scalar and Vector Fields

2.1 FIELDS

A physical quantity can be expressed as a continuous function of the position of a region of space. If the cartesian coordinates of a point P are x, y, z and the physical quantity at this is A, then if

$$A = A(x, y, z) \tag{2.1}$$

This function is called a point function and the region in which it specifies the physical quantity is known as a *'field'*.

Fields are of two kinds: (i) *Scalar Field* and (ii) *Vector Field*, depending upon the nature of the physical quantity concerned.

2.2 SCALAR FIELDS

When a 'scalar' physical quantity is expressed from point to point in a region of space by a continuous point function,

$$\phi = \phi(x, y, z)$$

which gives the value of the quantity at each point, then the region is a '*scalar field*' and the function ϕ is a 'scalar point function'. Distribution of density, temperature, electric potential, gravitational potential in space etc. are few familiar examples of scalar fields.

One can represent scalar quantities graphically by drawing such surfaces at which the value of the quantity remains a constant. We must note that one point in space corresponds to only one value of the quantity, i.e. scalar fields are single valued functions at each point, i.e. two level surfaces cannot intersect (other wise ϕ should have two values at the point of intersection which contradicts our definition) [Fig. 2.1]. These surfaces are called *equiscalar surfaces*. If we put a point charge at any place, then the electric potential around the charge will depend on the position of the point. Electric potential is a scalar quantity and hence the field around the charge is known as *scalar potential field*. If we join all such points at which the potential is constant, then such a surface is called a *equipotential surface*. Fig. 2.1 shows S_1 and S_2 as equipotential surfaces, with constant potentials ϕ_1 and ϕ_2 respectively.

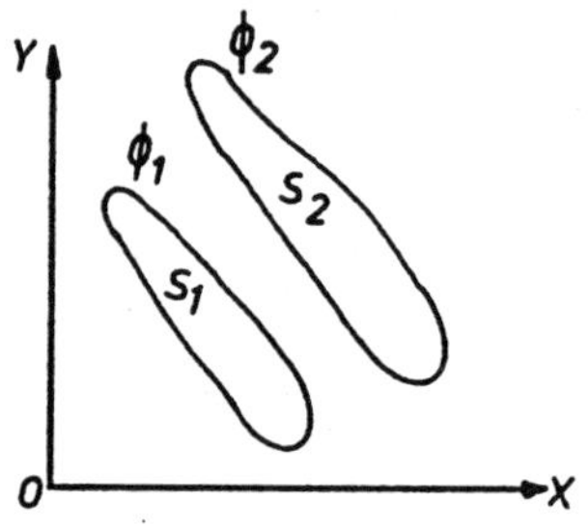

Figure 2.1. Graphical representation of scalar fields

2.3 VECTOR FIELDS

When a vector physical quantity is expressed from point to point in a region of space by a continuous vector function $\mathbf{A}(x, y, z)$ then the region is said to be a 'vector field' and the function **A** is called a 'vector point function'. At any given point in the field the function **A** is represented by a vector of definite magnitude and direction which changes continuously from point to point. Distribution of velocity in a fluid and distribution of gravitational or magnetic field intensity in space are few examples of vector fields.

Vector fields are graphically represented by lines and these lines are called *field lines* or *flux lines*. These lines are drawn in the field in such a way that the tangent at any point of the line represent the direction of the vector field at that point. The magnitude of **A** at any point on a flux line is given by the number of flux lines crossing unit area perpendicular to their direction drawn at that point. Two flux lines cannot intersect (otherwise the directions of the vector would become indefinite at the point of intersection).

Let us consider that the number of field lines, or flux going out of an area element ds perpendicular to field lines at any point is $d\phi$ (Fig. 2.2).

Obviously, the magnitude of the intensity of the vector field will be

$$|\mathbf{A}| = \frac{d\phi}{ds} \tag{2.2}$$

$$\therefore \quad \mathbf{A} = \left(\frac{d\phi}{ds}\right)\hat{n} \tag{2.3}$$

where $\hat{n}$ is unit vector perpendicular to the area ds. Thus

$$\hat{n}\, ds = \mathbf{ds}$$

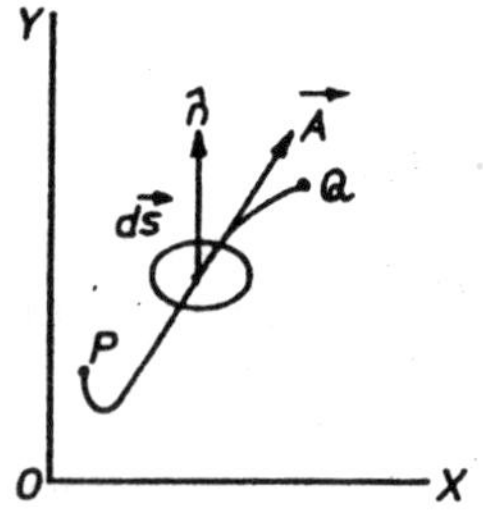

Figure 2.2.

Now,

$$\mathbf{A}\cdot\hat{n} = \frac{d\phi}{ds}\cdot\hat{n}\cdot\hat{n} = \frac{d\phi}{ds}$$

or

$$d\phi = (\mathbf{A}\cdot\hat{n})ds = \mathbf{A}\cdot\mathbf{ds} \tag{2.4}$$

We must remember that point function $\phi(x,y,z)$ or $\mathbf{A}(x,y,z)$ at every point of the field is single valued.

2.4 PARTIAL DERIVATIVE

Let us consider that a physical quantity f is a positional function of the coordinates x, y and z of the point of observation, where x, y and z are independent variables, then

$$f = f(x,y,z) \tag{2.5}$$

Keeping y and z constant, let us change the value of x by an infinitesimal amount Δx and the value of f changes by an amount Δf, then for constant y and z the rate of change of f with respect to x, $\left(\frac{\Delta f}{\Delta x}\right)$ is known as partial derivative of f with respect to x. Usually it is expressed as $\left(\frac{\partial f}{\partial x}\right)_{y,z}$ or simply $\frac{\partial f}{\partial x}$. Obviously, Δx is considered infinitely small, i.e. $\Delta x \to 0$ and hence

$$\frac{\partial f}{\partial x} = Lt_{\Delta x\to 0}\frac{f(x+\Delta x,y,z)-f(x,y,z)}{\Delta x} \tag{2.6}$$

Similarly if one keeps x and z constant and the value of y is changed by an infinitesimal amount Δy, then

$$\frac{\partial f}{\partial y} = Lt_{\Delta y\to 0}\frac{f(x,y+\Delta y,z)-f(x,y,z)}{\Delta y} \tag{2.7}$$

and if one keeps x and y constant and the value of z is changed by an infinitesimal amount Δz, then

$$\frac{\partial f}{\partial z} = Lt_{\Delta z \to 0} \frac{f(x,y,z+\Delta z) - f(x,y,z)}{\Delta z} \tag{2.8}$$

Now, consider that all the three independent variables x, y, z are simultaneously changed by infinitesimal amounts, say δx, δy and δz respectively and the value of f is changed by an amount δf. When $\delta x \to 0$, $\delta y \to 0$ and $\delta z \to 0$, i.e. all the three vanish simultaneously $\delta f \to 0$, i.e. δf will also vanish. Obviously, in this state, the total change in the value of f, i.e. δf is known as total derivative. Thus

$$\delta f = \left(\frac{\partial f}{\partial x}\right)\delta x + \left(\frac{\partial f}{\partial y}\right)\delta y + \left(\frac{\partial f}{\partial z}\right)\delta z$$

One can write the total derivative for the limiting change as

$$df = \left(\frac{\partial f}{\partial x}\right)dx + \left(\frac{\partial f}{\partial y}\right)dy + \left(\frac{\partial f}{\partial z}\right)dz \tag{2.9}$$

The partial derivatives $\frac{\partial f}{\partial x}, \frac{\partial f}{\partial y}$ and $\frac{\partial f}{\partial z}$ are called *first partial derivatives* or partial derivatives of the first order. By differentiating these derivatives once more, one can obtain the *second partial derivatives* (or partial derivatives of the second order) e.g., $\frac{\partial^2 f}{\partial x^2}, \frac{\partial^2 f}{\partial y \partial x}, \frac{\partial^2 f}{\partial x \partial y}, \frac{\partial^2 f}{\partial y^2}, \frac{\partial^2 f}{\partial z^2}$ etc. Obviously, $\frac{\partial^2 f}{\partial x^2}$ is the partial derivative of $\frac{\partial f}{\partial x}$ with respect to x, $\frac{\partial^2 f}{\partial x \partial y}$ is the partial derivative of $\frac{\partial f}{\partial x}$ with respect y etc. If all the derivatives concerned are continuous, then $\frac{\partial^2 f}{\partial x \partial y} = \frac{\partial^2 f}{\partial y \partial x}$, i.e. then the order of differentiation is immaterial. By differentiating the second partial derivatives again with respect to x, y and z respectively, one obtains the third partial derivatives and so on.

If $\mathbf{A}$ is vector depending on more than one scalar variables such as cartersion coordinates x, y, z of a point in space. Then $\frac{\partial \mathbf{A}}{\partial x}, \frac{\partial \mathbf{A}}{\partial z}$ and $\frac{\partial \mathbf{A}}{\partial z}$ are partial derivatives of $\mathbf{A}$ with respect to x (when y and z remain constant), y (when x and z remain constant) and z (when x and y remain constant respectively, If now x, y and z change simultaneously by small increments dx, dy and dz, then the total change or total derivative of **A** will be

$$\begin{aligned} d\mathbf{A} &= \frac{\partial \mathbf{A}}{\partial x}dx + \frac{\partial \mathbf{A}}{\partial y}dy + \frac{\partial \mathbf{A}}{\partial z}dz \\ &= \left(\frac{\partial}{\partial x}dx + \frac{\partial}{\partial y}dy + \frac{\partial}{\partial z}dz\right)\mathbf{A} \\ &= \left(\frac{\partial}{\partial x}\hat{i} + \frac{\partial}{\partial y}\hat{j} + \frac{\partial}{\partial z}\hat{k}\right)\cdot(dx\hat{i} + dy\hat{j} + dz\hat{k})\mathbf{A} \end{aligned} \tag{2.10}$$

where $\hat{i}$, $\hat{j}$ and $\hat{k}$ are unit vectors in the directions of x, y and z axes respectively. $\frac{\partial}{\partial x}\hat{i} + \frac{\partial}{\partial y}\hat{j} + \frac{\partial}{\partial z}\hat{k}$ is defined as the *vector differential operator*, written as ∇ and read as '*del*'. Thus

$$\nabla = \frac{\partial}{\partial x}\hat{i} + \frac{\partial}{\partial y}\hat{j} + \frac{\partial}{\partial z}\hat{k} \tag{2.11}$$

Further if **r** is the position vector of the point (x, y, z) with respect to origin, then

$$\mathbf{r} = \hat{i}x + \hat{j}y + \hat{k}z \tag{2.12}$$

so that

$$d\mathbf{r} = \hat{i}dx + \hat{j}dy + \hat{k}dz \tag{2.13}$$

The Eq. (2.10) can be expressed as

$$d\mathbf{A} = (\nabla \cdot d\mathbf{r})\mathbf{A} \tag{2.14}$$

Here ∇ is a vector quantity possessing all the properties of an ordinary vector. Since it is an '*operator*', its magnitude has no physical significance.

The differential operator $\nabla \cdot \nabla = \nabla^2$ is defined as

$$\nabla^2 = \frac{\partial^2}{\partial x^2} + \frac{\partial^2}{\partial y^2} + \frac{\partial^2}{\partial z^2} \tag{2.15}$$

∇^2 is a scalar and is known as '*Laplacian Operator*'.

2.5 GRADIENT OF A SCALAR FUNCTION

A scalar field can be mapped out by a series of level surfaces. Let us consider that in a scalar field two surfaces S_1 and S_2 represent surfaces with constant scalar quantities ϕ and $\phi + \delta\phi$ respectively (Fig. 2.3). **r** is the position vector of any point P on S, relative to the origin O.

The position vector of another point Q on S_2 is $\mathbf{r}+\delta\mathbf{r}$. Obviously, the displacement of Q with respect to P is $\mathbf{PQ}=\delta\mathbf{r}$. The rate of change of scalar function in the direction of **PQ** is $\frac{\delta\phi}{\delta r}$. Now, draw a perpendicular PR from P on surface S_2.

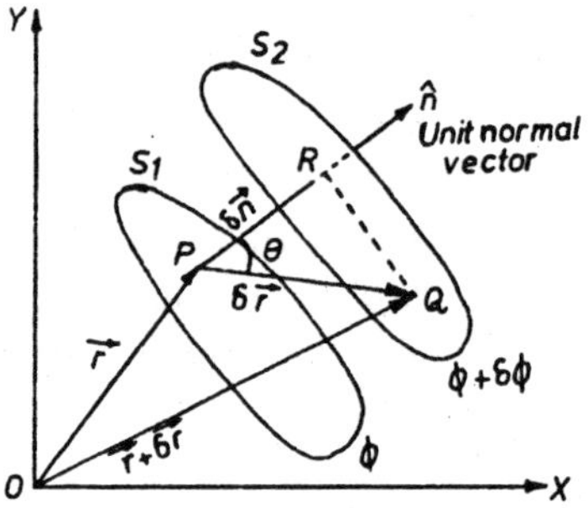

Figure 2.3.

The minimum distance between the surfaces S_1 and S_2, i.e. $PR=\delta n$, will be in the direction of their normal. Therefore, the rate of change of scalar function ϕ in the direction of $PR=\frac{\delta\phi}{\delta n}$. If the distance δr is infinitesimal then one can consider ΔPQR, a right angled triangle. If Q is the angle between $\delta\mathbf{r}$ and $\delta\mathbf{n}$, then

$$PR = PQ\cos\theta \quad \text{or} \quad \delta n=\delta r\cos\theta$$

$$\therefore \quad PQ = \frac{\delta\phi}{\delta n}\cos\theta \tag{2.16}$$

obviously, (2.16) represents the rate of change of ϕ in the direction of PQ. Since the maximum value of $\cos\theta$ is 1 and hence the maximum rate of change of $\phi=\left(\frac{\delta\phi}{\delta r}\right)_{\max}=\left(\frac{\delta\phi}{\delta n}\right)$

$$\therefore \text{Lt}\ \delta r\to 0\ \left(\frac{\delta\phi}{\delta r}\right)_{\max}=\frac{d\phi}{dn}$$

obviously, the rate of change of a scalar function is maximum in a direction perpendicular to the equiscalar surface.

Thus in a *scalar field, the maximum rate of change of the scalar function is known as the gradient of the scalar field.*

Gradient is a vector quantity and its direction is perpendicular to the equiscalar surface, i.e. the direction in which rate of change is maximum. If $\hat{n}$ is the unit vector normal to the equiscalar surface, then

$$\text{grad}\,\phi = \left(\frac{d\phi}{dn}\right)\hat{n} = \left(\frac{d\phi}{dr}\right)_{\max}\hat{n} \tag{2.17}$$

obviously, the gradient of a scalar field ϕ depends only (i) on the way the values of ϕ are distributed in space and (ii) entirely independent of the position and direction of the coordinate axes. Thus grad ϕ is an invariant vector field with respect to transformation of coordinates.

To make the above statement more clear, let us consider an example. Suppose ϕ is the potential in an electric field. The intensity of the field at any point is in the direction of the greatest rate of fall of potential, i.e. normal to the equipotential surface and is equal to the rate, i.e.

$$\mathbf{E} = -\frac{\partial\phi}{\partial n}\hat{n} = -\text{grad}\,\phi \tag{2.18}$$

The negative sign is used because the direction of the field is opposite to the direction of the increase of potential.

The Gradient of a Scalar field in Rectangular Cartesian Coordinates

Suppose the scalar field ϕ is a function of (x, y, z), then with the use of partial differentiation, one can write

$$d\phi = \frac{\partial\phi}{\partial x}\cdot dx + \frac{\partial\phi}{\partial y}\cdot dy + \frac{\partial\phi}{\partial z}\cdot dz \tag{2.19}$$

Now, we have the gradient of ϕ is a vector given by

$$\nabla\phi = \frac{\partial\phi}{\partial n}\hat{n}$$

where $\hat{n}$ is a unit vector. Let us take its scalar product with an element of radius vector $d\mathbf{r}$, one obtains

$$\begin{aligned}
(\text{grad}\,\phi)\cdot d\mathbf{r} &= \frac{\partial\phi}{\partial n}\hat{n}\cdot d\mathbf{r} \\
&= \frac{\partial\phi}{\partial n}dr\cos\theta \quad (\because \quad \hat{n}\cdot d\mathbf{r} = dr\cos\theta) \\
&= \frac{\partial\phi}{\partial n}\cdot dn \\
&\qquad (\because \quad \frac{\partial\phi}{\partial n}\text{is the rate of change of } \phi \text{ in the direction } \hat{n}) \\
&= d\phi
\end{aligned}$$

$$= \left(\frac{\partial\phi}{\partial x}\right)dx + \left(\frac{\partial\phi}{\partial y}\right)dy + \left(\frac{\partial\phi}{\partial z}\right)dz$$

$$= \left(\frac{\partial\phi}{\partial x}\hat{i} + \frac{\partial\phi}{\partial y}\hat{j} + \frac{\partial\phi}{\partial z}\hat{k}\right)\cdot(dx\hat{i} + dy\hat{j} + dz\hat{k})$$

$$= \left(\frac{\partial\phi}{\partial x}\hat{i} + \frac{\partial\phi}{\partial y}\hat{j} + \frac{\partial\phi}{\partial z}\hat{k}\right)\cdot d\mathbf{r}$$

$$\therefore \quad \nabla\phi = \frac{\partial\phi}{\partial x}\hat{i} + \frac{\partial\phi}{\partial y}\hat{j} + \frac{\partial\phi}{\partial z}\hat{k} = \text{grad}\ \phi \tag{2.20}$$

Obviously, $\nabla\phi$ is such a vector which represents the maximum rate of change of scalar function ϕ.

It is worthwhile to note that the gradient of a scalar field is always a vector field, however, its converse is always not true. All vector fields cannot be expressed as gradient of scalar fields.

A vector field which can be expressed as gradient of a scalar field is called a *lamellar* or *non-curl field.*

A pure electrostatic field is a lamellar vector field, because it can be expressed as (negative) gradient of a scalar potential field. If V is the electric potential, then one can write the electric field as

$$\mathbf{E} = -\nabla V \tag{2.21}$$

A scalar field $\phi(x,y,z)$ evaluated at a particular point is independent of the coordinates of that point. For example, the temperature at a point is not dependent on whether co-ordinates (x,y,z) or (x',y',z') are used. This means a scalar field is invariant with respect to the transformation of coordinates. Its gradient which is a vector field, must therefore be also invariant with respect to the transformation.

2.6 LINE INTEGRAL OF VECTOR FIELD

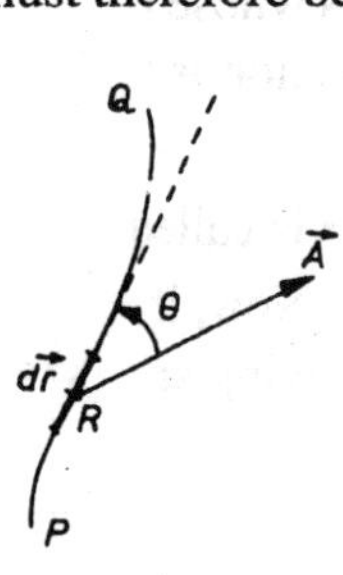

Figure 2.4.

Let us consider two points P and Q in a vector field. One can divide the path between P and Q into displacement elements of lengths $dr_1, dr_2, dr_3, \ldots$ etc. (not shown in Fig. 2.4) respectively. Since these are very small and one can consider them as straight lines. $d\mathbf{r}$ is one of such element shown in Fig. 2.4. Let $\mathbf{A}$ be the vector at R in a direction making an angle θ with $d\mathbf{r}$. If A varies in magnitude and direction from point to point along the curve PQ, then the integral

$$\int_A^B \mathbf{A}\cdot d\mathbf{r} = \int_P^Q (A\cos\theta)dr = \int_P^Q (A_x dx + A_y dy + A_z dz) \qquad (2.22)$$

is defined as the line integral of vector **A** along the curve AB. Few familiar examples of line integral are:

(i) If **A** denotes a force acting on a particle which moves along the curve PQ then the integral $\int_P^Q \mathbf{A}\cdot d\mathbf{r}$ represents the work done by the force in moving from P to Q.

(ii) If **A** represents the electric field intensity at any point, then the line integral represents the potential difference between P and Q.

(iii) If **A** represents the velocity at any point in a fluid then the line integral of **A** round a closed path is known as circulation of the fluid.

If the value of the line integral depends only upon the coordinates of the two points in the vector field and not upon the actual path taken between them, then the vector field $\mathbf{A}(x,y,z)$ is called a '*conservative*' or '*non curl*' or '*lamellar field*'.

2.7 SURFACE INTEGRAL

Let S be a surface drawn in a vector field, and ds an infinite small element of the surface (Fig. 2.5). Let $\hat{n}$ be a Unit positive vector normal to ds. The surface element of area ds is represented by a vector $d\mathbf{s}$ whose magnitude is $|ds|$ and direction is that of $\hat{n}$. Thus

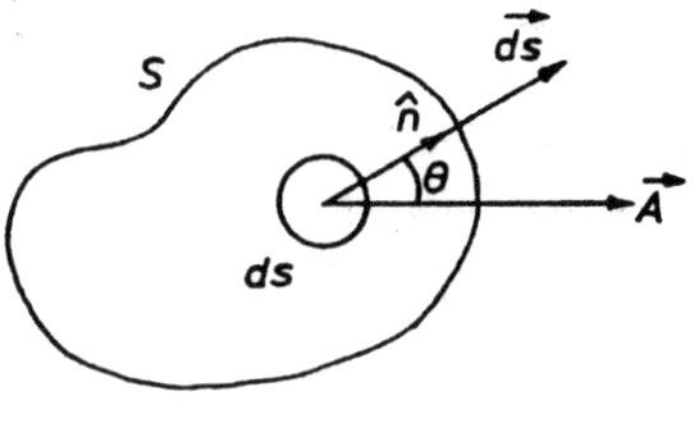

Figure 2.5.

$$d\mathbf{s} = \hat{n}ds$$

$d\mathbf{s}$ is called the **area vector**.

Now, if **A** be a vector at the middle of the element ds in a direction making an angle θ with $\hat{n}$, then the integral

$$\iint \mathbf{A}\cdot d\mathbf{s} \quad \text{or} \quad \int_S \mathbf{A}\cdot\hat{n}ds = \int_S (A\cos\theta)ds \qquad (2.23)$$

is defined as the 'surface integral' or 'Flux' of **A** across the surface S.

If **A** denotes the velocity of a moving fluid in which a fixed surface S is drawn, then the surface integral $\int_S \mathbf{A}\cdot d\mathbf{s}$ gives the amount of fluid flowing per unit time normally through the surface S. In case of a closed surface, a positive surface integral means that the fluid is flowing outwards, if negative then inwards.

Other examples of surface integrals are $\int_S \mathbf{A} \times ds$ and $\int_S \phi ds$, where ϕ is a scalar function.

2.8 VOLUME INTEGRAL

Let $dV = dx\,dy\,dz$ denote the element of volume V. Volume element dV is a scalar. If $\mathbf{A}$ be a vector function inside it then the integral $\int_V \mathbf{A}dV$ is called the volume integral, $\iiint_V \mathbf{A}dV$ (we will write it as $\int_V \mathbf{A}dV$) of $\mathbf{A}$ over the volume V, i.e.

$$\int_V \mathbf{A}dV = \int_V (A_x\hat{i} + A_y\hat{j} + A_z\hat{k})dx\,dy\,dz \tag{2.24}$$

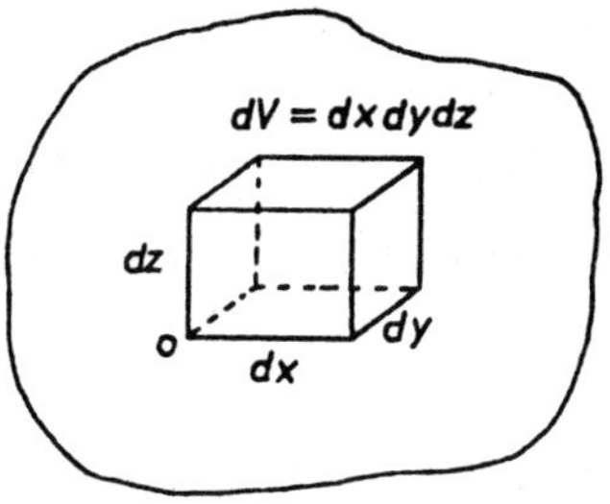

Figure 2.6.

In case of a scalar function ϕ, the integral $\int_V \phi dV$ represents the volume integral over V.

2.9 DIVERGENCE OF A VECTOR FIELD

The divergence of a vector field at any point is defined as the amount of flux per unit volume diverging from that point. Since the divergence is the amount of flux, it is essentially a scalar. One can express the divergence of a vector field $\mathbf{A}(x,y,z)$ as

$$\nabla \cdot \mathbf{A} = \frac{\partial A_x}{\partial x} + \frac{\partial A_y}{\partial y} + \frac{\partial A_z}{\partial z} = \text{div}\,\mathbf{A} \tag{2.25}$$

Eq. (2.25) is the expression for divergence in cartesian coordinates. Divergence of a vector field has important applications in hydrodynamics. Now, we derive the expression for divergence in cartesian coordinates.

Let us consider a small rectangular parallelopiped whose centre is $C(x,y,z)$ and whose sides have lengths dx, dy and dz parallel to the coordinate axes x, y, and z respectively as shown in Fig. 2.7.

Suppose the vector $\mathbf{A}$ has components $\mathbf{A}_x$, $\mathbf{A}_y$ and $\mathbf{A}_z$ along x, y and z axes respectively.

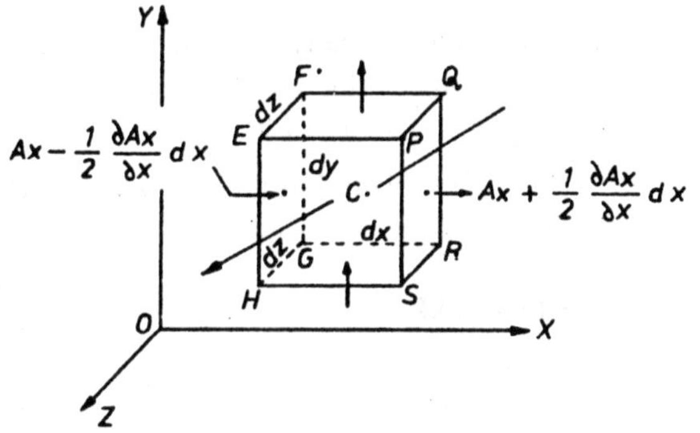

Figure 2.7.

The value of the x component of **A** at the centre of the face $EFGH$ is

$$A_x - \frac{1}{2}\frac{\partial A_x}{\partial x}dx \tag{2.26}$$

and that at the centre of the face $PQRS$ is

$$A_x + \frac{1}{2}\frac{\partial A_x}{\partial x}dx \tag{2.27}$$

Now, the volume of the fluid flowing per unit time through a face is given by the product of the area of the face and the normal component of the velocity upon it. We call it the 'flux' through the face. The area of each of faces $EFGH$ and $PQRS$ is $dydz$, one finds,

$$\text{flux entering the face } EFGH = \left(A_x - \frac{1}{2}\frac{\partial A_x}{\partial x}dx\right)dydz \tag{2.28}$$

and

$$\text{flux leaving the face } PQRS = \left(A_x + \frac{1}{2}\frac{\partial A_x}{\partial x}dx\right)dydz \tag{2.29}$$

obviously, the excess of the flux leaving the parallelopiped over that entering it in the x-direction is

$$\left(A_x + \frac{1}{2}\frac{\partial A_x}{\partial x}dx\right)dydz - \left(A_x - \frac{1}{2}\frac{\partial A_x}{\partial x}dx\right)dydz = \frac{\partial A_x}{\partial x}dxdydz \tag{2.30}$$

Similarly, one obtains the net flux leaving the parallelopiped in the y and z directions are $\frac{\partial A_y}{\partial y}dxdydz$ and $\frac{\partial A_z}{\partial z}dxdydz$ respectively. Therefore the total net flux diverging from (i.e. leaving the) parallelopiped is

$$\left(\frac{\partial A_x}{\partial x}+\frac{\partial A_y}{\partial y}+\frac{\partial A_z}{\partial z}\right)dxdydz \tag{2.26}$$

Since $dxdydz$ is the volume of the elementary parallelopiped and hence the amount of flux diverging per unit volume, which is defined as divergence of **A** is:

$$\nabla\cdot\mathbf{A}=\text{div}\mathbf{A}=\frac{\partial A_x}{\partial x}+\frac{\partial A_y}{\partial y}+\frac{\partial A_z}{\partial z} \tag{2.27}$$

2.10 PHYSICAL MEANING OF DIVERGENCE OF A VECTOR

$\nabla\cdot\mathbf{A}$ at a point gives the amount of flux diverging per unit volume from that point. If **A**, denotes the velocity of the moving fluid, then $\nabla\cdot\mathbf{A}$ tells the rate at which the fluid is diverging from the point per unit volume.

If the divergence of a vector is positive at a point in a fluid, then either the fluid is expanding and its density at that point is falling with time or the point as a source of fluid. In case, if the divergence of a vector is negative, then either the fluid is contracting and its density is rising or the point is a negative source, i.e. *sink*.

If the flux, entering any element of field space is exactly balanced by that leaving it, the quantity $\nabla\cdot\mathbf{A}=0$. Obviously, there is no source or sink, nor its density is changing, i.e. the fluid is incompressible. If the flux enerting and leaving an element are equal, the lines of flow of vector **A** should form either closed curves (e.g., in case of magnetic field due to a current) or extend to infinity. A vector which satisfies this condition is called *Solenoidal*.

In the case of non-material fluxes (e.g. magnetic or electric flux) the existence of divergence means the presence of a source or sink of flux at that point.

2.11 GAUSS'S DIVERGENCE THEOREM

This theorem states that the volume integral of the divergence of a vector field **A** taken over any volume V is equal to the surface integral of A taken over the closed surface surrounding the volume, i.e.

$$\iiint_v(\nabla\cdot\mathbf{A})dV=\iint_s\mathbf{A}\cdot ds \tag{2.28}$$

Proof: In Cartesian coordinates

$$\begin{aligned}\nabla\cdot\mathbf{A} &= \left(\hat{i}\frac{\partial}{\partial A}+\hat{j}\frac{\partial}{\partial y}+\hat{k}\frac{\partial}{\partial z}\right)\cdot(A_x\hat{i}+A_y\hat{j}+A_z\hat{k})\\ &= \frac{\partial A_x}{\partial x}+\frac{\partial A_y}{\partial y}+\frac{\partial A_z}{\partial z}\end{aligned}$$

and $dv = dxdydz$

Also

$$\begin{aligned}\mathbf{A}\cdot d\mathbf{s} &= (A_x\hat{i}+A_y\hat{j}+A_z\hat{k})\cdot(dS_x\hat{i}+dS_y\hat{j}+dS_z\hat{k})\\ &= A_x dS_x + A_y dS_y + A_z dS_z\\ &= A_x dydz + A_y dxdz + A_z dxdy\end{aligned}$$

Using the above results, one can express Gauss's theorem in cartesian coordinates as

$$\iiint_V \left(\frac{\partial A_x}{\partial x}+\frac{\partial A_y}{\partial y}+\frac{\partial A_z}{\partial z}\right) dxdydz = \iint_s (A_x dS_x + A_y dS_y + A_z dS_z) \qquad (2.29)$$

Let us consider the first part of the L.H.S. integral of (2.29). As shown in Fig. 2.8, integrate it w.r.t. x along a stripe of cross section $dydz$ from $P_1(x_1,y,z)$ to $P_2(x_2,y,z)$, i.e.

$$\begin{aligned}\iiint_v \frac{\partial A_x}{\partial x}dxdydz &= \iint_s |A_x|_{P_1}^{P_2} dydz\\ &= \iint_s [A_x(P_2)-A_x(P_1)]dydz\\ &= \iint_s [A_x(x_2,y,z)-A_x(x_1,y,z)dydz\end{aligned}$$

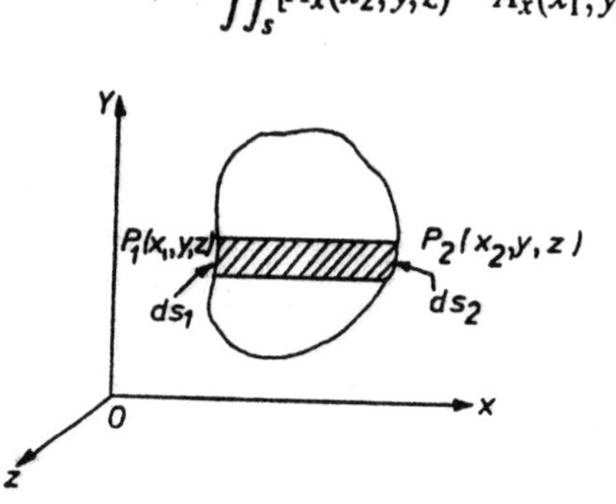

Figure 2.8.

Evidently, at P_1, $A_x dydz = -A_x dS_x$ and at P_2,

$$A_x dydz = A_x dS_x$$

($\because$ at P_1 and P_2, the x components of area have opposite directions)

$$\therefore \quad \iiint_v \frac{\partial A_x}{\partial x} dxdydz = \iint_s A_x dS_x \tag{2.30}$$

We should not confuse that we have written for

$$\iiint_v A_x(P_2) - A_x(P_1)dydz = \iint_s (A_x dS_x + A_x dS_x)$$

only $\iint_s A_x dS_x$, because if we take all the stripes then each integral $\iint_s A_x dS_x$ will account for the half integral. Hence $\iint_s A_x dS_x$ accounts for the total surface integral for x component. Similarly, one obtains

$$\iiint_v \frac{\partial A_y}{\partial y} dxdydz = \iint_s A_y dS_y \tag{2.31}$$

and

$$\iiint_v \frac{\partial A_z}{\partial z} dxdydz = \iint_s A_z dS_z \tag{2.32}$$

Adding (2.30) to (2.32), one obtains

$$\iiint_v \left(\frac{\partial A_x}{\partial x} + \frac{\partial A_y}{\partial y} + \frac{\partial A_z}{\partial z} \right) dxdydz = \iint_s (A_x dS_x + A_y dS_y + A_z dS_z)$$

or

$$\iiint_v (\nabla \cdot \mathbf{A} dv = \iint_s \mathbf{A} \cdot ds$$

or

$$\iiint_v (\text{div}\mathbf{A}) dv = \iint_s \mathbf{A} \cdot ds \tag{2.33}$$

This proves the *Gauss's divergence theorem.*

Now we apply this theorem to an electric field **E**. Let us consider that there exist throughout the volume V enclosed by the surface S, a charge distribution of varying density ρ, then

$$\frac{1}{\varepsilon_0} \int_v \rho dv = \int_s \mathbf{E} \cdot ds = \int_v (\nabla \cdot \mathbf{E}) dv \tag{2.34}$$

The two volume integrals in (34) are equal, whatever the volume over which the integration takes place. It therefore follows that integrands themselves must be equal, i.e.

$$\nabla \cdot \mathbf{E} = \frac{\rho}{\varepsilon_0} \tag{2.35}$$

Eq. (2.35) is sometimes referred to as the *differential from of Gauss's law in electrostatics.* According to this, the divergence of the electric field at a point equals ρ/ε_0 in the immediate neighbourhood of that point. If the point is situated in a material of dielectric constant K, then R.H.S. of (2.35) must be divided by K.

The differential or point form (2.35) is valid only where the derivatives exist. The integral form (2.34) is valid everywhere.

2.12 THE LAPLACIAN OPERATOR

Let ϕ be a scalar potential function. If ϕ represents an electric field **E**, then

$$\mathbf{E} = -\nabla\phi = -\text{grad}\,\phi \tag{2.36}$$

According to Gauss's law in electrostatics, the divergence of an electric field emerging from a closed surface is equal to $\frac{1}{\varepsilon}$ times the charge density enclosed in the surface, i.e.

$$\nabla\cdot\mathbf{E} = \frac{\rho}{\varepsilon_0} \tag{2.37}$$

Here ε is the permittivity of the vacuum. Substituting (2.36) in (2.37), one obtains

$$-\nabla\cdot\nabla\phi = \frac{\rho}{\varepsilon_0} \tag{2.38}$$

But

$$\begin{aligned}
\nabla\cdot\nabla\phi &= \text{div}(\text{grad}\phi)\\
&= \left(\hat{i}\frac{\partial}{\partial x} + \hat{j}\frac{\partial}{\partial y} + \hat{k}\frac{\partial}{\partial z}\right)\cdot\left(\hat{i}\frac{\partial}{\partial x} + \hat{j}\frac{\partial}{\partial y} + \hat{k}\frac{\partial}{\partial z}\right)\phi\\
&= \left(\frac{\partial^2}{\partial x^2} + \frac{\partial^2}{\partial y^2} + \frac{\partial^2}{\partial z^2}\right)\phi\\
&= \nabla^2\phi
\end{aligned} \tag{2.39}$$

Thus $\nabla^2 = \frac{\partial^2}{\partial x^2} + \frac{\partial^2}{\partial y^2} + \frac{\partial^2}{\partial z^2}$ is known as *Laplacian operator* or simply, the *Laplacian*. $\nabla^2 = \nabla\cdot\nabla$ is usually pronounced as del squared.

2.13 POISSON'S AND LAPLACE'S EQUATIONS

Substituting (2.39) in (2.38), one obtains

$$\nabla^2\phi = -\frac{\rho}{\varepsilon_0} \tag{2.40}$$

Eq. (2.40) is known as *Poisson's equation.* One can calculate the electric potential at a point due to a charge distribution with the help of this equation. In a medium of dielectric constant K, Eq. (2.40) takes the form

$$\nabla^2\phi = -\frac{\rho}{K\varepsilon_0} \tag{2.41}$$

In the charge free region, volume charge density $\rho = 0$, so the Eq. (2.40) or (2.41) become

$$\nabla^2\phi = 0 \tag{2.42}$$

Eq. (2.42) is known as *Laplace's equation.* By solving this equation, one can calculate the potential in an electric field, which is devoid of the source charge.

In *Spherical polar Co-ordinates,* Poisson's equation becomes

$$\frac{1}{r^2}\frac{\partial}{\partial r}\left(r^2\frac{\partial\phi}{\partial r}\right)+\frac{1}{r^2}\frac{1}{\sin\theta}\frac{\partial}{\partial\theta}\left(\sin\theta\frac{\partial\phi}{\partial\theta}\right)+\frac{1}{r^2\sin^2\theta}\frac{\partial^2\phi}{\partial\theta^2}=\frac{\rho}{k\varepsilon_0} \tag{2.43}$$

In *cylindrical polar co-ordinates,* Poisson's equation becomes

$$\frac{1}{r}\frac{\partial}{\partial r}\left(r\frac{\partial\phi}{\partial r}\right)+\frac{1}{r^2}\frac{\partial^2\phi}{\partial\theta^2}+\frac{\partial^2\phi}{\partial z^2}=\frac{\rho}{k\varepsilon_0} \tag{2.44}$$

2.14 THE CURL OF A VECTOR FUNCTION

We have seen that divergence of a vector field is a scalar. Now, we define a vector called curl **A**, Also written as $\nabla \times \mathbf{A}$, associated with a vector field **A**.

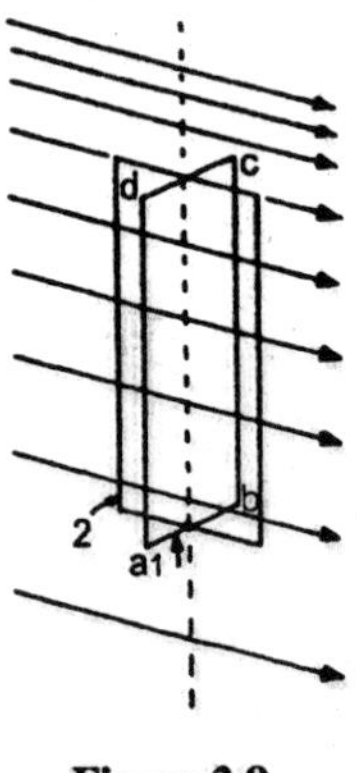

Figure 2.9.

If a vector field is derived as the gradient of a scalar field then the line integral of the vector around any closed path in the field is zero. Such a field is called '*lamellar*' or *non curl field.* There are, however, such vector fields for which the line-integral round any closed path is non-zero and have a finite value. Such vector fields cannot be derived as gradient of any scalar field and they exhibit the property of curl.

Suppose a non-lamellar vector field is represented by several lines of flow as shown in Fig. (2.9). Let us consider a plane rectangular area in this field. When the plane *abcd* is perpendicular to the field (position 1) none line of the flow lies along its boundary so that the line integral round it is zero. When the plane, *abcd* is parallel to the field (position 2) lines of flow lie along the upper and lower edges of the plane. Since the value of the vector at the upper edge is different from that at the lower edge, the line integral round the boundary has a finite value. This value will be different for different orientations of the plane, and shall be maximum for a particular orientation. This maximum line integral computed per unit area along the boundary is called the curl of the vector field. Thus one can define the curl of a vector field as:

The curl of a (non-lamellar) vector field at any point is defined as the maximum line integral of the vector component per unit area along the boundary of an infinitesimal area at that point. Obviously, a curl of a vector field is a vector.

Curl in Cartesian Coordinates

We now calculate curl **A**, where **A** is function of cartesian coordinates (x,y,z). Let us consider a rectangular shaped infinitely small element of area *dxdy* in the (x,y) plane as shown in Fig. 2.10.

Let the magnitude of the vector **A** at a point O is $|\mathbf{A}|$ and A_x, A_y and A_z be its components along x, y and z axes respectively. Let the sides dx, dy of the rectangle be parallel to the x and y axes respectively. So the normal to the area is along the z-axis. The arrow heads on the sides of the rectangle show the directions in which the component of **A** act.

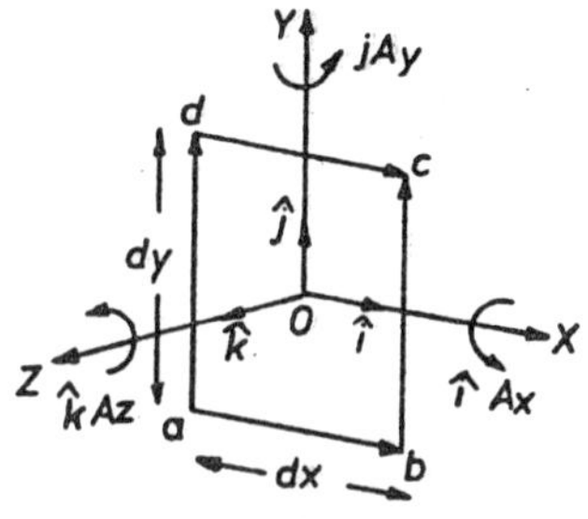

Figure 2.10.

Since the rectangle is infinitely small, the average value of **A** along any side of it may be taken as the value at the middle point of that side. Obviously, the average values along the four sides will be as follows:

$$\text{along } ab; \quad A_x - \frac{1}{2}\frac{\partial A_x}{\partial y}dy$$

$$\text{along } bc; \quad A_y + \frac{1}{2}\frac{\partial A_y}{\partial x}dx$$

$$\text{along } dc; \quad A_x + \frac{1}{2}\frac{\partial A_x}{\partial x}dy$$

and

$$\text{along } ad; \quad A_y - \frac{1}{2}\frac{\partial A_y}{\partial x}dx$$

Here we have considered only the first order of small quantities dx, dy in the Taylor's expansion.

Therefore, the closed path line integral around the boundary $abcd$ of the rectangle is given by

$$= \left[\left(A_x - \frac{1}{2}\frac{\partial A_x}{\partial y}dy\right) - \left(A_x + \frac{1}{2}\frac{\partial A_x}{\partial y}dy\right)\right]dx$$

$$+ \left[\left(A_y + \frac{1}{2}\frac{\partial A_y}{\partial x}dx\right) - \left(A_y - \frac{1}{2}\frac{\partial A_y}{\partial x}dx\right)\right]dy$$

$$= \left(\frac{\partial A_y}{\partial x} - \frac{\partial A_x}{\partial y}\right)dxdy$$

on dividing the value of the circulation by the area $dxdy$ of the rectangle, one obtains the magnitude of the component of curl **A** along z axis as

$$(\text{curl } \mathbf{A})_z = \left(\frac{\partial A_y}{\partial x} - \frac{\partial A_x}{\partial y}\right)\hat{k}$$

Similarly, one obtains the components of curl **A** along x and y axes as

$$(\text{Curl } \mathbf{A})_x = \left(\frac{\partial A_z}{\partial y} - \frac{\partial A_y}{\partial z}\right)\hat{i}$$

and

$$(\text{Curl } \mathbf{A})_y = \left(\frac{\partial A_x}{\partial z} - \frac{\partial A_z}{\partial x}\right)\hat{j}$$

Adding the three components, one obtains

$$\text{Curl } \mathbf{A} = \left(\frac{\partial A_z}{\partial y} - \frac{\partial A_y}{\partial z}\right)\hat{i} + \left(\frac{\partial A_x}{\partial z} - \frac{\partial A_z}{\partial x}\right)\hat{j} + \left(\frac{\partial A_y}{\partial x} - \frac{\partial A_x}{\partial y}\right)\hat{k} \tag{2.45}$$

One can express $\nabla \times \mathbf{A}$ in the form of a determinant as

$$\text{Curl } \mathbf{A} = \begin{vmatrix} \hat{i} & \hat{j} & \hat{k} \\ \frac{\partial}{\partial x} & \frac{\partial}{\partial y} & \frac{\partial}{\partial z} \\ A_x & A_y & A_z \end{vmatrix} \tag{2.46}$$

We must note that (2.46) is independent of axes. The name curl indicates that a vector field at a point in space has circulation. The determinant in (2.46) is equal to the vector product of del operator (∇) and vector **A**, i.e.

$$\text{Curl}\,\mathbf{A} = \nabla \times \mathbf{A} \tag{2.47}$$

2.15 STOKE'S CURL THEOREM

This states that *the flux of the curl of a vector field* **A** *over any surface of any shape is equal to the line integral of the vector field* **A** *over the boundary of that surface.* We must note that the sense of line integral is related with the positive normal to the surface.

Mathematically one Can express it as

$$\iint_s \text{Curl}\mathbf{A} \cdot ds = \oint \mathbf{A} \cdot d\mathbf{l} \tag{2.48}$$

Here the line integration is being carried out for the boundary of the surface.

The path C bounds the surface S as shown in Fig. 2.11. Let the surface S be divided into large number of small areas $\mathbf{ds}_1, \mathbf{ds}_2, \mathbf{ds}_3, \ldots$ etc. Since all the areas are to be traversed by the curves bounding them in the same sense. The sum of the line integrals around these areas is equal to the line integral around the closed path C, i.e.

$$\oint \mathbf{A} \cdot d\mathbf{l} = \Sigma_i \oint_c \mathbf{A} \cdot d\mathbf{l}$$

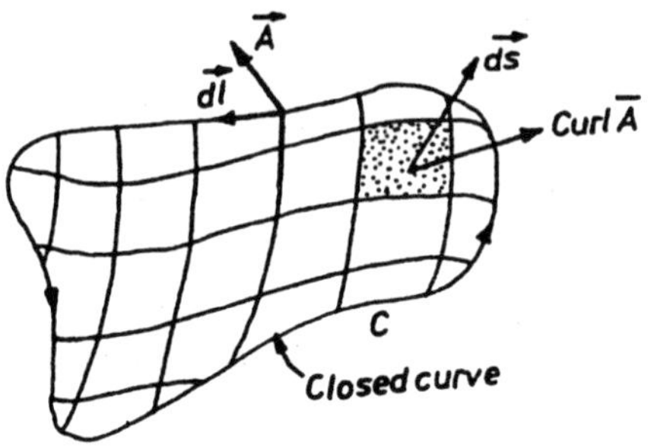

Figure 2.11.

From the definition of curl A, at a point inside ds_i, we have

$$\text{Curl}\mathbf{A} \cdot \mathbf{ds_i} = \oint_{c_i} \mathbf{A} \cdot d\mathbf{l}$$

Since the number of these areas is very large and hence in the limiting case

$$\oint \mathbf{A} \cdot d\mathbf{l} = \left[\lim \partial s \to 0 \quad \Sigma_i \frac{\partial}{\partial s_i} \oint \mathbf{A} \cdot d\mathbf{l} \right] \partial s_i \tag{2.49}$$

Since the curl is the line integral per unit area in the limiting case, when area tends to zero, we have

$$\oint \mathbf{A} \cdot d\mathbf{l} = \iint_s \text{Curl } \mathbf{A} \cdot d\mathbf{s} \tag{2.50}$$

Eq. (2.50) represents the *Stoke's curl theorem.* Obviously, *the line integral of a vector field along the limiting boundary of the area enclosed by a closed path is equal to the surface integral of the curl of the vector field.* This theorem is used in converting line integral into surface integral and vice versa.

2.16 PHYSICAL MEANING OF CURL OF A VECTOR FIELD

The existence of the curl of a vector field is always due to its rotation or circulation. A vector field having non-zero curl has circulation and has curly or rotational field lines about the point of non-zero curl. In German literature curl is written as 'rot' which is derived from rotation. The following examples illustrates clearly the physical meaning of curl.

(i) In an electrostatic field **E**, the curl **E** is zero every where. The line integral of the field **E** between two points A and B is independent of the path. Therefore line integral along A to B is negative of the line integral along B to A or in other words

$$\int_A^B \mathbf{E} \cdot d\mathbf{l} = -\int_B^A \mathbf{E} \cdot d\mathbf{l}$$

Hence

$$\int_A^B \mathbf{E} \cdot d\mathbf{l} + \int_B^A \mathbf{E} \cdot d\mathbf{l} = 0$$

Obviously, the line integral of the field **E** around any closed path is zero. Also if the circulation is zero around any closed path, then from the stoke's theorem, the surface integral of curl **E** is over an element of any size and shape. This clearly reveals that curl **E** should also be zero everywhere, i.e.

$$\text{Curl } \mathbf{E} = 0 \tag{2.51}$$

The above condition is for a field to be conservative. Since the line integral of a vector around the closed path is zero when the vector is derived by taking the gradient of a scalar function. Therefore one conclude that electric field can be described as the gradient of some potential function.

(ii) Let us consider an example of a vector field having non-zero curl. In the subsequent chapters we shall see that the curl of the magnetic induction vector is proportional too the current density vector at the given point, i.e.

$$\nabla \times \mathbf{B} = \mu_0 \mathbf{J} \tag{2.52}$$

where μ_0 is the permeability of free space ($\mu_0 = 4\pi \times 10^{-7}\ NA^{-2}$). However, we must note that (2.52) hold good only for the magnetic field in a vacuum in the absence of time-varying electric fields.

A comparison of (2.51) and (2.52) shows that electrostatic and magnetic fields are of an appreciably different nature. The curl of an electrostatic field is zero, i.e. an electrostatic field is conservative and can be characterized by the scalar potential ϕ. The curl of a magnetic field at points where a current differs from zero. Accordingly the circulation of **B** is proportional to the current enclosed by a loop. This is why we cannot ascribe to a magnetic field a scalar potential that would be related to **B** by an equation similar to (2.36). This potential would not be unique-upon each circum-vention of the look and return to the initial point it would receive an increment equal to $\mu_0 I$. A field whose curl differs from zero is called a *Vortex* or a *Solenoidal* one.

Since $\nabla \cdot \mathbf{B}$ is zero every-where and hence **B** can be represented as the curl of another vector **A**. Thus

$$\mathbf{B} = \nabla \times \mathbf{A} \tag{2.53}$$

The divergence of a curl is always zero. **A** is called the *vector potential.*

2.17 GREEN'S THEOREM

Green's theorem is a corollary of the divergence theorem. The theorem states that if ϕ and ψ are two scalar point functions such that these functions and their first derivatives are continuously differentiable, then

$$\iiint_v (\phi\nabla^2\psi + \nabla\phi\cdot\nabla\psi)dv = \iint_s (\phi\nabla\psi)\cdot ds \tag{2.54}$$

and

$$\iiint_v (\phi\nabla^2\psi - \psi\nabla^2\phi)dv = \iint_s (\phi\nabla\psi - \psi\nabla\phi)\cdot ds \tag{2.55}$$

(2.54) and (2.55) are respectively known as first and second form of Green's theorem. One can derive it as follows:

We start with Gauss's divergence theorem which states

$$\iiint_v \nabla \cdot \mathbf{A} dv = \iint_s \mathbf{A} \cdot ds \tag{2.56}$$

Let us take $\mathbf{A} = \phi \nabla \psi$ or

$$\hat{i}A_x + \hat{j}A_y + \hat{k}A_z = \phi \left(\hat{i}\frac{\partial \psi}{\partial x} + \hat{j}\frac{\partial \psi}{\partial y} + \hat{k}\frac{\partial \psi}{\partial z} \right)$$

Obviously,

$$A_x = \phi \frac{\partial \psi}{\partial x}, \quad A_y = \phi \frac{\partial \psi}{\partial y} \quad \text{and} \quad A_z = \phi \frac{\partial \psi}{\partial z}$$

Now,

$$\begin{aligned}
\nabla \cdot \mathbf{A} &= \frac{\partial A_x}{\partial x} + \frac{\partial A_y}{\partial y} + \frac{\partial A_z}{\partial z} \\
&= \frac{\partial}{\partial x}\left(\phi \frac{\partial \psi}{\partial x}\right) + \frac{\partial}{\partial y}\left(\phi \frac{\partial \psi}{\partial y}\right) + \frac{\partial}{\partial z}\left(\phi \frac{\partial \psi}{\partial z}\right) \\
&= \left(\phi \frac{\partial^2 \psi}{\partial x^2} + \frac{\partial \phi}{\partial x}\frac{\partial \psi}{\partial x}\right) + \left(\phi \frac{\partial^2 \psi}{\partial y^2} + \frac{\partial \phi}{\partial y}\frac{\partial \psi}{\partial y}\right) \\
&\quad + \left(\phi \frac{\partial^2 \psi}{\partial z^2} + \frac{\partial \phi}{\partial z}\frac{\partial \psi}{\partial z}\right) \\
&= \phi \left(\frac{\partial^2 \psi}{\partial x^2} + \frac{\partial^2 \psi}{\partial y^2} + \frac{\partial^2 \psi}{\partial z^2} \right) + \frac{\partial \phi}{\partial x}\frac{\partial \psi}{\partial x} + \frac{\partial \phi}{\partial y}\frac{\partial \psi}{\partial y} + \frac{\partial \phi}{\partial z}\frac{\partial \psi}{\partial z} \\
&= \phi \nabla^2 \psi + \nabla \phi \cdot \nabla \psi
\end{aligned}$$

Making use of these results in (2.56), one obtains

$$\iiint_v (\phi \nabla^2 \psi + \nabla \phi \cdot \nabla \psi) dv = \iint_s (\phi \nabla \psi) \cdot ds \tag{2.57}$$

This is Green's first equation.

Interchanging ϕ and ψ in (2.57), one obtains

$$\iiint_v (\psi \nabla^2 \phi + \nabla \psi \cdot \nabla \psi) dv = \iint_s (\psi \nabla \phi) \cdot ds \tag{2.58}$$

Subtracting (2.58) from (2.57), one obtains

$$\iiint_v (\phi \nabla^2 \psi - \psi \nabla^2 \phi) dv = \iint_s (\phi \nabla \psi - \psi \nabla \phi) \cdot ds \tag{2.59}$$

This is Green's second equation. This is also referred as *symmetrical theorem*.

In the R.H.S. of (2.59), we have

$$\nabla\psi\cdot ds = \frac{\partial\psi}{\partial n}ds$$

and

$$\nabla\phi\cdot ds = \frac{\partial\phi}{\partial n}ds$$

where $\frac{\partial\psi}{\partial n}$ and $\frac{\partial\phi}{\partial n}$ are the directional derivatives of ψ and ϕ respectively along the outward normal to ds. In the light of these results (2.59) can also be expressed as

$$\iiint_v(\phi\nabla^2\psi-\psi\nabla^2\phi)dv=\iint_s\left(\phi\frac{\partial\psi}{\partial n}-\psi\frac{\partial\phi}{\partial n}\right)ds \qquad (2.60)$$

(2.60) is an alternative from of Green's Symmetrical theorem.

One can easily derive Gauss's divergence theorem from Green's theorem. We have Green's symmetrical theorem (2.59). If ϕ is constant, then we have

$$\phi\iiint_v\nabla^2\psi dv = \phi\iint_s\nabla\psi\cdot ds$$

or

$$\iiint_v(\nabla\cdot\nabla\psi)dv = \iint_s\nabla\psi\cdot ds$$

Putting $\nabla\psi=\mathbf{A}$, one obtains

$$\iiint_v(\nabla\cdot\mathbf{A})dv=\iint_s\mathbf{A}\cdot ds \qquad (2.61)$$

Which is *Gauss's divergence theorem.*

EXAMPLE 2.1: A field is expressed by the potential function $V = 3x^2z-xy^3+z$. Calculate the potential at the point $(1,-2,+2)$.

Solution: We have

$$V=3x^2z-xy^3+z$$

Potential at the point $(1,-2,+2)$ is obtained as

$$V_{(1,-2,+2)}=3(1)^2\times2-(1)(-2)^3+2=16 \text{ Volts}$$

EXAMPLE 2.2: A vector field **A** is represented by the function $\mathbf{A}=\hat{\imath}(2x^2y-x^4)+\hat{\jmath}(yz^2)-\hat{k}(xy^2)$, then obtain the values of $\frac{\partial\mathbf{A}}{\partial x}$, $\frac{\partial^2\mathbf{A}}{\partial x^2}$ and $\frac{\partial^2\mathbf{A}}{\partial y\partial z}$.

Solution: We have

$$\mathbf{A} = \hat{i}(2x^2y - x^4) + \hat{j}yz^2 - \hat{k}xy^2$$

$$\therefore \quad \frac{\partial \mathbf{A}}{\partial x} = \frac{\partial}{\partial x}[\hat{i}(2x^2y - x^4) + \hat{j}yz^2 - \hat{k}xy^2]$$

$$= \hat{i}(4xy - 4x^3) + \hat{j}0 - \hat{k}y^2$$

Now
$$\frac{\partial^2 \mathbf{A}}{\partial x^2} = \frac{\partial}{\partial x}[\hat{i}(4xy - 4x^3) - \hat{k}y^2]$$

$$= \hat{i}(4y - 12x^2) - \hat{k}0$$

and
$$\frac{\partial^2 \mathbf{A}}{\partial y \partial z} = \frac{\partial}{\partial z}\left(\frac{\partial \mathbf{A}}{\partial y}\right)$$

$$= \frac{\partial}{\partial z}\left[\frac{\partial}{\partial y}\{\hat{i}(2x^2y - x^4) + \hat{j}yz^2 - \hat{k}xy^2\}\right]$$

$$= \frac{\partial}{\partial z}[\hat{i}(2x^2) + \hat{j}z^2 - \hat{k}2xy] = 2\hat{j}z.$$

EXAMPLE 2.3: Find the value of $\text{div}(r^n\,\hat{r})$, where $\hat{r}$ is the unit vector.

Solution: We have $\mathbf{r} = \hat{i}x + \hat{j}y + \hat{k}z$

$$\therefore |\mathbf{r}| = \sqrt{x^2 + y^2 + z^2}$$

$$\therefore \text{div}(r^n\hat{r}) = \nabla \cdot (r^n\hat{r}) = \left(\hat{i}\frac{\partial}{\partial x} + \hat{j}\frac{\partial}{\partial y} + \hat{k}\frac{\partial}{\partial z}\right) \cdot (r^{n-1}\, r\,\hat{r})$$

$$= \left(\hat{i}\frac{\partial}{\partial x} + \hat{j}\frac{\partial}{\partial y} + \hat{k}\frac{\partial}{\partial z}\right) \cdot (r^{n-1}\,\mathbf{r})$$

$$= \left(\hat{i}\frac{\partial}{\partial x} + \hat{j}\frac{\partial}{\partial y} + \hat{k}\frac{\partial}{\partial z}\right) \cdot r^{n-1}(\hat{i}x + \hat{j}y + \hat{k}z)$$

$$= \frac{\partial}{\partial x}(xr^{n-1}) + \frac{\partial}{\partial y}(yr^{n-1}) + \frac{\partial}{\partial z}(zr^{n-1})$$

$$= \left[r^{n-1} + x\frac{\partial}{\partial x}(x^2 + y^2 + z^2)^{n-1/2}\right]$$

$$+ \left[r^{n-1} + y\frac{\partial}{\partial y}(x^2 + y^2 + z^2)^{n-1/2}\right]$$

$$+ \left[r^{n-1} + z\frac{\partial}{\partial z}(x^2 + y^2 + z^2)^{n-1/2}\right]$$

$$= r^{n-1} + x\left(\frac{n-1}{2}\right)(x^2+y^2+z^2)^{n-3/2}2x + \cdots$$

$$= 3r^{n-1} + (n-1)[x^2+y^2+z^2]^{n-1/2}$$

$$= 3r^{n-1} + (n-1)r^{n-1} = (n+2)r^{n-1}$$

EXAMPLE 2.4: Find (a) $\nabla \cdot \mathbf{r}$, where $\mathbf{r}$ is the position vector and (b) $\nabla \cdot [\hat{i}(x+y) + \hat{j}(x-y) + \hat{k}(4z)]$

Solution: (a) $\nabla \cdot \mathbf{r} = \left(\hat{i}\frac{\partial}{\partial x} + \hat{j}\frac{\partial}{\partial y} + \hat{k}\frac{\partial}{\partial y}\right) \cdot (\hat{i}x + \hat{j}y + \hat{k}z)$

$$= \frac{\partial x}{\partial x} + \frac{\partial y}{\partial y} + \frac{\partial z}{\partial z} = 3$$

(b) $\nabla \cdot [\hat{i}(x+y) + \hat{j}(x-y) + \hat{k}(4z)]$

$$= \left(\hat{i}\frac{\partial}{\partial x} + \hat{j}\frac{\partial}{\partial y} + \hat{k}\frac{\partial}{\partial z}\right) \cdot [\hat{i}(x+y) + \hat{j}(x-y) + \hat{k}(4z)]$$

$$= \frac{\partial}{\partial x}(x+y) + \frac{\partial}{\partial y}(x-y) + \frac{\partial}{\partial z}(4z)$$

$$= 1 - 1 + 4 = 4$$

EXAMPLE 2.5: If $\mathbf{r}$ is the position vector of a point, calculate the gradient of $\frac{1}{r}$.

Solution:

$$\mathbf{r} = \hat{i}x + \hat{j}y + \hat{k}z$$

$$\therefore \quad |r| = \sqrt{x^2+y^2+z^2}$$

Now,

$$\nabla\left(\frac{1}{r}\right) = \nabla(x^2+y^2+z^2)^{-1/2}$$

or

$$\nabla\left(\frac{1}{r}\right) = \left(\hat{i}\frac{\partial}{\partial x} + \hat{j}\frac{\partial}{\partial y} + \hat{k}\frac{\partial}{\partial z}\right)(x^2+y^2+z^2)^{-1/2}$$

$$= \hat{i}\frac{\partial}{\partial x}(x^2+y^2+z^2)^{-1/2}$$

$$+ \hat{j}\frac{\partial}{\partial y}(x^2+y^2+z^2)^{-1/2}$$

$$+ \hat{k}\frac{\partial}{\partial z}(x^2+y^2+z^2)^{-1/2}$$

$$\because \quad \frac{\partial}{\partial x}(x^2+y^2+z^2)^{-1/2} = -\frac{1}{2}(x^2+y^2+z^2)^{-3/2}(2x) = \frac{-x}{r^3}$$

Similarly,

$$\frac{\partial}{\partial y}(x^2+y^2+z^2)^{-1/2} = \frac{-y}{r^3}$$

and

$$\frac{\partial}{\partial z}(x^2+y^2+z^2)^{-1/2} = \frac{-z}{r^3}$$

$$\therefore \quad \nabla\left(\frac{1}{r}\right) = -\frac{\hat{i}x+\hat{j}y+\hat{k}z}{r^3} = \frac{-\mathbf{r}}{r^3}$$

Example 2.6: The electric field due to a point charge is expressed as $\mathbf{E} = \frac{Q}{r^2}\hat{r}$. Show that the divergence of electric field due to that point charge is zero.

Solution:

$$\begin{aligned}
\nabla\cdot\mathbf{E} &= \nabla\cdot\left(\frac{Q}{r^2}\hat{r}\right) = \nabla\cdot\left(\frac{Q}{r^3}\hat{r}r\right) = \nabla\cdot\left(\frac{Q}{r^3}\mathbf{r}\right) \\
&= \left(\hat{i}\frac{\partial}{\partial x}+\hat{j}\frac{\partial}{\partial y}+\hat{k}\frac{\partial}{\partial z}\right)\cdot Q\left(\frac{\hat{i}x+\hat{j}y+\hat{k}z}{(x^2+y^2+z^2)^{3/2}}\right) \\
&\quad \left(\because \quad \mathbf{r} = \hat{i}x+\hat{j}y+\hat{k}z \quad \therefore \quad |\mathbf{r}| = \sqrt{x^2+y^2+z^2}\,\right) \\
&= Q\left[\frac{\partial}{\partial x}\left\{\frac{x}{(x^2+y^2+z^2)^{3/2}}\right.\right] + Q\left[\frac{\partial}{\partial y}\left\{\frac{y}{(x^2+y^2+z^2)^{3/2}}\right.\right] \\
&\quad +Q\left[\frac{\partial}{\partial z}\left\{\frac{z}{(x^2+y^2+z^2)^{3/2}}\right.\right] \\
&= Q\left[\frac{1}{(x^2+y^2+z^2)^{3/2}} - \frac{3x^2}{(x^2+y^2+z^2)^{5/2}} + \frac{1}{(x^2+y^2+z^2)^{3/2}}\right. \\
&\quad \left. - \frac{3y^2}{(x^2+y^2+z^2)^{5/2}} + \frac{1}{(x^2+y^2+z^2)^{3/2}} - \frac{3z^2}{(x^2+y^2+z^2)^{5/2}}\right] \\
&= 0
\end{aligned}$$

Example 2.7: Show that the potential function $V = q(x^2+y^2+z^2)^{1/2}$ does not satisfy the Laplace's equation.

Solution: The Laplace's equation is $\nabla^2 V = 0$ i.e.

$$\frac{\partial^2 V}{\partial x^2}+\frac{\partial^2 V}{\partial y^2}+\frac{\partial^2 V}{\partial z^2} = 0 \tag{1}$$

Using $V = q(x^2+y^2+z^2)^{1/2}$, L.H.S. of (1) becomes

$$\begin{aligned}
\text{L.H.S.} &= q\frac{\partial}{\partial x}\left[\frac{\partial}{\partial x}(x^2+y^2+z^2)^{1/2}\right] + q\frac{\partial}{\partial y}\left[\frac{\partial}{\partial y}(x^2+y^2+z^2)^{1/2}\right] \\
&\quad + q\frac{\partial}{\partial z}\left[\frac{\partial}{\partial z}(x^2+y^2+z^2)^{1/2}\right] \\
&= q\frac{\partial}{\partial x}\left[x(x^2+y^2+z^2)^{1/2}\right] + q\frac{\partial}{\partial y}\left[y(x^2+y^2+z^2)^{1/2}\right] \\
&\quad + q\frac{\partial}{\partial z}\left[z(x^2+y^2+z^2)^{1/2}\right] \\
&= q\left\{-\frac{1}{4}(x^2+y^2+z^2)^{-3/2}4x + (x^2+y^2+z^2)^{-1/2}\right. \\
&\quad -\frac{1}{4}(x^2+y^2+z^2)^{-3/2}4y + (x^2+y^2+z^2)^{-1/2} \\
&\quad \left. -\frac{1}{4}(x^2+y^2+z^2)^{-3/2}4z + (x^2+y^2+z^2)^{-1/2}\right\} \\
&= q\left\{(x^2+y^2+z^2)^{-3/2}(x^2+y^2+z^2) + 3(x^2+y^2+z^2)^{-1/2}\right\} \\
&= 2q(x^2+y^2+z^2)^{-1/2} \neq 0
\end{aligned}$$

Obviously, the given potential function does not satisfy the Laplace's equation.

EXAMPLE 2.8: Show that curl $\mathbf{r} = 0$, where $\mathbf{r}$ is position vector.

Solution:

$$\begin{aligned}
\nabla \times \mathbf{r} &= \left(\hat{i}\frac{\partial}{\partial x} + \hat{j}\frac{\partial}{\partial y} + \hat{k}\frac{\partial}{\partial z}\right) \times (\hat{i}x + \hat{j}y + \hat{k}z) \\
&= \begin{vmatrix} \hat{i} & \hat{j} & \hat{k} \\ \frac{\partial}{\partial x} & \frac{\partial}{\partial y} & \frac{\partial}{\partial z} \\ x & y & z \end{vmatrix} = 0
\end{aligned}$$

EXAMPLE 2.9: A potential function in an electric field is given by, $\phi = (2x^2 - y^2 + 3z^2)$ Volts. Show that the intensity of electric field at point $(1, -1, 0)$m is $-(4\hat{i} - 2\hat{j})$ V/m.

Solution: $\text{E} = -\nabla\phi = -\left(\hat{i}\frac{\partial\phi}{\partial x} + \hat{j}\frac{\partial\phi}{\partial y} + \hat{k}\frac{\partial\phi}{\partial z}\right)$

Now,

$$\frac{\partial \phi}{\partial x} = 4x, \ \frac{\partial \phi}{\partial y} = -2y \quad \text{and} \quad \frac{\partial \phi}{\partial z} = 6z$$

$$\therefore \mathbf{E} = -(\hat{i}4x - \hat{j}2y + \hat{k}6z)$$

$\therefore$ **E** at the point (1,-1,0)

$$= -(\hat{i}4 - 2\hat{j}) = (-4\hat{i} + 2\hat{j})\,\text{V/m}$$

EXAMPLE 2.10: Show that curl grad $\phi = 0$, where ϕ is any scalar function.

Solution:

$$\begin{aligned}
\nabla \times \nabla\phi &= \left(\hat{i}\frac{\partial}{\partial x} + \hat{j}\frac{\partial}{\partial y} + \hat{k}\frac{\partial}{\partial z}\right) \times \left(\hat{i}\frac{\partial \phi}{\partial x} + \hat{j}\frac{\partial \phi}{\partial y} + \hat{k}\frac{\partial \phi}{\partial z}\right) \\
&= \begin{vmatrix} \hat{i} & \hat{j} & \hat{k} \\ \frac{\partial}{\partial x} & \frac{\partial}{\partial y} & \frac{\partial}{\partial z} \\ \frac{\partial \phi}{\partial x} & \frac{\partial \phi}{\partial y} & \frac{\partial \phi}{\partial z} \end{vmatrix} \\
&= \hat{i}\left(\frac{\partial^2 \phi}{\partial y \partial z} - \frac{\partial^2 \phi}{\partial z \partial y}\right) + \hat{j}\left(\frac{\partial^2 \phi}{\partial z \partial x} - \frac{\partial^2 \phi}{\partial x \partial z}\right) \\
&\quad + \hat{k}\left(\frac{\partial^2 \phi}{\partial x \partial y} - \frac{\partial^2 \phi}{\partial y \partial x}\right) \\
&= 0\left(\because \quad \frac{\partial^2 \phi}{\partial y \partial z} = \frac{\partial^2 \phi}{\partial z \partial y} \quad \text{and so on}\right)
\end{aligned}$$

EXAMPLE 2.11: Show that div grad $(R^n) = n(n+1)r^{n-2}$, where R is position vector.

Solution: div grad $(R^n) = \nabla \cdot \nabla(R^n)$ Now

$$\begin{aligned}
\nabla(R^n) &= \left(\hat{i}\frac{\partial}{\partial x} + \hat{j}\frac{\partial}{\partial y} + \hat{k}\frac{\partial}{\partial z}\right) R^n \\
&= \hat{i}nR^{n-1}\frac{\partial R}{\partial x} + \hat{j}nR^{n-1}\frac{\partial R}{\partial y} + \hat{k}nR^{n-1}\frac{\partial R}{\partial z} \\
&= nR^{n-1}\left(\hat{i}\frac{\partial R}{\partial x} + \hat{j}\frac{\partial R}{\partial y} + \hat{k}\frac{\partial R}{\partial z}\right)
\end{aligned}$$

$$\because \quad \mathbf{R} = \hat{i}x + \hat{j}y + \hat{k}z$$
$$\therefore \quad R^2 = \mathbf{R}\cdot\mathbf{R} = x^2 + y^2 + z^2$$
$$\therefore \quad 2R\frac{\partial R}{\partial x} = 2x$$
$$\therefore \quad \frac{\partial R}{\partial x} = \frac{x}{R}$$

Similarly $\dfrac{\partial R}{\partial y} = \dfrac{y}{R}$ and $\dfrac{\partial R}{\partial z} = \dfrac{z}{R}$

$$\therefore \nabla R^n = nR^{n-1}\left(\hat{i}\frac{x}{R} + \hat{j}\frac{y}{R} + \hat{k}\frac{z}{R}\right)$$

$$\therefore \nabla\cdot\nabla R^n = \left(\hat{i}\frac{\partial}{\partial x} + \hat{j}\frac{\partial}{\partial y} + \hat{k}\frac{\partial}{\partial z}\right)\cdot nR^{n-2}(\hat{i}x + \hat{j}y + \hat{k}z)$$

$$= \frac{\partial}{\partial x}(nR^{n-2}x) + \frac{\partial}{\partial y}(nR^{n-2}y) + \frac{\partial}{\partial z}(nR^{n-2}z)$$

$$= n\left\{R^{n-2} + x(n-2)R^{n-3}\frac{\partial R}{\partial x} + R^{n-2}\right.$$
$$\left. + y(n-2)R^{n-3}\frac{\partial R}{\partial y} + R^{n-2} + z(n-2)R^{n-3}\frac{\partial R}{\partial z}\right\}$$

$$= 3nR^{n-2} + n(n-2)R^{n-3}\left\{x\frac{\partial R}{\partial x} + y\frac{\partial R}{\partial y} + z\frac{\partial R}{\partial z}\right\}$$

$$= 3nR^{n-2} + n(n-2)R^{n-3}\left\{x\frac{x}{R} + y\frac{y}{R} + z\frac{z}{R}\right\}$$

$$= 3nR^{n-2} + n(n-2)R^{n-3}\left(\frac{x^2 + y^2 + z^2}{R}\right)$$

$$= 3nR^{n-2} + n(n-2)R^{n-3}R$$

$$= 3R^{n-2}(3 + n - 2) = n(n+1)R^{n-2}$$

QUESTIONS

1. Describe Scalar and vector fields. How can they be represented graphically? **(Madras, 1998)**
2. What do you mean by the field of any physical quantity? What are scalar and vector fields? Give examples.

   (Calcutta, 1998; Delhi, 1997)

3. Define gradient of a scalar field. Show that the gradient of a scalar field is a vector. Give an example. **(MDS Ajmer, 1999)**
4. Prove that the gradient of a scalar field is normal to the surface, $\phi = \text{constant}$. **(Rajasthan, 1997; Patna, 1996)**
5. Express the gradient of a scalar field $\phi(x, y, z)$ in terms of cartesian coordinates. **(Rajasthan, 1992; MDSA, 1997)**
6. Show that when a vector field can be derived as the gradient of a scalar field the line integral of the vector taken round any closed path in the vector field is zero. **(MDSA, 1996; Mysore, 1998).**
7. Explain the meaning of line integral of a vector field using atleast one example. **(Rajasthan, 1999; Delhi, 1997)**
8. Explain the physical meaning of the divergence of a vector field **A**. Express divergence **A** in orthogonal coordinates and show that div $\mathbf{A} = \nabla \cdot \mathbf{A}$.
9. Define curl of a vector field and give its physical significance. Derive expression for a vector field and show that curl $\mathbf{A} = \nabla \times \mathbf{A}$.
10. What is a Laplacian operator? Express div grad ϕ in cartesion coordinates, where ϕ is a scalar quantity.
11. Derive Laplace' and Poisson's equations starting from the differential form of Gauss's law.
12. State and Prove Green's theorem.

PROBLEMS

1. For a position vector $\mathbf{R} = \hat{i}x + \hat{j}y + \hat{k}z$, and the values of (i) grad $\frac{1}{R}$ (ii) div $\left(\frac{\mathbf{R}}{R^3}\right)$ and (iii) curl $\left(\frac{\mathbf{R}}{R^3}\right)$

   [Ans. (i) $-\frac{\mathbf{R}}{R^3}$ (ii) 0 (iii) 0]

2. A potential field is represented by the equation, $\phi = 4yz^3 + 3xyz - z^2 + 2$. Calculate the potential at the points $(1, -1, -2)$.

   [Ans. -36V]

3. A vector field is represented by the equation, $\mathbf{A} = \hat{i}x^2yz - \hat{j}2xz^3 + \hat{k}xz^2$. Calculate the Values of the partial derivatives (i) $\frac{\partial \mathbf{A}}{\partial x}$, (ii) $\frac{\partial \mathbf{A}}{\partial z}$.

   [Ans. (i) $\hat{i}2xyz, -\hat{j}2z^3 + \hat{k}z^2$ (ii) $\hat{i}x^2z - \hat{j}6xz^2 + \hat{k}2xz$]

4. The temperature of a body at any point changes according to the relation, $T = 4x^2 + 3y^2 - 2z^2$. Show that the grad T at the point (1, 2, 3) is equal to $-8\hat{i} + 12\hat{j} - 12\hat{k}$.
5. Show that the function $\phi = x^2 - y^2$ satisfies the Laplace's equation.
6. Show that the potential functions
   (i) $V = x^2 - y^2 + z$; (ii) $V = x^2 + y^2 - 2z$;
   (iii) $V = ax + by + cz$ and; (iv) $V = ax^2 - ay^2 + 4z$,
   satisfy Laplace's equation.
7. Calculate the Curl of the following fields
   (a) $\mathbf{E} = a\{3x^2yz\hat{i} + [x^3z + yz^2(yz)^{1/2}]\hat{j} + [x^3y + y^2z(yz)^{1/2}]\hat{k}\}$
   (b) $\mathbf{B} = y\hat{i} - x\hat{j}$ and
   (c) $\mathbf{B} = a\left\{\dfrac{y\hat{i}}{(x^2+y^2)^{3/2}} - \dfrac{x\hat{k}}{(x^2+y^2)^{3/2}}\right\}$

$$\left[\textbf{Ans.}\,(a)0, \quad (b) - 2\hat{k} \quad (c)\frac{a\hat{k}}{(x^2+y^2)^{3/2}}\right]$$

8. Calculate the divergence at a point (x, y, z) for the fields in problem 7. [**Ans.** (a) $6axyz + \frac{3}{4}a(yz)^{1/2}$ (b) zero (c) Zero].
9. If $\mathbf{A} = xz^3\hat{i} - 2x^2yz\hat{j} + 2yz^4\hat{k}$, show that Curl $\mathbf{A}$ at the point $(1, -1, 1)$ is $(3\hat{i} + 4\hat{k})$.
10. Evaluate $\iint_s \mathbf{F} \cdot ds$, where $\mathbf{F} = \hat{i}4xz - \hat{j}y^2 + \hat{k}yz$, and S is the surface of the cube bounded by $x = 0, x = 1, y = 0, y = 1, z = 0$ and $z = 1$. [**Ans.** 3]
11. Evaluate $\iint_S \mathbf{r} \cdot \hat{n} ds$, where S is a closed surface and V is the volume enclosed by S. [**Ans.** 3V]
12. Show that $\iint_S \hat{n} ds = 0$ for any closed surface S.
13. Show that $\iint_S$ curl $\mathbf{F} \cdot d\mathbf{S} = 0$ for any closed surface S.
14. A fluid of density $\rho(x, y, z, t)$ moves with velocity $\mathbf{v}(x, y, z, t)$. If there are no sources or sinks, prove that

$$\nabla \cdot \mathbf{J} + \frac{\partial \rho}{\partial t} = 0, \quad \text{where} \quad \mathbf{J} = e\mathbf{V}$$

15. Show that the necessary and sufficient condition for $\oint_c \mathbf{A} \cdot d\mathbf{r} = 0$ to be true for every curve c is that curl $\mathbf{A} = 0$
16. Prove Stoke's theorem for the vector $\mathbf{A}(x + y, 2x - z, y + z)$ taken over the triangle cut from the plane $3x + 2y + z = 6$ by the Coordinate planes.

SHORT ANSWER QUESTIONS

1. How many kinds of fields are there?

[**Ans.** Two: (i) Scalar (ii) vector]

2. Give two examples of vector field.

[**Ans.** (i) Distribution of velocity in a fluid (ii) Distribution of gravitational, magnetic or electric field in space]

3. If $\phi(x,y,z)$ is a scalar field, is $\hat{i}\frac{\partial\phi}{\partial x}+2\hat{j}\frac{\partial\phi}{\partial y}+3\hat{k}\frac{\partial\phi}{\partial z}$ a vector field?

[**Ans.** We know that a scalar field $\phi(x,y,z)$ evaluated at a particular point is independent of the coordinates of that point. Obviously, scalar field, is invariant with respect to the transformation of coordinates. Its gradient, which is a vector field, must also be invariant with respect to the transformation. Now, on applying coordinate transformation, the vector $\hat{i}\frac{\partial\phi}{\partial x}+2\hat{j}\frac{\partial\phi}{\partial y}+3\hat{k}\frac{\partial\phi}{\partial z}$ does not retain its form on applying a coordinate transformation. Thus it cannot be a vector field corresponding to given scalar field $\phi(x,y,z)$.]

4. What are lamellar fields?

[**Ans.** Certain vector fields can be expressed as gradient of a scalar field ϕ and such fields are known as lamellar fields.]

5. What is name of the theorem which provides a method of reducing volume integrals to surface integrals?

[**Ans.** Gauss's theorem of divergence.]

OBJECTIVE QUESTIONS

1. In two dimension, the divergence transforms as:
(a) a scalar, (b) a vector,
(c) a scalar and as a Vectòr, (d) none of the above. [a]

2. The curl of a vector function **A** is:
(a) Scalar, (b) Vector,
(c) vector as well as scalar, (d) none of the above. [b]

3. The divergence of a curl is always:
(a) infinite, (b) Zero, (c) finite, (d) none of the above. [b]

4. The curl of a gradient is always:

   (a) Zero, (b) infinite, (c) finite, (d) none of the above. **[a]**

5. If **E** is the intensity of electric field and *V* is the electric potential, then the relation between these two is:

   (a) $\mathbf{E} = \nabla \times V$, (b) $\mathbf{E} = \nabla \cdot V$, (c) $\mathbf{E} = \nabla V$, (d) $\mathbf{E} = \nabla^2 V$ **[c]**

6. The theorem which relates the line integral of a vector field along the limiting boundary of the area enclosed by a closed path to the surface integral of the curl of the vector field is known as:

   (a) Green's theorem, (b) Gauss's divergence theorem,
   (c) Stoke's curl theorem, (d) none of the above. **[b]**

Chapter 3

Electrostatics

3.1 INTRODUCTION

Physics is based on the concept that all matter is made of atoms and that these atoms in turn are made of electric charged particles whose behaviour is governed by electrodynamics.

There are two kinds of electric charges: *positive* and *negative*. Like charges repel each other while unlike charges attract each other. The algebric sum of electric charges remains constant in a closed system. This is known as the *law of conservation of electric charges.*

The electric charge of any body consists of a whole number of elementary charges each of which equals 1.6×10^{-19}C charge. The smallest stable subatomic particle having a negative elementary charge is called the *electron.* The mass of an electron is 9.1×10^{-31} kg. The smallest stable subatomic particle having a positive charge is the proton. The mass of the proton is 1.67×10^{-27} kg. Electrons and protons are found in the atoms of any substance. A neutral body contains charges of opposite sign which are equal in absolute value. Electric charges are said to be point charges if the linear dimensions of the bodies on which the charges are concentrated are very much smaller than any other lengths pertitent to the problem under consideration. A body exhibiting positive electrification has a positive electric charge, and one with negative electrification has a negative electric charge. The net charge of a body is the algebraic sum of its positive and negative charges. A particle having a nonzero net charge is often called an *ion.* Since matter in bulk does not exhibit gross electrical forces. One may assume that it is composed of equal amounts of positive and negative charges. ***The net or total charge does not change for any process occurring within an isolated system.*** **No exception has been found to this rule, i.e.** ***principle of conservation of charge.***

3.2 COULOMB'S LAW

The results of the electric interaction between two charged particles at rest in the observer's inertial frame of reference or, at most, moving with a very small velocity constitute what is called *electrostatics*. The electrostatic interaction for two charged particles is given by *Coulomb's law*.

Charles A. de Coulomb (1736–1806), from experimental observations concluded that the *electrostatic interaction between two charged particles is proportional to the square of the distance between them, and its direction is along the lines joining the two charges.*

Let '1' and '2' are two particles carrying charges q_1, and q_2 respectively and separated by a distance r_{12} (Fig. 3.1). According to Coulomb's law, the electric force exerted by the particle '1' on the particle '2' is

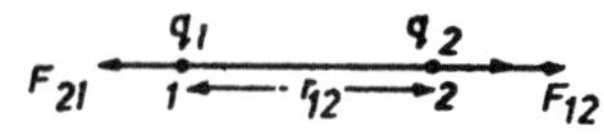

Figure 3.1. Coulomb's Law

$$F_{12} \propto \frac{q_1 q_2}{r_{12}^2} \quad \text{i.e.,} \quad F_{12} = k\frac{q_1 q_2}{r_{12}^2} = -F_{12} \tag{3.1}$$

where k is the constant of proportionality and F_{21} is the force exerted by the particle '2' on particle '1'. Expressing (3.1) in vector notation, one obtains

$$\mathbf{F}_{12} = k\frac{q_1 q_2}{|\mathbf{r}_{12}|^2}\hat{r}_{12} = k\frac{q_1 q_2}{|\mathbf{r}_{12}|^3}\mathbf{r}_{12} \tag{3.2}$$

where $\hat{r}_{12}$ is the unit vector along $\mathbf{r}_{12}$.

Coulomb's law is very similar to the law of gravitational interaction. The sign of charges in (3.2) decides whether the force is *attractive* or *repulsive*. When both the charges are similar (i.e., positive or negative, F_{12} is positive and represents the force of repulsion; while if one is positive and other is negative, $\mathbf{F}_{12}$ is negative and is the force of attraction.

The numerical value of the constant K in the MKSC system is equal to $10^{-7}C^2 = 8.9874 \times 10^9$, where C is the Velocity of light. For practical purposes, we may say that $k = 9 \times 10^9$. Then, when the distance r_{12} is measured in meters and the force in newtons, Eq. (3.1) becomes

$$F_{12} = 9 \times 10^9 \frac{q_1 q_2}{r_{12}^2} = -F_{21} \tag{3.3}$$

once we have decided on the value of k, the unit charge is fixed. This unit is called a Coulomb and is designated by C. *The Coulomb is that charge which, when it is placed one meter from an equal charge in Vacuum, repels it with a force of* 8.987×10^9 *newtons.* Formula (3.3) hold only for two charged particles in vacuum; i.e., for two charged particles in the absence of any other charge or matter. We must that according to (3.2), the unit of k is $N - m^2 - C^{-2}$ or m^3 kg $s^{-2}C^{-2}$. The constant k is then Equal to

$$k = \frac{1}{4\pi\varepsilon_0}$$

where ε_0 is the vacuum permeability or *permittivity of free space.* It is also some times referred as Capacitivity in vacuo, or the dielectric constant of a vacuum. Thus

$$\varepsilon_0 = \frac{10^7}{4\pi c^2} = 8.854 \times 10^{-12} N^{-1} m^{-2} C^2 \quad (\text{or } m^{-3} kg^{-1} s^2)$$

Now, the Coulomb's is written as

$$\mathbf{F}_{12} = \frac{1}{4\pi\varepsilon_0} \frac{q_1 q_2}{|\mathbf{r}_{12}|^3} \mathbf{r}_{12} \quad (\text{newtons}) \tag{3.4}$$

The forces of electrostatic interaction are central and long range. One can obtain an idea about the relative magnitudes of electrostatic and gravitational forces by comparing the force of attraction between two electrons due to gravitation and the force of repulsion due to electrostatic repulsion.

$$\frac{\text{Electrostatic repulsion between two electrons}}{\text{Gravitational attraction between two electrons}}$$

$$= \frac{1}{4\pi\varepsilon_0} \frac{e^2}{r^2} \frac{r^2}{Gm^2} = \frac{e^2}{4\pi\varepsilon_0 Gm^2}$$

$$= 9 \times 10^9 \times \frac{(1.6 \times 10^{-19})^2}{(6.67 \times 10^{-11})(9.1 \times 10^{-31})^2}$$

$$= 4.17 \times 10^{42}$$

Obviously, the electrical interaction is of the order of magnitude required to produce the binding between atoms to form molecules, or the binding between the electrons and protons to form atoms.

This reveals that the chemical processes (and in general the behaviour of matter in bulk) are due to electrical interactions between atoms and molecules.

Coulomb's Law is based primarily on experiments. At this stage, one can raise a question whether the law is exactly that of inverse square, i.e., if the force is proportional to $\frac{1}{r^n}$, is n exactly equal to 2? Cavendish found that $n = 2 \pm 0.02$. Plimpton and Lawton in 1936 reported that n differs from 2 by not more than one part in 10^9. Lamb-and Rutherford in 1947 found from their measurements of energy levels of the hydrogen atom that the exponent in Coulomb's law is correct to one part in 10^9 at distances of the order of 10^{-10}m. Evidence from nuclear experimental observation has shown that the electrostatic forces vary approximately according to inverse square law even at distances of the order of 10^{-15} m.

3.3 ELECTRIC FIELD

Any region where an electric charge experiences a force is called an *electric field*. The force is due to the presence of other charges in that region. For example, a charge q placed in a region where there are other charges q_1, q_2, q_3, etc. (Fig. 3.2) experiences a force $\mathbf{F} = \mathbf{F}_1 + \mathbf{F}_2 + \mathbf{F}_3 + \cdots$, and we say that it is in an electric field produced by the charges $q_1, q_2, q_3, \ldots$. We must note that charges q of course also exerts forces on $q_1, q_2, q_3, \ldots$ but we are not concerned with them now. Since the force that each charge $q_1, q_2, q_3, \ldots$ produces on the charge q is proportional to q, the resultant force $\mathbf{F}$ is also proportional to q. Obviously, the force on a particle placed in an electric field is proportional to the charge of the particle.

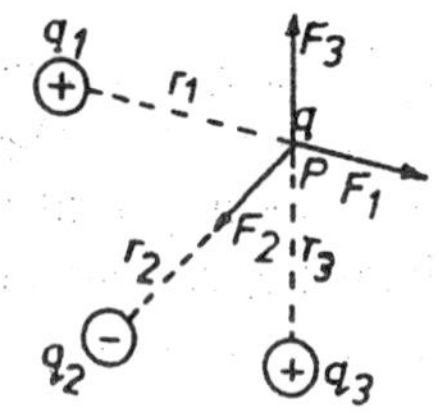

Figure 3.2. Resultant electric field at P due to the presence of several charges in the region.

The intensity of the electric field at a point is equal to the force per unit charge placed at that point. Thus

$$\mathbf{E} = \frac{\mathbf{F}}{q} \quad \text{or} \quad \mathbf{F} = q\mathbf{E} \qquad (3.5)$$

The unit of electric field intensity is newton/Coulomb or NC

If q is positive, the force $\mathbf{F}$ acting the charge has the same direction as the field $\mathbf{E}$, but if q is negative, the force $\mathbf{F}$ has the direction Opposite to $\mathbf{E}$ (Fig. 3.3).

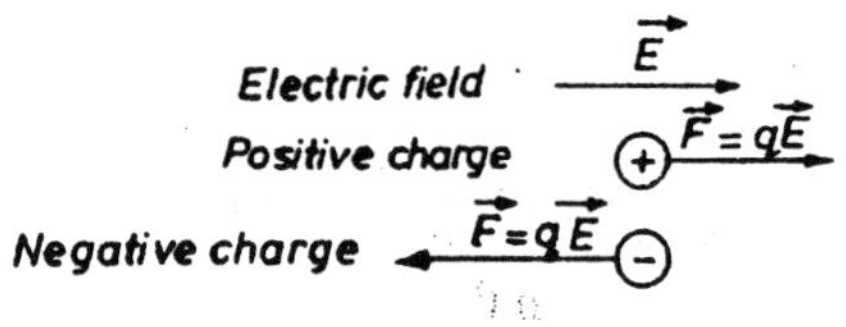

Figure 3.3. Direction of the force produced by an electric field on a positive and a negative charge.

Obviously, if we apply an electric field to a region where positive and negative particles or ions are present, the field will tend to move the positively and negatively charged bodies in opposite directions, resulting in a charge separation, an effect sometimes referred as *polarization*.

Let us write eq. (3.3) in the form $F = q_2\left(\frac{q_1}{4\pi\varepsilon_0 r_{12}^2}\right)$. This gives the force produced by the charge q_1 on the charge q_2 placed at a distance r_{12} from q_1. Using (3.5), we may say that the electric field E at the point where q_2 is placed is such that $F = q_2E$. Obviously, by comparing both expressions of F, we conclude that the electric field at a distance r_{12} from the charge q_1, is $E = \frac{q_1}{4\pi\varepsilon_0 r_{12}^2}$, or in the vector form,

$$\mathbf{E} = \frac{q_1}{4\pi\varepsilon_0 r_{12}^2}\hat{a}_r \qquad (3.6)$$

where $\hat{a}_r$ is the unit vector in the radial direction away from the charge q, since $\mathbf{F}$ is along this direction. Eq. (3,6) is valid for both positive and negative charges with the direction of $\mathbf{E}$ relative to $\hat{a}_r$ given by the sign of q_1. Obviously, $\mathbf{E}$ is directed away from a positive charge and toward a negative charge (Fig. 3.4).

Let $q_1 = q_0$ is sufficiently small so that it does not disort the field whose intensity is to be measured. The resultant force $\mathbf{F}$ exerted on a test charge q by a field set up by a system of fixed charges $q_1, q_2, q_3,$

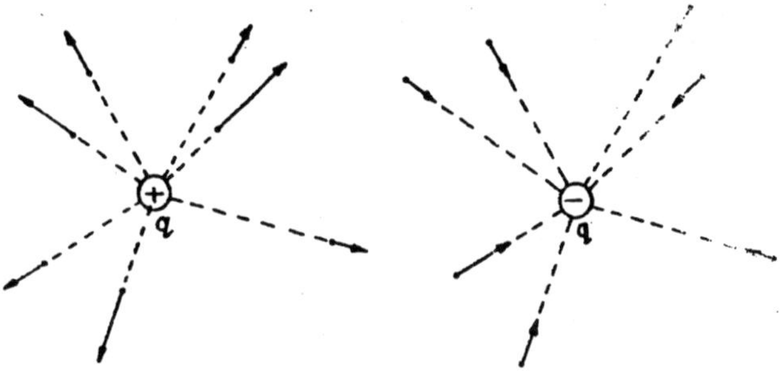

Figure 3.4. Electric field produced by (a) a positive and (b) a negative charge.

$\ldots q_n$ is equal to the vector sum of the forces $\mathbf{F}_i$ exerted on the test charge by each of the fields of charges q_1, i.e.

$$\mathbf{F} = \sum_{i=1}^{n} \mathbf{F}_i \tag{3.7}$$

This leads to the principle of superposition of electric fields:

$$\mathbf{E} = \sum_{i}^{n} \mathbf{E}_i \tag{3.8}$$

3.4 PRINCIPLE OF SUPERPOSITION

If there are more than two particles present with charges, say $q_1, q_2, q_3, \ldots$, the total force on any one particle is the vector sum of forces it experiences due to all other particles present there separately. This is called the principle of superposition. There are three charges q_1, q_2 and q_3 as shown in Fig. 3.5. The force on q_3 is

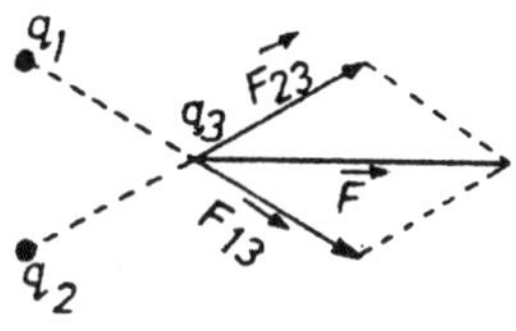

Figure 3.5.

$$\begin{aligned}\mathbf{F} &= \mathbf{F}_{13} + \mathbf{F}_{23} \\ &= \frac{1}{4\pi\varepsilon_0}\frac{q_1 q_3}{|\mathbf{r}_{13}|^3}\mathbf{r}_{13} + \frac{1}{4\pi\varepsilon_0}\frac{q_2 q_3}{|\mathbf{r}_{23}|^3}\mathbf{r}_{23}\end{aligned}$$

Generalising, one finds the force acting on a charge q_j due to a number of other charges present in the region is

$$\mathbf{F}_J = \frac{1}{4\pi\varepsilon_0}\sum_{i\neq j}\frac{q_i q_j}{|r_{13}^3|}\mathbf{r}_{ij} \tag{3.9}$$

Let $\mathbf{r}_i$ and $\mathbf{r}_j$, are the vectors representing the location of charges q_i and q_j respectively (Fig. 3.6). One can write the Eq. (3.9) in terms of these position vectors as

$$\mathbf{F}_j = \frac{1}{4\pi\varepsilon_0} \sum_{i \neq j} \frac{q_i q_j}{|\mathbf{r}_j - \mathbf{r}_i|^3} (\mathbf{r}_j - \mathbf{r}_i) \tag{3.10}$$

The experimentally observed linear superposition of forces due to many charges means that one can write the electric field at **X** due to a system of point charges q_i, located at $\mathbf{X}_i$, $i = 1, 2, \ldots, n$, as a vector sum

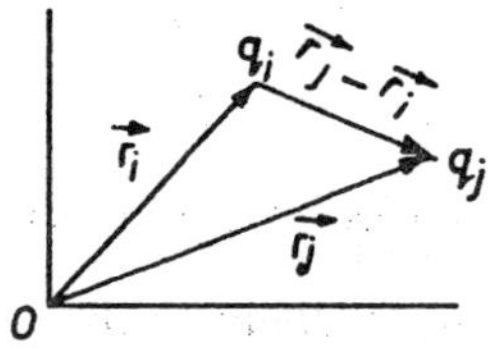

Figure 3.6.

$$\mathbf{E}(X) = \frac{1}{4\pi\varepsilon_0} \sum_{i=1}^{n} q_i \frac{\mathbf{X} - \mathbf{X}_i}{|\mathbf{X} - \mathbf{X}_i|^3} \tag{3.11}$$

If the charges are so small and so numerous that they can be described by a charge density $\rho(\mathbf{x}')$ [if Δq is the charge in a small volume $\Delta x \Delta y \Delta z$ at the point $\mathbf{X}'$, then $\Delta q = \rho(\mathbf{X}')\Delta x \Delta y \Delta z$], the sum in (3.11) is replaced by the integral

$$E(X) = \frac{1}{4\pi\varepsilon_0} \int \rho(X') \frac{X - X'}{|X - X'|^3} d^3x' \tag{3.12}$$

where $d^3x' = dx'dy'dz'$ is a three-dimensional volume element at $\mathbf{X}'$.

We have seen that the principle of superposition has facilitated considerably the mathematical handling of electromagnetic theory. However, this principle fails in nuclear interactions and this is one of the reasons why the nuclear theory is somewhat more complicated and troublesome than the theory of atomic interactions.

3.5 DIRAC DELTA FUNCTION

It is sometimes convenient to regard a point charge as a 'fictitious' continuous charge distribution. One can achieve this with the help of Dirac delta function, written $\delta(x - a)$, is a mathematically improper function having the following properties:

(i) $\delta(x - a) = 0$ for $x \neq a$, and

(ii) $\int \delta(x - a) dx = 1$ if the region of integration includes $x = a$, and is zero otherwise.

(iii) $\int f(x)\delta(x-a)dx = f(a)$

(iv) $\int f(x)\delta'(x-a)dx = -f'(a)$ where prime denotes differentiation with respect to the argument.

(v) $\delta(f(x)) = \sum_i \frac{1}{\left|\frac{df}{dx}(x_i)\right|}\delta(x-x_i)$ where $f(x)$ is assumed to have only simple zeros, located at $x = x_i$. In more than one dimension, one merely take products of delta functions in each dimension. In three dimensions, for e.g., with cartesian coordinates,

(vi) $\delta(x-X) = \delta(x_1-X_1)\delta(x_2-X_2)\delta(x_3-X_3)$ is a delta function that vanishes everywhere except at $x = X$, and is such that

(vii) $\int_{\Delta V}\delta(x-X)d^3x = \begin{cases} 1 & \text{if } \Delta V \text{ contains } x = X \\ 0 & \text{if } \Delta V \text{ does not contain } x = X \end{cases}$

We must remember that a delta function has the dimensions of an inverse volume in whatever number of dimensions the space has.

A discrete set of point charges can be described with a charge density by means of delta functions. For example,

$$\rho(x) = \sum_{i=1}^{n} q_i(x-x_i) \tag{3.13}$$

represents a distribution of n point charges q_i, located at the points x_1. Using (3.13) in (3.12) and integration, using the properties of the delta function, yields the discrete sum (3.11).

3.6 LINES AND TUBES OF FORCE

Just as in the case of a gravitational field, an electric field may be represented by lines of force, which are lines that, at each point, are tangent to the direction of the electric field at that point, i.e. *lines of force* or *field lines* are parallel to the direction of the field every where. Since there is a single direction for the electric field at every point of the field and hence there is usually just one field line through any given point, i.e. two field lines can never intersect. The lines of force in Fig. 3.7(a) depict the electric field of a positive charge, and those in Fig. 3.7(b) show the electric field of a negative charge. We can see that they are straight lines passing through the charges.

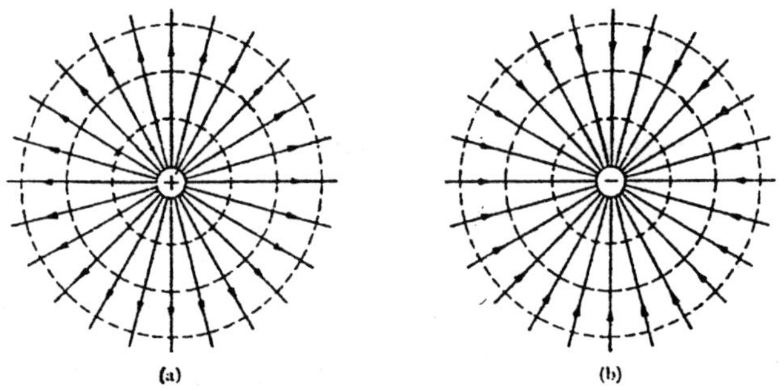

Figure 3.7. Lines of force and equipotential surfaces of the electric field of a positive and negative charge.

Figs. 3.8 and 3.9 shows the lines of force near a pair of charges of equal magnitude one positive and one negative and for two equal positive charges respectively.

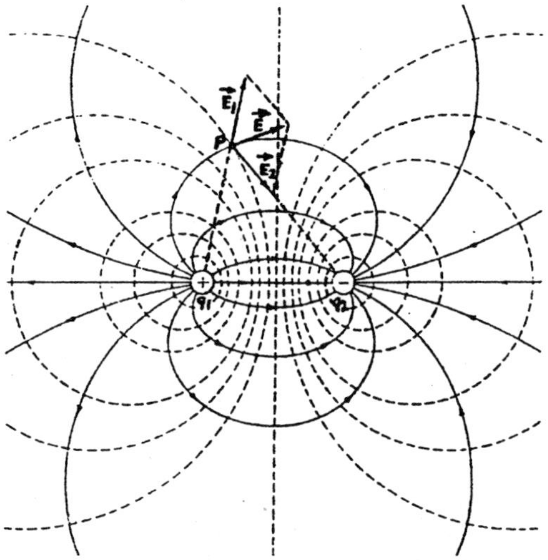

Figure 3.8. Lines of force and equipotential surfaces of the electric field of two equal but opposite charges.

In both figures the lines of force of the resultant electric field produced by the two charges have also been represented.

Let us consider an element of area *ds* in the field (small enough). Suppose that the electric intensity to be the same in magnitude and direction at every point of *ds*, so that the field of lines or lines of force at all points on the boundary will be approximately all parallel.

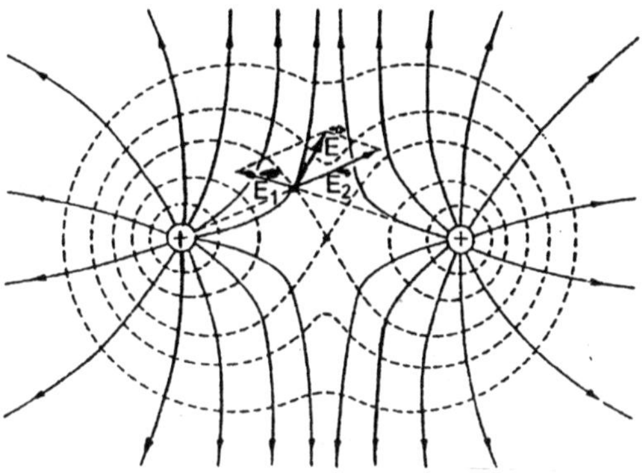

Figure 3.9. Lines of force and equipotential surfaces of the two identical positive charges.

One obtains the "tubular" surface as Fig. 3.10 shown in and it is called a '*tube of force*' we must note that the normal cross-section of the tube of force forms a part of an *equipotential surface*.

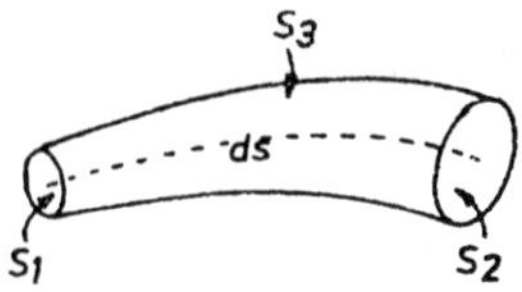

Figure 3.10. Tube of flux.

3.7 ELECTRIC FLUX

The concept of electric flux is of great usefulness in many physical problems. To understand the concept of electric flux let us imagine that a fluid is flowing with a speed V through a small flat surface ds in a direction normal to the surface as shown in Fig. 3.11 (a). The rate of flow of the fluid, i.e. the volume of the fluid crossing the area per unit time is termed as the flux of fluid and it is equal to vds. Now, if the normal to the surface is not parallel to the direction of flow of the fluid and makes an angle with the surface as shown in Fig. 3.11(b), then one finds the projected area in a plane perpendicular to v is $ds\cos\theta$. Thus the flux F is

$$F = vds\cos\theta \tag{3.14}$$

Let us represent the surface by vector $ds\hat{a}_n$ having magnitude ds and direction along the normal to the surface, $\hat{a}_n$ being the unit vector in this direction. One can express the flux as

$$\mathbf{F} = \mathbf{v}\cdot\hat{a}_n ds \tag{3.15}$$

In an electrostatic field, there is nothing actually flowing, one can mathematically define a quantity analogous to the flux of a fluid. This

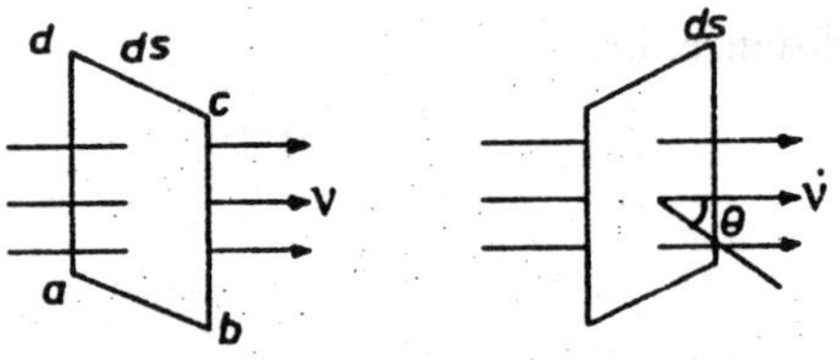

Figure 3.11.

is called as *electric flux* and as defined as

$$\text{Electric flux}, \phi = \mathbf{E} \cdot \hat{a}_n ds \tag{3.16}$$

The concepts of electric flux and tubes are sometimes useful in drawing the electric field lines about charges.

3.8 SOLID ANGLE

When every point of the boundary of a surface element, e.g. CD of area ds is joined to a given point O, then one finds that as a result of it, a cone is formed. Let us imagine a spherical surface with center at O and of radius r (= distance of surface element from O), then the above cone will intersect this surface in the form of another surface AB of area da. We can easily see that the area da is the projection of the area ds perpendicular to OP, i.e. $da = ds\cos\theta$, where θ is the angle between the surfaces AB and CD (Fig. 3.12). Obviously, if we consider the surface element as vector whose direction is along the outward normal to the surface element and O is considered as origin. Then the angle between **ds** and **r** will be θ. We have

$$da = \hat{r} \cdot \mathbf{ds} = ds\cos\theta$$

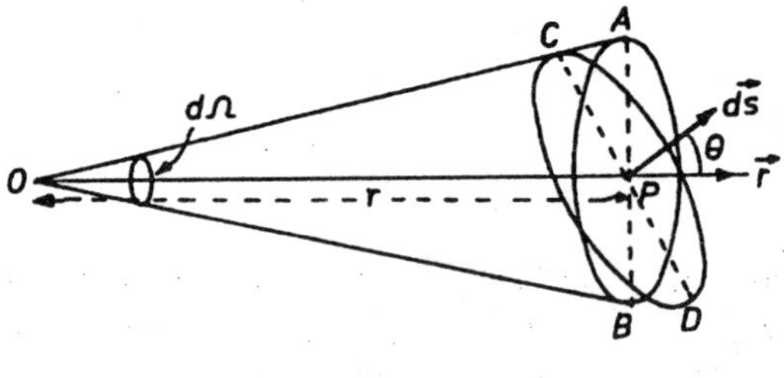

Figure 3.12. Concept of a solid angle.

The angle formed by this surface element da at the point O in three dimensions is termed as *solid angle*. The solid angle formed by the area ds at a point O is obtained as the ratio of the normal area to the square of the distance, i.e.

$$d\Omega = \frac{da}{r^2} = \frac{ds\cos\theta}{r^2} \tag{3.16(a)}$$

Solid angle has no dimensions like an angle but it is represented by the unit, *Steradian.* If $r = 1$, then $d\Omega = da$. Thus *the solid angle formed by any surface at point O is equal in magnitude to the area intersected by the cone or pyramid formed by joining every point of that surface to the point O, on a spherical surface of unit radius and having centre at O.*

Let us now determine the solid angle formed by the element of the area *PQRS* situated on the spherical surface of radius *r*, at the (centre *O* (Fig. 3.13). Let us draw planes *TOM* and *TON* making angles ϕ and $\phi + d\phi$ respectively from the plane *ZOX*, i.e. the angle formed by the side *SP* of area element *PQRS* on the Z axis is $d\phi$. From Fig. 3.13 it is evident that lines *OP* and *OQ* make angles θ and $\theta + d\theta$ from the Z-axis and hence the angle formed at the centre *O* by the side *PQ* of the area element is $d\theta$ as shown in Fig. 3.13.

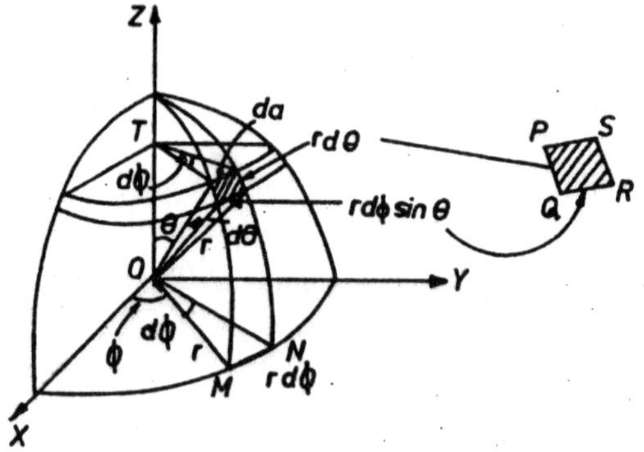

Figure 3.13. Concept of a solid angle

The radius of the spherical surface *PQRS* is *r*, and hence

$$PQ = rd\theta$$

From ΔOPT, we have

$$\frac{TP}{OP} = \sin\theta$$

$$\therefore \quad TP = OP\sin\theta = r\sin\theta \tag{3.16b}$$

or

$$SP = (r\sin\theta)d\phi$$

$\therefore$ Area of element $PQRS = PQ \times PS = (rd\theta)r\sin\theta d\phi$

or

$$da = r^2 \sin\theta d\theta d\phi \tag{3.16c}$$

$$\therefore \quad \text{Solid angle } d\Omega = \frac{da}{r^2} = \frac{r^2 \sin\theta d\theta d\phi}{r^2}$$
$$= \sin\theta d\theta d\phi \tag{3.16d}$$

Now, the solid angle formed at the centre by the entire spherical surface is equal to the sum of solid angles formed at the centre by all the area elements situated on the surface. Since the values of θ and ϕ vary from 0 to π and 0 to 2π respectively for the entire spherical surface and hence solid angle subtended by spherical surface at its centre is

$$\Omega = \int_0^{2\pi}\int_0^{\pi} \sin\theta d\theta d\phi = 2\pi \int_0^{\pi} \sin\theta d\theta$$
$$= 2\pi(-\cos\theta)_0^{\pi} = 4\pi \tag{3.16e}$$

3.9 GAUSS'S LAW FOR THE ELECTRIC FIELD (INTEGRAL FORM)

One can use relation (3.16) to obtain an important law in electrostatics i.e. Gauss's law. This law relates the electric flux (ϕ) through any closed surface to the net amount of charge within the surface.

Let us now imagine a closed surface S surrounding a charge q_i as shown in Fig. 3.14.Let ds be an element of area around the point P on the surface and $\hat{a}_n$ an outward unit vector normal to it. Let θ be the angle between the electric field at P and unit vector $\hat{a}_n$. The electric flux through the element of area ds is

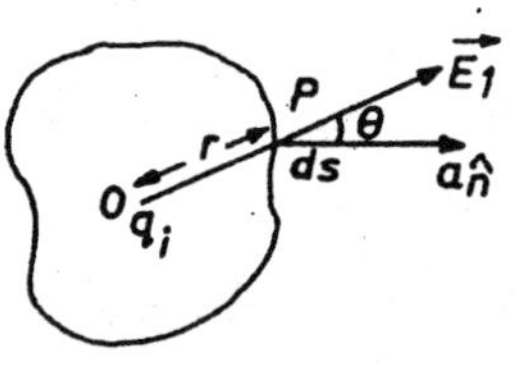

Figure 3.14.

$$d\phi = \mathbf{E}_i \cdot \hat{a}_n ds = E_i \cos\theta ds$$
$$= \frac{1}{4\pi\varepsilon}\frac{q_i}{r^2}\cos\theta ds$$

where r is the distance of the element ds from charge q_i.

Now $d\Omega = \frac{ds\cos\theta}{r^2}$ is the solid angle subtended by ds at q_i. Thus

$$d\phi = \mathbf{E}_i \cdot \hat{a}_n ds = \frac{1}{4\pi\varepsilon_0} q_i d\Omega \tag{3.17}$$

Thus the total flux through the entire surface S is

$$\phi = \int_S \mathbf{E}_i \cdot \hat{a}_n ds = \frac{1}{4\pi\varepsilon_0} q_i \int d\Omega = \frac{q_i}{\varepsilon_0} \tag{3.18}$$

Obviously, the electric flux through the surface S, then is proportional to the charge and independent of the distance r (in case of a spherical surface r be radius). Therefore, if we draw several concentric spherical surfaces $S_1, S_2, S_3, \ldots$ (Fig. 3.15) around the charge q_i the electric flux through all of them is the same and equal to $\frac{q}{\varepsilon_0}$. The result is due to $\frac{1}{r^2}$ dependence of the field. The same is also true in the case of gravitational field.

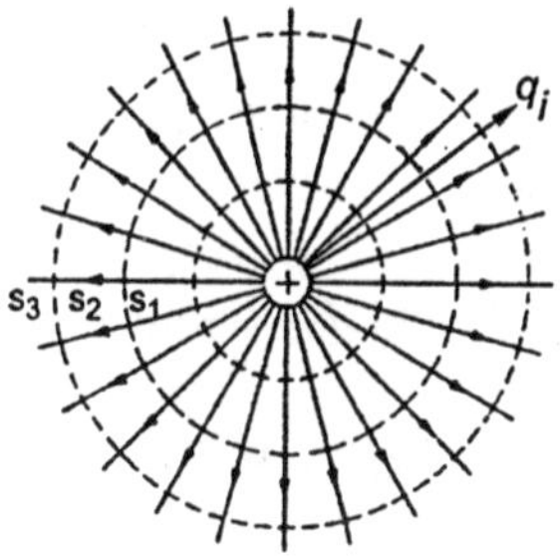

Figure 3.15. The electric flux through concentric spheres surrounding the same charge is same.

If there is any arbitrary distribution of charges within the surface S, by the principle of superposition

$$\mathbf{E} = \sum \mathbf{E}_i \tag{3.19}$$

(3.19) leads to the relation

$$\oint_s \mathbf{E}_i \cdot \hat{a}_n ds = \frac{\Sigma q_i}{\varepsilon_0} = \frac{Q}{\varepsilon_0} \tag{3.20}$$

where Q is the total charge $(= \Sigma q_i)$ within the closed surface S. This result is valid for any closed surface, irrespective of the position of the charge within the surface.

In the case of a continuous distribution of charge within the surface S, the relation (3.20) changes to

$$\oint_s \mathbf{E}_i \cdot \hat{a}_n ds = \frac{1}{\varepsilon_0} \int \rho dv \tag{3.21}$$

If no charges are present inside the closed surface S, or if the net charge is zero, the total electric flux through it is zero. The charges outside the closed surface do not contribute to the total flux. One can easily versify it. Let us suppose that charge q_i is outside the closed surface S (Fig. 3.16). The electric field vector $\mathbf{E}$ passes through two elements of

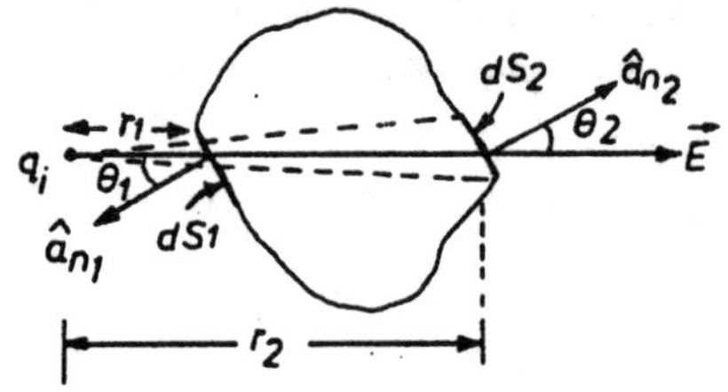

Figure 3.16.

area ds_1, and ds_2 cut out by a cone with its apex at charge q_i as shown in Fig. 3.16. $\hat{a}_{n_1}$ and $\hat{a}_{n_2}$ are the outward drawn unit vectors normal to ds_1, and ds_2 making angles θ_1 and θ_2 respectively with the field. The net outward flux through ds_1 and ds_2 area is

$$d\phi = \frac{1}{4\pi\varepsilon_0}\left[\frac{q_i}{r_1^2}\cos(180° + \theta_1)ds_1 + \frac{q_i}{r_2^2}\cos\theta_2 ds_2\right]$$

where r_1 and r_2 are the distances of the elements ds_1, ds_2 from charge q_i. Thus

$$d\phi = \frac{q_i}{4\pi\varepsilon_0}\left[-\frac{\cos\theta_1}{r_1^2}ds_1 + \frac{\cos\theta_2}{r_2^2}ds_2\right]$$

Now, the solid angle subtended by ds_1 and ds_2 at q_i are the same, i.e.

$$\frac{\cos\theta_1}{r_1^2}ds_1 = \frac{\cos\theta_2}{r_2^2}ds_2 = d\Omega$$

Hence $d\phi = 0$.

One can see that this holds good for all the cones drawn from q_i through the surface and, hence, the total flux $\phi = 0$.

These results constitute what is known as Gauss's law for the electric field. One can state it as

$$\left.\begin{aligned} \oint_s \mathbf{E}\cdot\hat{a}_n\,ds &= \frac{Q}{\varepsilon_0} \text{ for charge } Q \text{ inside } S \\ &= 0 \text{ for charge } Q \text{ outside } S \end{aligned}\right\} \tag{3.22}$$

Relation (3.22) is the *integral form* of Gauss's law for the electric field.

Gauss's law is particularly useful when we wish to compute the electric field produced by charge the electric field produced by charge distribution having certain geometrical symmetries.

3.10 GAUSS'S LAW FOR THE ELECTRIC FIELD (DIFFERENTIAL FORM)

One can also express the Gauss's law in a differential form. We have the relation

$$\text{div}\,\mathbf{A} = \lim_{dv\to 0} \frac{\oint_s \mathbf{A}\cdot\hat{a}_n\,ds}{dV} \tag{3.23}$$

where S is the surface enclosing the volume element. Integrating over the finite volume, one obtains

$$\int_v \text{div}\,\mathbf{A}dV = \oint \mathbf{A}\cdot\hat{a}_n ds \tag{3.24}$$

we are familiar with (3.24), which is the well-Known *Divergence theorem* usually used for transforming a volume integral into a surface in-. tegral and vice versa. Using (3.24), the Gauss's law can be expressed as

$$\oint_s \mathbf{E}\cdot\hat{a}_n\,ds = \oint_v \text{div}\,\mathbf{E}dV = \frac{1}{\varepsilon_0}\int_v \rho\,dV$$

i.e.

$$\int_v \left(\text{div}\mathbf{E} - \frac{\rho}{\varepsilon_0}\right) dV = 0$$

The above is true for any arbitrary volume V. Thus

$$\text{div}\mathbf{E} - \frac{\rho}{\varepsilon_0} = 0$$

or

$$\nabla\cdot\mathbf{E} = \frac{\rho}{\varepsilon_0} \tag{3.25}$$

Relation (3.25) is the differential form of Gauss's law for the electric field. In this form, the law shows that the sources of electric displacement are free electric charges (since $\rho \neq 0$), i.e. the electric flux lines either leave or enter a closed surface, therefore, there must be either a source or sink of flux lines. We know from convention that flux lines start from positive charge as sources and terminate in negative charges as sinks. This shows that positive and negative charges exist independently. Obviously, Gauss's law in differential form expresses a local relation between **E** and ρ. Thus we may say that electric charges are the sources of the electric field, and that their distribution and magnitude determine the electric field at each point of space.

We have adopted Coulomb's law in electrostatics as the fundamental law. We have also seen that the Gauss's law in electrostatics is a cosequence of the fact that the electric force between charged particles is inversely proportional to the square of the distance between them, i.e. Coulomb's inverse square law. One can easily see that any other law, say, $\frac{1}{r^n}$ with $n \neq 2$ would not give Gauss's law. Obviously, we can take Gauss's law as also the fundamental law of electrostatics.

3.11 APPLICATIONS OF GAUSS'S LAW

This law provides a powerful method for evaluating the electric field intensity **E** and potential variation due to simple charge distribution whose *symmetry* is such that the field is constant and normal over the entire Gaussian surface. We shall discuss some useful applications of Gauss's law in this section. We must remember that *symmetry* is crucial to the application of Gauss's law. There are only three kinds of symmetry which are sufficient:

(i) *Spherical symmetry.* One can make Gaussian surface a concentric sphere

(ii) *Cylindrical Symmetry.* One can make Gaussian surface a coaxial cylinder (Fig. 3.17).

(iii) *Plane Symmetry.* One can make use of Gaussian 'pillbox', which straddles the surface. (Fig. 3.18).

Figure 3.17.

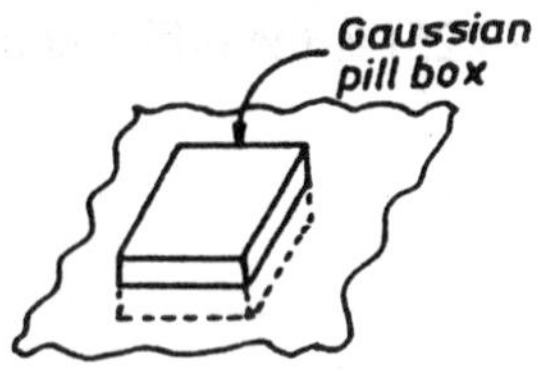

Figure 3.18..

(i) Field due to an Infinite Layer of Positive charge with Uniform surface density (σ)

Let us Consider that *ABCD* be a plane as shown in Fig. 3.19. Symmetry leads that the field lines are perpendicular to the plane. Now, we consider a right circular cylinder *PQR* with cross section *ds*.

One obtains from Gauss's law

$$(E_1 - E_2)ds = \frac{\sigma ds}{\varepsilon_0}$$

$$\therefore E_1 - E_2 = \frac{\sigma}{\varepsilon_0} \qquad (3.26)$$

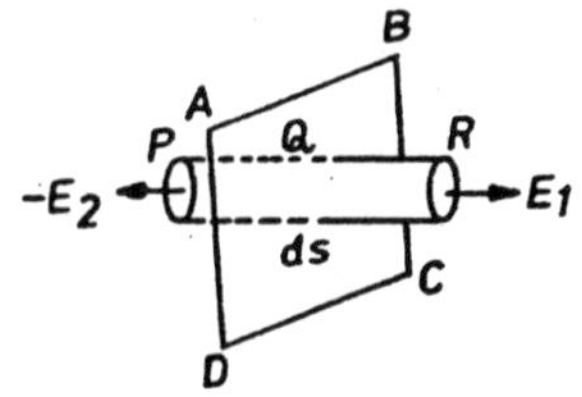

Figure 3.19.

Obviously, the field changes across a charge layer and the change is equal to $\frac{\sigma}{\varepsilon_0}$.

(ii) Field outside an isolated charged Sphere

Let us suppose that a sphere *A* is filled with a uniform charge distribution (Fig. 3.20). We are interested in finding the electric field at a point *P* outside the sphere.

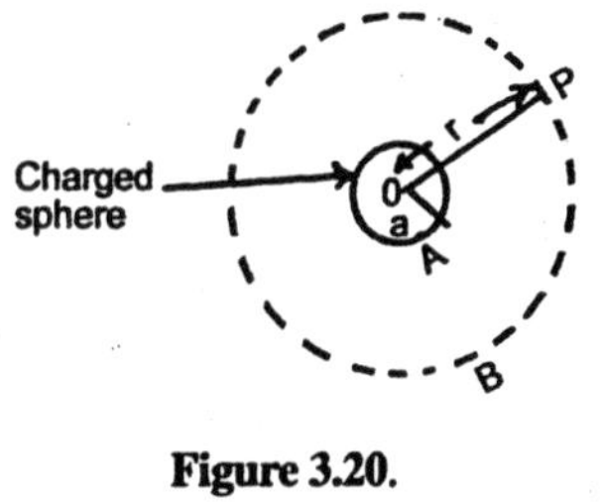

Figure 3.20.

Construct a spherical surface *B* concentric with the charged sphere *A* and passing through the point *P*. Let *r* is the radius of the imaginary sphere *B* and its surface area is $4\pi r^2$. By symmetry, we can say that the electric intensity *E* is the same at every point of the surface. Obviously, the flux outwards through the surface is

$$\oint_s \mathbf{E} \cdot \hat{a}_n ds = 4\pi r^2 E = \frac{Q}{\varepsilon_0} \quad \text{(Guass's law)}$$

where Q is the total charge within the sphere. Thus

$$E = \frac{1}{4\pi\varepsilon_0}\frac{Q}{r^2} \tag{3.27}$$

Eq. (3.27) is the same as the field produced by the point charge Q at the centre of the sphere. If there is continuous charge distribution within the sphere, we have

$$Q = \frac{4}{3}\pi a^3 \rho$$

where ρ is the charge density and a the radius of the sphere. Thus

$$E = \frac{1}{4\pi\varepsilon_0}\frac{4\pi a^3 \rho}{3r^2} = \frac{a^3\rho}{3\varepsilon_0 r^2} \tag{3.28}$$

(iii) Electric Field due to a Spherical Shell of charge

Following the arguments applied in the case of a field outside an isolated charged sphere, one can see that the field outside a thin spherical shell of charge is the same as if the total charge on the shell is concentrated at the centre. Now, we are interested in finding out the field at a point O inside the shell.

Let us imagine a cone with apex at O and extending on either side to cut surface elements ds_1, and ds_2 as shown in Fig. 3.21. Let r_1 and r_2 be the distances of ds_1 and ds_2 from O respectively. Let σ is the surface density of charge, the fields at O due to elements are $\frac{\sigma s_1}{r_1^2}$ and $\frac{\sigma ds_2}{r_2^2}$ and these will act in opposite direction. However,

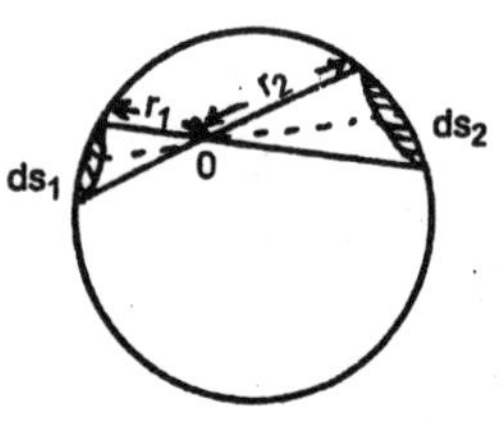

Figure 3.21.

$$\frac{ds_1}{r_1^2} = \frac{ds_2}{r_2^2} = d\Omega$$

($d\Omega$ is the solid angle subtended at O by ds_1, and ds_2).

Thus the contribution to the electric field due to the two elements being equal and opposite and hence cancel exactly. Extending this to cover the entire shell by balancing off field contributions of opposite differential areas, one can see that each pair of differential areas gives a zero contribution. Thus net field at O is zero.

(iv) Electric Field Strength at an Internal point

Let P be an internal point at a distance r from the centre of the charge distribution O as shown in Fig. 3.22. Imagine a sphere of radius $(OP =)$ r concentric with the spherical charge. Let ρ be the volume charge density (i.e., charge per unit volume), i.e.

$$\rho = \frac{\text{Charge}}{\text{Volume}} = \frac{Q}{\frac{4}{3}R^3}$$

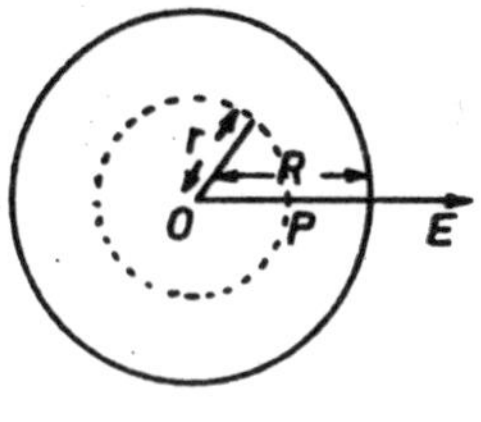

Figure 3.22.

We must note that the charge in the shell of thickness $R-r$ does not contribute to the field at P since the point P lies inside the shell (see case iii). If E is the field at P, the flux through the imaginary or Gaussian surface is

$$4\pi r^2 E = \frac{1}{\varepsilon_0}\frac{4}{3}\pi r^3 \rho \quad \text{(Gauss law)}$$

or

$$E = \frac{r\rho}{3\varepsilon_0} = \frac{1}{4\pi\varepsilon_0}\frac{Qr}{R^3} \qquad (3.29)$$

Thus the electric field strength at point P inside a spherically symmetric charge distribution is directly proportional to the distance of the point P from the centre of the spherical charge.

Eqs. (3.28) and (3.29) give the field outside and inside the sphere respectively. One can see that the two forms match, as they should, when $r = a$.

The variation of magnitude of electric field strength with distance from the centre of the spherically charge distribution is represented by the curve as shown in Fig. 3.23.

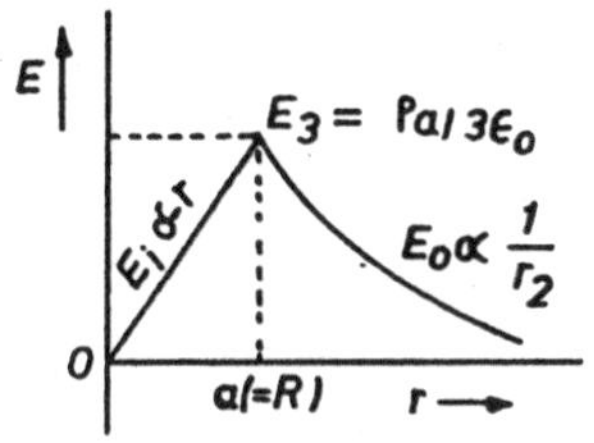

Figure 3.23.

(v) Electric Field Strength due to Uniformly Charged infinite Cylinder

Let us consider a charged cylinder of infinite length of radius r as shown in Fig. 3.24. We are interested in finding the field at a point P at a distance 'd' from the axis. For this, let us imagine a closed cylindrical surface coaxial with the cylinder and passing through P, its two end faces being perpendicular to the axis of the cylinder. One can easily see by symmetry, that the field will be normal to the axis directed away from the axis and is the same at equal distances from the axis. If 'l' is the height of the cylindrical Gaussian surface, the flux through this surface is $= 4\pi l E$. We must remember that the end faces do not contribute to the field since E is tangential to these faces. Applying Gauss's law we have

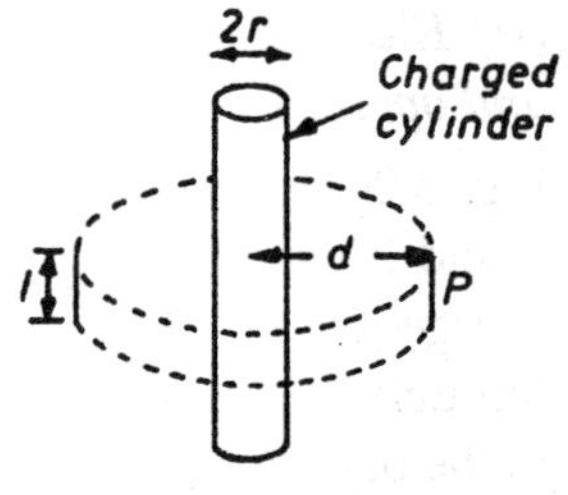

Figure 3.24.

$$2\pi l E = \frac{\lambda l}{\varepsilon_0}$$

where λ is the charge per unit length. Thus

$$E = \frac{\lambda}{2\pi\varepsilon_d} = \frac{1}{4\pi\varepsilon_0}\frac{2\lambda}{d} \tag{3.30}$$

Obviously, the electric field strength at a distance d from an infinite cylinder uniformly charged with linear charge density λ has magnitude $\left(\frac{\lambda}{2\pi\varepsilon_0}\right)$ which is inversely proportional to d. We can further see that the result is independent of the radius of charged cylinder and holds also for the rectilinear distribution of charges.

One could obtain the above result by direct integration. Let us consider a filament of finite length of constant line density of charge λ Coulomb/m as shown in Fig. 3.25. Now, the field at a point P due to a line element dz is

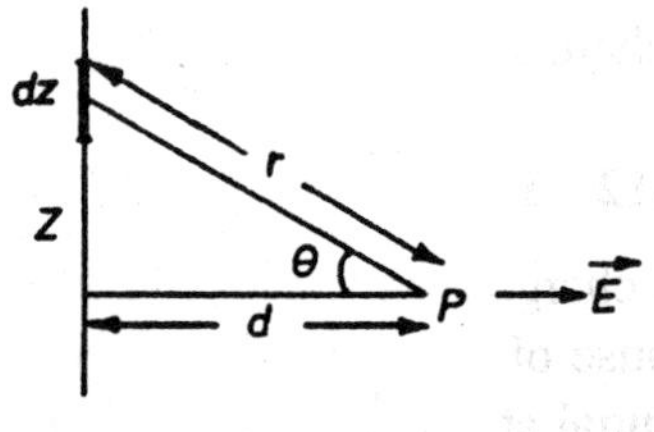

Figure 3.25.

$$dE_d = \frac{\lambda dz \cos\theta}{4\pi\varepsilon_0 r^2}$$

Here we have assumed that the filament coincide with the z-axis. The distance of the point P from the element is dz and the perpendicular distance from the z-axis is 'd'. One obtains the total field as

$$E_d = \frac{\lambda}{4\pi\varepsilon_0}\int_{-\infty}^{\infty}\frac{\cos\theta}{r^2}dz$$

we have $z = d\tan\theta$ and $r = d\sec\theta$

$$\therefore E_d = \frac{\lambda}{4\pi\varepsilon_0}\int_{-\pi/2}^{\pi/2}\frac{\cos\theta}{d}d\theta = \frac{\lambda}{2\pi\varepsilon_0 d}$$

(vi) Electrostatic Field between two Concentric spheres with equal and opposite charge

Fig. 3.26 exhibit two concentric spheres 1 and 2 which have equal and opposite charge. Imagine a Gaussian surface drawn through P. Obviously, the field at P due to the outer sphere is zero and that due to the inner sphere is $E = \frac{1}{4\pi\varepsilon_0}\frac{q}{r^2}$.

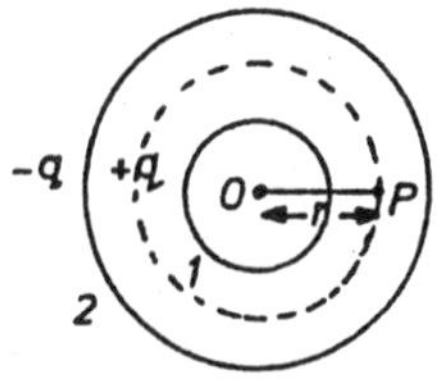

Figure 3.26.

Let us now consider the case of two nonconcentric charged spheres as shown in Fig. 3.27. Figure shows the redistribution of charge and field lines. Draw a cone with P as the apex. We can see that the two charge densities on ds_1, and ds_2 are not equal $dE_2 > dE_1$, and hence the two different areas yield a net dE in the negative direction. One can find a net E_- due to the outer sphere in the negative x-direction by covering the entire sphere. The field due to the inner sphere at P is E_+. Thus the effective field at P, $E = E_+ + E_-$.

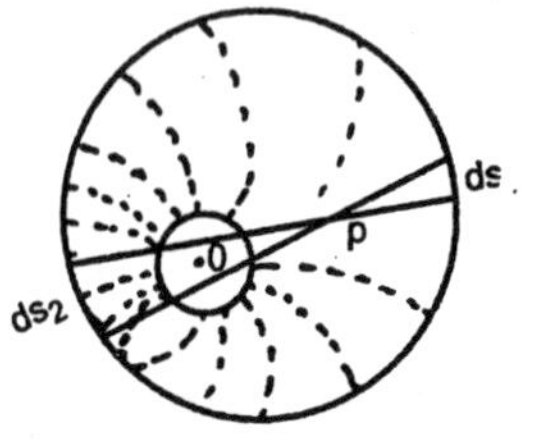

Figure 3.27.

3.12 ELECTROSTATIC POTENTIAL

A charged particle placed in an electric field has potential energy because of its interaction with the field. One can define the electric potential at a point as the potential energy per unit charge placed at that point. The electric potential is usually designated by V and the potential energy of a charge q by E_p, we have

$$V = \frac{E_p}{q} \quad \text{or} \quad E_p = qV \tag{3.31}$$

The electric potential is measured in Joules/Coulomb or JC^{-1}, a unit called a Volt, abbreviated V. In terms of international units, $V = m^2kgs^{-2}C^{-1}$.

Let us consider the field produced by a stationary point charge q. At any point of this field, the point charge q' experiences the force

$$\mathbf{F} = \frac{1}{4\pi\varepsilon_0}\frac{qq'}{r^2}\hat{a}_r = F(r)\hat{a}_r \tag{3.32}$$

Here $F(r)$ is the magnitude of the force $\mathbf{F}$, and $\hat{a}_r$ is the unit vector of the position vector $\mathbf{r}$ determining the position of the charge q' relative to the charge q.

The force (3.32) is a central one. *A central field of forces is conservative.* Consequently, the work done by the forces of the field on the charge q' when it is moved from one point to another does not depend on the path. This work is

$$W_{12} = \int_1^2 F(r)\,\hat{a}_r d\mathbf{l} \tag{3.33}$$

where $d\mathbf{l}$ is the elementary displacement of the charge q'. Fig. 3.28 shows that the scalar product $\hat{a}_r d\mathbf{l}$ equals the increment of the magnitude of the position vector $\mathbf{r}$, i.e. dr. Eq (3.33) can therefore be written in the form

$$W_{12} = \int_1^2 F(R)dr$$

Using (3.32), one obtains

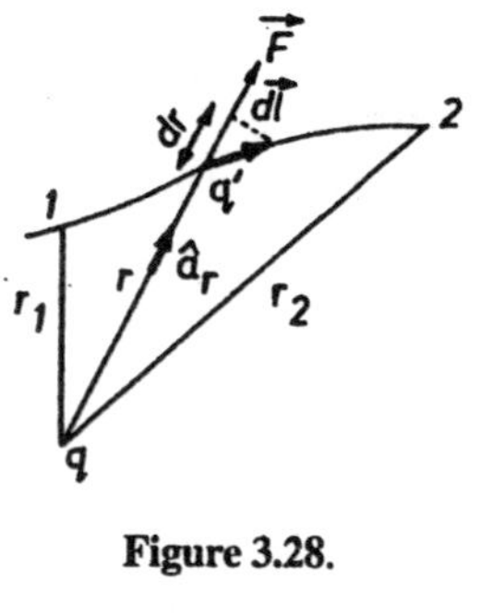

Figure 3.28.

$$W_{12} = \frac{qq'}{4\pi\varepsilon_0}\int_{r_1}^{r_2}\frac{dr}{r^2} = \frac{1}{4\pi\varepsilon_0}\left(\frac{qq'}{r_1} - \frac{qq'}{r_2}\right) \tag{3.34}$$

The work of the forces of the conservative field can be represented as a decrement of the potential energy

$$W_{12} = V_{p,1} - V_{p,2} \tag{3.35}$$

A comparison of eqs. (3.34) and (3.35) leads to the following expression for the potential energy of the charge q' in the field of charge q:

$$V_p = \frac{1}{4\pi\varepsilon_0}\frac{qq'}{r} + \text{Const.}$$

The value of the constant in the expression for the potential energy is usually chosen so that when the charge moves away to infinity (i.e.

when $r = \infty$), the potential energy vanishes. When this condition is observed, one obtains

$$V_p = \frac{1}{4\pi\varepsilon_0}\frac{qq'}{r} \tag{3.36}$$

Let us use the charge q' as a test charge for studying the field. By Eq. (3.36), the potential energy which the test charge has depends not only on its magnitude q', but also on the quantities q and r determining the field. Obviously one can use this energy to describe the field just like we used the force acting on the test charge for this purpose.

Different test charges q_t', q_t'', etc. will have different energ·es V_p', V_p'' etc. at the same point of the field. But the ratio $\frac{V_p}{q_t}$ will be the same for all the charges [see Eq. (3.36)]. The quantity

$$V \text{ or } \phi = \frac{V_p}{q_t} \tag{3.37}$$

is called the *field potential* at a given point and is used together with the field strength E to describe electric fields.

It can be seen from (3.37) that the potential numerically equals the potential energy which a unit positive charge would have at the given point of the field. Substituting for the potential energy in Eq. (3.37) its value from (3.36), one obtains the expression for the potential of a point charge

$$V \text{ or } \phi = \frac{1}{4\pi\varepsilon_0}\frac{q}{r} \tag{3.38}$$

Let us consider the field produced by a system of N point charges $q_1, q_2, \ldots q_N$. Let $r_1, r_2, \ldots r_N$ be the distances from each of the charges to the given point of the field. The work done by the forces of this field on the charge q' will equal the algebraic sum of the work done on the charge q' will equal the algebraic sum of the work done by the forces set up by each of the charges separately.

$$W_{12} = \sum_{i=1}^{N} W_i$$

By eq. (3.34), each work W_i equal

$$W_i = \frac{1}{4\pi\varepsilon_0}\left(\frac{q_i q'}{r_{i,1}} - \frac{q_i q'}{r_{i,2}}\right)$$

where $r_{i,1}$ is the distance from the charge q_i to the initial position of the charge q' and $r_{i,2}$ is the distance from q_i to the final position of the charge q'. Thus

$$W_{12} = \frac{1}{4\pi\varepsilon_0}\sum_{i=1}^{N}\frac{q_i q'}{r_{i,1}} - \frac{1}{4\pi\varepsilon_0}\sum_{i=1}^{N}\frac{q_i q'}{r_{i,2}}$$

Comparing this equation with eq. (3.35), one obtains the following expression for the potential energy of the charge q' in the field of a system of charges:

$$V_p = \frac{1}{4\pi\varepsilon_0}\sum_{i=1}^{N}\frac{q_i q'}{r_i}$$

from which it can be seen that

$$V \text{ or } \phi = \frac{1}{4\pi\varepsilon_0}\sum_{i=1}^{N}\frac{q_i}{r_i} \tag{3.39}$$

In general for a set of point charges q_j the potential at a point r_i is given by the algebraic sum of the individual potentials, i.e.

$$V(r_i) = \frac{1}{4\pi\varepsilon_0}\sum_{j}\frac{q_j}{r_{ij}} \tag{3.39a}$$

In case of a continuous distribution of charges

$$V(r_i) = \frac{1}{4\pi\varepsilon_0}\int\frac{\rho d\tau_j}{r_{ij}} \tag{3.39b}$$

Comparing (3.39) with (3.38), we arrive at the conclusion that the *potential of the field produced by a system of charges equals the algebraic sum of the potentials produced by each of the charges separately.* Where as the field strengths are added vectorially in the superposition of fields, the potentials are added algebraical This is why it is usually much simpler to calculate the potentials than the electric field strengths.

Examination of Eq. (3.37) shows that the charge q at a point of the field with the potential V or ϕ has the potential energy

$$V_P = qV = q\phi \tag{3.40}$$

Hence, the work of the field forces on the charge q can be expressed through the potential difference:

$$W_{12} = V_{P,1} - V_{P,2} = q(V_1 - V_2) = q(\phi_1 - \phi_2) \tag{3.41}$$

obviously, the work done on a charge by the forces of a field equals the product of the magnitude of the charge and the difference between the potentials at the initial and final points (i.e. the potential decrement).

If the charge q is removed from a point having the potential V or ϕ to infinity (where by convention the potential vanishes), then the work of the field forces will be

$$W_\infty = q\phi = qV \tag{3.42}$$

obviously, the *potential numerically equals the work done by the forces of the field on a unit positive charge when the latter is removed from the given point to infinity.* Work of the same magnitude must be done against the electric field forces to move a unit positive charge from infinity to the given point of a field.

3.13 RELATION BETWEEN ELECTRIC FIELD STRENGTH AND POTENTIAL

An electric field can be described either with the aid of the vector quantity **E**, or with the aid of the scalar quantity V or ϕ. There must evidently be a definite relation between these quantities. If we bear in mind that **E** is proportional to the force acting on a charge and ϕ to the potential energy of the charge, it is easy to see that this relation must be similar to that between the potential energy and the force.

The force **F** is related to the potential energy by the expression

$$F = -\nabla V_p \tag{3.43}$$

For a charged particle in an electrostatic field, we have

$$\mathbf{F} = q\mathbf{E}$$

and

$$V_p = q\phi = qV$$

Introducing these values in (3.43),one obtains

$$q\mathbf{E} = -\nabla(q\phi) = -q\nabla(\phi)$$

or

$$\mathbf{E} = -\nabla\phi = -\text{grad}\,\phi \tag{3.44}$$

This is the relation between the field strength and potential. Since **E** has a finite value at any point of the field, the potential ϕ is a continuous function of the coordinates of points of the field. The rectangular components of the electric field **E** are given by

$$E_x = -\frac{\partial \phi}{\partial x},\ E_y = -\frac{\partial \phi}{\partial y} \ \text{ and } \ E_z = -\frac{\partial \phi}{\partial z} \tag{3.45}$$

Thus

$$\mathbf{E} = -\left[\hat{i}\frac{\partial}{\partial x} + \hat{j}\frac{\partial}{\partial y} + \hat{k}\frac{\partial}{\partial z}\right]\phi \tag{3.45(a)}$$

In general, the component along the direction corresponding to a displacement ds is

$$E_s = -\frac{\partial \phi}{\partial s} \tag{3.46}$$

Eqs. (3.45) or (3.46) are used to find the electric potential ϕ or V when the field **E** is known, and conversely.

Let us consider a simple case of a uniform electric field (Fig. 3.29).

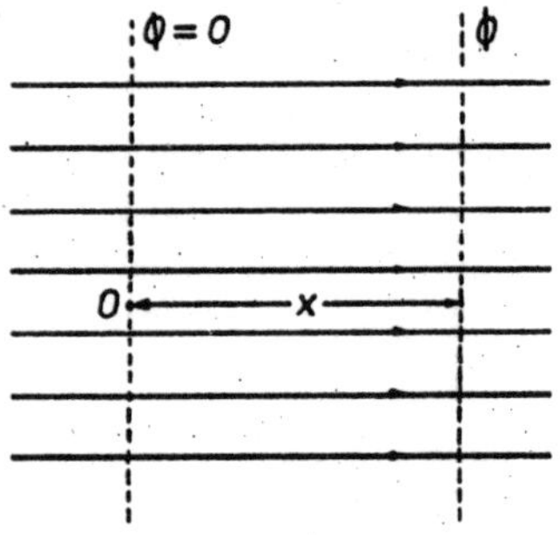

Figure 3.29. Uniform electric field.

The first of eqs. (3.45) gives, for an x-axis which is parallel to the field, $E = -\frac{d\phi}{dx}$. Since E is constant, and we assume $\phi = 0$ at $x = 0$, by integration, one obtains

$$\int_0^{\phi} d\phi = -\int_0^x E dx = -E\int_0^x dx \text{ or } \phi = -Ex \tag{3.47}$$

Eq. (3.47) is very useful relation represented graphically in Fig. 3.30.

We may note that, because of the negative sign in (3.46) or (3.47), the electric field points in the direction in which the electric potential decreases. When we consider two points x_1 and x_2, Eq. (3.47) gives $\phi_1 = -Ex_1$ and $\phi_2 = -Ex_2$. Subtracting we have $\phi_2 - \phi_1 = -E(x_2 -$

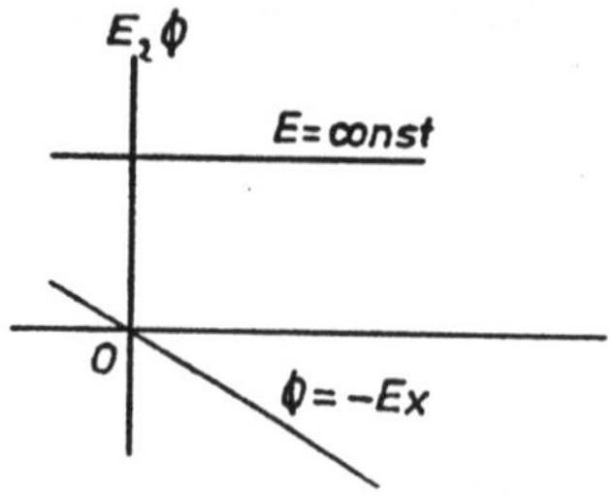

Figure 3.30. Variation of E and ϕ for a uniform electric field.

x_1); or, calling $d = x_2 - x_1$, we obtain

$$E = -\frac{\phi_2 - \phi_1}{d} = \frac{\phi_1 - \phi_2}{d} \tag{3.48}$$

Although (3.48) is valid only for uniform electric fields, it can be used to estimate the electric field between two points separated by a distance d, if the potential difference $\phi_1 - \phi_2$ between them is known. If the potential difference $\phi_1 - \phi_2$ is positive, the field points in the direction from x_1 to x_2, and if it is negative it points in the opposite direction. Eq. (3.48) [or in fact also (3.45) or (3.46)] indicates that the electric field can also be expressed in volts/meter, a unit which is equivalent to the newtons/Coulomb given before. One can easily verify it in the following way.

$$\frac{\text{Volt}}{\text{meter}} = \frac{\text{Joule}}{\text{Coulomb} - \text{m}} = \frac{\text{newton} - \text{m}}{\text{Coulomb} - \text{m}} = \frac{\text{newton}}{\text{Coulomb}}$$

By common usage, the term volt/m, abbreviated Vm^{-1} is preferred to NC^{-1}.

We have seen that eq. (3.44) allows us to find the field strength at every point from the known values of ϕ. One can also solve the reverse problem, i.e. find the potential difference between two arbitrary points of a field according to the given values of E. For this purpose, one can take advantage of the circumstance that the work done by the forces of a field on the charge q when it is moved from point 1 to point 2 can be calculated as (Fig. 3.28)

$$W_{12} = \int_1^2 qE\,dl$$

At the same time in accordance with eq. (3.41), this work can be written as

$$W_{12} = q(\phi_1 - \phi_2)$$

Equating these two expressions, one obtains

$$\phi_1 - \phi_2 = \int_1^2 \mathbf{E} \cdot d\mathbf{l} \qquad (3.49)$$

The integral can be taken along any line joining points 1 and 2 because the work of the field forces is independent of the path for circumvention along closed contour $\phi_1 = \phi_2$,and Eq. (3.49) becomes

$$\oint \mathbf{E} \cdot d\mathbf{l} = 0 \qquad (3.50)$$

(the circle on the integral sign indicates that integration is performed over a closed contour). In vector calculus the line integral of a vector round a closed curve is called the circulation or the curl of the vector field. Eq. (3.50) indicates that the circulation of the electric field is zero. Its sources are given by (3.25). Other examples of such fields are the gravitational field, the magnetic field in region free from electric current and the velocity field of the flow of incompressible fluids free from viscosity. Such fields are termed as *irrotational fields*. We must note that relation (3.50) is not applicable to the field of moving charges (i.e. a field changing with time), i.e. this field is not a potential one.

Now, we take curl of vectors in (3.44) one obtains

$$\nabla \times \mathbf{E} = -\nabla \times \nabla\phi = 0 \qquad (3.50a)$$

Using stokes' theorem, one can transform line integral in (3.50) into a surface integral

$$\oint \mathbf{E} \cdot d\mathbf{l} = \int_s \nabla \times \mathbf{E} \cdot \hat{a}_n dS$$

$$\therefore \quad \int_s \nabla \times \mathbf{E} \cdot \hat{a}_n dS = 0$$

and since S is arbitrary $\nabla \times \mathbf{E} = 0$. Thus, one obtains the following two important laws of electrostatics:

(i) $$\nabla \cdot \mathbf{E} = \frac{\rho}{\varepsilon_0} \qquad (3.51)$$

(ii) $$\nabla \times \mathbf{E} = 0 \qquad (3.52)$$

(3.51) follows from Coulomb's inverse square law. (3.52) does not depend upon this law. All predictions of electrostatics follow from these laws and hence (3.51) and (3.52) can be considered fundamental laws of electrostatics.

3.14 EQUIPOTENTIAL SURFACES

A locus of the points in an electrostatic field having the same potential is called an equipotential surface. Its equation has the form

$$\phi(r) = c$$

where c is a constant, represents an equipotential surface. Different values of c generate a family of such surfaces.

The potential does not change in movement along an equipotential surface over the distance $dr(d\phi = 0)$. Hence, according to Eq. (3.46), the tangential component of the E to the surface equals zero. Let us consider a displacement $d\mathbf{r}$ on an equipotential surface

$$\phi(\mathbf{r}) - \phi(\mathbf{r} + d\mathbf{r}) = 0$$

i.e.

$$\phi(\mathbf{r}) - \left\{\phi(\mathbf{r}) + \frac{\partial \phi}{\partial r} \cdot d\mathbf{r}\right\} = 0$$

$$\therefore \frac{\partial \phi}{\partial r} \cdot d\mathbf{r} = 0$$

i.e.

$$\nabla\phi \cdot d\mathbf{r} = 0$$

or

$$\mathbf{E} \cdot d\mathbf{r} = 0 \tag{3.53}$$

This shows that if $d\mathbf{r}$ is on the surface $\mathbf{E}$ must be normal to the surface, i.e. the field lines are normal or orthogonal to the equipotential surface.

An equipotential surface can be drawn through any point of a field consequently one can construct an infinitely great number of such surfaces. They are conventionally drawn so that the potential difference for two adjacent surfaces is the same everywhere. Thus the density of the equipotential surfaces allows us to assess the magnitude of the field strength. Indeed the denser are the equipotential surfaces, the more rapidly does the potential change when moving along a normal to the surface. Hence $\nabla\phi$ is at the given place, and therefore $\mathbf{E}$ is greater too.

Figure 3.7 shows equipotential surfaces (more exactly, their intersections with the plane of the drawing) for the field of a point charge. Figs. 3.8 and 3.9 shows the equipotential surfaces along with the lines of force for two equal but opposite charges and two identical charges respectively. In accordance with the nature of the dependence of E or r equipotential surfaces become the denser, the nearer we approach a charge.

Equipotential surfaces for a homogeneous field are a collection of equispaced planes at right angles to the direction of field.

3.15 ELECTROSTATIC ENERGY

The energy of electrostatic systems is solely the potential energy arising from the interaction between the charges.

Let us consider a charge q_1, be situated at a certain point in space and charge q_2 which was initially at infinity, be brought up to a distance r_{12} from q_1. We have the potential of the charge q_1 at r_{12} is $\frac{q_1}{4\pi\varepsilon_0 r_{12}}$. Obviously, this is the work done in bringing a unit charge from infinity a the charge q_2 from infinity to the point specified by r_{12} is $\frac{q_1 q_2}{4\pi\varepsilon_0 r_{12}}$. Now a third charge q_3 is added to the system. We will have to perform work against the field of q_1 and q_3. If r_{13}, r_{23} are the distances of q_3 from q_1 and q_2 respectively, then the additional contribution to the potential energy is as follows,

$$\frac{q_1 q_3}{4\pi\varepsilon_0 r_{13}} + \frac{q_2 q_3}{4\pi\varepsilon_0 r_{23}}$$

Proceeding to build up the assembly in this way, one finds the total energy of the assembly is

$$\begin{aligned} W &= \frac{q_2}{4\pi\varepsilon_0}\left(\frac{q_1}{r_{12}}\right) + \frac{q_3}{4\pi\varepsilon_0}\left(\frac{q_1}{r_{13}} + \frac{q_2}{r_{23}}\right) \\ &\quad + \frac{q_4}{4\pi\varepsilon_0}\left(\frac{q_1}{r_{14}} + \frac{q_2}{r_{24}} + \frac{q_3}{r_{34}}\right) + \cdots \\ &= \frac{1}{4\pi\varepsilon_0}\sum_i q_i \sum_{j<i} \frac{q_i}{r_{ij}} \end{aligned} \tag{3.54}$$

The restriction $j < i$ ensures that the interaction between every pair is counted only once. One can write (3. 54) in a different form, i.e.

$$\begin{aligned} W &= \frac{1}{4\pi\varepsilon_0}\frac{1}{2}q_1\left(\frac{q_2}{r_{12}} + \frac{q_3}{r_{13}} + \cdots\right) + \frac{1}{4\pi\varepsilon_0}\frac{1}{2}q_2\left(\frac{q_1}{r_{21}} + \frac{q_3}{r_{23}} + \cdots\right) \\ &\quad + \frac{1}{4\pi\varepsilon_0}\frac{1}{2}q_3\left(\frac{q_1}{r_{31}} + \frac{q_2}{r_{32}} + \frac{q_4}{r_{34}} + \cdots\right) + \cdots \\ &= \frac{1}{4\pi\varepsilon_0}\frac{1}{2}\sum_{i=1} q_i \sum_{j=1} \frac{q_j}{r_{ij}}; \quad i \neq j \end{aligned} \tag{3.55}$$

The factor $\frac{1}{2}$ appears in (3.55) because in this expression each pair is counted twice.

Since $\frac{1}{4\pi\varepsilon_0}\sum_{j\neq i}\frac{q_j}{r_{ij}}$ is the potential ϕ_i, produced by all the charges of the system except i[th] one at the point at which q_i is situated, we have

$$W = \frac{1}{2}\sum_i q_i\phi_i \tag{3.56}$$

When there is a continuous distribution of charges and charged are not localized, we have

$$W = \frac{1}{2}\int \phi\rho\, d\tau \tag{3.57}$$

3.16 ELECTRIC DIPOLE

An electric dipole is defined as a system of two point charges $+q$ and $-q$ identical in value and opposite in sign, the distance between which is much smaller than that to the points at which the field of the system is being determined. The straight line passing through both the charges is called the *dipole axis* (Fig. 3.31). There are several familiar examples of such systems in physics. For example:

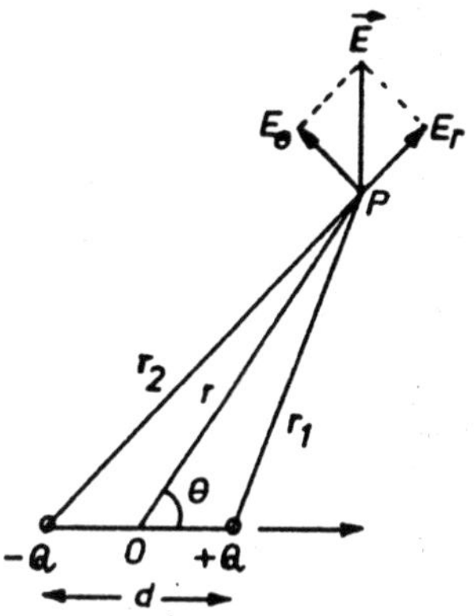

Figure 3.31. An electric dipole

(i) When we place an atom or a molecule in an electric field, the positive and negative charges feel opposite forces and are displaced slightly forming dipole.

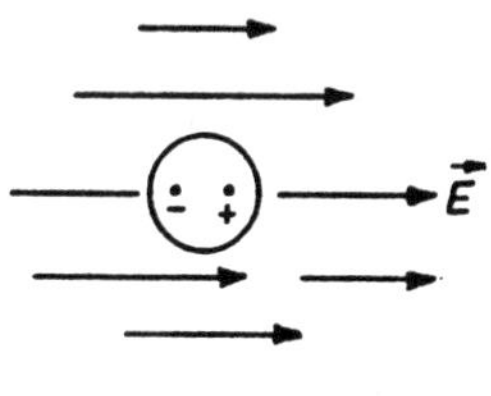

Figure 3.32.

(ii) The structure of some molecules, e.g. H_2O is such that even though there is no external field the positive charge, is slightly separated from the negative charge (Fig 3.33). One can easily see that though this structure is not equivalent to the one we defined as dipole, however, one can treat them as such when one is considering a field they produce at a very long distance.

Let us first calculate the potential and then the field strength of a dipole. Suppose the two point charges $+q$ and $-q$ are placed on the z-axis, one at $\frac{d}{2}$ and other at $-\frac{d}{2}$ from the origin O.

When the point P at which the magnitude of the potential and field due to dipole are to be determined is close to either of the charges, the potential is almost the same as if the charges were independent of one another and equipotential surfaces are nearly spherical. If the point $P(r,\theta,\phi)$ is at a distance r from the origin and $r \gg d$ the potential of the positive charge at P is

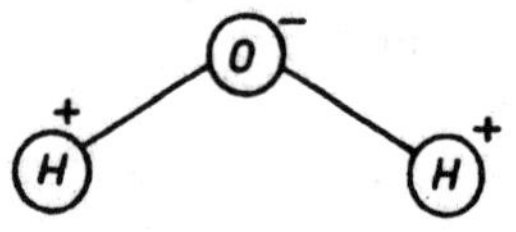

Figure 3.33. Structure of water molecule.

$$\phi_1(r_1) = \frac{1}{4\pi\varepsilon_0}\frac{Q}{r_1}$$

The potential of the negative charge at P is

$$\phi_2(r_2) = \frac{1}{4\pi\varepsilon_0}\frac{-Q}{r_2}$$

The total potential ϕ at P is then

$$\phi(r) = \phi_1(r_1) + \phi(r_2) = \frac{Q}{4\pi\varepsilon_0}\left(\frac{1}{r_1} - \frac{1}{r_2}\right) \tag{3.58}$$

where r_1 and r_2 are the distances of P from $+Q$ and $-Q$ respectively.

From fig. 3.31, we have

$$r_1^2 = r^2 + \left(\frac{d}{2}\right)^2 - 2\left(\frac{d}{2}\right) r\cos\theta$$

where θ is the angle between the z-axis and the position vector of P. Therefore

$$r_1^2 = r^2\left[1 + \frac{d^2}{4r^2} - \frac{d\cos\theta}{r}\right]$$

$$\therefore \quad \frac{1}{r_1} = \frac{1}{r}\left[1 + \frac{d^2}{4r^2} - \frac{d\cos\theta}{r}\right]^{-1/2} = \frac{1}{r}\left[1 + \frac{d\cos\theta}{2r}\cdots\right]$$

$$\approx \frac{1}{r} + \frac{d\cos\theta}{2r^2}$$

we have neglected terms in d^2 and higher order terms. The said approximation is justified if we let $\left(\frac{d}{r}\right) \to 0$ keeping the product qd

finite. Similarly, one can show

$$\frac{1}{r^2} \approx \frac{1}{r} - \frac{d\cos\theta}{2r^2}$$

Substituting the values of $\frac{1}{r_1}$ and $\frac{1}{r_2}$ in (3.58), one obtains

$$\phi(r) = \frac{Q}{4\pi\varepsilon_0} \frac{d\cos\theta}{r^2} \tag{3.59}$$

It is interesting to note that:

(i) the potential due to a dipole falls off as the square of the distance r, where as the potential from a single point charge varies only as $\frac{1}{r}$. This comes from the fact that the charges in a dipole appear close together for an observer at some distance away and that their fields cancel more and more as the distance r is increased

(ii) the potential of the dipole is also a function of the angle θ and this suggests that at a fixed radius 'r' the potential is a maximum on the axis of the dipole ($\theta = 0$) and is zero normal to the axis ($\theta = 90°$). One can easily that when $\theta = 90°$ the point P is exactly equidistant from the two charges so that their effects cancel.

The vector along the axis of the dipole pointing in the direction $-Q$ to $+Q$ and having magnitude Qd is termed as the *dipole moment* and represented by $\mathbf{P}$. Since

$$\mathbf{P}\cdot\mathbf{r} = Pr\cos\theta = Qdr\cos\theta$$

$$\therefore \quad \phi(r) = \frac{P\cos\theta}{4\pi\varepsilon_0 r^2} = \frac{\mathbf{P}\cdot\mathbf{r}}{4\pi\varepsilon_0 |r|^3} \tag{3.60}$$

One can also express dipole potential as a gradient. Since

$$\nabla\left(\frac{1}{r}\right) = -\frac{r}{|r|^3}$$

Therefore

$$\phi(r) = -\frac{\mathbf{P}}{4\pi\varepsilon_0}\cdot\nabla\left(\frac{1}{r}\right) = -\mathbf{P}\cdot\nabla(\phi_0) \tag{3.61}$$

where $\phi_0 = \frac{1}{4\pi\varepsilon_0 r}$ is the potential of a unit charge.

Knowing potential, one can obtain electrostatic field by taking the gradient of ϕ. Thus for a electric dipole oriented along the z-axis (Fig. 3.31), one obtains the three components of the field strength of a dipole as

$$E_z = -\frac{\partial\phi}{\partial z} = \frac{P}{4\pi\varepsilon_0}\frac{\partial}{\partial z}\left(\frac{r\cos\theta}{r^3}\right) = -\frac{P}{4\pi\varepsilon_0}\frac{\partial}{\partial z}\left(\frac{z}{r^3}\right)$$

$$= -\frac{P}{4\pi\varepsilon_0}\left(\frac{1}{r^3}-\frac{3z^2}{r^5}\right) = \frac{P}{4\pi\varepsilon_0 r^3}\left(\frac{3z^2}{r^2}-1\right)$$

$$= \frac{P}{4\pi\varepsilon_0 r^3}(3\cos^2\theta - 1)$$

$$E_x = \frac{P}{4\pi\varepsilon_0}\frac{3zx}{r^5} \quad \text{and} \quad E_y = \frac{P}{4\pi\varepsilon_0}\frac{3xy}{r^5} \tag{3.62}$$

It is sometimes convenient to express the field of a dipole in spherical polar coordinates. We have

$$\mathbf{E} = -\left[\hat{a}_r\frac{\partial\phi}{\partial r} + \hat{a}_\theta\frac{\partial\phi}{\partial\theta} + \hat{a}_\phi\frac{1}{r\sin\theta}\frac{\partial\phi}{\partial\Phi}\right]$$

Using (3.60), one obtains

$$E_r = -\frac{\partial\phi}{\partial r} = \frac{2P\cos\theta}{4\pi\varepsilon_0 r^3} \tag{3.63}$$

$$E_\theta = -\frac{1}{r}\frac{\partial\phi}{\partial\theta} = \frac{P}{4\pi\varepsilon_0}\frac{\sin\theta}{r^3} \tag{3.64}$$

and

$$E_\phi = -\frac{1}{r\sin\theta}\frac{\partial\phi}{\partial\Phi} = 0 \tag{3.65}$$

$$\therefore \mathbf{E} = \hat{a}_r\frac{2Qd\cos\theta}{4\pi\varepsilon_0 r^3} + \hat{a}_\theta\frac{Qd\sin\theta}{r^3} \tag{3.66}$$

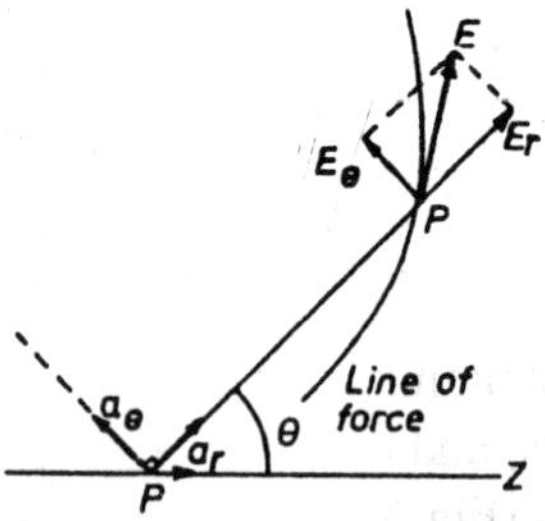

Figure 3.34.

The two components E_r and E_θ of the field of an electric dipole are shown in Fig. 3.34. The lines of force are indicated in Fig. 3.35. Although in an electric dipole the two charges are equal and opposite, giving a zero net charge, the fact that they are slightly displaced is enough to produce a nonvanishing electric field.

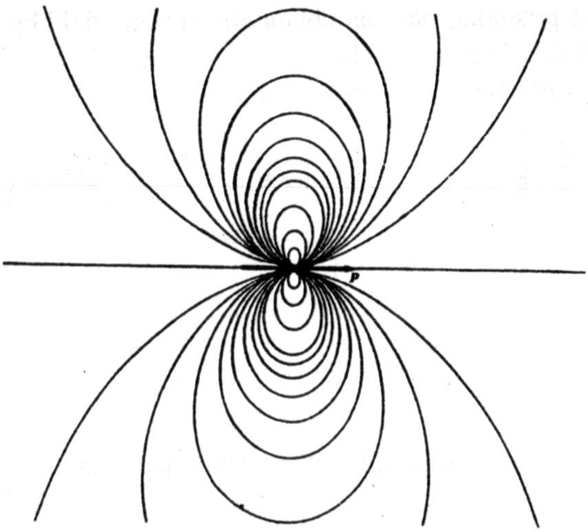

Figure 3.35. Lines of force the electric field of an electric dipole.

One can also express the field of an electric dipole in the following way,

$$
\begin{aligned}
\mathbf{E} = -\nabla\phi &= -\frac{1}{4\pi\varepsilon_0}\nabla\left(\frac{\mathbf{P}\cdot\mathbf{r}}{|\mathbf{r}|^3}\right) \\
&= -\frac{1}{4\pi\varepsilon_0}\left[\frac{1}{|\mathbf{r}|^3}\nabla(\mathbf{P}\cdot\mathbf{r}) + \mathbf{P}\cdot\mathbf{r}\nabla\left(\frac{1}{|\mathbf{r}|^3}\right)\right]
\end{aligned}
$$

Since

$$
\begin{aligned}
\nabla(\mathbf{P}\cdot\mathbf{r}) &= \mathbf{P} \text{ and } \nabla\left(\frac{1}{|\mathbf{r}|^3}\right) = -\frac{3r}{|\mathbf{r}|^5} \\
\therefore \mathbf{E} &= \frac{1}{4\pi\varepsilon_0}\left[\frac{3(\mathbf{P}\cdot\mathbf{r})\mathbf{r}}{|\mathbf{r}|^5} - \frac{\mathbf{P}}{|\mathbf{r}|^3}\right] \\
&= \frac{1}{4\pi\varepsilon_0 r^3}\left[\frac{3(\mathbf{P}\cdot\mathbf{r})\mathbf{r}}{r^2} - \mathbf{P}\right] \qquad (3.67)
\end{aligned}
$$

In atoms the center of mass of the electrons coincides with the nucleus, and therefore the average electric dipole moment of the atom is zero (Fig 3.36(a)). But if an external field is applied, the electronic motion is distorted, and the center of mass of the electrons is displaced through the distance x relative to the nucleus (Fig 3.36(b)). The atom is thus polarized and becomes an electric dipole of moment **P**. This moment is proportional to the external field **E**.

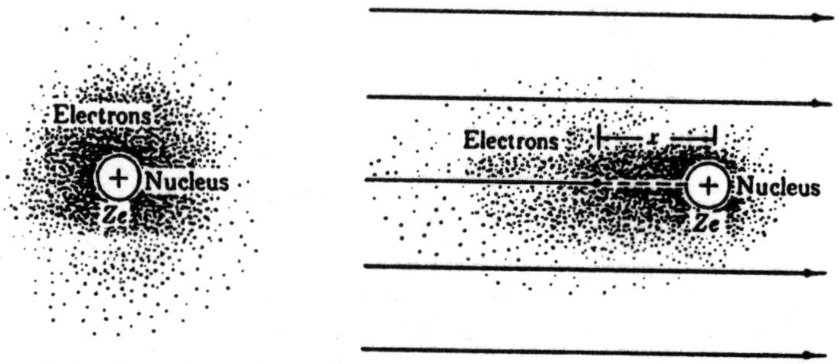

(a) No external field (b) External field

Figure 3.36. Polarization of an atom under an external field

Molecules, on the other hand, may have a permanent electric dipole moment. Such molecules are called *polar*. For example, in the HCl molecule (Fig 3.37), the electron of the H atom spends more time moving around the Cl atom than around the H atom. Therefore the center of the negative charges does not coincide with that of the positive charges, and the molecule has a dipole moment directed from the Cl atom to the H atom. That is, one may write $H^+ Cl^-$. The electric dipole of the HCl molecule is $P = 3.43 \times 10^{-30}$C-m. In the CO molecule (Fig. 3.37) the charge distribution is only slightly asymmetric and the electric dipole moment is relatively small, about 0.4×10^{-30} C-m, with the carbon atom corresponding to the positive and the oxygen atom to the negative end of the molecule.

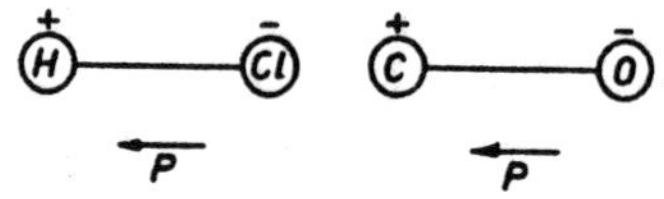

Figure 3.37. Polar diatomic molecules.

In the H_2O molecule, where the two H —O bonds are at an angle slightly over 90° (Fig. 3.38), the electrons try to crowd around the oxygen atom, which there upon becomes slightly negative relative to the H atoms.

Each H—O bond thus contributes to the electric dipole moment, whose resultant, because of symmetry, lies along the axis of the molecule and has a value equal to 6.2×10^{-30} C-m. But in the all CO_2 molecule, the atoms are in a straight line (Fig. 3.39), and the resultant electric dipole moment is zero because of the symmetry. Obviously, the electric dipole moments can provide useful information about the structure of molecules. The values of P for several polar molecules are

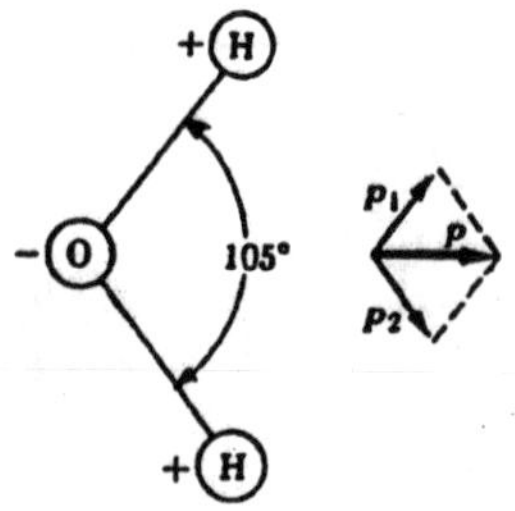

Figure 3.38. Electric dipole of H_2O molecule.

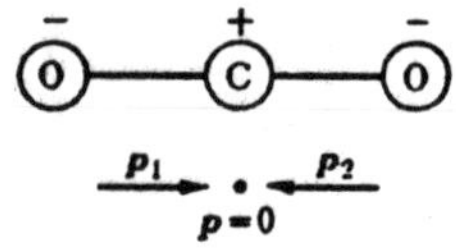

Figure 3.39. The CO_2 molecule has no electric dipole.

given in Table 3.1. We must note that dipole moments of CO_2, H_2, CH_4, C_2H_6 and CCl_4 molecules are zero.

Table 3.1. Electric Dipole moments for selected polar molecules

Molecule	P (C-m)
HCl	3.43×10^{-30}
HBr	2.60×10^{-30}
HI	1.26×10^{-30}
CO	0.40×10^{-30}
H_2O	6.2×10^{-30}
H_2S	5.3×10^{-30}
SO_2	5.3×10^{-30}
NH_3	5.0×10^{-30}
C_2H_5OH	3.66×10^{-30}

3.17 ELECTRIC DIPOLE IN UNIFORM ELECTRIC FIELD

When an electric dipole consisting, of two charges $-q$ and $+q$ distance d apart is placed in a uniform electric field (Fig. 3.40), a force is produced on each charge of the dipole. The resultant force is

$$\mathbf{F} = q\mathbf{E} - q\mathbf{E}' = q(\mathbf{E} - \mathbf{E}')$$

Considering that electric field is along the x-axis and the dipole is oriented parallel to the field. Now, considering magnitudes only, $E - E' = \frac{dE}{dx}d$, and therefore $F = P\frac{dE}{dx}$. This shows that *an electric dipole oriented parallel to the electric field tends to move in the direction in*

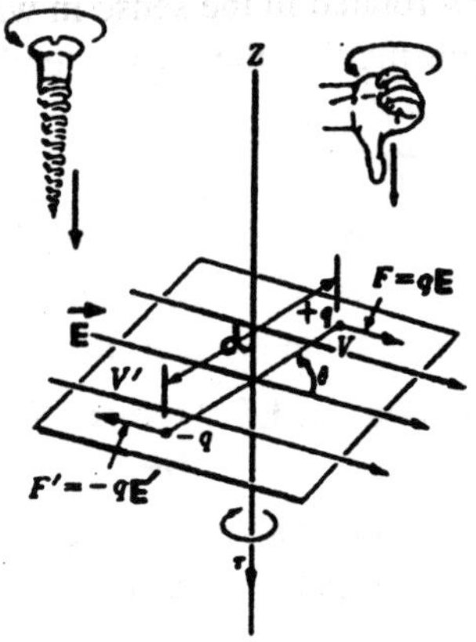

Figure 3.40. Electric dipole in an external uniform electric field

which the, field increases. One obtains the opposite result if the dipole is oriented antiparalled to the field. We must note that if the electric field is uniform, the resultant force on the electric dipole is zero.

The potential energy of the dipole is

$$V_P = q\phi - q\phi' = q(\phi - \phi') = -qd\left(-\frac{\phi - \phi'}{d}\right)$$

If θ is the angle between the dipole and the electric field, then the component $E_d = E\cos\theta$ of the field **E** parallel to d. Therefore

$$E_P = -PE\cos\theta = -\mathbf{P}\cdot\mathbf{E} \tag{3.68}$$

The potential energy is a minimum when $\theta = 0$, indicating that the dipole is in equilibrium when it is oriented parallel to the field. If we neglect the slight difference between **E** and **E**′, the force $q\mathbf{E}$ and $-q\mathbf{E}'$ on the charges comprising the dipole form essentially a couple whose torque is

$$\tau = \mathbf{d} \times (q\mathbf{E}) = (q\mathbf{d}) \times \mathbf{E} = \mathbf{P} \times \mathbf{E} \tag{3.69}$$

The torque of the electric field tends to align the dipole parallel to the field. The magnitude of the torque is $\tau = PE\sin\theta$, and its direction is indicated in Fig. 3.40. One can easily see that $\tau_z = -\frac{\partial E_P}{\partial\theta}$. Using (3.68), one obtains $\tau_z = -PE\sin\theta$. The difference in sign for τ is due to the fact that τ gives the magnitude of the torque while τ_z gives the component of the torque along a direction Z, perpendicular to the plane in which the angle θ is measured, and oriented in the sense of advance of a right-handed screw rotated in the sense in which θ increases. The negative sign in τ_z confirms that the torque tends to decrease the angle θ.

These properties of a dipole placed in an electric field have very important applications. For example, the electric field of an ion in solution polarizes the molecules of the solvent (electrolysis) which surrounds the ions and they become oriented in the form indicated in the Fig. 3.41.

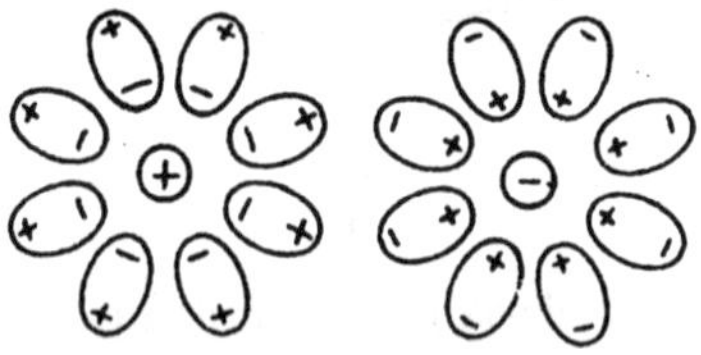

Figure 3.41. Polarization effects of an ion in Solution (electrolyse).

These oriented molecules more or less attached to the ion, increasing its effective mass and decreasing its effective charge, which is partially screened by the molecules. The net effect is that the mobility of the ion in an external field is decreased. Also when a gas or a liquid whose molecules are permanent dipoles is placed where an electric field exists, the molecules, as a result of the torques due to the electric field, tend to align with their dipoles parallel. We say that the substances has been polarized.

3.18 ELECTRIC DIPOLE IN A NON-UNIFORM FIELD

Let us consider a dipole in a non-uniform electric field (Fig. 3.42). A translational force is exerted on the dipole. We are interested in finding the expression for this force.

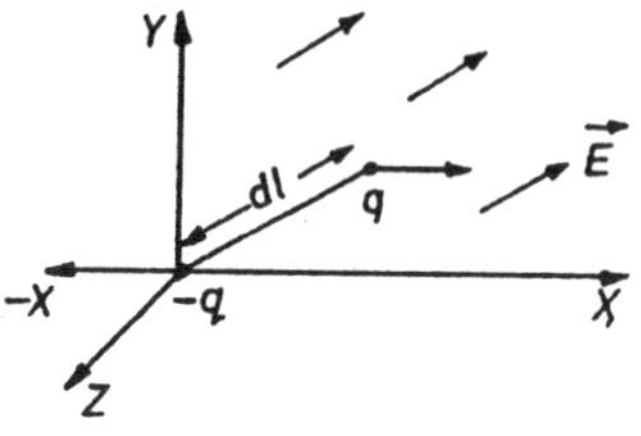

Figure 3.42.

Suppose the origin of the coordinate system coincides with the charge $-q$. The force acting on the dipole in the positive x-direction is

$$F_x^{(+)} = q(E_x + dE_x)$$

The force acting on the dipole in the negative x direction is

$$F_x^{(-)} = qE_x$$

The net force component in the positive x-direction is

$$F_x = qdE_x$$

As the field is non-uniform and hence

$$\begin{aligned} dE_x &= \left(\frac{\partial E_x}{\partial x}\right) dx + \left(\frac{\partial E_x}{\partial y}\right) dy + \left(\frac{\partial E_x}{\partial z}\right) dz \\ \therefore \quad F_x &= q\left(dx\frac{\partial E_x}{\partial x} + dy\frac{\partial E_x}{\partial y} + dz\frac{\partial E_x}{\partial z}\right) \\ &= q\left(dx\frac{\partial}{\partial x} + dy\frac{\partial}{\partial y} + dz\frac{\partial}{\partial z}\right) E_x \\ &= q(d\mathbf{l}\cdot\nabla)E_x = (\mathbf{P}\cdot\nabla)E_x \end{aligned} \tag{3.70}$$

Because

$$d\mathbf{l} = \hat{a}_x dx + \hat{a}_y dy + \hat{a}_z dz$$

and

$$\nabla = \hat{a}_x\frac{\partial}{\partial x} + \hat{a}_y\frac{\partial}{\partial y} + \hat{a}_z\frac{\partial}{\partial z}$$

One can obtain the other components is a similar manner. In general, the force on the electric dipole, is

$$\mathbf{F} = (\mathbf{P}\cdot\nabla)\mathbf{E} \tag{3.71}$$

3.19 ELECTRIC DOUBLE LAYERS

When one studies biological and Celloid problems, then it is observed that on a given surface there is a double layer of surface charges consisting of two neighbouring layers of charges of equal magnitude but opposite sign separated by an infinitesimal distance dx as shown in fig. 3.43(a). If the surface charge density is σ, then the strength of the layer

$$D = \sigma dx$$

Fig. 3.43(b) shows two surfaces with two small elemental areas on it. The two charges $\pm\sigma ds$ separated by dx are equivalent to dipole moment $p = \sigma ds dx$. Obviously, one may think it as a layer of dipoles rather than two layers with positive and negative charges. The direction of the dipole moment of the layer is normal to the surface. One obtains

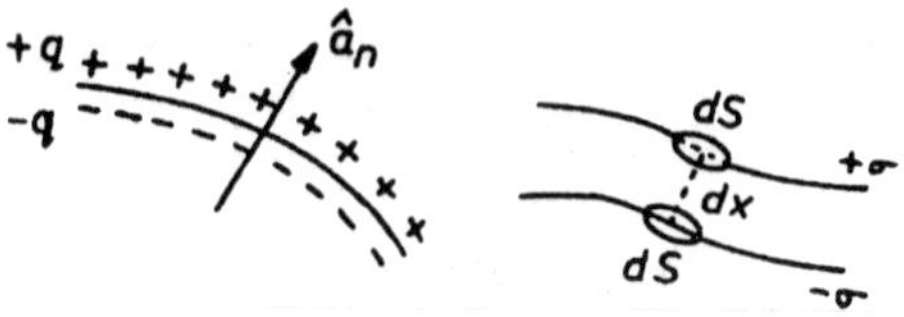

Figure 3.43.

the potential of the dipole layer at a point distance r as

$$\phi(r) = \frac{1}{4\pi\varepsilon_0}\int \hat{a}_n \frac{Dds \cdot \mathbf{r}}{|\mathbf{r}|^3} = \frac{D}{4\pi\varepsilon_0}\int \frac{\hat{a}_n ds \cdot \mathbf{r}}{|\mathbf{r}|^3} = \frac{D}{4\pi\varepsilon_0} = d\Omega. \quad (3.72)$$

where $d\Omega$ is the element of the solid angle subtended at the point by ds.

3.20 HIGHER ELECTRIC MULTIPOLES

It is possible to define higher-order or multipole electric moments. For example, a charge distribution such as that shown in Fig. 3.44 constitutes an *electric quadrupole*. Its total charge Σq_i and electric dipole moment **p** are zero and hence gives a zero value of potential. Actually, however, the field of a quadrupole, although it is much weaker than that of a dipole (with the same values of q and d), differs from zero. Let us consider the case of a *linear quadrupole* as shown in Fig. 3.45.

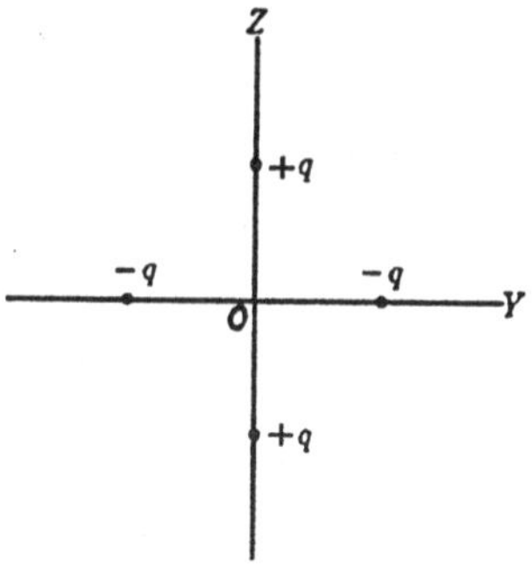

Figure 3.44. Electric quadrupole.

One can regard the distribution of charge of a linear quadrupole (Fig. 3.45) as derived from two equal and opposite dipole moments **P** placed in a line and separated by a distance d. The potential due to the quadrupole at a point P is given by

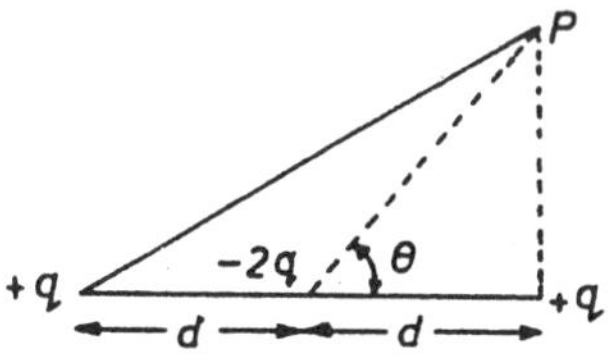

Figure 3.45. Linear quadrupole.

$$\phi_P = \phi_1 - \left(\phi_1 + d\frac{\partial \phi_1}{\partial x}\right) = -d\frac{\partial \phi_1}{\partial x}$$

where ϕ_1 is the potential due to the first dipole. Therefore

$$\begin{aligned}\phi_P &= -d\frac{\partial}{\partial x}\left(\frac{p\cos\theta}{4\pi\varepsilon_0 r^2}\right) = -\frac{pd}{4\pi\varepsilon_0}\frac{\partial}{\partial x}\left(\frac{\cos\theta}{r^2}\right)\\ &= -\frac{pd}{4\pi\varepsilon_0}\left[-\frac{1}{r^2}\sin\theta\frac{\partial\theta}{\partial x}-\frac{2\cos\theta}{r^3}\frac{\partial r}{\partial x}\right]\\ &= -\frac{pd}{4\pi\varepsilon_0}\left[\frac{\sin^2\theta}{r^3}-\frac{2\cos^2\theta}{r^3}\right] \quad \left(\because \ \frac{\partial\theta}{\partial x}=-\frac{\sin\theta}{r}, \frac{\partial r}{\partial x}=\cos\theta\right)\\ &= \frac{pd}{4\pi\varepsilon_0 r^3}(3\cos^2\theta-1) \qquad (3.73)\end{aligned}$$

The quadrupole moment, defined analogously to the dipole moment, is given by

$$Q = pd = qd^2$$

$$\therefore \quad \phi_p = \frac{Q}{4\pi\varepsilon_0 r^3}(3\cos^2\theta-1) \qquad (3.74)$$

obviously the potential of the field set up by a quadrupole is proportional to $\frac{1}{r^3}$, where as the field decreases as $\frac{1}{r}$.

The electric quadrupole moment is zero for a spherical distribution of charge positive for an elongated or prolate charge distribution, and negative for an oblate or flattened charge distribution (Fig. 3.46).

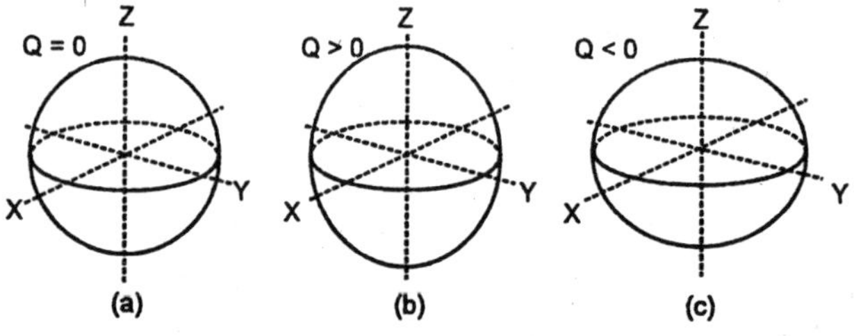

Figure 3.46. Electric quadrupole of elliposidal charge distribution

Therefore the electric quadrupole moment gives an indication of the degree of departure from the spherical form of a charge distribution. We generally consider atomic nuclei spherical. However, careful measurements indicate that certain nuclei have relatively large electric quadrupole moments. This has been interpreted as indicating that such nuclei are greatly deformed, and thus the electric field they produce departs from a point charge. This in turn affects the energy of the electronic motion.

We must note that the potential of a point charge decreases as r^{-1} and the field as r^{-2}. Similarly for an electric dipole the potential decreases as r^{-2} and the field is r^{-3}. For an electric quadrupole the potential varies as r^{-3} and the field as r^{-4}. Similar results are obtained for higher order multipoles. We conclude then that the higher the multipole order the smaller the range within which its electric field has any noticeable effect.

3.21 ELECTRIC POTENTIAL DUE TO AN ARBITRARY DISTRIBUTION OF CHARGE

Let us consider that an given assembly of point charges distributed within a region in a complicated manner as shown in Fig. 3.47. We are interested in determining the potential due to this distribution at a point P at a distance R from the origin O.

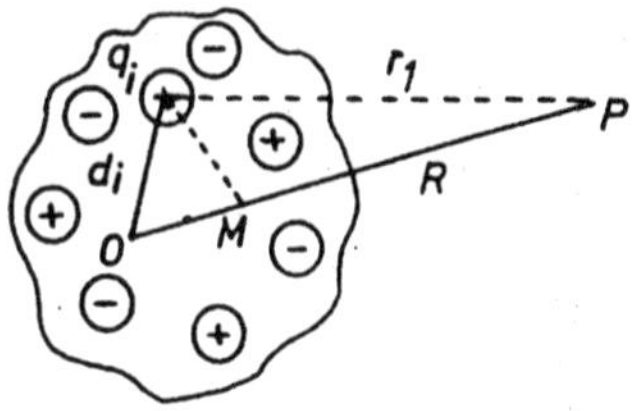

Figure 3.47.

Let us consider that a charge q_i be at a distance d_i from the origin. The potential at P due to this charge is

$$\phi_i(r_i) = \frac{q_i}{4\pi\varepsilon_0 r_i} = \frac{q_i}{4\pi\varepsilon_0 |R - d_i|} \tag{3.75}$$

where r_i is the distance of q_i from P. If $R \gg d_i$ one may write to the first approximation

$$\phi_i(r_i) = \frac{q_i}{4\pi\varepsilon_0 R}$$

for each charge q_i.

Thus the total potential due to entire distribution of charges is

$$\phi(R) = \frac{\Sigma q_i}{4\pi\varepsilon_0 R} = \frac{Q}{4\pi\varepsilon_0 R} \tag{3.76}$$

obviously the potential is same as that obtained by a point charge Q at the origin, where

$$Q = \Sigma_i q_i$$

Let us consider the case when there are equal number of positive and negative charges in the region. Clearly, in this case $Q = 0$, i.e. the object under consideration is neutral. At once, one may conclude that there will not be anything like potential around the object. However, there may be objects which are neutral and yet the charges are not all at one point and hence, one should expect some effect of these separate charges at points not far away from the object. we are interested in finding the potential due to such a charge distribution. For this purpose one can use Eq. (3.75) without making the approximation that d_i being much smaller than R. However, we can introduce better approximation. Let us assume that r_i differs from R by projection of d_i on R. i.e. by OM as shown in Fig. 3.47. Thus

$$r_i = R - \mathbf{d}_i \cdot \hat{a}_R \tag{3.77}$$

where $\hat{a}_R$ is the unit vector in the direction of R. Thus

$$\frac{1}{r_i} = \frac{1}{R\left(1 - \frac{\mathbf{d}_i \cdot \hat{a}_R}{R}\right)} = \frac{1}{R}\left(1 + \frac{\mathbf{d}_i \cdot \hat{a}_R}{R}\right)$$

Here we have neglected higher terms

$$\begin{aligned} \therefore \quad \phi(R) &= \Sigma_i \frac{q_i}{4\pi\varepsilon_0 R}\left(1 + \frac{\mathbf{d}_i \cdot \hat{a}_R}{R}\right) \\ &= \frac{Q}{4\pi\varepsilon_0 R} + \Sigma_i q_i \frac{\mathbf{d}_i \cdot \hat{a}_R}{4\pi\varepsilon_0 R^2} \end{aligned} \tag{3.78}$$

we must note that the first term is same as in (3.77) and is equal to zero, since $Q = 0$. Therefore

$$\phi(R) = \Sigma_i q_i \frac{\mathbf{d}_i \cdot \hat{a}_R}{4\pi\varepsilon_0 R^2} \tag{3.79}$$

Defining $\mathbf{P} = \Sigma q_i d_i$ as the dipole moment of charge distribution, one obtains

$$\phi(R) = \frac{\mathbf{P} \cdot \hat{a}_R}{4\pi\varepsilon_0 R^2} \tag{3.80}$$

Obviously for any distribution of charges that is as a whole netural, the potential is a dipole potential. Examples of such charge distributions are H_2O and CO_2 molecules.

EXAMPLE 3.1: A charge q coulomb is distributed uniformly throughout a non-conducting spherical volume of radius R metre. Show that the potential at a distance r form the centre $r \leq R$ is given by

$$\phi = \frac{1}{4\pi\varepsilon_0} \frac{q(3R^2 - r^2)}{2R^3}$$

Solution: Let us consider that charge q Coulomb be distributed uniformly throughout a non-conducting sphere of radius R meter. We have the field strength at point distance $r > R$ and $r < R$ from the centre of spherical charge are, respectively, given by

$$\mathbf{E}_0 = \frac{1}{4\pi\varepsilon_0}\frac{q}{r^3}\mathbf{r} \quad \text{and} \quad \mathbf{E}_i = \frac{1}{4\pi\varepsilon_0}\frac{q}{R^3}\mathbf{r}$$

The potential at distance $r < R$ from the centre is given by

$$\begin{aligned}
\phi &= -\int_{\infty}^{r}\mathbf{E}\cdot d\mathbf{r} = -\left[\int_{\infty}^{R}\mathbf{E}_0\cdot d\mathbf{r} + \int_{R}^{r}\mathbf{E}_i\cdot d\mathbf{r}\right] \\
&= -\left[\int_{\infty}^{R}\frac{1}{4\pi\varepsilon_0}\frac{q\mathbf{r}}{r^3}\cdot d\mathbf{r} + \int_{R}^{r}\frac{1}{4\pi\varepsilon_0}\frac{q}{R^3}\mathbf{r}\cdot d\mathbf{r}\right] \\
&= -\left[\frac{1}{4\pi\varepsilon_0}q\int_{\infty}^{R} r^{-2}dr + \frac{1}{4\pi\varepsilon_0}\int_{R}^{r}\frac{q}{R^3}\mathbf{r}\cdot d\mathbf{r}\right] \\
&= -\frac{1}{4\pi\varepsilon_0}q\left[\left\{-\frac{1}{r}\right\}_{\infty}^{R} + \left\{\frac{r^2}{2R^3}\right\}_{R}^{r}\right] \\
&= -\frac{1}{4\pi\varepsilon_0}q\left[-\frac{1}{R} + \frac{r^2}{2R^2} - \frac{1}{2R}\right] \\
&= -\frac{1}{4\pi\varepsilon_0}q\left[\frac{r^2}{2R^3} - \frac{3}{2R}\right] = \frac{q}{4\pi\varepsilon_0}\left[\frac{3R^3 - r^2}{2R^3}\right]
\end{aligned}$$

EXAMPLE 3.2: Two particles each of mass m and having a charge Q are suspended by string of length l from a common point, show that the angle θ, which each string makes with the vertical, is given by the relation

$$\frac{\tan^3\theta}{1+\tan^2\theta} = \frac{Q^2}{16\pi\varepsilon_0 mgl^2}$$

Solution: The two particles carry like charges and hence the force F between the charges will be repulsive (Fig. 3.48). The magnitude of the force F is given by

$$F = \frac{Q\times Q}{4\pi\varepsilon_0(AB)^2} = \frac{Q^2}{4\pi\varepsilon_0(AB)^2} \qquad \text{(i)}$$

Now

$$AB = 2AM = 2l\sin\theta \qquad \text{(ii)}$$

Using (ii), (i) becomes

$$F = \frac{Q^2}{4\pi\varepsilon_0(2l\sin\theta)^2} = \frac{Q^2}{16\pi\varepsilon_0 l^2\sin^2\theta} \qquad \text{(iii)}$$

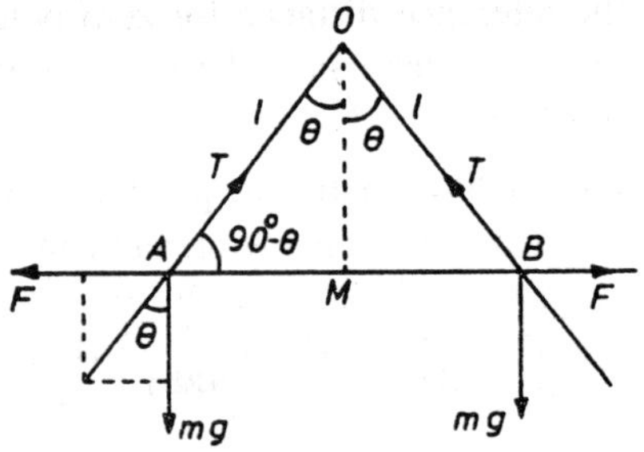

Figure 3.48.

Applying Lame's theorem at point A, one obtains

$$\frac{F}{\sin(180-\theta)} = \frac{mg}{\sin(90+\theta)}$$

or

$$\frac{F}{\sin\theta} = \frac{mg}{\cos\theta} \tag{iv}$$

Substituting the value of F from (iv) in (iii), one obtains

$$\frac{Q^2}{16\pi\varepsilon_0 l^2 \sin^2\theta} = \frac{mg}{\cos\theta}\sin\theta \tag{v}$$

or

$$\frac{Q^2}{16\pi\varepsilon_0 l^2 mg} = \frac{\sin^3\theta}{\cos\theta} = \frac{\tan^3\theta}{\sec^2\theta} = \frac{\tan^3\theta}{1+\tan^2\theta}$$

EXAMPLE 3.3: Show that the potential difference between two points distance r_1 and r_2 from an infinitely long charge of linear charge density (i.e. charge per unit length) λ is given by

$$\nabla\phi = \frac{1}{4\pi\varepsilon_0} 2\lambda \log_e\left(\frac{r_2}{r_1}\right)$$

Solution: The field strength at a distance r from an infinite linear charge of density λ is given by

$$E = \frac{1}{4\pi\varepsilon_0}\frac{2\lambda}{r} \tag{i}$$

The potential at a distance r from the axis is

$$\phi_r = -\int_{r_{\text{res}}}^{r} \mathbf{E}\cdot d\mathbf{r}$$

Here r_{ref} denotes the reference distance for zero potential. We must note that the reference distance cannot be taken as infinity since the wire itself extends to infinity.

As per the problem, we have to find the potential difference between two points distance r_1 and r_2 from the wire, and hence

$$\nabla\phi = -\int_{r_2}^{r_1} \frac{1}{4\pi\varepsilon_0}\frac{2\lambda}{r}dr = \frac{1}{4\pi\varepsilon_0}2\lambda\log_e\frac{r_2}{r_1}$$

EXAMPLE 3.4: Determine the potential at a point (i) outside and (ii) inside a uniformly charged sphere of radius a.

Solution:

(i) $r > a$ we have the field produced at a point r outside the sphere is

$$E = -\frac{\partial\phi}{\partial r} = \frac{a^3\rho}{3\varepsilon_0 r^2}$$

where e is the charge density

$$\therefore \phi = \int_{\infty}^{r}\frac{a^3\rho}{3\varepsilon_0 r^2}dr = \frac{a^3\rho}{3\varepsilon_0 r}$$

(ii) $r < a$

$$\because \quad E = -\frac{\partial\phi}{\partial r}$$

$$\begin{aligned}\therefore \quad \phi &= \int_{\infty}^{r}\frac{\partial\phi}{\partial r}dr = \int_{\infty}^{r}\frac{\partial\phi}{\partial r}dr + \int_{a}^{r}\frac{\partial\phi}{\partial r}dr \\ &= -\int_{\infty}^{a}\frac{a^3\rho}{3\varepsilon_0 r^2}dr - \int_{a}^{r}\frac{\rho r}{3\varepsilon_0}dr \\ &= \frac{a^3\rho}{3\varepsilon_0} - \frac{\rho}{6\varepsilon_0}(r^2 - a^2) = \frac{\rho}{6\varepsilon_0}(3a^2 - r^2)\end{aligned}$$

EXAMPLE 3.5: Show that the potential at a point distance r_1 and r_2 respectively from centres of a long parallel pair of wires of negligible cross section and having equal and opposite linear charge density λ Coulomb-metre is given by

$$\phi = \frac{\lambda}{4\pi\varepsilon_0}\log_e\left(\frac{r_2}{r_1}\right)$$

Solution: The electric field strength at distance r from the line charge is

$$E = \frac{1}{4\pi\varepsilon_0}\frac{2\lambda}{r} \tag{i}$$

By definition, we have electric potential

$$\phi = -\int_{\text{ref}}^{r} \mathbf{E}\cdot d\mathbf{r}$$

where r_{ref} is the reference point for zero potential.

In this problem, we can not choose reference point r_{ref} at infinity as charge itself extends to infinity. Therefore we choose reference point at r_0 where potential is ϕ_0. Thus potential difference between r_0 and r is

$$\begin{aligned}\phi - \phi_0 &= -\int_{r_0}^{r} \mathbf{E}\cdot d\mathbf{r} = -\int_{r_0}^{r} \frac{1}{4\pi\varepsilon_0}\frac{2\lambda}{r}dr \\ &= \frac{1}{4\pi\varepsilon_0}2\lambda \log_e \frac{r_0}{r} = \frac{\lambda}{2\pi\varepsilon_0}\log_e \frac{r_0}{r}\end{aligned}$$

Thus the potential at distance r is

$$\phi = \frac{\lambda}{2\pi\varepsilon_0}\log_e \frac{r_0}{r} + \phi_0 \tag{ii}$$

Let us consider the case of two lines of charge of linear charge densities λ and $(-\lambda)$ if the point under consideration (say P) lies at distance r_1 and r_2 from wires then potential (ϕ_1) at P due to one line charge of charge density $(+\lambda)$ will be

$$\phi_1 = \frac{\lambda}{2\pi\varepsilon_0}\log_e\left(\frac{r_0}{r_1}\right) + \phi_0 \tag{iii}$$

The potential due to other line charge of charge density $(-\lambda)$ will be

$$\phi_2 = \frac{-\lambda}{2\pi\varepsilon_0}\log_e \frac{r_0}{r_2} + \phi_0 \tag{iv}$$

$\therefore$ The net potential due to both the wires

$$\phi = \phi_1 + \phi_2 = \frac{\lambda}{2\pi\varepsilon_0}\log_e \frac{r_2}{r_1} + 2\phi_0 \tag{v}$$

Let the potential at reference point r_0 is zero. This is possible in this case if the reference point r_0 is at equal distance from both the wires, i.e. we have to choose $\phi_0 = 0$. Thus

$$\phi = \frac{\lambda}{2\pi\varepsilon_0} \log_e \frac{r_0}{r_1}$$

EXAMPLE 3.6: Show that in free space the normal component of field on any surface is discontinuous to the extent of $\frac{\rho}{\varepsilon_0}$ where ρ is the surface charge density.

Solution: Construct a volume element in the shape of a pill (pill box) across the surface (Fig 3.49). Now apply Gauss's law to the volume of the pill box as the flat faces approach the surface. The flux through the lateral surface approaches zero and E_{n1}, E_{n2} approach the normal components of the field on the two sides of the surface, we have

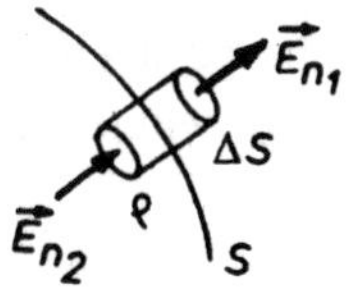

Figure 3.49.

$$(E_{n1} - E_{n2})\Delta S = \frac{\rho}{\varepsilon_0}\Delta S$$

or

$$E_{n1} - E_{n2} = \frac{\rho}{\varepsilon_0}$$

EXAMPLE 3.7: Show that the energy of a sphere of charge of radius R in which the charge is uniformly distributed is

$$W = \frac{4\pi\rho^2}{15\varepsilon_0} R^5$$

where ρ is the density of charge.

Solution: We know that the energy is equal to the workdone in bringing charges from infinity to the point (here it is sphere) under consideration. Let us imagine that sphere is formed by assembling various thin shells of charge. Now we consider a small sphere of charge of radius r as shown in Fig. 3.50. ρ is the density of charge. The total charge on the surface on the sphere is $\frac{4}{3}\pi r^3 \rho$. Imagine that a small layer of charge dq in the form of thin shell of thickness dr is deposited on the

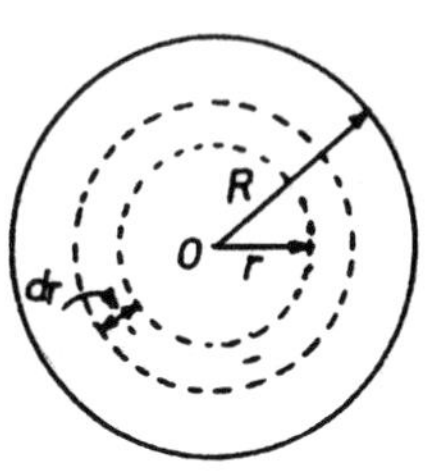

Figure 3.50.

sphere. Therefore

$$dq = \rho 4\pi r^2 dr$$

Now, the workdone in bringing this charge from infinity is

$$\begin{aligned} dW &= \text{Potential at } r \times dq \\ &= \frac{1}{4\pi\varepsilon_0}\frac{\left(\frac{4}{3}\right)\pi r^3\rho}{r}\rho 4\pi r^2 dr \\ &= \frac{4\pi\rho^2}{3\varepsilon_0}r^4 dr \end{aligned}$$

One can obtain the total energy required to assemble charges so as to build up a sphere of radius R is

$$W = \frac{4\pi\rho^2}{3\varepsilon_0}\int_0^R r^4 dr = \frac{4\pi\rho^2}{15\varepsilon_0}R^5$$

EXAMPLE 3.8: Write down the interaction potential energy of two short electric dipoles separated by a distance. If one of the dipoles is inclined at an angle θ_1 to the radius vector joining them, show that in the state of equilibrium, the other dipole will make an angle θ_2 with it given by

$$\tan\theta_2 = -\frac{1}{2}\tan\theta_1$$

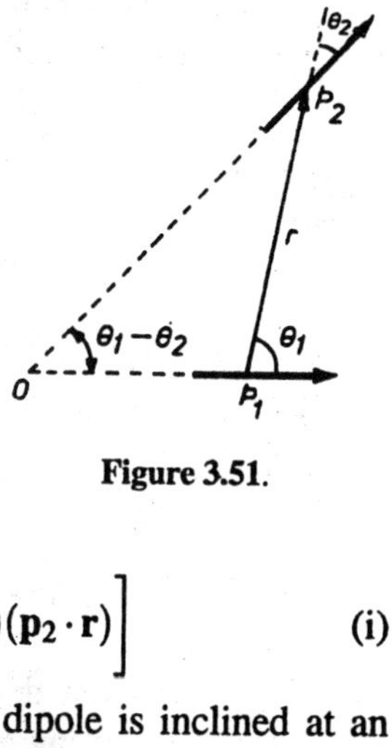

Figure 3.51.

Solution: The interaction potential energy of two short dipoles is

$$U = \frac{1}{4\pi\varepsilon_0}\left[\frac{\mathbf{p}_1\cdot\mathbf{p}_2}{r^3} - \frac{3}{r^5}(\mathbf{p}_1\cdot\mathbf{r})(\mathbf{p}_2\cdot\mathbf{r})\right] \qquad \text{(i)}$$

Let the dipole moment, $\mathbf{p}_1$, of one of the dipole is inclined at an angle θ_1 to the radius vector r and dipole moment $\mathbf{p}_2$ of second dipole is inclined at angle θ_2 with the same radius vector r (Fig 3.51), then the angle between the two poles would be θ_1. Now the expression (i) for U takes the form

$$\begin{aligned} U &= \frac{1}{4\pi\varepsilon_0}\left[\frac{p_1p_2\cos(\theta_1-\theta_2)}{r^3} - \frac{3}{r^5}(p_1r\cos\theta_1)(p_2r\cos\theta))\right] \\ &= \frac{1}{4\pi\varepsilon_0}\frac{p_1p_2}{r^3}[\cos(\theta_1-\theta_2) - 3\cos\theta_1\cos\theta_2] \qquad \text{(ii)} \end{aligned}$$

We must note that in the state of equilibrium of dipoles the force must be zero and potential energy must be extremum. Now for applying the condition of extremum potential energy we differentiate (ii) w.r.t. θ_2 and then equal to zero one obtains

$$\frac{\partial U}{\partial \theta_2} = \frac{1}{4\pi\varepsilon_0}\frac{p_1 p_2}{r^3}[\sin(\theta_1 - \theta_2) + 3\cos\theta_1 \sin\theta_2] = 0$$

Since $\frac{1}{r^3}$ is not equal to zero, we have

$$\sin(\theta_1 - \theta_2) + 3\cos\theta_1 \sin\theta_2 = 0$$

$$\sin\theta_1 \cos\theta_2 - \cos\theta_1 \sin\theta_2 + 3\cos\theta_1 \sin\theta_2 = 0$$

or

$$\sin\theta_1 \cos\theta_2 + 2\cos\theta_1 \sin\theta_2 = 0$$

$$\therefore \quad \tan\theta_2 = -\frac{1}{2}\tan\theta_1$$

EXAMPLE 3.9: Calculate the resultant dipole moment for the system of charges as shown in Fig. 3.52 and hence determine the electric potential and field at a point quite far away from the charges.

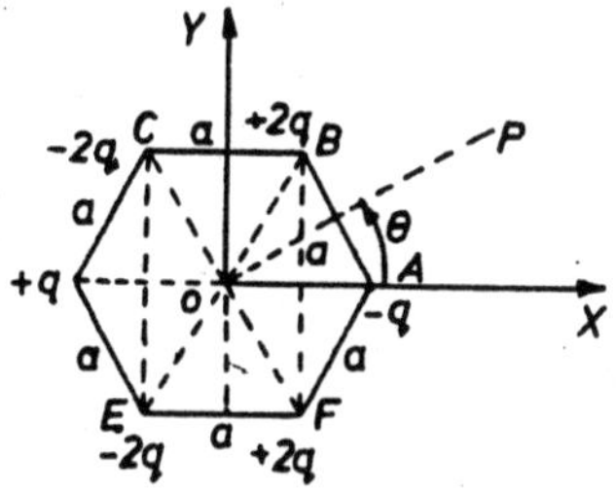

Figure 3.52.

Solution: We have the x-component of dipole moment due to charge $-q$ at point $A(a, 0)$ is

$$\int_v \rho x dv = -q \cdot \mathbf{a} = -qa \tag{i}$$

similarly, the x component of dipole moment due to charge

$$+2q \text{ at } B\left(\frac{a}{2}, \frac{a\sqrt{3}}{2}\right) = 2q \cdot \frac{a}{2} = +qa \tag{ii}$$

The x-component of dipole moment due to charge

$$-2q \text{ at } C\left(-\frac{a}{2},\frac{a\sqrt{3}}{2}\right) = -2q\left(\frac{-q}{2}\right) = +qa \qquad \text{(iii)}$$

The x-component of dipole moment due to charge

$$+q \text{ at } D(-a,0) = +q(-a) = -qa \qquad \text{(iv)}$$

The x-component of dipole moment due to charge

$$-2q \text{ at } E\left(-\frac{a}{2},\frac{-a\sqrt{3}}{2}\right) = -2q\left(-\frac{a}{2}\right) = +qa \qquad \text{(v)}$$

and lastly, the x-component of dipole moment due to charge

$$+2q \text{ at } F\left(+\frac{a}{2},\frac{-q\sqrt{3}}{2}\right) = 2q\cdot\frac{a}{2} = +qa \qquad \text{(vi)}$$

Thus the x-component of the dipole moment for the given system of charges $p_x = +2qa$.

Similarly, one can obtain the contributions of all the charges to y-component of dipole moment. One obtains

$$P_y = \int_v \rho dv = 0 + 2q\left(\frac{a\sqrt{3}}{2}\right) - 2q\left(\frac{a\sqrt{3}}{2}\right) + 0$$

$$-2q\left(-\frac{a\sqrt{3}}{2}\right) + 2q\left(-\frac{a\sqrt{3}}{2}\right) = 0$$

∴ Total dipole moment

$$\mathbf{p} = \hat{i}p_x + \hat{j}p_y = 2qa\hat{i}$$

∴ The potential at a distant point p due to dipole moment

$$\phi_p = \frac{1}{4\pi\varepsilon_0}\frac{\mathbf{p}\cdot\mathbf{r}}{r^2} = \frac{1}{4\pi\varepsilon_0}\frac{2qa\cos\theta}{r^2}$$

The field components are

$$E_r = -\frac{\partial\phi}{\partial r} = \frac{1}{4\pi\varepsilon_0}\frac{4qa\cos\theta}{r^3}$$

$$E_\theta = -\frac{1}{r}\frac{\partial\theta}{\partial r} = \frac{1}{4\pi\varepsilon_0}\frac{2qa\sin\theta}{r^3}$$

$$\therefore \quad \mathbf{E} = \hat{r}E_r + E_\theta$$
$$= \frac{1}{4\pi\varepsilon_0}\frac{2qa}{r^3}[2\cos\theta\hat{r} + \sin\theta\hat{\theta}]$$

EXAMPLE 3.10: An infinite circular cylindrical volume carries a uniform charge density except for the cylindrical cavity which is also infinitely long. The cavity carries no charge. Determine the field inside the cavity and show that it is constant.

Solution: Applying Gauss's law, one can write (Fig 3.53):

$$\mathbf{E}(P) = \frac{\rho(O\mathbf{P})}{2\varepsilon_0} - \frac{\rho}{2\varepsilon_0}(\mathbf{BP})$$
$$= \frac{\rho}{2\varepsilon_0}(\mathbf{AB}) = \text{constant.}$$

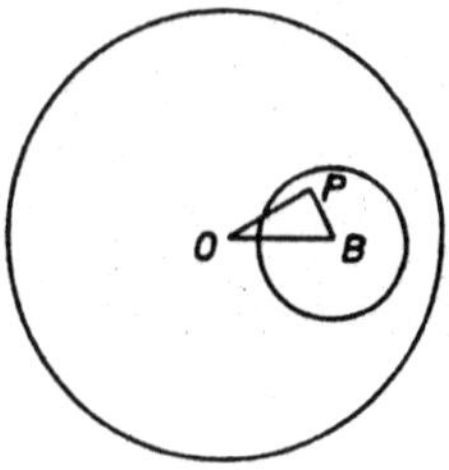

Figure 3.53. Field inside a cylindrical cavity

EXAMPLE 3.11: Four equal charges of magnitude Q of sign indicated are arranged in air as shown in Fig. 3.54 forming an quadrupole show that at large distance $r(r \gg l$ and $r \gg s)$ the potential due to this quadrupole configuration is

$$V = \frac{Qls\sin\theta\cos\theta}{2\pi\varepsilon_0 r^3}$$

where $r \to$ radial distance and $\theta \to$ angle from axis to radial line.

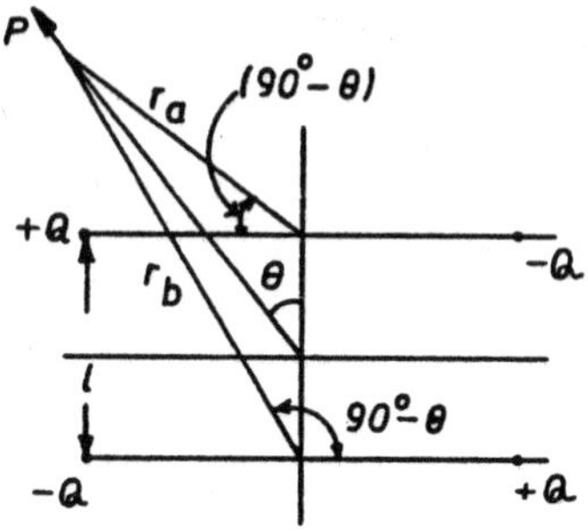

Figure 3.54.

Solution: Figure 3.54 shows a combination of two dipoles and the potential due to the quadrupole may be evaluated by superposition of the potentials due to each dipole system taken separately. We have the potential due to an electric dipole as

$$V = \frac{Ql\cos\theta}{4\pi\varepsilon_0 r^2} \qquad \text{(i)}$$

As shown in Fig 3.54 the horizontally placed charges $+Q$ and $-Q$ at the top $-Q$ and $+Q$ at the bottom may be considered to be two sets of dipoles which constitute quadrupole.

Let P is situated at such a distance that r_a, r_b, r are infinitely large as compared to l and s, then r_a, r_b and r may be considered to be parallel and the angle 'α' measure the positive charge of each dipole, namely the top and the bottom - dipole will $90-\theta$ and $90+\theta$ respectively. Also

$$r_a = r - \frac{l}{2} \quad \text{and} \quad r_b = r + \frac{l}{2}\cos\theta \qquad \text{(ii)}$$

Using (i) the potential due to the top dipole,

$$V_1 = \frac{Qs\cos(90-\theta)}{4\pi\varepsilon_0\left(r-\frac{l}{2}\cos\theta\right)^2} = \frac{Qs\sin\theta}{4\pi\varepsilon_0\left(r-\frac{l}{2}\cos\theta\right)^2} \qquad \text{(iii)}$$

and the potential due to the bottom dipole, in a similar way, will then be

$$V_2 = \frac{Qs\cos(90+\theta)}{4\pi\varepsilon_0\left(r+\frac{l}{2}\cos\theta\right)^2} = -\frac{Qs\sin\theta}{4\pi\varepsilon_0\left(r+\frac{l}{2}\cos\theta\right)^2} \qquad \text{(iv)}$$

The resultant potential at P,

$$\begin{aligned} V &= V_1 + V_2 = \frac{Qs\sin\theta}{4\pi\varepsilon_0}\left[\frac{1}{\left(r-\frac{l}{2}\cos\theta\right)^2} - \frac{1}{\left(r+\frac{l}{2}\cos\theta\right)^2}\right] \\ &= \frac{Qs\sin\theta}{4\pi\varepsilon_0}\left[\frac{\left(r+\frac{l}{2}\cos\theta\right)^2 - \left(r-\frac{l}{2}\cos\theta\right)^2}{r^2 \cdot r^2}\right] \\ &= \frac{Qs\sin\theta}{4\pi\varepsilon_0 r^4} \quad \left[\because r^2 \gg \left(\frac{l}{2}\cos\theta\right)^2\right] \\ &= \frac{Qls\sin 2\theta}{4\pi\varepsilon_0 r^3} \end{aligned} \qquad \text{(v)}$$

We have

$$E = -\nabla V$$

$$\therefore \quad \nabla V = a_r \frac{\partial V}{\partial r} + a_\theta \frac{1}{r}\frac{\partial V}{\partial \theta}$$

$$\therefore \quad \mathbf{E} = \frac{3Ql \sin 2\theta}{4\pi\varepsilon_0 r^4}\hat{a}_r - \frac{Ql s \cos 2\theta}{2\pi\varepsilon_0 r^4}\hat{a}_\theta \qquad \text{(vi)}$$

If $\theta = 0$, point P lies along the axis which represents zero potential line ($\theta = 0$, eq(v) gives $V = 0$) eq. (vi) shows that the radial component of electric field vanishes but the angular component, E_θ, persists and is directed to the right. In case $\theta = 45°$, E_θ vanishes but E_r persists.

EXAMPLE 3.12: A point charge has a charge $Q = 10^{-10}$C. Arbitrarily taking 10^{20} flux lines per Coulomb of charge, determine the number of lines per square metre at a distance of 10 metres. Also find the density D at this distance

Solution:

$$\text{Total number of lines} = 10^{-10} \times 10^{20} = 10^{10}$$

$$\text{Number of lines per } m^2 \text{ at a distance of } 10\ m = \frac{10^{10}}{4\pi r^2} = 7.95 \times 10^6$$

$$\begin{aligned} \therefore \quad \text{Density } D &= 7.95 \times 10^6 \times \text{Charge/line} \\ &= 7.95 \times 10^6 \times 10^{-20} \\ &= 7.95 \times 10^{-14}\ \text{C/m}^2 \end{aligned}$$

QUESTIONS

1. Define the flux of a vector field and intensity of electric field.
2. State the principle of superposition as applied to electric potential.
3. State Gauss's law in electrostatics. Apply it to calculate the electric field intensity due to a uniformly charged sphere (non-conducting) at points (i) outside the sphere (ii) at the surface of the sphere and (iii) inside the sphere.
4. Apply Gauss's law to find the electric field intensity at a point near a uniformly charged wire of infinite length. Hence calculate the force per unit length on a similarly charged wire parallel to it and placed at a distance 'd' from the first one.

5. (a) Plot the magnitude of electric field with distance for a conducting sphere of radius 'a' carrying charge Q.
   (b) In what region does this field differ from its value for a uniformly charged sphere of the same radius and charge?
6. Define dipole moment of an electric dipole Derive an expression for the potential at a point (r, θ) due to a dipole of moment p.
7. Define potential at a point in electrostatic field show that $\mathbf{E} = -\nabla V$.
8. Derive an expression for the electric field due to an electric dipole and show that the absolute of the electric field of an electric dipole decreases monotonically as the angle θ between the dipole moment and radius to the observer increases from 0 to $\frac{\pi}{2}$ radians when the distance of the observer is constant.
9. Define electric dipole, quadrupole and octopole. Show that the potential of charge distribution over a finite volume can always be expressed as a series of multipole potentials, referred of a suitable origin of coordinates.
10. What do you understand by electric dipole and quadrupole moment? Obtain expressions for electric potential and field at a point in space due to (i) a dipole and (ii) a quadrupole.
11. An electric quadrupole is formed by a charge $-q$ at the origin and charge $+q$ at the points $(\pm a, 0, 0)$. Show that the potential ϕ at a distance r large compared with a is approximately given by

$$\phi = \frac{1}{4\pi\varepsilon_0} \frac{qa^2(3\cos^2\theta - 1)}{r^3}$$

where θ is the angle between r and the line through the charge and that the value of electric field E on the axis of quadrupole for point distance r from its centre (assume $r \gg a$) is given by

$$E = \frac{1}{4\pi\varepsilon_0} \frac{6qa^2}{r^4}$$

12. Show that the interaction energy of two dipoles of moments $\mathbf{p}_1$ and $\mathbf{p}_2$ is given by

$$U = \frac{1}{4\pi\varepsilon_0} \left[\frac{\mathbf{p}_1 \cdot \mathbf{p}_2}{r^2} - \frac{3}{r^5} (\mathbf{p}_1 \cdot \mathbf{r})(\mathbf{p}_2 \cdot \mathbf{r}) \right]$$

where r is the radius vector joining the centres of the two dipoles.

PROBLEMS

1. The potential at a certain distance from a point charge is 600 V and the electric field is 200 N/C. (a) what is the distance of the point charge (b) What is the magnitude of the point charge.

   [**Ans.** (a) 2×10^{-7}C (b) 3 m]

2. Two particles each of mass m and charge q are suspended by strings each of length l from the same point. Show that the inclination θ of each string to the vertical is given by

$$\frac{\cos\theta}{\sin^3\theta} = 16\pi\varepsilon_0 mgl^2/q^2$$

3. The flux entering a closed surface is 2×10^3 Nm^2/C and leaving the closed surface is 8×10^3 Nm^2/C. Calculate the charge enclosed by the surface. [**Ans.** 5.31×10^{-8}C]

4. An infinite line charge has a charge density 2 micro Coulomb/m. Calculate the electric field at a point which is at a normal distance of 10 cm from the line charge. [**Ans.** 36×10^4N/C]

5. Show that the rectangular components of the electric field produced by a charge q at the distance r are $E_x = \dfrac{qx}{4\pi\varepsilon_0 r^3}$, etc.

6. Four equal small spheres, each of mass m and carrying a charge q are suspended by light strings each of length l, from the same point show that the side of the square that the spheres will from due to the force of repulsion is given by

$$\frac{a}{\sqrt{2}l} = \frac{q^2}{4\pi\varepsilon_0 mga^2}(\sqrt{2}+0.5)\frac{\sqrt{l^2-a^2/2}}{l}$$

   If the charge on each sphere and length of each string is increased by a factor k, what should be the mass of each sphere so as to make the side of the square ka? [**Ans.** $m = m'$]

7. Consider a conducting spherical shell of radius a and charge ρ per unit area. Find by integration the potential at a point distant r from its centre. Then show that

$$\begin{aligned} E &= \frac{\rho}{\varepsilon_0}\frac{a^2}{r^2} \quad \text{for} \quad r > a \\ &= 0 \qquad\quad \text{for} \quad r < a \end{aligned}$$

8. Find the electric potential and field along the points on the axis of a disk of radius R carrying a charge σ per unit area. [Hint: Divide the disk into rings and add the contribution of all rings]

$$\left[\textbf{Ans.} \left(\frac{\sigma}{2\varepsilon_0}\right)(\sqrt{R^2+r^2}-r);\ \left(\frac{a}{4\varepsilon_0}\right)\log\left(\frac{R^2}{r^2}+1\right)\right]$$

9. A spherical region of space of radius 'a' is filled with positive charge of volume density ρ which varies with the distance r from the centre of the sphere but is constant for a give r. Find the total charge in the sphere if

(a) $\dfrac{\rho}{\rho_0} = \dfrac{a}{r}$ and (b) $\dfrac{\rho}{\rho_0} = \left(1-\dfrac{r^2}{a^2}\right)$

[**Ans.** (a) $2\pi\rho_0 \cdot a^3$ (b) $\dfrac{8\pi}{15}\rho_0 a^3$]

10. A wire of length L carries a charge of λ per unit length (Fig. 3.55) (a) show that the electric field at a point a distance R from the wire is given by

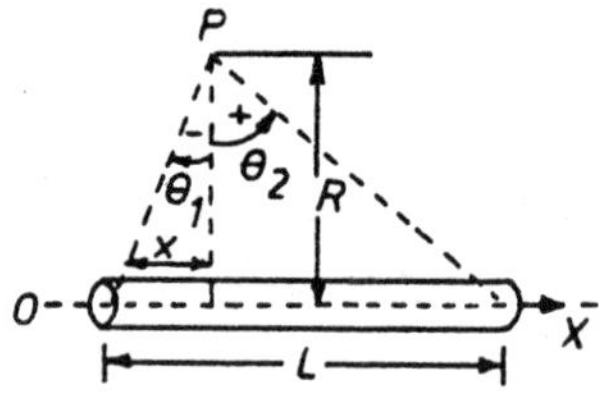

Figure 3.55.

$$E_\perp = \left(\frac{\lambda}{4\pi\varepsilon_0 R}\right)(\sin\theta_2 - \sin\theta_1)$$

$$E_\parallel = -\left(\frac{\lambda}{4\pi\varepsilon_0 R}\right)(\cos\theta_2 - \cos\theta_1)$$

where $E_\perp$ and $E_\parallel$ are the components of E perpendicular and parallel to the wire and θ_1 and θ_2 are the angles that the lines from the point to the ends of the wire make with the perpendicular to the wire. Find the field when the point is equidistant from both ends. (c) The signs of angles θ_1 and θ_2 are as indicated in the figure.

[**Ans.** (b) $E_\perp = \left(\dfrac{2}{4\pi\varepsilon_0 R}\right)\sin\theta;\ E_\parallel = 0$]

11. Two identical spheres of radius r are separated by a distance $d \gg r$. (a) What is the potential energy of the system if we put charge

$\frac{Q}{2}$ on each sphere (b) what is the potential energy if we put all the charge Q on one sphere (c) What are the electrical potentials on each sphere in cases (a) and (b)?

[**Ans.** (a) $\frac{kQ^2}{4d}$ (b) $\frac{KQ^2}{2r}$ (c) $\frac{1}{2}\frac{kQ}{r}$ for case (a) and $\frac{kQ}{r}$, O for case (b), where $K = \frac{1}{4\pi\varepsilon_0}$]

12. A square sheet is uniformly charged with charge density σ. Show that the potential at the centre of the square is $\phi_0 = \frac{\sigma a}{\pi\varepsilon_0}\log(1+\sqrt{2})$, where a is the length of the side of the square.
13. Two electric dipoles A and B are placed in such a way that the direction of A passes through B while the direction of B is perpendicular to that of A. Show that the actual force exerted by A to B is not in the same direction as the actual force exerted by B on A. Does this not contradict well known Newton's third law of motion?
14. Obtain the interaction energy between two electric dipoles. Use the result to estimate the interaction energy between two water molecules. Also discuss relative orientation effects.

SHORT ANSWER QUESTIONS

1. Can the net or total charge for any process occuring within an isolated system change? [**Ans.** No.]
2. Can an electric field may be represented by lines of force. [**Ans.** Yes]
3. When several charges are present, what will be the resultant electric field?

   [**Ans.** The vector sum of the electric fields produced by each charge]
4. How many electrons are there in one Coulomb charge?

   [**Ans.** $\frac{1}{1.602 \times 10^{-19}} = 6.25 \times 10^{18}$]
5. Does Coulomb's law of electric force obey Newton's third law of motion? [**Ans.** Yes]

6. Write the dimensional formula for ε_0. [**Ans.** $[M^{-1}L^{-3}T^4A^2]$]
7. Is the electric force between two charges greater than the gravitational force between them? If so, by what factor?

   [**Ans.** $\frac{F_e}{F_g} \approx 10^{39}$, i.e. electric force ($F_e$) is about 10^{39} times greater than the gravitational interaction]
8. Which is bigger -1C or charge on the electron?

   [**Ans.** 1 Coulomb]
9. Why do two electric lines of force not intersect each other?

   [**Ans.** Let us suppose that two lines of force interest each other. Then two tangents can be drawn from the point of intersector making for two directions of the resultant electric force, which is not possible].
10. What is an electric dipole?

    [**Ans.** It is a system of two equal and opposite electric charges (in which total charge, i.e. $q_1 + q_2 = 0$) separated by a distance. Obviously an electric dipole or just a dipole, consists of a pair of equal and opposite point charges q and $-q$ (where $q > 0$) separated by a very small distance a The strength of the dipole moment $|\mathbf{p}| = qa$ is a vector quantity.]
11. Does the work needed to move a charge from one point to another depends on the path it follows between them?

    [**Ans.** No. The work done to move a charge between two points depends on the position of the points only, otherwise the principle of conservation of energy would be violated.]
12. Two identical metallic spheres of exactly equal masses are taken. One of them is given a positive charge $+Q$ Coulomb and the other an equal negative charge. What will be their masses after charging?

    [**Ans.** Different. When positive charge is given it involves removal of electrons where as a negative charge is given adding electrons.]
13. Can a point in space have a non-zero potential even when the electric field in the space is zero?

    [**Ans.** Yes. Electric intensity at a given point is given by $E = -\frac{d\phi}{dx}$. If potential ϕ is constant in space, the potential gradient in any direction is zero, hence the intensity is zero. Obviously, potential can exist where there is no electric field.]

OBJECTIVE QUESTIONS

1. The number of lines of force that radiate outward from one Coulomb of positive charge is:
   (a) infinite, (b) $\varepsilon_0(= 8.85 \times 10^{-12})$,
   (c) $\frac{1}{\varepsilon_0}(= 1.13 \times 10^{11})$, (d) Zero. **[c]**
2. A positively charged particle of mass m kg and charge Q Coulomb travels from rest through a $p \cdot d$ of V volt. Its kinetic energy in joule is:
   (a) $m\frac{Q}{V}$, (b) $\frac{m}{QV}$, (c) mQV, (d) QV. **[d]**
3. Two charges of 1C and 5C are placed at some distance in air. The ratio of the forces acting on them is:
   (a) 1 : 1, (b) 5 : 1, (c) 1 : 5, (d) 1 : 25. **[a]**
4. The work done in moving a positive charge on an equipotential surface is:
   (a) finite and positive, (b) infinite,
   (c) finite and negative, (d) zero. **[d]**
5. A and B are two points in an electric field. If 8J of work is done in taking 2 Coulomb of electric charge from A to B, then the potential difference between A and B is:
   (a) 16V; (b) 4V, (c) 2V, (d) Zero. **[b]**
6. Two point charges +2C and +6C repel each other with a force of 12 N. If a charge of –4C is given to each of these charges the force will be:
   (a) 4N (attractive), (b) 4N (repulsive),
   (c) 8N (repulsive), (d) 8N (attractive). **[a]**
7. Two thin infinite parallel plates have uniform charge densities $+\sigma$ and $-\sigma$. The electric field in the space between them is:
   (a) $\frac{\sigma}{\varepsilon_0}$, (b) $\frac{\sigma}{2\varepsilon_0}$, (c) $\frac{2\sigma}{\varepsilon_0}$, (d) Zero. **[a]**
8. A hollow metal sphere is charged so that the potential on its surface is 10 V. The potential at the centre of the sphere is:

(a) 10 v.
(b) 0 v.
(c) same as at point 5 cm away from the surface.
(d) same as at point 25 cm away from the surface. [a]

9. When placed in a uniform field a dipole experiences:
   (a) only a torque.
   (b) only a net force.
   (c) both a net force and a torque.
   (d) neither a net force nor a torque. [a]

10. An electric dipole placed in a uniform electric field will have minimum potential energy when the dipole moment is inclined to the field at an angle:
   (a) $\frac{\pi}{2}$, (b) $\frac{3\pi}{2}$, (c) π, (d) 0. [d]

11. Three charges $2q, -q, -q$ are located at the vertices of an equilateral triangle. At the circumcentre of the triangle:
   (a) field is non zero but potential is zero.
   (b) both field and potential are zero.
   (c) both field and potential are non-zero.
   (d) field is zero but potential is non-zero. [b]

12. The work done in carrying a charge q once round a circle of radius r with a charge Q at the centre is:
   (a) Zero, (b) $\frac{qQ}{4\pi\varepsilon_0}\left(\frac{1}{2\pi r}\right)$, (c) $\frac{qQ}{4\pi\varepsilon_0 r}$, (d) $\frac{qQ}{4\pi\varepsilon_0}\frac{1}{\pi r}$. [a]

13. A charge q is placed at the centre of the line joining two equal charges Q. The system of the three charges will be in equilibrium if q is equal to:
   (a) $+\frac{Q}{2}$, (b) $-\frac{Q}{2}$, (c) $-\frac{Q}{4}$, (d) $+\frac{Q}{4}$. [c]

14. The electric potential in a region along the x-axis varies with x according to the relation $V(x) = 5 + 4x^2$.
   (a) The force experienced by the above charge is along the +ve x-axis.
   (b) A uniform electric field exists in this region along the x-axis.
   (c) The force experienced by a charge of 1C placed at $x = -1$ m is 8N.
   (d) Potential difference between the points $x = 1$ and $x = -3$ is 32V. [a, c, d]

15. Bond length of HCl molecule is 1.28 A. Its electric dipole moment will be:
    (a) 16×10^{-19} C-m, (b) 1.28×10^{-10} C-m,
    (c) 2.05×10^{-29} C-m, (d) 2.05×10^{-10} C-m. **[c]**
16. The radius of two spheres are R_1 and R_2 and their charges are q_1 and q_2 respectively. They are at the same potential. The ratio $\frac{q_1}{q_2}$ will be equal to:
    (a) $\frac{R_1}{R_2}$, (b) $\frac{R_2}{R_1}$, (c) $\frac{R_1^2}{R_2^2}$, (d) $\frac{R_2^2}{R_1^2}$. **[a]**
17. The S-I units of electric dipole moment are:
    (a) C-m, (b) C/m, (c) C/m^2, (d) N/C. **[a]**
18. State which of the following is correct:
    (a) Joule = volt + amp, (b) Joule = volt × amp,
    (c) Joule = Coulomb × volt, (d) Joule = Coulomb + volt. **[a]**

Chapter 4

Electrostatics in Dielectrics

4.1 INTRODUCTION

So far, we have limited our discussion to electrostatic fields in free space and the field produced was exclusively by the free charges. Now, we shall examine the phenomena which results from the introduction of a dielectric materiel into an electrostatic field. However, if matter is present, the magnitude of electric field decreases. The medium plays an important part in determining the absolute value of the field. The electromagnetic fields that result in matter are, in general, very complex displaying considerable variation from particle to particle over distances of the order of separation of atoms. In dealing with this phenomena that range over large distances, we shall not be concerned with the detailed fluctuations in the field on the atomic scale. We shall try to present a consistent macroscopic model of matter that does predict in every instance the correct fields. We must note that the macroscopic fields within matter are not necessarily the fields actually exerted on any of the particles in matter.

On the basis of electric behaviour, media can be classified into two distinct groups:

(i) *Conductors*: We are all familiar with conductors which contain some electrons which are free to move in the presence of a field. These free electrons can be imagined as charge carriers. The conductivity of these substances is of the order of 10^8 mho/m. Examples of conductors are metals, etc. There are among these some substances, both elements and compounds, the resistance of which vanishes at a certain temperature, known as critical temperature (T_c). These substances are known as *superconductors*. For example, (T_c) for mercury is, 4.2 K. The highest known T_c so far is 134 K for mercury cuprate.

(ii) *Insulators or Dielectrics*: These are the substances in which the electrons are strongly bound to the atoms or molecules composing the material and cannot be detached by the application of an electric field to these materials. The forbidden energy gap ($\sim$ 7 eV) is very large compared to the thermal energy and hence the excitation of electrons from valence band to conduction band is normally not possible and free charge carriers are not available. The conductivity of these substances is very small of the order of 10^{-16} mho/m. In the absence of an external field, the dipole moments of the molecules of a dielectric usually either equal zero (non polar molecules) or are distributed in space by directions chaotically (polar molecules). In both cases, the total dipole moment of a dielectric equals zero. However, there are substances that can have a dipole moment in the absence of an external field. These substances are called *ferroelectrics*. These substances exhibit spontaneous polarization in the absence of an external field, e.g. Rochelle salt. Obviously, under the action of external electric field, no steady motion of electrons is possible at the most they may be slightly displaced from their positions in response to the field applied. Such insulating media are also known as dielectric media. Sulphur, mica, porcelain etc. are few examples of dielectrics. We must note that if the applied field is too large, catastrophic breakdown may occur in these substances. A good insulator breaks down in fields of about 10^9 volts/m.

In addition to conductors and insulators, we have another class of materials the so called *semi-conductors* which have properties intermediate between conductors and insulators. In these substances electrons in the conduction band and holes in the valence band act as charge carriers. There is a forbidden gap between the conduction and valence bands of the order of 1 eV. The conductivity of these substances is less than that of conductors and is of the order of $10^{-1} - 10^{-2}$ mho/m. The present day technology depends on such materials.

4.2 CONDUCTORS IN ELECTROSTATIC FIELD

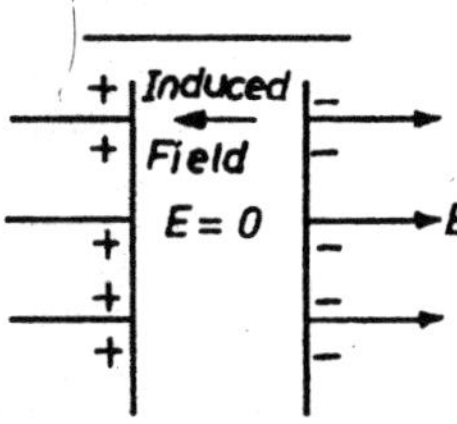

Figure 4.1. A conductor in an electric field.

Let us consider that a conductor is placed in an electric field (Fig. 4.1). Figure 4.1 shows that a slab of conducting material is placed in an electric field. One can see that under the influence of the field, the electrons can move freely in the material but cannot leave the surface because strong forces act to prevent them from leaving the surface. This means that they leave a net charge on the surface. The charges thus induced on the surface of the slab generate a electric field inside it (conductor) which is in a direction opposite to the external electric field. We must note that the electrons continue to move so long as there is any field left in the conductor and until the induced charge ensures that the resultant field inside the conductor is zero. Obviously, no electrostatic field can exist within the body of a conductor, i.e. $E = 0$ inside the conductor. Since $E = -\nabla\phi = 0$, shows that the conductor is an *equipotential region*. Further, because $\mathbf{E} = 0$, $\nabla = \frac{\rho}{\varepsilon} = 0$, i.e. $\rho = 0$. This means, there cannot be a volume distribution of charges within the conductor, i.e. the charges must reside entirely on the surface of a conductor. As stated earlier, the surface of a conductor upon which charges are at rest, is an equipotential surface. Since the surface being an equipotential surface, the field is normal to the surface of the conductor. If this was not the case, there would be an electrostatic field along the surface resulting into a current.

4.3 ELECTRIC FIELD AT THE SURFACE OF A CHARGED CONDUCTOR

Let us imagine a small disc enclosing an area *ds* on the surface of a conductor as shown in Fig. 4.2. From figure it is evident that the volume enclosed by the outer surface of the disc is half in the conductor and half in the air. Further, the flat surfaces of the disc are parallel to the surface of the conductor. If *E* is the electric field then the flux through the surface outside = Eds. The flux through the other surface is zero since the field inside a conductor is zero. Applying Gauss's law, we have

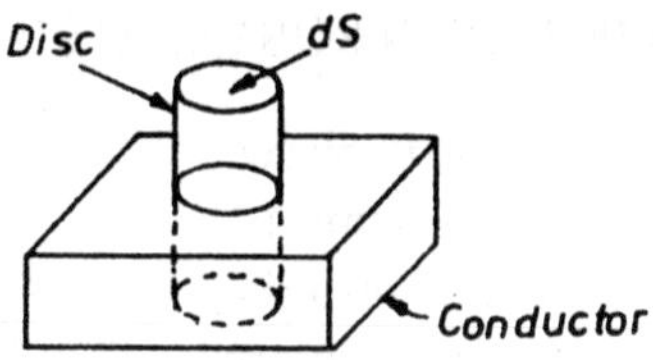

Figure 4.2.

$$\text{Eds} = \frac{\sigma}{\varepsilon_0} ds$$

or

$$E = \frac{\sigma}{\varepsilon_0} \tag{4.1}$$

where σ is surface charge density.

4.4 DIELECTRIC AND PERMITTIVITY

An important Characteristics of a dielectric is its permittivity or dielectric constant ε. If two point charges are placed in an electrically homogeneous and isotropic medium then according to Coulomb's law, the force between the two point charges is given by

$$\mathbf{F} = \frac{1}{4\pi\varepsilon} \frac{q_1 q_2}{r^2} \hat{r} \tag{4.2}$$

Here ε is called the permittivity of the medium. We must note that ε (permittivity) is not always a constant. The permittivity of a dielectric is always greater than the permittivity of vacuum, hence it is often convenient to use the relative permittivity ε_r of the dielectric expressed as the ratio of its permittivity to that of vacuum. Thus

$$\varepsilon_r = \frac{\varepsilon}{\varepsilon_0} = \frac{\text{Permittivity of the medium}}{\text{Permittivity of the vaccum}} \tag{4.3}$$

The value of relative permittivity (ε_r) or dielectric constant or specific inductivity capacity is different for different media. One can find the explanation of this effect in terms of the electric nature of dielectric media.

4.5 CAPACITOR

It is a device used for storing energy. It is a system of two oppositely charged conductors having charges of equal magnitude and having such shapes and relative position that the field set up by the system is concentrated (localized) in a limited region of space. In the case of a parallel plate capacitors, the conductors are called plates.

Let us suppose that the two conductors in a capacitor carry charges $\pm Q$. These charges are distributed over the surfaces with the surface densities σ_+ and σ_- respectively. We must note that these may not be constant over the surfaces of the conductors, but the potential ϕ_+ and ϕ_- are, since the surfaces of the conductors are equipotential surfaces. Let V be the potential difference between the two conductors, then

$$V = \phi_+ - \phi_- \tag{4.4}$$

In an isolated charged conductor, not subject to the action of external fields, potential difference is always proportional to the charge Q, i.e.

$$Q = CV \tag{4.5}$$

where C is a constant and is called the capacitance (or capacity) of the capacitor. The capacitance of the capacitor is the mutual capacitance of its plates (or conductors in the general case) and is related to the ability of a capacitor to hold charges. The capacitance depends on the geometry, i.e. the shape, size of the conductors. Capacitors are made in the form of two conductors placed close to each other. The conductors forming a capacitor are called its plates. To prevent external bodies from influencing the capacitance of a capacitor, the plates are shaped and arranged relative to each other so that the field set up by the charges accumulating on them is concentrated inside the capacitor. This condition is satisfied by two plates arranged close to each other, two coaxial (plane), cylindrical, and spherical capacitors are encountered. Since the field is confined inside a capacitor, the electric displacement lines begin on one plate and terminate on the other. Consequently, the extraneous charges produced on the plates have the same magnitude and are opposite in sign. Now we will calculate the capacitance of a few capacitors.

(i) Parallel plate capacitor

It is the most commonly used type of capacitor. The conductors forming a parallel plate capacitor are called its plates. Let us consider two plates, each of area A, separated by a small distance 'd' as shown in Fig. 4.3. Let equal and opposite charges $\pm Q$ be put on the plates. One can ignore the edge effect, if the lateral dimensions of the plates are much larger than this separation and the field between the plates be

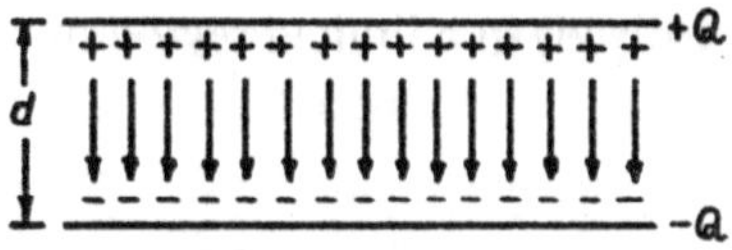

Figure 4.3. A parallel plate capacitor

assumed to be uniform. We know that the potential difference between the plates is the work required to take a unit charge from one plate to the other. We have

$$V = Ed = \frac{\sigma}{\varepsilon_0}d = \frac{Qd}{\varepsilon_0 A} \qquad \left(\because \quad \sigma = \frac{Q}{A}\right)$$

$$\therefore \quad C = \frac{Q}{V} = \frac{\varepsilon_0 A}{d} \tag{4.6}$$

It must be noted that the accuracy of determining the capacitance of a real parallel-plate capacitor by Eq. (4.6) is the greater, the smaller is the separation distance d in comparison with the linear dimensions of the plates.

The formula for the capacitance of a multi-plate capacitor differs from that for a parallel-plate capacitor in that the term A is replaced by $A(n-1)$, where n is the number of plates.

(ii) Capacitance of an isolated Sphere

Let us consider that a capacitor is formed from two spheres. Let one sphere have a radius a and carry a charge Q. Let us imagine that another sphere is of infinite radius carrying a charge $-Q$. The potential (V) of the sphere is given by

$$V = -\int_{\infty}^{a} Edr = -\int_{\infty}^{a} \frac{Q}{4\pi\varepsilon_0 r^2}dr = \frac{Q}{4\pi\varepsilon_0 a} \tag{4.7}$$

$$\therefore \quad C = \frac{Q}{V} = 4\pi\varepsilon_0 a \tag{4.8}$$

(iii) "Co-axial cable" or cylindrical Capacitor

Consider a cylindrical capacitor, i.e. two coaxial cylinders of infinite length. Let their radius be r_1 and r_2, as shown in Fig. 4.4.

If the charge per unit length on the inner cylinder be $+\lambda$, then the E is given by the relation

$$E = \frac{\lambda}{2\pi\varepsilon_0 r}$$

The potential difference between the two co-axial cylinders is given by

$$V = \phi_{r_1} - \phi_{r_2} = -\int_{r_2}^{r_1} E dr = -\frac{\lambda}{2\pi\varepsilon_0}\int_{r_2}^{r_1}\frac{dr}{r}$$

$$= -\frac{\lambda}{2\pi\varepsilon_0}(\log r_1 - \log r_2) = \frac{\lambda}{2\pi\varepsilon_0}\log\frac{r_2}{r_1} \quad (4.8)$$

If we consider that l be the length of cylinders then the capacitance of a cylindrical capacitor':

$$C = \frac{\lambda l}{\frac{\lambda}{2\pi\varepsilon_0}\log\frac{r_2}{r_1}} = \frac{2\pi\varepsilon_0 l}{\log\frac{r_2}{r_1}} \quad (4.9)$$

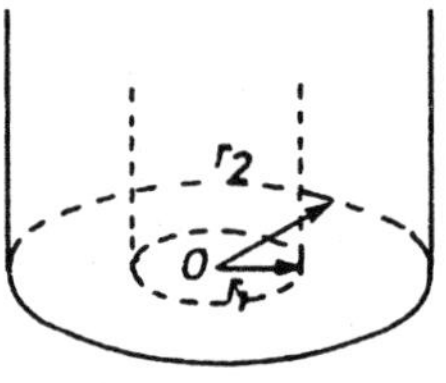

Figure 4.4.

The accuracy of determining the capacitance of a real capacitor by Eq. (4.9) is greater, the smaller is the separation distance $d = r_2 - r_1$ in comparison with l and r_1.

Apart from the capacitance every capacitor is characterized by the maximum voltage U_{max} that may be applied across its plates without the danger of a breakdown. When this voltage is exceeded, a spark jumps across the space between the plates. The result is the failure of the capacitor. This voltage is called the '*breakdown voltage*'. The breakdown voltage depends upon the thickness of the dielectric, its properties and the shape of the conductors in the capacitance.

The capacitance can be increased by connecting the conductors of like charge of two or more capacitors in *parallel*. The equivalent capacitance for such a connection is

$$C = \sum_{i=1}^{n} C_i \quad (4.10)$$

where C_i is the capacitance of the ith capacitor.

When capacitors are connected in *series*, their oppositely charged conductors are connected together. In this case

$$\frac{1}{C} = \sum_{i=1}^{n}\frac{1}{C_i} \quad (4.11)$$

The equivalent capacitance is always less than the minimum capacitance of any capacitor in the network.

4.6 ENERGY OF A CHARGED CAPACITOR

We have seen that the charge q on a conductor can be considered as a system of point charges Δq. Moreover, the charge in the case of conductors resides only on the surfaces, the potential energy of a conductor can be found from

$$W = \frac{1}{2}\int \sigma\phi ds \tag{4.12}$$

(we have replaced summation by integration in Chapter 3).

Assume that the potential of a capacitor plate carrying the charge $+q$ is ϕ_+, and that of a plate carrying the charge $-q$ is ϕ_-. Consequently, each of the elementary charges Δq into which the charge $+q$ can be divided is at a point with the potential ϕ_+, and each of the charges into which the charge $-q$ can be divided is at a point with the potential ϕ_-. By equation (4.12), the potential energy of such a system of charges is

$$W = \frac{1}{2}\int \sigma_+\phi_+ ds + \frac{1}{2}\int \sigma_-\phi_- ds \tag{4.13}$$

where σ_+, σ_- are the surface densities of charges on the two conductors. Therefore

$$W = \frac{1}{2}Q(\phi_+ - \phi_-) = \frac{1}{2}QV \tag{4.14}$$

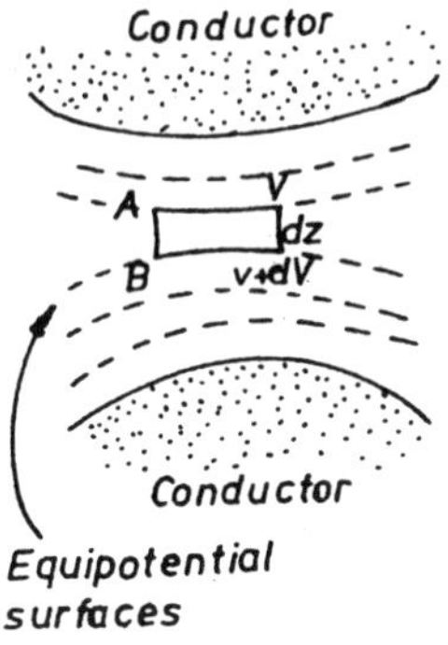

Figure 4.5.

One can also express (4.14) in terms of capacity as

$$W = \frac{1}{2}CV^2 = \frac{1}{2}\frac{Q^2}{C} \tag{4.15}$$

one can also express the energy expression (4.15) in a different way by considering it to be distributed over the space between the charges occupied by their electric field. Let us consider two equipotential surfaces close enough together between two conductors (Fig. 4.5).

Let us suppose that the two surfaces differ in their potential by a small amount dV. If now two conducting surfaces with potentials exactly as on the equipotential surfaces are inserted so as to coincide with the surfaces, one finds that the condition of the problem remain unaltered. Thus the two conducting surfaces will form a capacitor with capacity $\varepsilon_0 \frac{ds}{dz}$ where ds is the area of each plate and dz their separation. Thus the energy of the capacitor

$$\begin{aligned} dW &= \frac{1}{2}CV^2 = \frac{1}{2}\varepsilon_0\frac{ds}{dz}(dV)^2 = \frac{1}{2}\varepsilon_0\frac{ds}{dz}(Edz)^2 \\ &= \frac{1}{2}\varepsilon_0 E^2 ds\,dz = \frac{1}{2}\varepsilon_0 E^2 d\tau \end{aligned}$$

where $d\tau$ is the volume occupied by the capacitor. Thus

$$W = \frac{1}{2}\varepsilon_0 \int E^2 d\tau$$

Obviously,

$$W = \frac{1}{2}\varepsilon_0 \int E^2 d\tau$$

This shows that one can regard the energy as distributed throughout the field and hence the energy density at a point being $\frac{1}{2}\varepsilon_0 E^2$.

One can derive the above result in a more rigorous way. Let us consider an assembly of conductors enclosed by a surface so sufficiently far away that the field on it is negligible (Fig. 4.6). Let us suppose that there exists within the surface a volume distribution of charge of density ρ and surface distribution of density σ over the conductors. One obtains the electrostatic energy as

$$W = \frac{1}{2}\int \rho\phi\, d\tau + \frac{1}{2}\int \sigma\phi\, ds_1 \tag{4.16}$$

The surface integral is taken over the surfaces of all the conductors and the volume integral over the entire region occupied by the field, i.e.

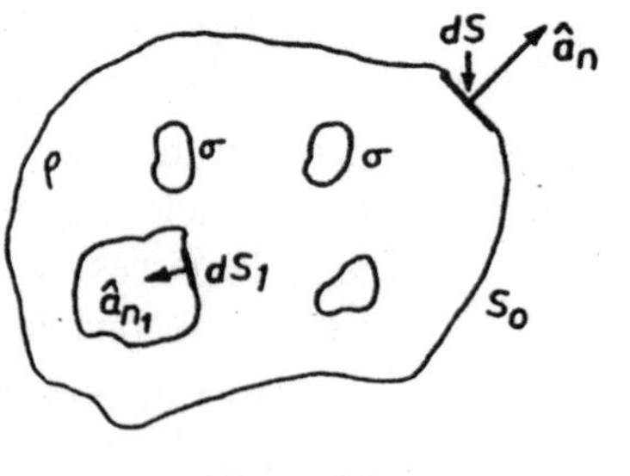

Figure 4.6.

the volume bounded by the surface so as shown in Fig. 4.6. Applying Gauss's differential form $\nabla \cdot \mathbf{E} = \frac{\rho}{\varepsilon_0}$, one obtains using Divergence theorem,

$$\frac{1}{2}\int \rho\phi d\tau = \frac{1}{2}\int \varepsilon_0(\nabla \cdot \mathbf{E})\phi d\tau = \frac{\varepsilon_0}{2}\int \{\nabla \cdot (\phi\mathbf{E}) - \mathbf{E} \cdot \nabla\phi\} d\tau$$

$$= \frac{\varepsilon_0}{2}\int \nabla \cdot (\phi E) d\tau + \frac{\varepsilon_0}{2}\int E^2 d\tau = \frac{\varepsilon_0}{2}\int \phi E \cdot \hat{a}_n ds + \frac{\varepsilon_0}{2}\int E^2 d\tau$$

The first integral is to be performed over the closed surface s_0 enclosing the volume as well as over the surfaces of the conductors. Thus

$$\frac{1}{2}\int \rho\phi d\tau = \frac{\varepsilon_0}{2}\int \phi\mathbf{E} \cdot \hat{a}_n ds_0 + \underbrace{\frac{\varepsilon_0}{2}\int \phi\mathbf{E} \cdot \hat{a}_{n_1} ds_1}_{\text{over the surface of the conductor}} + \frac{\varepsilon_0}{2}\int E^2 d\tau$$

Here $\hat{a}_{n_1}$ and $\hat{a}_{n_2}$ are unit vectors normal to the respective surfaces.

We know that ϕ varies as r^{-1} and E as r^{-2} while ds increases as r^2. The first integral is proportional to $1/r$ and hence, vanishes when r is very large, i.e. on the surface s_0.

The integration in the second integral is over the surfaces of the medium $\hat{a}_n ds_1$ is the vector drawn outwards from the medium and hence into the conducting surface (Fig. 4.6). We have $E = \frac{-\sigma}{\varepsilon_0}$ in this direction,

$$\frac{1}{2}\int \rho\phi d\tau = -\frac{1}{2}\int \sigma\phi ds_1 + \frac{\varepsilon_0}{2}\int E^2 d\tau$$

Thus one obtains the total energy as

$$W = -\frac{1}{2}\int \sigma\phi ds_1 + \frac{\varepsilon_0}{2}\int E^2 d\tau + \frac{1}{2}\int \sigma\phi ds_1 = \frac{\varepsilon_0}{2}\int E^2 d\tau \quad (4.17)$$

Since $E = 0$ within a conductor, one can regard this energy as distributed throughout the surrounding medium with density $\varepsilon_0 \frac{E^2}{2}$.

Eq. (4.17) has certain limitations. For example, let us compute the electrostatic energy of a point charge q with the help of (4.17). The energy will be

$$W = \frac{\varepsilon_0}{2}\int \frac{q^2}{(4\pi\varepsilon_0)^2 r^4} 4\pi r^2 dr = \frac{q^2}{8\pi\varepsilon_0}\int_0^\infty \frac{dr}{r^2} = \frac{q^2}{8\pi\varepsilon_0}\left[-\frac{1}{r}\right]_0^\infty = \infty$$

This shows that the self energy of a point charge is infinite, which is a result physically absurd. This means that the concept of energy density is not compatible with the concept of point charge.

4.7 NON-CONDUCTING MEDIUM IN AN ELECTRIC FIELD

How non-conducting medium i.e. an dielectric behave in an electric field? If a dielectric is introduced into an electric field, then the field and the dielectric itself undergo appreciable changes.

Cavandish and later Faraday, studied the effects of introducing a dielectric material in between the plates of a capacitor. They observed that the capacity of a conductor depends not only on shape and size of the conducting plates but also on the nature of the dielectric material by which they are separated. It is found that the capacitance of a capacitor increases on inserting a slab of dielectric between its plates. We have, $C = \frac{Q}{V}$. Since the capacity C increases on inserting a dielectric, where Q remaining the same. Obviously, the potential difference V falls, i.e. the field is weakened. If the capacity is increased by a factor ε_r, the field is weakened by a factor $\frac{1}{\varepsilon_r}$. Thus the electric field in a dielectric medium is given by

$$\mathbf{E} = \frac{q}{4\pi\varepsilon_0\varepsilon_r}\frac{\mathbf{r}}{|\mathbf{r}|^3} \tag{4.18}$$

The parameter ε_r, the *relative permittivity* of the dielectric is called the dielectric constant of the medium. ε_r depends on the nature of the dielectric medium. The value of ε_r for vacuum is 1, for water it is 81, for air it is 1.00057 and for most substances it varies from 1 to 10.

4.8 POLARIZATION

Let us study the effect of an electric field on a piece of matter. We know that atoms do not have permanent electric dipole moments because of their spherical symmetry, but when they are placed in an electric field, they become polarized, acquiring induced electric dipole moments in the direction of the field. This results from the perturbation of the motion of the electrons produced by the applied electric field.

On the other hand, many molecules do have permanent electric dipole moments. When a molecule has a permanent electric dipole moment, it tends to be oriented parallel to the applied electric field because of the torque it experiences. As a consequence of either of these two effects, a piece of matter placed in an electric field becomes electrically *polarized*. That is, its molecules (or atoms) become electric dipoles oriented in the direction of the local electric field (Fig. 4.7), either because of the distortion of the electronic motion or the orientation of their permanent dipoles. A medium that can be polarized by an electric field is called a '*dielectric*'. The polarization gives rise to a net positive charge on one side of the piece of matter and a net negative charge on the positive side. The piece of matter then becomes a large electric dipole that tends to move in the direction in which the field increases. The polarization of a dielectric material due to some externally applied electric field may occur as a consequence of three effects:

(i) *Electron polarization*: This occurs when the externally applied field causes in the atom's positive and negative charges.

(ii) *Ionic polarization*: This occurs in molecules composed of positively and negatively charged ions.

(iii) *Orientational polarizations*: This occurs in materials that possess permanent, microscopic separation of charge centres and such materials possess persistent polarization. In the absence of an applied field, these permanent dipoles are randomly ori-

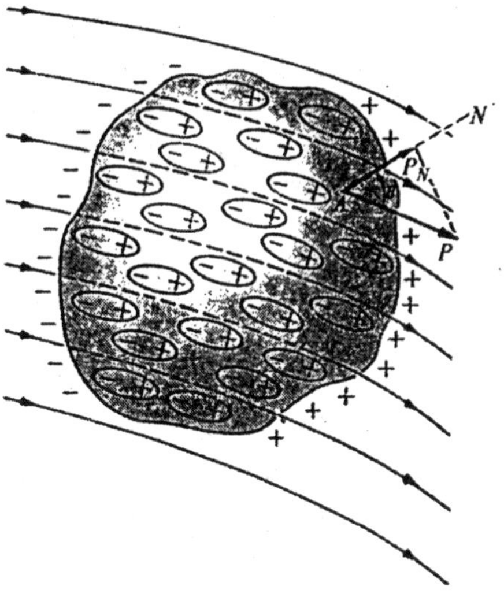

Figure 4.7. Polarization of matter by an electric field

ented. However, in the presence of an applied electric field, these permanent dipoles tend to rotate to align with the direction of the applied field. Materials (such as water) possessing these permanent dipoles in the absence of a external applied field are termed as *polar* molecules. Obviously, there are two types of materials: (i) *Polar* and (ii) *non-polar*. In a non-polar dielectric the centres of positive charge and negative charge normally coincide where as in a polar dielectric the centres of charges are always displaced from each other.

Now, we are interested in finding how the dielectric constant ε_r is related to the intrinsic properties of the materials?

When an electric field is applied on an atom, a dipole moment is induced in it. In a non-polar atom the induced dipole moment as in the direction of the applied field and directly proportional to it, i.e.

$$\mathbf{P} \propto \mathbf{E}$$

or

$$\mathbf{P} = \alpha\mathbf{E} \tag{4.19}$$

Here α is a constant known as atomic polarizability. We have

$$\alpha = \frac{P}{E} = \frac{\text{Coulomb-metre}}{\text{Volt/metre}} = \text{farad-m}^2$$

The above statement is not true for all materials especially crystals for which α is a matrix. For dielectric materials used as insulators α is a constant. Such materials when placed in an external field get polarized. The polarized matter gives rise to its own field which modifies the external field which in turn modifies polarization. Under equilibrium conditions we have the polarized matter and the original external field modified by polarization.

The polarization P per unit volume of the dielectric is defined as

$$\mathbf{P} = \frac{1}{\Delta V}\sum_{1}^{N}\mathbf{p}_i \tag{4.20}$$

where ΔV is the volume of the dielectric containing a large number of polarized molecule and $\mathbf{p}_i$ is the dipole moment of the ith molecule. Obviously, the polarized matter is now replaced by a "jelly" of dipoles.

For a homogeneous dielectric with nonpolar molecules in a homogeneous electric field

$$\mathbf{P}_e = n_0\mathbf{p}_e \tag{4.21}$$

where $\mathbf{p}_e$ is the induced dipole moment of one molecule and n_0 is the number of molecules per unit volume.

In an isotropic dielectric, dipole moment **p** is induced in the direction of the electric field **E**. In such a medium the polarization **P** is directly proportional to the intensity of electric field **E**, i.e.

$$\mathbf{P} \propto \mathbf{E}$$

or

$$\mathbf{P} = \varepsilon_0 \chi \mathbf{E} \tag{4.22}$$

The quantity x is a dimensionless scalar constant i.e. independent of **E** and known as *electrical susceptibility*. For most substances x is a positive quantity.

In anisotropic or *nonisotropic* dielectrics, the directions of **P** and **E**, generally speaking, do not coincide. In this case, the relation between **P** and **E** is described by the equations

$$\begin{aligned} P_x &= \varepsilon_0(\chi_{xx}E_x + \chi_{xy}E_y + \chi_{xz}E_z) \\ P_y &= \varepsilon_0(\chi_{yx}E_x + \chi_{yy}E_y + \chi_{yz}E_z) \\ P_z &= \varepsilon_0(\chi_{zx}E_x + \chi_{yz}E_y + \chi_{zz}E_z) \end{aligned} \tag{4.23}$$

The combination of the nine quantities forms a symmetrical tensor of rank two called the *tensor of dielectric susceptibility*. This tensor characterizes the electrical properties of an anisotropic dielectric.

Let us consider a slab of material of thickness l and surface s placed perpendicular to to a uniform field **E** (Fig. 4.8). The polarization **P**, being parallel to **E**, is also perpendicular to s. The volume of the slab is ls and therefore its total electric dipole moment is $P(ls) = (Ps)l$. But l is just the separation between the positive and negative charges that appear on the two surfaces. Since by definition the electric dipole moment is equal to charge times distance, we conclude that the total electric charge that appears on each of the surfaces is Ps, and therefore the charge per unit area σ_p on the faces of polarized slab is P, or $\sigma_p = P$. Although this result has been obtained for a particular geometrical arrangement, it is of general validity, and

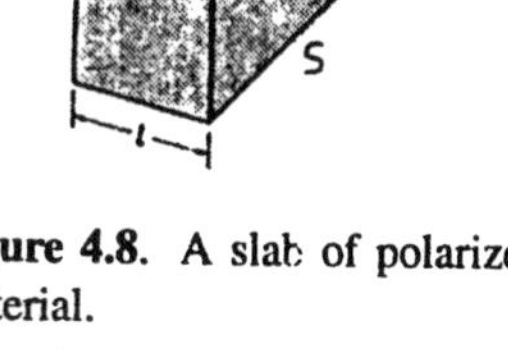

Figure 4.8. A slab of polarized material.

the charge per unit area on the surface of a polarized piece of matter is equal to the component of the polarization P in the direction of the normal to the surface of the body.

So, in Fig. 4.7, the charge per unit area on the surface at A is $P_N = P\cos\theta$.

8(a) Electric Displacement

We have seen that a polarized dielectric has certain charges on its surface (and also, unless the polarization is uniform, throughout its volume). These polarization charges however, are "frozen" in the sense that they are bound to specific atoms or molecules and are not free to move through the dielectric. In other materials such as metal or an ionized gas, there may be electric charges capable of moving the material, and therefore we call them *free* charges.

To understand a clear distinction between free charges and polarization charges, let us consider a slab a dielectric material placed between two conducting parallel plates (Fig. 4.9), carrying equal and opposite free charges. The surface charge density on the left hand plate is $+\sigma_{\text{free}}$ and on the right hand plate is $-\sigma_{\text{free}}$. These charges produce an electric field that polarizes the slab so that polarization charges appear on each surface of the slab. These polarization charges have a sign opposite to that on the plates. Therefore the polarization charges on the faces of the dielectric slab partially balance the free charges on the conducting

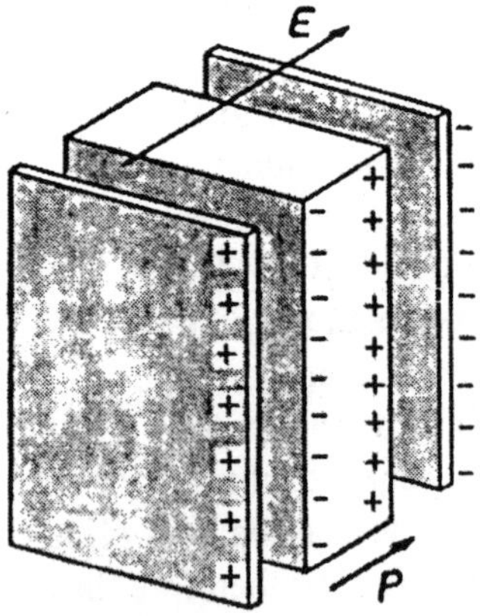

Figure 4.9. Dielectric placed between oppositely charged plate. Charges on the plate are free charges and charges on the dielectric surface are polarization charges.

plates. If P is the magnitude of the polarization in the slab, the surface charge density on the left face of the slab is $-P$, while on the right face it is $+P$. The effective, or net, surface charge density on the left is $\sigma = \sigma_{\text{free}} - P$, with an equal and opposite result on the right. These net surface charges give rise to a uniform electric field, given by $E = \dfrac{\sigma}{\varepsilon_0}$. Thus, using the effective value of σ, we have

$$E = \frac{1}{\varepsilon_0}(\sigma_{\text{free}} - P) \quad \text{or} \quad \sigma_{\text{free}} = \varepsilon_0 E + P \tag{4.23a}$$

an expression that gives the free charges on the surface of a conductor surrounded by a dielectric in terms of the electric field in the dielectric and the polarization of the dielectric. When **E** and **P** are vectors in the same direction, the above result suggests the introduction of a new vector field, which is called the *electric displacement*, and which is defined as

$$\mathbf{D} = \varepsilon_0 \mathbf{E} + \mathbf{P} \tag{4.24}$$

obviously, **D** is expressed in $C\text{-}m^2$, since those are the units of the two terms that appear on the right-hand side of Eq. (4.24). Obviously, $(\varepsilon_0 E + P)$ or D is a physical quantity which depends on free charge density only.

For isotropic dielectrics, the polarization vector **P** is proportional to the field intensity **E** and coincides with **E** in direction. Hence

$$\mathbf{D} = K\varepsilon_0 \mathbf{E} \tag{4.25}$$

where K is the relative permeability (ε_r) or dielectric Constant and is a scalar quantity, i.e.

$$\begin{aligned} K &= \frac{\text{Permittivity of medium}(\varepsilon)}{\text{Permittivity of vaccum}(\varepsilon_0)} \\ &= \frac{\text{Electric field in vaccum}(E_0)}{\text{Electric field in the medium}(E)} \end{aligned}$$

Using Eq. (4.22), Eq. (4.23) can be written as

$$\sigma_{\text{free}} = \varepsilon_0 E + \varepsilon_0 \chi E$$

or

$$\sigma_{\text{free}} = \varepsilon_0 E(1+\chi)$$

or

$$E = \frac{\sigma_{\text{free}}}{\varepsilon_0(1+\chi)} \tag{4.26}$$

Eq. (4.26) shows that the field is reduced by a factor $1/(1+\chi)$.

Using (22), Eq. (4.24) can be expressed as $\mathbf{D} = \varepsilon_0 \mathbf{E} + \chi \varepsilon_0 \mathbf{E} = \varepsilon_0 \mathbf{E}(1+\chi) = \varepsilon \mathbf{E}$ where the coefficient $\varepsilon = \frac{D}{E} = (1+\chi)\varepsilon_0$, is called the permittivity of the medium. The relative permittivity or the dielectric constant (K or ε_r) is defined as

$$k \quad \text{or} \quad \varepsilon_r = \frac{\varepsilon}{\varepsilon_0} = 1 + \chi \tag{4.26a}$$

This equation represents the relation between the dielectric constant ε_r and electrical susceptibility of the medium.

$$\text{or} \quad \varepsilon = \varepsilon_0(1+\chi) \tag{4.27}$$

For most substances K is larger than one and it is a pure number (Table 4.1). We must note that the quantities ε_r, χ, ε and P are all a measure to the extent of polarization produced in a dielectric medium under the action of the external field.

Table 4.1. Values of ε_r at room temperature

Material	ε_r
Carbon tetrachloride	2.24
Transformer oil	2.4
Paraffin wax	2.25
Polythelene	2.30
Nylon	3.50
Porcelain	6.00
Mica	7.00
Water, 0°C	88.00
Water, 100°C	48.00

Let us consider a parallel plate capacitor (Fig. 4.10) with a negative charge on the top and positive charge on the bottom plate. let A be the area of each plate, σ the surface charge density and d their separation. We have

$$E = \frac{\sigma}{\varepsilon_0}, \quad V = \frac{\sigma}{\varepsilon_0} d, \quad C = \frac{Q}{V} \frac{\varepsilon_0 A}{\delta}$$

Now, suppose we introduce a dielectric slab between the plates of the capacitor (Fig. 4.10). Fig. 4.10 exhibit a gap between the plates and the walls of the dielectric. We must note that such a gap will always be present though its width may be only of molecular dimensions. As

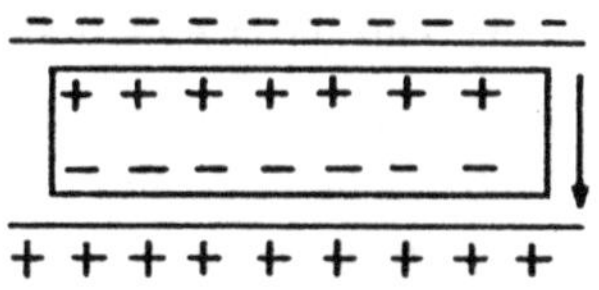

Figure 4.10.

shown in the Figure 4.10, a positive charge will be induced in the top and negative on the bottom side of the slab. This will produce a macroscopic field inside the material which is opposed to the field between the plates, thus, effectively reducing it. Thus, we have the effective field

$$E = \frac{\sigma - P}{\varepsilon_0}$$

and
$$V = Ed = \frac{\sigma d}{\varepsilon_0(1+\chi)} \tag{4.28}$$

and
$$C = \frac{Q}{V} = \frac{\sigma A}{\sigma d/\varepsilon_o(1+\chi)}$$

$$= \frac{\sigma A}{d}(1+\chi) = \frac{\varepsilon_0 \varepsilon_r A}{d} \tag{4.29}$$

Thus the capacity is increased by a factor ε_r.

4.9 LAWS OF ELECTROSTATICS IN THE PRESENCE OF DIELECTRICS

We have seen in the preceding sections that the field in a dielectric medium is not the same as would be expected from Coulomb's law. Obviously, the law, therefore, is not of universal validity. Experimentally, it has been proved for air and is found not to be true for dielectrics. This means that the laws of electrostatics so far established needs modification in dielectrics.

So far, we have considered uniform polarization. However, there may be situations where the polarization is not uniform. This may be due either to dielectric being non-uniform or due to the variation of the field even inside a uniform dielectric. Now, we consider a situation in which the polarization P is not the same everywhere.

Let us consider that an unpolarized slab of a material is placed in electric field. Due to polarization process, a certain amount of net charge will pass through a given amount of area ds due to polarization process. The amount of the charge $= \mathbf{P}\cdot\hat{a}_n\, ds$ (Product of the component of **P** normal to **ds** and the magnitude of ds). If P is not uniform, then the displacement of charges can result in a volume charge density. The total charge displaced out of any volume V by the polarization is obtained as

$$\int_s \mathbf{P}\cdot\hat{a}_n\, ds$$

where s is the surface that bounds the volume. Let q' be the equal excess charge of the opposite sign left behind, then

$$q' = -\int_s \mathbf{P}\cdot\hat{a}_n\, ds \tag{4.30}$$

If ρ' be the density of the polarization charges within the volume, then

$$q' = \int_v \rho'\, d\tau$$

$$\therefore \quad \int_v \rho'\, d\tau = -\int_s \mathbf{P}\cdot\hat{a}_n\, ds \tag{4.31}$$

Making use of the Divergence theorem, one obtains

$$\int_v \rho'\, d\tau = -\int_v \nabla\cdot\mathbf{P}\, d\tau \tag{4.32}$$

Thus

$$\rho' = -\nabla\cdot\mathbf{P} \tag{4.33}$$

obviously, the polarization charge density ρ' (C-m^3) is given by the negative divergence of Polarization density **P**.

We have Gauss's law in free space

$$\nabla\cdot\mathbf{E} = \frac{\rho}{\varepsilon_0}$$

one can write an equivalent law for dielectric medium by remembring that besides free charges there are also polarization charges present. One obtains

$$\nabla\cdot\mathbf{E} = \frac{\rho+\rho'}{\varepsilon_0} = \frac{\rho}{\varepsilon_0} - \frac{\nabla\cdot\mathbf{P}}{\varepsilon_0}$$

$$\therefore \quad \nabla\cdot\left(\mathbf{E} + \frac{\mathbf{P}}{\varepsilon_0}\right) = \frac{\rho}{\varepsilon_0}$$

or

$$\nabla\cdot(\varepsilon_0 E + \mathbf{P}) = \rho \tag{4.34}$$

or

$$\nabla\cdot\mathbf{D} = \rho \tag{4.35}$$

where $\mathbf{D} = \varepsilon_0\mathbf{E} + \mathbf{P}$.

Eq. (4.35) is the modified form of Gauss's law and also called as Gauss's theorem in differential form, which also includes the effects of polarization charges. Obviously, the flux of **D** out of a closed surfaces = total free charge enclosed by s. Obviously **D** has the same dimensions as **P**. Eq. (4.35) is also the Maxwell's first equation, which suggests that the evolution of electric displacement vector is due to free charges. Since

$$\begin{aligned} \mathbf{P} &= \varepsilon_0 \chi \mathbf{E} \\ \therefore \quad \mathbf{D} &= \varepsilon_0 \mathbf{E} + \varepsilon_0 \chi E = (1+\chi)\varepsilon_0 \mathbf{E} = \varepsilon_0 \varepsilon_r \mathbf{E} = \varepsilon \mathbf{E} \end{aligned} \tag{4.36}$$

where $\varepsilon = \varepsilon_0 \varepsilon_r$ is the *permittivity* of the medium. We must remember that $\nabla \times \mathbf{E} = 0$ remains unchanged in dielectrics.

4.10 ENERGY OF THE ELECTROSTATIC FIELD IN THE PRESENCE OF A DIELECTRIC

We have already obtained the expression for the potential energy, when the charges are distributed with the volume density ρ and surface density σ as

$$W = \frac{1}{2}\int_v \rho\phi\, dV + \frac{1}{2}\int_{s_1} \sigma\phi\, ds_1$$

The above result is not affected by the presence of a dielectric. We have from Gauss's theorem in the presence of a dielectric, $\nabla \cdot \mathbf{D} = \rho$. Thus

$$\begin{aligned} \frac{1}{2}\int_v \rho\phi\, d\tau &= \frac{1}{2}\int_v \phi(\nabla\cdot\mathbf{D})\, d\tau = \frac{1}{2}\int_v \{\nabla\cdot(\phi\mathbf{D}) - \mathbf{D}\cdot\nabla\phi\}\, d\tau \\ &= \frac{1}{2}\int_s \phi\mathbf{D}\cdot\hat{a}_n\, ds + \frac{1}{2}\int_v \mathbf{D}\cdot\mathbf{E}\, d\tau \end{aligned} \tag{4.37}$$

In obtaining the above result, we have used the following vector identity

$$\nabla\cdot(\phi\mathbf{D}) = \phi\nabla\cdot\mathbf{D} + \mathbf{D}\cdot\nabla\phi \tag{4.38}$$

and the divergence theorem. We can see that the first integral in (4.37) must be taken over a closed surface bounding the whole volume (which is zero) and also over the surfaces of the conductors. Since the integration is over the surface of the medium, the vector $\hat{a}_n\, ds$ is drawn outwards from the medium, i.e. into the conducting surface (see Fig. 4.6).

One can also show that the normal component of **D**, i.e.

$$\mathbf{D}\cdot\hat{a}_n ds_1 = -\sigma ds_1$$

$$\therefore \quad \frac{1}{2}\int_v \rho\phi d\tau = -\frac{1}{2}\int_{s_1}\sigma\phi ds_1 + \frac{1}{2}\int_v \mathbf{D}\cdot\mathbf{E}d\tau$$

$$\therefore \quad W = \frac{1}{2}\int_v \rho\phi d\tau + \frac{1}{2}\int_{s_1}\sigma\phi ds_1 = \frac{1}{2}\int_v \mathbf{D}\cdot\mathbf{E}d\tau \quad (4.39)$$

Eq. (4.39) shows that one may regard the energy as being distributed throughout all space. The energy density is $\frac{1}{2}\mathbf{D}\cdot\mathbf{E}$.

4.11 BOUNDARY CONDITIONS

In a single medium the electric field is continuous. In case the field is not constant, there occurs an infinitesimal amount of change in the field in an infinitesimal distance. However, at the boundary between two different media the electric field may change abruptly both in magnitude and direction. Thus it becomes necessary to establish the field relations at such boundaries. Moreover, the solution of Poisson's or Laplace's equation should always satisfy certain conditions at the boundaries between different media in the field. These conditions pertaining to potential, tangential component of the electric field intensity, normal component of the displacement density vector and refraction lines of forces are called boundary conditions.

Let us imagine a disc enclosing a part of the boundary surface between two media and whose axis is normal to the boundary as shown in Fig. 4.11.

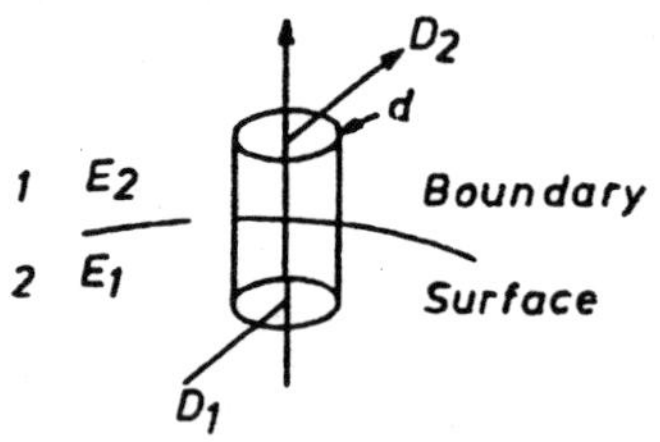

Figure 4.11.

Let ε_1 and ε_2 be the relative permittivities of two media marked as 1 and 2 respectively. Assuming that the thickness of the disc is very small, the only contribution to the outward flux from the disc comes from its flat surfaces. We have from Gauss's theorem

$$\nabla\cdot\mathbf{D} = \rho$$

$$\therefore \quad \int_v \nabla\cdot\mathbf{D}d\tau = \int_v \rho d\tau \quad (4.40)$$

where ρ is the density of free charges. Since there is no free charge on the surface, we have

$$\int_v \nabla \cdot \mathbf{D}\, d\tau = 0$$

By Divergence theorem, one obtains

$$\int_v \nabla \cdot \mathbf{D}\, d\tau = \int_s \mathbf{D} \cdot \hat{a}_n\, ds \tag{4.41}$$

$$\therefore \quad \int_s \mathbf{D} \cdot \hat{a}_n\, ds = \mathbf{D}_1 \cdot \hat{a}_{n_1}\, ds + \mathbf{D}_2 \cdot \hat{a}_{n_2}\, ds = 0$$

where D_1 and D_2 are the values of D in the two media and $\hat{a}_{n_1}$ and $\hat{a}_{n_2}$ are the unit vectors. Also

$$\hat{a}_{n_1} = -\hat{a}_{n_2} = \hat{a}_n$$

Thus

$$(\mathbf{D}_1 \cdot \hat{a}_n - \mathbf{D}_2 \cdot \hat{a}_n) ds = 0 \quad \text{or} \quad \mathbf{D}_1 \cdot \hat{a}_n = \mathbf{D}_2 \cdot \hat{a}_n \tag{4.42}$$

Obviously, *the normal component of* **D** *in two media are equal.*

Now we construct at the boundary a rectangular path ABCD (Fig. 4.12) with sides $AB = CD = dl$ parallel to the surface and $BC = DA = dt$ perpendicular to it.

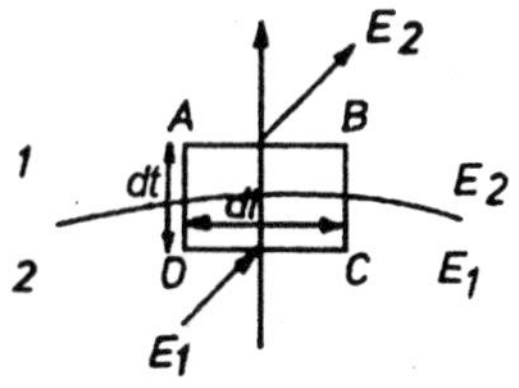

Figure 4.12.

Let $\mathbf{E}_1$, and $\mathbf{E}_2$ be the electric fields in the two media. We know that electric field is conservative and hence no net work is done by taking a unit charge round the rectangular path *ABCDA*. Obviously,

$$\oint \mathbf{E} \cdot \mathbf{dl} = E_{\parallel(1)} dl - E_{\parallel 2} dl + \text{Contribution from } BC \text{ and } DA$$
$$= 0 \tag{4.43}$$

where $E_{\parallel(1)}$ and $E_{\parallel(2)}$ are the components of electric field **E** in the two media parallel to the surface. Let us assume that dt, i.e. BC, DA is vanishingly small, one finds

$$E_{\parallel(1)} dl = E_{\parallel(2)} dl = 0$$
$$\therefore \quad E_{\parallel(1)} = E_{\parallel(2)} \tag{4.44}$$

Obviously, *the tangential component of electric field* **E**, *i.e.* $E_{\parallel}$ *is continuous across the boundary.*

Thus we have obtained two conditions represented by eqs. (4.42) and (4.44) to be satisfied by the displacement vector **D** and electric field **E**.

The results we have obtained signify that when passing through the interface between two dielectrics, the normal component of the vector **D** and the tangential component of vector **E** change continuously. The tangential component of the vector **D** and the normal component of the vector **E**, however, are disrupted when passing through the interface.

One can also express the above conditions in terms of potential. Let we have two media in contact with potentials ϕ_1 and ϕ_2 respectively, at all connecting points of the two media 1 and 2

$$\phi_1 = \phi_2$$

and
$$\varepsilon_1 \frac{\partial \phi_1}{\partial n} = \varepsilon_2 \frac{\partial \phi_2}{\partial n} \tag{4.45}$$

We must note that condition (4.42) for the displacement vector was arrived at on the assumption that there is no free charge on the boundary. However, if there is any free charge present on the boundary, say with surface density σ, one finds that (4.42) changes to

$$\mathbf{D}_1 \cdot \hat{a}_n - \mathbf{D}_2 \cdot \hat{a}_n = \sigma \tag{4.46}$$

Obviously, the normal component of D abruptly changes by σ.

4.12 POTENTIAL AND FIELD DUE TO A POLARIZED SPHERE

Consider a sphere of dielectric material which has been polarized along the positive z-axis (Fig. 4.13). Due to polarization, the positive charges are concentrated at the top and the negative charges at the bottom. Thus a polarized sphere is equivalent to a combination of two spheres, one having $+Q$ charge and other having $-Q$ charge such that the centre of gravity of the positive charge has been displaced slightly upward while the centre of the negative charge has been displaced slightly downward. If d be the distance between these centres (Fig. 4.14), then the moment of the dipole is $\mathbf{P}_0 = Q\mathbf{d}$ (considering the polarized sphere to be equivalent to dipole). If r be the radius of the sphere, then volume of the sphere $= \frac{4}{3}\pi r^3$ and if **P** is the density of polarization, then dipole moment of the sphere $p_0 = \frac{4}{3}\pi r^3 p$.

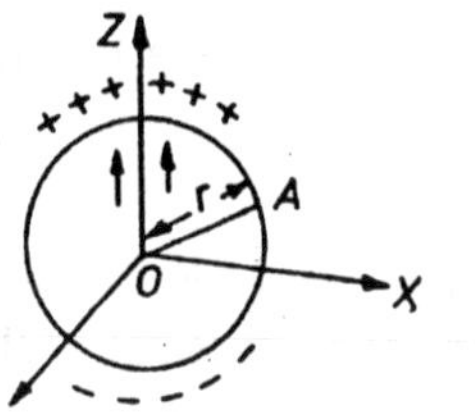

Figure 4.13.

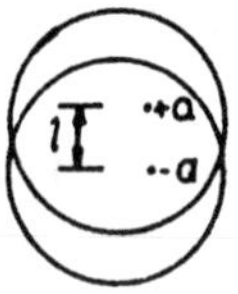

Figure 4.14.

Obviously, the dipole moment of a polarized sphere is equivalent to the vector sum of the dipole moments of atoms or molecules situated inside the sphere.

The potential V at a point A on the surface of the sphere is

$$\begin{aligned} V &= \frac{1}{4\pi\varepsilon_0}\frac{p_0\cos\theta}{r^2} \\ &= \frac{1}{4\pi\varepsilon_0}\frac{\frac{4}{3}\pi r^3 p\cos\theta}{r^2} = \frac{1}{3\varepsilon_0}pr\cos\theta \end{aligned} \tag{4.47}$$

Now $r\cos\theta = z_A$, the z-coordinate of the point A

$$\therefore \qquad V = \frac{1}{4\pi\varepsilon_0}\frac{4}{3}\pi p z_A = \frac{1}{3}\varepsilon_0 p z_A = K z_A \tag{4.48}$$

where $K = \dfrac{1}{3\varepsilon_0 p}$ is a constant. Obviously, the nature of the field is same as due to a dipole (Fig. 4.15).

The potential inside the sphere must satisfy the Laplace's equation, $\nabla^2 V = 0$, i.e.

$$\frac{\partial^2 V}{\partial x^2} + \frac{\partial^2 V}{\partial y^2} + \frac{\partial^2 V}{\partial z^2} = 0$$

Putting the value of $V = K z_A$, one finds that the equation is satisfied. Then according to *uniqueness theorem*, Laplace's equation can have only one solution, i.e. $V = K z_A$. This relation gives the value of potential everywhere inside the polarized sphere. The electric field inside the sphere, i.e. along z-axis is

$$E_z = -\frac{\partial^2 V}{\partial z} = -\frac{\partial}{\partial z}(Kz) = -K = -\frac{p}{3\varepsilon_0} \tag{4.49}$$

Eq. (4.49) shows that the field inside the sphere is uniform and constant, and is directed opposite to the direction of polarization as shown in Fig. 4.16.

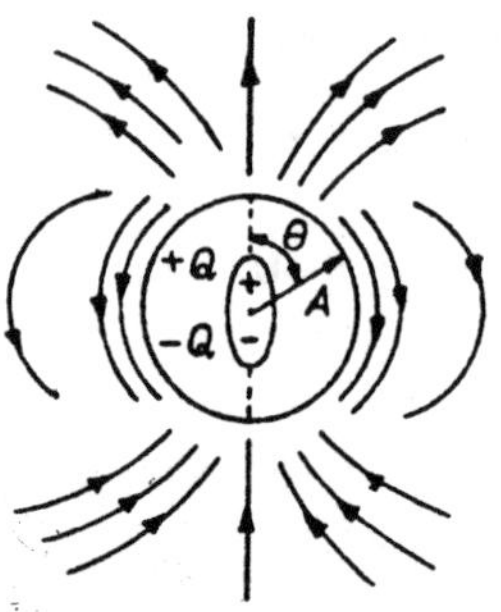

Figure 4.15.

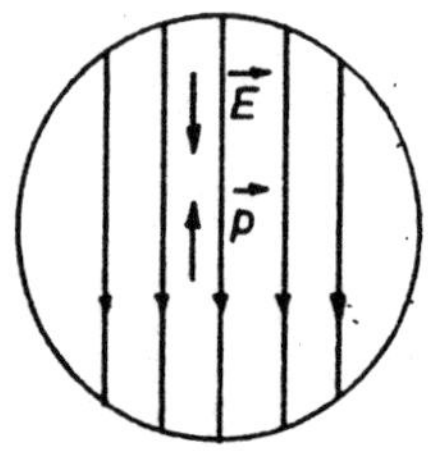

Figure 4.16.

Since $E_x E_y = 0$ and hence $\mathbf{E} = -\dfrac{P}{3\varepsilon_0}$. Obviously, the electric field inside the polarised sphere is constant.

Potential and Field due to a polarized sphere when point is outside the sphere ($R > r$)

We have seen that a polarised sphere behaves like a dipole of dipole moment p. Thus the potential at a point, having coordinates (R, θ) situated outside the polarized sphere, will be

$$V = \frac{p\cos\theta}{4\pi\varepsilon_0 R^2} = \frac{\frac{4}{3}\pi r^3 p\cos\theta}{4\pi\varepsilon_0 R^2} = \frac{pr^3\cos\theta}{3\varepsilon_0 R^2} \tag{4.50}$$

The components of the electric field intensity at (R, θ) will be

$$E_R = -\frac{\partial V}{\partial R} = -\frac{\partial}{\partial R}\left(\frac{pr^3\cos\theta}{3\varepsilon_0 R^2}\right) = \frac{2pr^3\cos\theta}{3\varepsilon_0 R^3} \tag{4.51}$$

or $$E_\theta = -\frac{1}{R}\frac{\partial V}{\partial \theta} = -\frac{1}{R}\frac{\partial}{\partial \theta}\left(\frac{pr^3\cos\theta}{3\varepsilon_0 r^2}\right) = \frac{pr^3\sin\theta}{3\varepsilon_0 R^3} \tag{4.52}$$

$\therefore$ Total intensity

$$\mathbf{E} = \left[\frac{2pr^3\cos\theta}{3\varepsilon_0 R^3} + \frac{pr^3\sin\theta}{3\varepsilon_0 R^3}\hat{\theta}\right]$$

$$= \frac{pr^3}{3\varepsilon_0 R^3}\left[(2\cos\theta)\hat{R} + (\sin\theta)\hat{\theta}\right] \tag{4.53}$$

where $\hat{R}$ and $\hat{\theta}$ are unit vectors.

Point on the surface of the polarised sphere (r = R)

Putting $r = R$ in (4.50), (4.51) and (4.52), one obtains

$$\text{Potential } V = \frac{pR\cos\theta}{3\varepsilon_0} = \frac{pz}{3\varepsilon_0} \qquad (\because \quad R\cos\theta = z) \tag{4.54}$$

Obviously, the potential of a polarized sphere depends on the z-coordinates only. The components of the intensity of electric field on the surface of the intensity of electric field on the surface of polarised sphere are

$$E_R = \frac{2p\cos\theta}{3\varepsilon_0} \tag{4.55}$$

or

$$E_\theta = \frac{p\sin\theta}{3\varepsilon_0} \tag{4.56}$$

$\therefore$ Total intensity

$$\mathbf{E} = \frac{2p\cos\theta}{3\varepsilon_0}\hat{r} + \frac{p\sin\theta}{3\varepsilon_0}\hat{\theta} \tag{4.57}$$

4.13 DIELECTRIC SPHERE IN A UNIFORM ELECTRIC FIELD

Let us consider that a sphere of radius R and dielectric constant (k or ε_r is placed in a uniform field $\mathbf{E}_0$. The centre of the sphere is at the origin and the electric field is along z-axis as shown in Fig. 4.17.

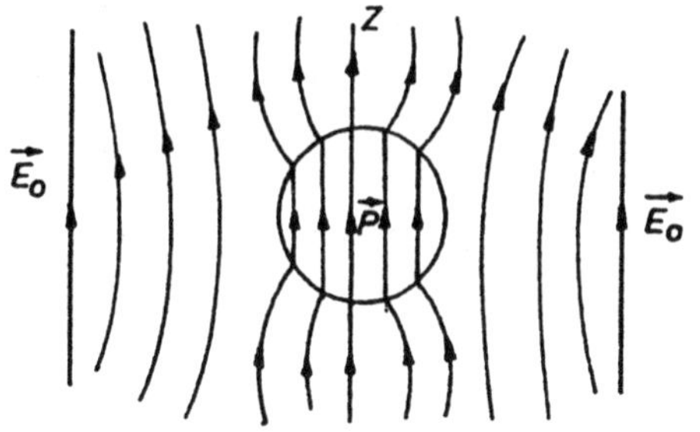

Figure 4.17.

Due to this, field polarisation is induced in the sphere along z-axis which alters the field intensity inside and outside the sphere. However, the field at a large distance from the sphere remains same, i.e. equal to $\mathbf{E}_0$.

(i) Field inside the dielectric sphere

We can easily see that the resultant electric field $\mathbf{E}$ inside the dielectric sphere is equal to the vector sum of the superposed electric fields, i.e. external field $\mathbf{E}_0$ and the field produced due to the polarisation $\mathbf{P}$, i.e. $\mathbf{E}_P$. Thus

$$\mathbf{E}_{\text{in}} = \mathbf{E}_0 + \mathbf{E}_P \tag{4.58}$$

The electric field inside the dielectric sphere due to induced polarisation $\mathbf{P}$ will be

$$\mathbf{E}_P = -\frac{\mathbf{P}}{3\varepsilon_0} \tag{4.59}$$

$$\therefore \quad \mathbf{E}_{\text{in}} = \mathbf{E}_0 - \frac{\mathbf{P}}{3\varepsilon_0} \tag{4.60}$$

Since the induced polarisation $\mathbf{P}$ in a dielectric is proportional to $\mathbf{E}_{\text{in}}$, therefore, one obtains

$$\mathbf{P} = \varepsilon_0 \chi \mathbf{E}_{\text{in}}$$

where for a dielectric $\chi = \varepsilon_r - 1$, is the electrical susceptibility

$$\therefore \quad \mathbf{P} = \varepsilon_0(\varepsilon_r - 1)\mathbf{E}_{\text{in}} \tag{4.61}$$

Making use of (4.61), (4.60) becomes

$$\mathbf{E}_{\text{in}} = \mathbf{E}_0 - \frac{\varepsilon_0(\varepsilon_r - 1)\mathbf{E}_{\text{in}}}{3\varepsilon_0} = \frac{3}{2+\varepsilon_r}\mathbf{E}_0 \tag{4.62}$$

Obviously, the direction of electric field inside the dielectric will be the same as that of the external field $\mathbf{E}_0$. Moreover, for a dielectric ε_r is always greater than 1, the magnitude of E_{in} is always less than E_0. Using (4.62), (4.61) becomes

$$\mathbf{P} = \frac{3\varepsilon_0(\varepsilon_r - 1)}{(\varepsilon_r + 2)}\mathbf{E}_0 = \frac{3\varepsilon_0(k-1)}{(k+2)}\mathbf{E}_0 \tag{4.63}$$

The relation (4.63) suggests that polarisation $\mathbf{P}$ in the dielectric is induced along the direction of the applied field.

(ii) Electric Field outside the dielectric sphere

The resultant field outside the dielectric sphere $\mathbf{E}_{\text{out}}$ would be equal to the vector sum of electric field $\mathbf{E}_0$ and the electric field $\mathbf{E}$ produced due to polarized sphere, i.e.

$$\mathbf{E}_{\text{out}} = \mathbf{E}_0 + \mathbf{E}' \tag{4.64}$$

one can find the intensity of electric field at a point (R, θ) outside the polarized sphere with polarization **P** (using (4.53)) is

$$\mathbf{E} = \frac{2PR^3\cos\theta}{3\varepsilon_0 r^3}\hat{r} + \frac{PR^3\sin\theta}{3\varepsilon_0 r^3}\hat{\theta}$$

where $\hat{r}$ and $\hat{\theta}$ are unit vectors along r and θ directions respectively.

$$\begin{aligned} \therefore \qquad \mathbf{E}_{\text{out}} &= \mathbf{E}_0 + \frac{2PR^3\cos\theta}{3\varepsilon_0 r^3}\hat{r} + \frac{PR^3\sin\theta}{3\varepsilon_0 r^3}\hat{\theta} \\ &= \mathbf{E}_0 + \frac{2R^3(\varepsilon_r - 1)}{r^3(\varepsilon_r + 2)}E_0\cos\theta\,\hat{r} \\ &\quad + \frac{R^3(\varepsilon_0 - 1)}{r^3(\varepsilon_r + 2)}E_0\sin\theta\,\hat{\theta} \end{aligned} \tag{4.65}$$

4.14 GASEOUS NON-POLAR DIELECTRICS

A dielectric, in electromagnetic theory is considered to be a continuous medium and the macroscopic quantities such as charge density and electric polarization are defined as continuous functions. A dielectric is really an assembly of positive and negative charges. Therefore it is essential to relate the electrical properties of the constitutent discrete particles and the corresponding properties of the dielectric (continuous medium). We have mentioned earlier that if the field is not too large, the strength of the dipole induced in an atom, ion or molecule is proportional to the actual field acting on the particle. If we call this field E_{local}, then

$$\mathbf{p} = \alpha\varepsilon_0\mathbf{E}_{\text{local}} \tag{4.66}$$

where **p** is the induced dipole moment and $\alpha\varepsilon_0$ is the constant of proportionality.

We must note that the local field E_{local} acting on a particular particle is not the same as the macroscopic field, which is the field averaged over a region large enough to contain many particles, and calculated by taking the average of the vector sum of the electric fields set up by all the dipoles in the medium including that of the particle for which one require vector E_{local}. The local field E_{local} is equal to the maroscopic field **E** due to the dipoles more than a few atomic diameters away, which we can treat as a continuous medium with polarization **P**, plus the contribution from nearby dipoles excluding the field of the dipole at which one requires the value of E_{local}, since the field due to this dipole does not act on itself. We must note that this is particularly true in the case of liquids and solids. In the case of gases, the molecules, for most of the time, are far apart, and hence the short range fields are negligible. Thus E_{local} is the same as the macroscopic field **E**. Obviously, in the case of gases

$$\mathbf{p} = \alpha\varepsilon_0\mathbf{E}$$

and

$$\mathbf{P} = N\mathbf{p} = N\alpha\varepsilon_0\mathbf{E} = \varepsilon_0\chi\mathbf{E} \tag{4.67}$$

$$\therefore \quad \alpha = \frac{\chi}{N} \tag{4.68}$$

Here α is called the polarizability of the atom, ion or molecule. α measures the resistance of the particle to the displacement of its electron cloud. The scalar χ is the electrical susceptibility of the medium.

4.15 GASEOUS POLAR DIELECTRICS

We now study the dielectric behaviour of molecules which have a permanent dipole moment. We have mentioned in the previous section that the local field acting on a molecule in a gas is the same as the external field **E**. This field induces an extra dipole moment in the molecule giving it the same kind of polarizability as in non-polar molecules. Besides this, the field also tends to line up the individual dipoles. However, this does not happen effectively due to the thermal motion of molecules. The potential energy of a dipole when placed in an electric field is given by

$$W = -\mathbf{p}\cdot\mathbf{E} = -pE\cos\theta \tag{4.69}$$

The potential energy varies with the orientation of the dipole. The energy is minimum when $\theta = 0$, i.e. when the dipole is aligned along the field. One can get an idea about the order of magnitude of this energy by calculating the energy required to reverse the dipole, so as to make it point in the direction opposite to that of the field. Obviously, this energy is equal to $2pE$. We have atomic spacing $\sim 10^{-10}m$ and let $E = 10^6$ V/m, the maximum that can be achieved in gaseous dielectric. Therefore

$$p = e \times \text{ atomic spacing} = 1.6\times10^{-19}\times10^{-10} = 1.6\times10^{-29}\text{ C-m}$$

and

$$2pE = 2\times1.6\times10^{-19}\times10^{6}J = 2\times10^{-4}\text{ eV} \tag{4.70}$$

We know that the average kinetic energy of a gaseous molecule at room temperature is of the order of 0.04 eV. Obviously, the energy given by (4.70) is much less than the average kinetic energy. This reveals that the random thermal motion will not be affected much by the presence of the field **E** and there will still be molecules with their dipoles pointing in all directions. However, the dipoles, in general tend to line up in a direction in which their potential energy is minimum, i.e. in the direction of the field. Obviously, there will be a slight excess of molecules pointing in this direction. Let us calculate the net polarization caused by this polarization. One can do it by using the methods of statistical methods.

Let us consider that N be the number of molecules per unit volume. We know, from statiscal mechanics that in the state of thermal equilibrium, the relative number of molecules with the potential energy W is proportional to

$$N = N_0 \exp\left(-\frac{W}{kT}\right)$$

where k is Boltzmann constant and T is the absolute temperature.

We know that in the absence of an electric field all orientations of molecule are equally probable. Thus the number of molecules per unit volume having their dipoles within a solid angle $d\Omega$ will be equal to $N\frac{d\Omega}{4\pi}$. Now, the solid angle in the range of orientation between θ and $\theta + d\theta$ is given by (Fig. 4.18)

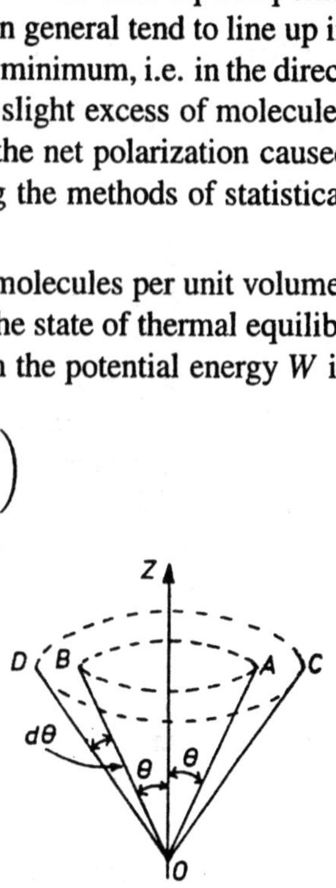

Figure 4.18.

$$d\Omega = \frac{2\pi r \sin\theta \; r d\theta}{r^2} = 2\pi \sin\theta d\theta \qquad (4.71)$$

$$\therefore \quad N\frac{d\Omega}{4\pi} = \frac{1}{2}N \sin\theta d\theta \qquad (4.72)$$

Now, consider that a field **E** is applied along the z-axis. Obviously, the dipoles making angle θ with the field acquire an additional potential energy $W = -pE\cos\theta$. The probability distribution will depend upon the Boltzmann factor $\exp\left(-\frac{W}{kT}\right)$. Thus the number of molecules per unit volume with their dipoles in the range between θ and $\theta + d\theta$ is given by

$$N(\theta)d\theta = C\exp\left(-\frac{W}{kT}\right)d\Omega \tag{4.73}$$

Since $\frac{W}{kT}$ is very small and hence one may write this as

$$N(\theta)\,d\theta = C\left(1-\frac{W}{kT}\right)d\Omega = C2\pi\sin\theta\left(1+\frac{pE\cos\theta}{kT}\right) \tag{4.74}$$

Therefore the total number of molecules per unit volume is obtained as

$$N = \int_0^\pi N(\theta)d\theta = 2\pi C\int_0^\pi \sin\theta\left(1+\frac{pE\cos\theta}{kT}\right)d\theta = 4\pi C$$

$$\therefore \quad C = \frac{N}{4\pi} \quad \text{and} \quad N(\theta) = \frac{N}{2}\sin\theta\left(1+\frac{pE\cos\theta}{kT}\right) \tag{4.75}$$

Eq. (4.75) clearly reveals that there will be more molecules in the field direction ($\cos\theta = 1$) than in the opposite direction. This is in agreement with our assumption that there will be net alignment when the field **E** is applied.

We find that the component of a dipole moment pointing in the direction θ, along the field is $p\cos\theta$. Thus the net dipole moment per unit volume along the field is

$$\begin{aligned} P &= \int_0^\pi N(\theta)p\cos\theta d\theta = \frac{N}{2}\int_0^\pi \sin\theta\left(1+\frac{pE\cos\theta}{kT}\right) \\ &= \frac{Np^2E}{3kT} \end{aligned} \tag{4.76}$$

This equation reveals the temperature dependence of the polarization taking place in dielectrics. As the temperature T increases, the polarization of the dielectrics is seen to decrease. This is physically visualized in terms of increase of orientational time with temperature which inhibits the increase of orientation of the electric dipoles along the applied field. Moreover, (4.36) also represent the contribution of the permanent dipole moments to the polarization. The quantity $\frac{p^2}{3kT}$ is usually called as *orientational polarizability*. We must note that, besides this, the molecules will also acquire an induced dipole moment along the field irrespective of the direction of the permanent dipole moment. This will lead to an additional polarization $N\alpha_0\varepsilon_0\mathbf{E}$, where α_0 is deformation polarizability. Therefore, the net polarization is obtained as

$$\mathbf{P} = \left(N\alpha_0\varepsilon_0 + \frac{Np^2}{3kT}\right)\mathbf{E} \tag{4.77}$$

The susceptibility is given by

$$\chi = N\left(\alpha_0 + \frac{p^2}{3kT\varepsilon_0}\right) \tag{4.78}$$

One can easily see from (4.78) that the measurement of susceptibilities at different temperatures will provide information about permanent and induced dipole moments. Further, we note that the plot of χ vs. $1/T$ will be a straight line. The slope of this plot will give the permanent dipole moment and intercept at $1/T = 0$ will give χ due to the induced polarization. We can also get some informations about the shape of molecules from the knowledge of susceptibilities and the dipole moments, e.g. CS_2 and H_2O both molecules contain two identical atoms bound to a common partner. One can see that there is electron transfer across C–S bond as well as across O–H bond and dipole moments associated with each bond. However, the dipole moment of H_2O is 6.2×10^{-30} C-m where as that of CS_2 is zero. One can conclude that since dipole moments of CS_2 cancel each other and hence it must be a linear molecule, i.e. S–C–S where as H_2O is not a linear molecule.

Eq. (4.78) Also depicts the density and temperature variation of the dielectric property of materials. In the case of polar molecules, it is found that the frequency response depends on the moment of inertia of the molecules. The moment of inertia of molecules determine the amount of time they take to turn toward the direction of the applied electric field. Obviously, if one applies electromagnetic fields in the microwave range, the polar contribution to the dielectric constant begins to fall because the heavy molecules do not respond to these high frequencies. In contrast to this, one finds that the electronic polarizability remains unaffected upto optical frequencies and the contribution to dielectric constant does not change with increasing frequency.

Note: If we consider the effect of collisions on the orientation time of the electric dipoles, the total polarization in the dielectric dipoles, the total polarization in the dielectric medium can be expressed as

$$\mathbf{P}_T = \frac{Np^2\mathbf{E}}{3kT(1 - j\omega\tau_r)} \tag{4.79}$$

where $\alpha(\omega)$ is the relaxation time of electric dipoles in dielectric medium. Thus

$$\frac{\mathbf{P}}{\mathbf{E}} = \frac{Np^2}{3kT(1 - j\omega\tau_r)} = N\varepsilon_0\alpha(\omega) \tag{4.80}$$

where $\alpha(\omega)$ is the atomic polarizability of the dielectrics.

$$\text{or} \qquad \alpha(\omega) = \frac{p^2}{3\varepsilon_0 kT(1 - j\omega\tau_r)} \tag{4.81}$$

This gives the complex form of *orientational polarizability*. The variation of real and imaginary parts of polarizability as a frequency is shown in Fig. 4.19. The real part of α (say $\alpha'(\omega)$) is constant at low frequencies and decreases steadily with increasing frequency. The slope becomes maximum at $\omega = 1/\tau_r$. From Fig. 4.19 it is evident that the maximum slope in $\alpha'(\omega)$ corresponds to maximum value of $\mathrm{I_m}\ \alpha(\omega)$, i.e. $\alpha''(\omega)$.

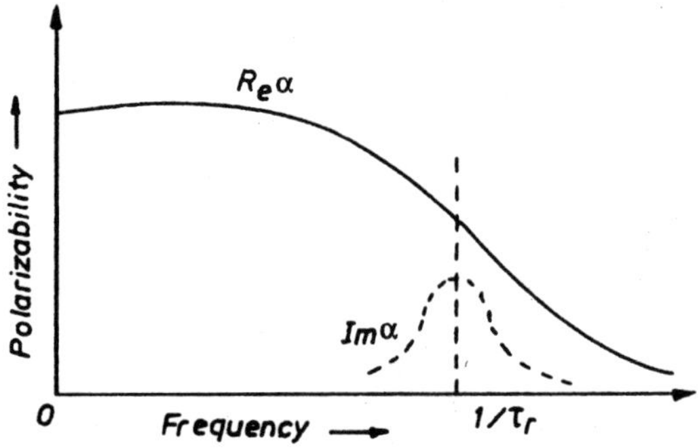

Figure 4.19.

4.16 NON-POLAR LIQUIDS

One cannot overlook the effects of the short range fields in solid and liquid dielectrics because the molecules in them are quite close together. When the polar molecules are present, the behaviour becomes more complicated because the fields generated by the permanent dipoles are sometimes much larger than any microscopic field. However, the short range fields in a non-polar liquid are not so strong and hence one can deduce $\mathbf{E}_{\text{local}}$ with the help of following procedure. Let us imagine a small spherical cavity carved out in a uniformly polarized material. The medium is polarized and hence the polarization charges will appear on the surface of the medium. The field at a point *A* inside the cavity, $\mathbf{E}_{\text{local}}$ is the sum of the macroscopic field $\mathbf{E}$ and the fields generated by the polarization charges on the surface of the cavity.

Let us consider that the macroscopic field is acting in the z-direction (Fig. 4.20). If σ_p is the surface density of polarization charge, then the polarization charge on a small element of area ds, on the surface of cavity ($BCED$),

$$\begin{aligned}\sigma ds = \mathbf{P}\cdot\hat{a}_n ds &= \varepsilon_0\chi\mathbf{E}\cdot\hat{a}_n ds \\ &= -\varepsilon_0\chi E\cos\theta\cos\theta ds\end{aligned} \tag{4.82}$$

where θ is the angle between the field and the normal to ds. The negative sign in (4.82) reveals that the outward normal to the dielectric is the inward normal to the cavity. The net field at A, the centre of cavity, generated by the polarization charges is, by symmetry, directed along the z-axis. One finds the magnitude of the z-component of the field at A due to this charge is

$$\frac{\varepsilon_0\chi E\cos\theta ds\cos\theta}{4\pi\varepsilon_0 R^2}$$

Here R is the radius of the cavity. Now, the field due to the entire band of area $ds = 2\pi R\sin\theta d\theta$ is

$$\frac{\varepsilon_0\chi E\cos^2\theta}{4\pi\varepsilon_0 R^2}2\pi R^2\sin\theta d\theta$$

and the total field at A due to all the polarization charges on the surface is obtained as

$$\frac{1}{2}\chi E\int_0^{\pi}\cos^2\theta\sin\theta d\theta = \frac{1}{2}\chi E$$

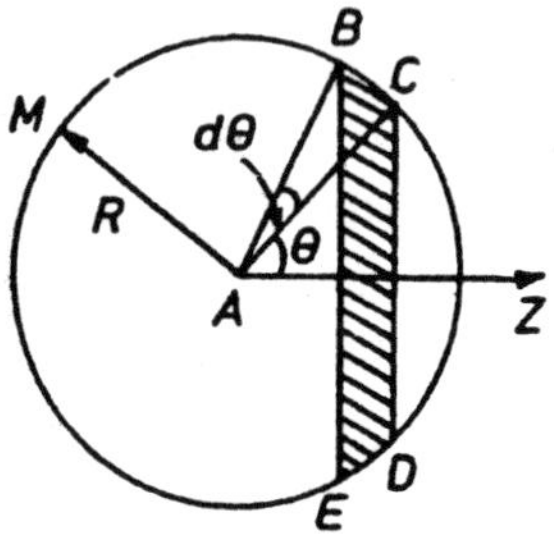

Figure 4.20.

Thus the effective field at *A*

$$\mathbf{E}_{\text{local}} = \mathbf{E}+\frac{1}{3}\chi\mathbf{E} = \left[1+\left(\frac{1}{3}\right)\chi\right]\mathbf{E} \tag{4.83}$$

$$\therefore \quad \mathbf{P} = N\alpha\varepsilon_0\mathbf{E}_{\text{local}} = N\alpha\varepsilon_0\mathbf{E}\left[1+\left(\frac{1}{3}\right)\chi\right] \tag{4.84}$$

or

$$\varepsilon_0\chi\mathbf{E} = N\alpha\varepsilon_0\mathbf{E}\left(1+\frac{1}{3}\chi\right) \tag{4.85}$$

$$\therefore \quad \chi = \frac{N\alpha}{1-\frac{N\alpha}{3}} \tag{4.86}$$

Since $\chi = \varepsilon_r - 1$ and $\mathbf{P} = \varepsilon_0(\varepsilon_r - 1)\mathbf{E}$, (4.85) can also be rewritten as

$$\varepsilon_0(\varepsilon_r-1)\mathbf{E} = N\alpha\varepsilon_0\mathbf{E}+\frac{N\alpha\varepsilon_0\mathbf{E}}{3}(\varepsilon_r-1)$$

or

$$\varepsilon_0(\varepsilon_r-1) = N\alpha\varepsilon_0\frac{2+\varepsilon_r}{3}$$

$$\therefore \quad \frac{\varepsilon_r-1}{\varepsilon_r+2} = \frac{N\alpha}{3} \tag{4.87}$$

This is called the *Clausius-Mossotti equation*, since Mossotti (in 1850) and Clausius independently (in 1879) established that for any given substance $\frac{(\varepsilon_r-1)}{(\varepsilon_r-2)}$ should be proportional to the density of the substance. (At optical frequencies, $\frac{\varepsilon}{\varepsilon_0} = n^2$ where n is the index of refraction. With n^2 replacing ε_r in (4.87), the equation is sometimes called the Lorentz-Lorentz equation). The relation holds best for dilute substances such as gases. For liquids and solids (4.87) is only approximately valid, especially if the dielectric constant is large. The Eq. (4.87) shows that the local field is larger than the macroscopic field. One can also find the size of the increase from it. We must remember that the formula is not very accurate, since it depends on the crude approximation, Table 4.2 provides of relative permittivity of some substances at N.T.P.

In terms of complex refractive index, one can write Eq. (4.87) as

$$\frac{M^2(\omega)-1}{M^2(\omega)+2} = \frac{N\alpha(\omega)}{3} \tag{4.88}$$

one can separate the real and imaginary parts of complex refractive

Table 4.2. Relative permittivity of some substances (at N.T.P.)

	Gas		Liquid	
Substance	ε_r	ρ(density)	ε_r	ρ(density)
O_2	1.0005	0.0014	1.51	1.190
CS_2	1.0029	0.0034	2.64	1.293
CCl_4	1.0030	0.0030	2.24	1.590
CO_2	1.0010	0.0019	1.61	0.975

index as

$$M^2 = (\mu + i\chi)^2 \tag{4.89}$$

With the help of Eq. (4.87), one can find the atomic polarizability and molar polarizability. If M_A is the mass number of the material of dielectric, ρ is density and N_A is Avogadro's number, then

$$\text{Number of atoms per unit volume } N = \frac{\rho N_A}{M_A}$$

putting N in Eq. (4.87), one obtains

$$\frac{N_A \alpha}{3} = \frac{M_A}{\rho}\left(\frac{\varepsilon_r - 1}{\varepsilon_r + 2}\right) \tag{4.90}$$

one can obtain α (atomic polarizability) and $N_A\alpha$ (molecular polarizability) from Eq. (4.90).

4.17 POLARIZATION IN CHANGING ELECTRIC FIELDS

So far we have considered the behaviour of dielectric materials in static electric fields. Now, we consider the effects of electric fields that are varying in time, like the field in a capacitor used in an alternating-current circuit. When the frequency of the applied field is low, i.e. the field changes slowly in time the polarization **P** follows **E** at every instant and the ratio of **P** to **E** is almost constant. We must note that slow variation in electric field is a relative term and depends on the physical process involved.

In dielectrics, the response to an electric field involves two processes (i) induced polarization and (ii) Orientation of permanent dipoles. We have seen that the induced polarization is due to the distortion of electric charge distribution. The motion of electrons in atoms and molecules are characterized by very high frequencies ($\sim 10^{16}$ Hz) or very low periods $\sim 10^{-16}$ s due to their very small mass and hence the period of the applied field (from *dc* to microwave) is comparatively quite large and these fields may be considered slowly varying fields. Thus in polar substances, in which polarization plays a predominant role, the polarization keeps in step with the applied field. Thus the susceptibility χ and hence the dielectric constant $\varepsilon_r = (1+\chi)$ is independent of frequency.

The orientation of a polar molecule is a process quite different from the mere distortion of the electron cloud. The whole molecular framework has to rotate. The frictional drag tends to make the rotation lag behind the torque and to reduce the amplitude of the resulting polarization, where on the time scale this effect sets in, varies enormously from one polar substance to another. In water, the "response time" for dipole reorientation is about 10^{-11} s. The dielectric constant remains around 80 upto frequencies of the order of 10^{10} cycles/s. Above 10^{11} C/s dielectric constant falls to a modest value typical of a nonpolar liquid. The dipoles simply cannot follow so rapid an alternation f the field. In other substances, especially solids, the characteristic time can be much longer. In ice just below the freezing point, the response time for electric polarization is around 10^{-5} s. Fig. 4.21 shows experimental curves of dielectric constant versus frequency for water and ice.

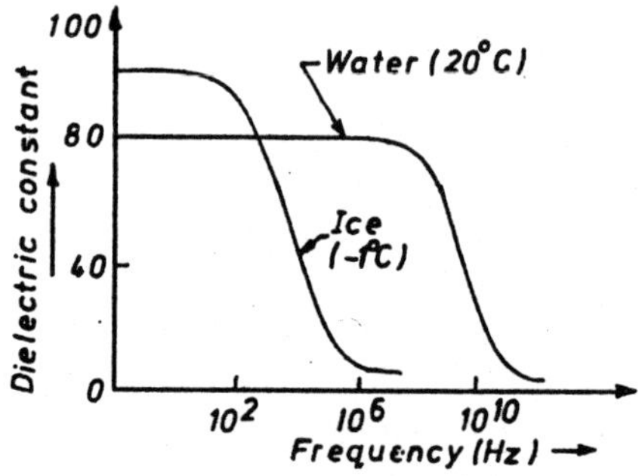

Figure 4.21.

4.18 THE BOUND CHARGE (POLARIZATION) CURRENT

The charges in the molecules of a dielectric are called *bound*. The action of a field can only cause bound charges to be displaced slightly from their equilibrium positions. They cannot leave the molecule containing them.

The charges that are within the boundaries of a dielectric, but are not inside its molecules and also charges outside a dielectric are called *extraneous* ones. Normally such charges are called *free* charges.

Bound charges differ from free charges only in that they cannot leave the confines of the molecules within which they are lying. Otherwise, they have the same properties as all other charges. In particular they are sources of an electric field. When a varying electric field is applied on a dielectric the centres of electron clouds start oscillating with the same frequency. Let the centre of electron cloud in an atom be displaced by a distance ds in time dt, then the change in dipole moment due to the displacement ds is given by

$$\mathbf{dp} = q\mathbf{ds} \tag{4.91}$$

where q is the charge of electron cloud.

If N is the number of atoms per unit volume in the dielectric, then the change in polarisation $\mathbf{P}$ in time dt is given by

$$\mathbf{dp} = N\mathbf{dp} = Nq\mathbf{ds} \tag{4.92}$$

Therefore the rate of polarization in dielectric due to changing electric field

$$\frac{\mathbf{dp}}{dt} = Nq\frac{\mathbf{ds}}{dt} = Nq\mathbf{v} \tag{4.93}$$

where $\mathbf{v} = \dfrac{ds}{dt}$ is the velocity.

A charge cloud of density Nq moving with velocity $\mathbf{v}$ produces a conduction current of density $\mathbf{J}$(C/s)/m^2 given by

$$\mathbf{J} = Nq\mathbf{v} \tag{4.94}$$

A comparison of (4.93) and (4.94) shows that current is produced due to change in polarization with respect to time, i.e. current is produced due to change in the induced bound charges on the surface of the dielectric. Such a current is source of magnetic field. This current is called bound charge current. Therefore, the bound charge current density,

$$\mathbf{J} = \frac{d\mathbf{P}}{dt} \tag{4.95}$$

The only difference between an "ordinary" conduction current density and current density $\dfrac{d\mathbf{P}}{dt}$ is that former involves free charges in motion while the latter involves bound charges in motion. For the development of magnetic field in dielectric medium three different current densities contribute: (i) Free current density $\mathbf{J}_{\text{free}}$, (ii) displacement current density due to change in electric field, $\mathbf{J}_d = \varepsilon_0 \frac{\partial \mathbf{E}}{\partial t}$ and (iii) bound charge current density produced due to change in polarisation as a result of change in electric field, $\mathbf{J}_p = \frac{d\mathbf{P}}{dt}$.

4.19 TRANSIENT BEHAVIOUR OF A R.C. CIRCUIT

The term transient is usually used to designate the result produced by making a sudden change in an electrical circuit, such as switching on or off an electromotive force, the release of the charge on a capacitor, the sudden change in the magnetic flux linking a circuit, etc. A study of the performance of a network produced by these typical changes may be effected by solving the network equations. Before discussing transient behaviour of a R.C. circuit, a brief idea of Kirchhoff's laws and their applications to any mesh (completely closed electric mesh) is essential.

Kirchhoff's Laws : The usual statement of Kirchhoff's laws is as follows:

1. The algebric sum of the currents which meet at a junction point in an electric circuit is zero.
2. The algebric sum of the electromotive forces and potential drops around any closed path or mesh of an electric circuit is zero.

With the application of Kirchhoff's second law, the electromotive forces in any mesh may be added to form the differential equation for the mesh. Alternatively, by the application of Kirchhoff's first law, the currents entering any branch point or junction in the network may be added to zero to form the differential equation for the branch point.

Now, we consider behaviour of a R.C. circuit.

Join a resistance R, Capacitor C and a Morse Key K in series with a battery of emf E as shown in Fig. 4.22. The capacitor can be charged and discharged with time by pressing or releasing the key K.

(i) Charging of a Capacitor

When Morse key K is pressed, then current flows through the resistance R and capacitor is charged. Let current I flow through the re-

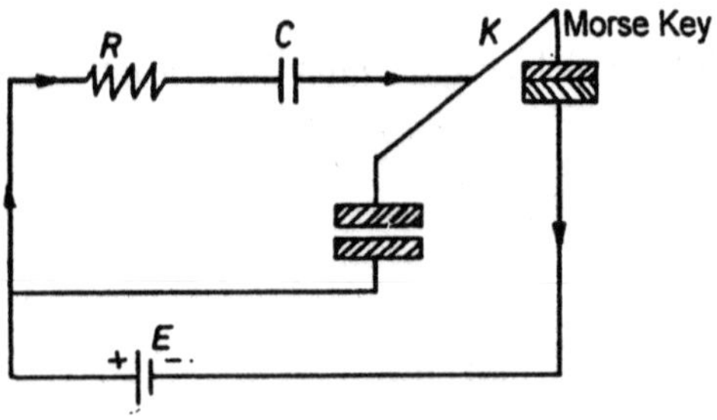

Figure 4.22.

sistance R for time t and accumulate a charge Q in the capacitor C. Due to storage of this charge, a potential difference, $V = \frac{Q}{C}$ is developed across the capacitor C and A potential difference IR across the resistance R, then at any instant for this circuit, we have

$$E = \frac{Q}{C} + IR \tag{4.96}$$

when the voltage across the capacitor becomes equal to the battery voltage E, a maximum charge Q_0. is stored in the capacitor and the current in the circuit becomes zero. Thus at $Q = Q_0$, we have

$$E = \frac{Q_0}{C} \qquad \text{and} \qquad I = 0$$

From Eq. (4.96), we have

$$\frac{Q_0}{C} = \frac{Q}{C} + IR$$

or
$$\frac{dQ}{dt} = \frac{Q_0 - Q}{RC} \qquad \left(\because I = \frac{dQ}{dt}\right)$$

or
$$\frac{dQ}{Q_0 - Q} = \frac{dt}{RC} \tag{4.96(a)}$$

Integrating the above equation, one obtains

$$\int \frac{dQ}{Q_0 - Q} = \int \frac{dt}{RC}$$

or
$$-\log_e(Q_0 - Q) = \frac{t}{RC} + A \tag{4.97}$$

where A is constant of integration. Initially, at $t = 0$, we have the charge on the capacitor is zero, i.e. $Q = 0$. Using this, Eq. (4.97) gives

$$A = -\log_e Q_0$$

$$\therefore \quad -\log_e(Q_0 - Q) = \frac{t}{RC} - \log_e Q_0$$

or $$\log_e\left(\frac{Q_0 - Q}{Q_0}\right) = -\frac{t}{RC}$$

or $$Q = Q_0(1 - e^{-\frac{t}{RC}}) \quad (4.98)$$

Eq. (4.98) represents the charging of the capacitor. The charge on the capacitor in this charging process increases exponentially with time. The product RC which controls the charging rate has the dimensions of time and is called "time constant" of the circuit.

Taking

$$CR = t, \quad \frac{t}{CR} = 1$$

$$\therefore \quad Q = Q_0(1 - e^{-1}) = 0.632 Q_0$$

Obviously, the time constant (RC) is the time taken by the condenser to get charged to 0.632 times its maximum value. Here C is measured in farads, R in Ohms and time in seconds.

Putting $RC = t$ in eq. 4.96(a), one obtains

$$\frac{dQ}{dt} = \frac{Q_0 - Q}{t}$$

or $$\frac{dQ}{dt} \propto \frac{1}{t} \quad (4.99)$$

Eq. (4.99) suggests that the rate of increase of charge is inversely proportional to t, the time constant. The circuit with large time constant will take more time in charging the capacitor to the maximum value and vice versa

Decay of Current

At the time of charging the capacitor, the current is given by

$$I = \frac{dQ}{dt}$$

$$= \frac{Q_0}{RC}\left(e^{-\frac{t}{RC}}\right) \quad \left[\because \; Q = Q_0\left(1 - e^{-\frac{t}{RC}}\right)\right]$$

$$\therefore \quad I = I_0 e^{-\frac{t}{RC}} \quad \left(I_0 = \frac{Q_0}{RC} \text{ is the maximum current which flows at } t = 0\right) \quad (4.100)$$

Obviously, the current decreases exponentially during charging and in

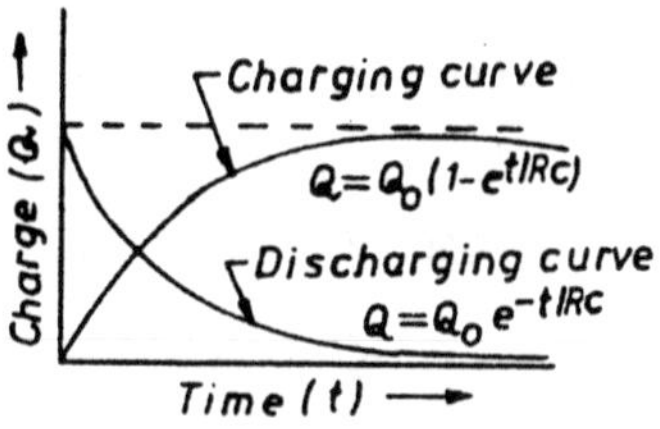

Figure 4.23.

time $t = RC$, the current reduces to $\frac{1}{e}$ times, i.e. 37% of the maximum current, I_0.

(ii) Discharging of a condenser

When a maximum charge Q_0. is accumulated in the capacitor, then if key K is released, the battery gets disconnected and the capacitor starts discharging through the resistance R, (Fig. 4.24). Obviously, during discharge, there is no source of e.m.f. in the circuit, we have

$$\frac{Q}{C}+RI = 0 \tag{4.101}$$

or
$$\frac{Q}{C}+R\frac{dQ}{dt} = 0 \qquad \left(\because I=\frac{dQ}{dt}\right)$$

$$\therefore \quad \frac{dQ}{Q} = -\frac{dt}{RC}$$

Integrating, one obtains

$$\log_e Q = -\frac{t}{RC}+B \tag{4.102}$$

where B is constant of integration. Initially, i.e. at time $t=0$, $Q=Q_0$,

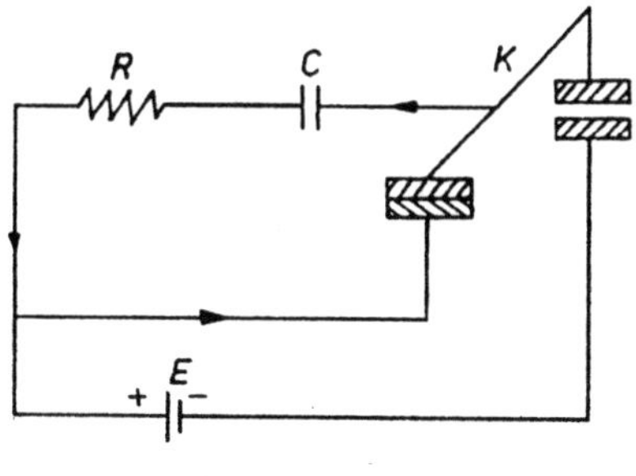

Figure 4.24. Discharging of a condenser.

we have

$$\log_e Q_0 = B$$

$$\therefore \quad \log_e Q = -\frac{t}{RC} + \log_e Q_0$$

or

$$\log_e \frac{Q}{Q_0} = -\frac{t}{RC}$$

$$\therefore \quad Q = Q_0 e^{-\frac{t}{RC}} \tag{4.103}$$

Eq. (4.103) shows that during discharging the capacitor, the charge on the capacitor decreases exponentially with time (Fig. 4.23).

Time Constant: If $t = RC$, then from Eq. (4.103)

$$Q = \frac{Q_0}{C} = 0.37Q_0$$

Obviously, *the time in which 63% of the capacitor decays or 37% remains on the capacitor is called the time constant of the circuit. In other words, the time in which the charge on the capacitor reduces to* $\frac{1}{e}$ *of its initial value* (Q_0) *is called the time constant.*

Decay constant: While discharging the capacitor, the current in the circuit

$$I = \frac{dQ}{dt} = -\frac{Q_0}{RC} e^{-\frac{t}{RC}}$$

$$\therefore \quad I = -I_0 e^{-\frac{t}{RC}} \tag{4.104}$$

At $t = 0$, the current is maximum but in opposite direction to the current at the start of charging the capacitor (Fig. 4.25). Further, again the current decreases in magnitude, exponentially with time.

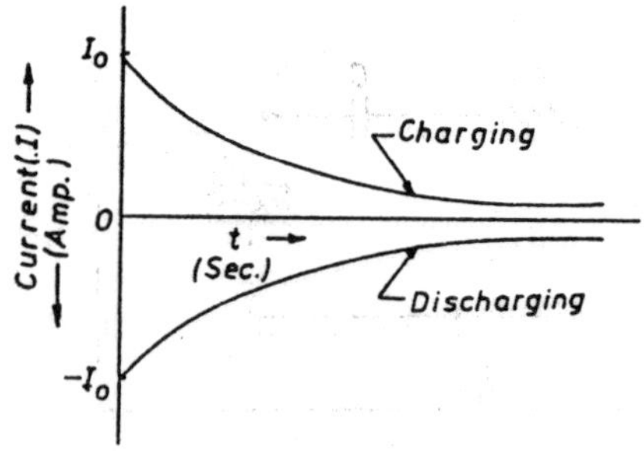

Figure 4.25.

Energy Stored in the capacitor during charging: We have seen that when an e.m.f. E is introduced suddenly into a RC circuit, the capacitor starts charging. At any time t, if the charge on the condenser is Q, then potential difference between the plates at that time will be $\frac{Q}{C}$. This potential difference will be opposed in polarity to the supply e.m.f.. Hence the current flows against the potential difference between the plates of the condenser and does external work. This work done to charge the condenser is stored between the plates of the condenser in the form of electrical potential energy U.

To obtain the qualitative expression for the storage of energy, consider an extra increment of charge dQ, the small amount of additional work needed will be

$$dU = \frac{Q}{C}dQ$$

Obviously, this work done will increase the potential energy of the system by $UdQ\left(=\frac{Q}{C}dQ\right)$. If the process is continued until a total charge Q_0 has been transferred. The total work done will be

$$U = \int dU = \int_0^{Q_0} \frac{Q}{C}dQ = \frac{1}{2}\frac{Q_0^2}{C} = \frac{1}{2}CE^2 \tag{4.105}$$

obviously, Eq. (4.105) gives the total energy of the charged condenser.

The Dissipation of Energy during charging of a Condenser through a Resistance: Let us examine the energy dissipated in the resistance R during the charging of a condenser. We have

$$Q = Q_0\left(1 - e^{-\frac{t}{RC}}\right)$$

$$I = \frac{dQ}{dt} = \frac{Q_0}{RC}\exp\left(-\frac{t}{RC}\right)$$

If δW is the energy dissipated in time δt, then

$$\delta W = I^2 R\delta t = \frac{Q_0^2 R}{C^2R^2}\exp\left(-\frac{t}{RC}\right)$$

Integrating the above, one obtains the total energy dissipated during charging when the condenser has charged to its maximum value.

$$W \doteq \int dW = \frac{Q_0^2}{RC}\int_0^{\infty}\exp\left(-\frac{2t}{RC}\right)dt$$

$$= \frac{1}{2}\frac{Q_0}{C} = \frac{1}{2}Q_0E \qquad (4.106)$$

which is independent of R. The energy supplied by the D.C. source is Q_0E, that stored in the condenser is $\frac{1}{2}Q_0E$. Obviously, the energy supplied by the source is accounted for.

4.20 SOLID DIELECTRICS (ELECTRETS)

One can prepare such materials in solid state which have a permanent polarization and exists even when a field is not applied, e.g. if some wax is melted and a strong electric field is applied to it when it is still in the liquid form, the dipoles get partly lined up and remain that way when the liquid freezes. The solid material thus formed has, like a magent, a permanent dipole moment. These materials are called solid dielectrics or electrets. Barium titanate is the most familiar example of electret. An electret, however, soon gets discharged, as it attracts free charges from the air.

It is also reported that a permanent internal polarization also exist in same complicated crystals. However, this is not noticed because the external fields are discharged, as in electrets.

One can work out a theory of the dielectric constant for crystals that do not have permanent dipole moment, on the same lines as for non-polar liquids.

EXAMPLE 4.1: Calculate the induced dipole moment per unit volume (density of polarization) of helium gas if placed in a field of 6×10^5 volt/m. The atomic polarizability of helium is 0.21×10^{-23} m^3 and density of helium is 2.6×10^{25} atoms/m^3. Also calculate the separation between centres of positive and negative charges in each atom.

Solution: The dipole moment induced in each atom is

$$p = \alpha E \qquad \left[\alpha = 0.21 \times 10^{-30}\ \text{m}^3,\ \frac{1}{4\pi\varepsilon_0 = 9 \times 10^9}\right]$$

$$= \frac{0.21 \times 10^{-30} \times 6 \times 10^5}{9 \times 10^9} = 1.4 \times 10^{-35}\ \text{C-m}$$

The dipole moment per unit volume is

$$P = Np = 2.6 \times 10^{25} \times 1.4 \times 10^{-35} = 3.64 \times 10^{-10}\ \text{C/m}^2$$

The atoms of helium contains two electrons. The separation between the centres of positive and negative charges in each helium atom is

$$d = \frac{p}{q} = \frac{p}{2e} = \frac{1.4 \times 10^{-35}}{2 \times 1.6 \times 10^{-19}} \text{ m} = 4.37 \times 10^{-17} \text{ m}$$

EXAMPLE 4.2: Assuming the negative charge in a hydrogen atom to be uniformly distributed in a spherical volume of radius 0.5×10^{10} m around the proton. Calculate the separation between its positive and negative charges when kept in external field of strength 3×10^6 V/m.

Solution: The separation is given by

$$d = \frac{4\pi\varepsilon_0 E a^3}{e} = \frac{1}{9 \times 10^9} \times \frac{3 \times 10^6 \times (0.5 \times 10^{-10})^3}{1.6 \times 10^{-19}}$$

EXAMPLE 4.3: An atom of oxygen on being polarized produces a dipole moment of 5×10^{-35} C-m. If the distance of the centre of negative charge cloud from the nucleus be 4×10^{-17} m, calculate the polarizability of oxygen atom.

Solution: The polarizing field E tries to push the negative cloud of 8 electrons away from the nucleus but the positive charges on the nucleus attract it. In the equilibrium state the two forces balance each other. If d be the equilibrium distance between positive and negative charges, then

$$8eE = \frac{1}{4\pi\varepsilon_0}\frac{8e \times Be}{d^2} \quad \text{or} \quad E = \frac{1}{4\pi\varepsilon_0}\frac{Be}{d^2}$$

$$\because \quad p = \alpha E$$

$$\therefore \quad \alpha = \frac{p}{E} = \frac{p \times 4\pi\varepsilon_0 d^2}{8e} = \frac{5 \times 10^{-35} \times (4 \times 10^{-17})^2}{8 \times 1.6 \times 10^{-19}} \text{ m}^3$$

$$= 12.5 \times 10^{-50} \text{ m}^3$$

EXAMPLE 4.4: The dielectric constant of a medium is 1.0005. Calculate its atomic and molecular polarizability. Given, Avogadro number $N_A = 6 \times 10^{23}$ per gm-mole and molar volume = 22.4 litre/mole.

Solution: From Clausius-Mossotti's relation

$$\frac{N_A \alpha}{3\varepsilon_0} = \frac{\varepsilon_{r-1}}{\varepsilon_{r+2}} V_A$$

$\varepsilon_r = 1.005$
$N_A = 6 \times 10^{23}$ atoms/mole
$V_A = 22.4$ litre/mole

Molar polarizability

$$\alpha_m = N_A\alpha = \frac{5\times10^{-4}\times22.4\times10^{-3}\times3\times8.85\times10^{-12}}{3.0005}$$
$$= 9.9\times10^{-18}\text{ farad-m}^2/\text{mole}$$

Atomic polarizability

$$\alpha_m = \frac{\alpha_m}{N_A} = \frac{9.9\times10^{-18}}{6\times10^{23}} = 1.65\times10^{-41}\text{ farad-m}^2$$

EXAMPLE 4.5: Calculate the dipole moment of a water molecule assuming that all the ten electrons in the molecule circulate symmetrically about the oxygen atom and the OH distance is 0.96*A* and the angle between OH bonds is 104°.

Solution: The effective centre of protons of Hydrogen atoms lies at the middle of the line joining the H-atoms (Fig. 4.26). Obviously, a water molecule can be assumed as a dipole of charge $2e$ and a separation distance 'a'. Thus the dipole moment of H_2O molecule.

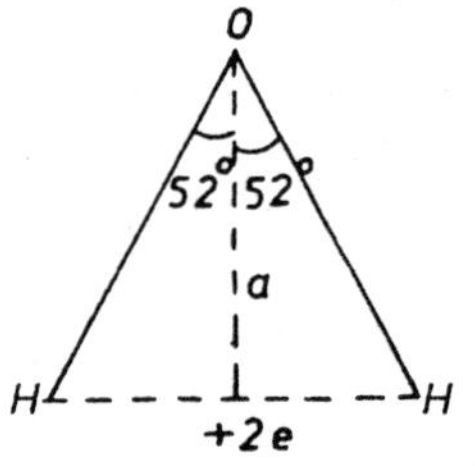

Figure 4.26.

Given $e = 1.6\times10^{-19}$ C ;
OH $= 0.96\times10^{-10}$ m ; $\theta = 52°$;
$\cos 52° = 0.616$
we have,

$$\begin{aligned} p &= (2e)a = (2e)\,(\text{OH})\cos\theta \\ &= 2\times1.6\times10^{-19}\times0.96\times10^{-19}\times\cos 52° \\ &= 2\times1.6\times10^{-19}\times0.96\times10^{-19}\times0.616 \\ &= 1.9\times10^{-29}\text{ C-m} \end{aligned}$$

EXAMPLE 4.6: A dielectric sphere is placed in a uniform electric field E_0. Polarization is uniform in the sphere. The field inside the sphere is 1/3 of the applied field. Determine dielectric constant and electrical susceptibility of the material.

Solution: In a dielectric sphere, we have

$$E_{in} = \frac{3}{2+\varepsilon_r}E_0 \qquad\qquad E_{in} = \frac{E_0}{3}$$

$$\therefore \quad \frac{E_0}{3} = \frac{3}{2+\varepsilon_r} E_0$$

or $$2+\varepsilon_r = 9 \quad \therefore \quad \varepsilon_r = 7$$

and $$\chi = \varepsilon_{r-1} = 7 - 1 = 6$$

EXAMPLE 4.7: An artificial dielectric of a large number of metal spheres of macroscopic size arranged in a three dimensional (lattice) structure as shown in Fig. 4.27. Show that the permittivity of the dielectric $= 1 + 3VN$, where V is the volume of the metal sphere and N number of spheres per unit volume.

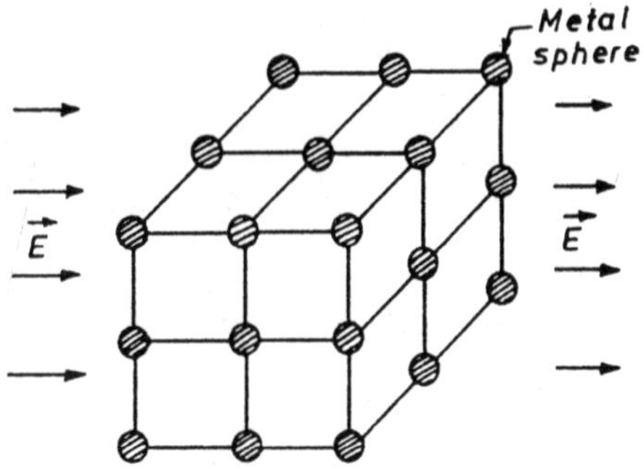

Figure 4.27.

Solution: Let **E** be a uniform field applied to the dielectric. This field will induce charges on the individual metallic spheres. (Fig. 4.28). Obviously, each sphere becomes analogous to a polarized atom and may be represented by a dipole moment $\mathbf{p} = q\mathbf{l}$ (Fig. 4.29). If N be the number of spheres per unit volume, then the polarization (P) is obtained as

$$\mathbf{P} = Nq\mathbf{l}$$

we have

$$\mathbf{D} = \varepsilon\mathbf{E} = \varepsilon_0\mathbf{E} + \mathbf{P}$$

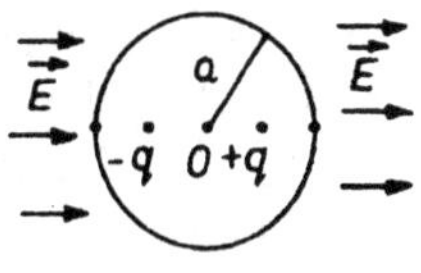

Figure 4.28.

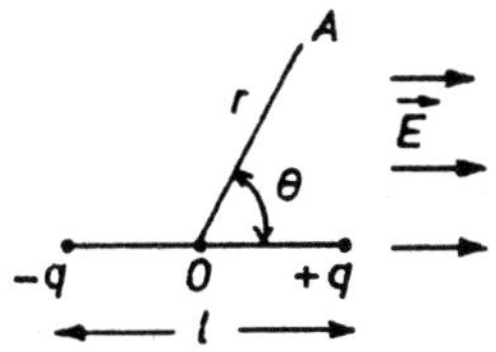

Figure 4.29.

$$\therefore \qquad \varepsilon = \varepsilon_0 + \frac{P}{E} = \varepsilon_0 + \frac{Nql}{E} \qquad \text{(i)}$$

The potential at a point due to the uniform field is

$$V = V_0 + V_d = -Er\cos\theta + \frac{ql\cos\theta}{4\pi\varepsilon_0 r^2} \qquad \text{(ii)}$$

Here r denotes the distance of the point from the origin which, for convenience is taken at the centre of the dipole and θ is the angle between the axis of the dipole and the radial line (Fig. 4.29). Now, the potential due to the dipole,

$$V_d = \frac{p\cos\theta}{4\pi\varepsilon_0 r^2} = \frac{ql\cos\theta}{4\pi\varepsilon_0 r^2} \qquad \text{(iii)}$$

The total potential at the point is

$$V = V_0 + V_d = -Er\cos\theta + \frac{ql\cos\theta}{4\pi\varepsilon_0 r^2} \qquad \text{(iv)}$$

We know that the metal sphere has equal amount of positive and negative charges on it and hence its potential is zero. If a is the radius of the sphere, then

$$0 = -Ea\cos\theta + \frac{ql\cos\theta}{4\pi\varepsilon_0 a^2} \qquad \text{(v)}$$

$$\therefore \qquad \frac{ql}{E} = 4\pi\varepsilon_0 a^3 \qquad \text{(vi)}$$

Using (vi), eq. (i) becomes

$$\varepsilon = \varepsilon_0 + 4\pi\varepsilon_0 N a^3$$

and

$$\varepsilon_r = \frac{\varepsilon}{\varepsilon_0} = 1 + 4\pi N a^3 = 1 + 3VN \qquad \text{(vii)}$$

where $V = \frac{4}{3}\pi a^3$ is the volume of the sphere. Relation (vii) shows that the permittivity (ε_r) of the dielectric under consideration depends on both the number of spheres and their sizes. Such dielectrics are called artificial dielectrics consisting of discrete metal particles of macroscopic size.

EXAMPLE 4.8: A dielectric slab of flat surface with relative permittivity 3 is disposed to a uniform field with its surface normal to the field. The slab has a flux density of 1.5 C/m^2 and occupies a volume of 0.8 m^3 and uniformly polarized. Determine (i) the polarization in the slab and (ii) the total dipole moment of the slab.

Solution: We have

$$\mathbf{D} = \varepsilon\mathbf{E} = \varepsilon_0\varepsilon_r\mathbf{E}$$

$$\therefore \quad |\mathbf{E}| = \frac{|\mathbf{D}|}{\varepsilon_0\varepsilon_r} = \frac{1.5}{3\varepsilon_0}$$

Now using

$$\mathbf{D} = \varepsilon_0\mathbf{E}+\mathbf{P}$$

or

$$\mathbf{P} = \mathbf{D}-\varepsilon_0\mathbf{E}$$

$$= \varepsilon_0\varepsilon_r\mathbf{E}-\varepsilon_0\mathbf{E} = \varepsilon_0(\varepsilon_r-1)\mathbf{E}$$

$$\therefore \quad P = 0.5(3-1) = 0.5\times 2 = 1 \text{ C/m}^2$$

$$\text{Total dipole moment} = \text{Polarization} \times \text{Volume}$$

$$= 1\times 0.08\times 0.08 \text{ C-m}$$

EXAMPLE 4.9: Show that at a change of dielectric medium field lines are refracted according to Snell's law $\frac{\varepsilon_1}{\varepsilon_2} = \frac{\tan\alpha_1}{\tan\alpha_2}$, where α_1 and α_2 are the angles between the directions of the field and common normal to the boundary (Fig. 4.30) and ε_1 and ε_2 are the relative permittivities of the two media.

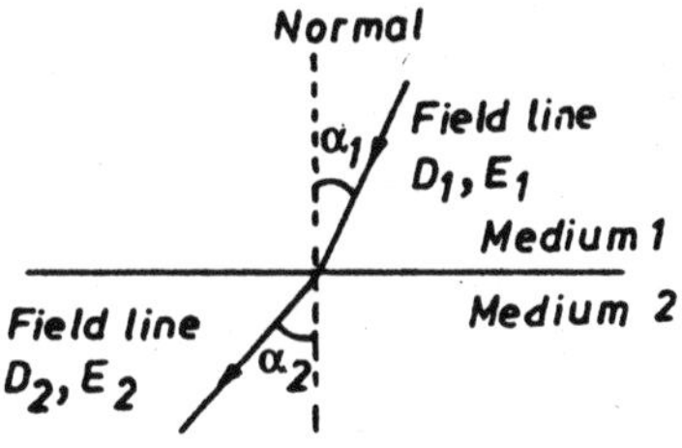

Figure 4.30.

Solution: By applying the first boundary condition, we have

$$\mathbf{D}_1\cdot\hat{a}_n = \mathbf{D}_2\cdot\hat{a}_n$$

or

$$D_1\cos\alpha_1 = D_2\cos\alpha_2$$

or

$$\varepsilon_1 E_1\cos\alpha_1 = \varepsilon_2 E_2\cos\alpha_2 \tag{1}$$

The second boundary condition gives

$$E_1\sin\alpha_1 = E_2\sin\alpha_2 \tag{2}$$

Dividing (2) by (1), one obtains

$$\frac{\varepsilon_1}{\varepsilon_2} = \frac{\tan\alpha_1}{\tan\alpha_2}$$

EXAMPLE 4.10: Find the angle of incidence of an electric field on the boundary between a conductor and a dielectric.

Solution: Medium 2 is a conductor and hence $D_2 = \varepsilon_2 E_2 = 0$ under static conditions. According to the boundary conditions, we have

$$D_{n_1} = \rho_s$$

or

$$E_1 E_{n_1} = \rho_s$$

and

$$E_{t_1} = = E_{t_2} = 0$$

$$\therefore \quad \alpha_1 = \tan^{-1}\left(\frac{E_{t_1}}{E_{t_2}}\right) = \tan^{-1}(0) = 0$$

obviously, a static electric field line or flux tube at a dielectric - conductor boundary is always perpendicular to the conductor surface (where no currents are present). This result is of fundamental importance in field mapping.

EXAMPLE 4.11: A water drop of volume 30 cubic mm is polarized uniformly. Calculate the maximum field strength due to the polarized drop at a distance of 10 cm. from its centre and polarization density. The dipole moment of water molecule is 6.1×10^{-30} C-m.

Solution: A uniformly polarized sphere is equivalent to a dipole of dipole moment **P** times the volume of the sphere for external points. The molecular weight of the water is 18 and Avogadro number is 6×10^{23}. Therefore the number of molecules in 1m^3 of water is = $\frac{6 \times 10^{23} \times 10^6}{18} = \frac{10^{29}}{3}$.

$$\text{Polarization density } P = \rho N = 6.1 \times 10^{-30} \times \frac{10^{29}}{3} = 2.03 \times 10^{-1}$$

$$= 0.203 \text{ C/m}^2$$

The dipole moment of the drop,

$$p = P \times \text{ volume}$$

$$= 0.203 \times 30 \times 10^{-9} \text{ C-m}$$

$$= 6.09 \times 10^{-9} \text{ C-m}$$

One can obtain the field components at any point (r, θ) due to a short dipole from

$$E_r = \frac{1}{4\pi\varepsilon_0}\left(\frac{2p\cos\theta}{r}\right) \quad \text{and} \quad E_\theta = \frac{1}{4\pi\varepsilon_0}\left(\frac{p\sin\theta}{r^3}\right)$$

The maximum field corresponds to $\theta = 0$ and is given by

$$E_{\max} = \frac{1}{4\pi\varepsilon_0}\frac{2p}{r^3} = \frac{9\times 10^9 \times 2 \times 6.09 \times 10^{-9}}{(-1)^3}[r = 0.1\text{m}]$$

$$= 1.09 \times 10^5 \text{ N/C.}$$

EXAMPLE 4.12: A slab of dielectric material with dielectric constant ε_r, thickness b and area a is to be kept in between two parallel plates of area A. The distance between the plates is d. Initially a potential difference of V_0 Volts is applied between two ends of the plates without dielectric. Determine (i) Capacity C_0 of the capacitor without dielectric (ii) Free charge on the plates Q_0, (iii) Electric field intensity in free space (E_0), (iv) Electric field in dielectric and (v) potential difference between the plates (V).

Solution:

(i) We have the electric field between the plates of the capacitor

$$E_0 = \frac{V_0}{d} = \frac{\sigma}{\varepsilon_0} = \frac{Q_0}{A\varepsilon_0}$$

$$\therefore \quad C_0 = \frac{Q_0}{V_0} = \frac{\varepsilon_0 A}{d}$$

(ii) Free charge on the plates, Q_0 is

$$Q_0 = \frac{A\varepsilon_0 V_0}{d}$$

(iii) Electric field intensity in the free space

$$E_0 = \frac{V_0}{d}$$

(iv) Electric field in the dielectric

$$E = \frac{E_0}{\varepsilon_r} = \frac{V_0}{\varepsilon_r d}$$

(v) Potential difference between the plates of the capacitor with dielectric

$$\begin{aligned} V &= E_0(d-b)+Eb \\ &= \frac{V_0}{d}(d-b)+\frac{V_0 b}{\varepsilon_r d} \\ &= \frac{V_0}{d}\left[(d-b)+\frac{b}{\varepsilon_r}\right] \end{aligned}$$

EXAMPLE 4.13: A coaxial cable contains two materials of different relative permittivities ε_1 and ε_2 [Fig 4.31]. The radii of inner and outer conductors are r_1 and r_2 respectively and the radius of the interface between the dielectric is r. Considering that both dielectric materials breakdown in an electric field greater than E_{r_2}, find the value of r which will allow the cable to carry the maximum possible voltage. Find this voltage.

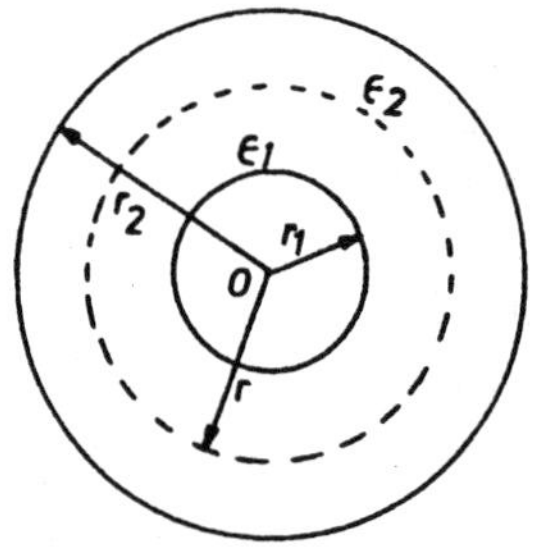

Figure 4.31.

Solution: We know that when the inner dielectric is on the point of breakdown the field on the surface of the inner conductor must be E_{r_2}. One can easily verify that the field $E_1(a)$ at a point in the region $r_1 < a < r_2$ is

$$E_1(a) = E_{r_2}\frac{r_1}{a}$$

According to boundary relations, we must see that the perpendicular component of the electric displacement be continuous at r_2. The field in the outer region, therefore, has to satisfy the condition

$$\varepsilon_2 E_2(r_2) = \varepsilon_1 E_1(r_2) = \varepsilon_1 E_r \frac{r_1}{r_2}$$

we know that the highest voltage occurs when both dielectrics are on the point of breakdown. Thus

$$\begin{aligned} E_2(r_2) &= E_r\frac{\varepsilon_1 r_1}{\varepsilon_2 r_2} = E_r \\ \therefore \quad r_2 &= \frac{\varepsilon_1}{\varepsilon_2} r_1 \end{aligned}$$

The field at a point 'a' in the outer region is

$$E_2(a) = E_r\frac{r_2}{a}$$

$$\therefore \quad V = -\int_{r_1}^{a} \mathbf{E}\cdot d\mathbf{a} = -\int_{r}^{r_2} E_r\frac{r_2}{r}da - \int_{r_2}^{r_1} E_r\frac{r_1}{a}da$$

$$= E_r\left[r_2\log\frac{r}{r_2} + r_1\log\frac{r_2}{r_1}\right]$$

EXAMPLE 4.14: Using the following data, calculate the individual dipole moment **p** of a carbon tetrachloride ($CC1_4$) molecule. Also calculate the average eectron displacement. Relative permittivity, $(\varepsilon_r) =$ 2.24, molecular weight = 156, density = 1.6 gm/cm^3 and field strength = 10^7 V/m.

Solution:

$$\text{Molecular Density}(N) = \frac{\text{Avogadro's number}}{\text{Molecular weight}} \times \text{Density}$$

$$= \frac{6.02\times10^{23}}{156}\times 1.6 = 6.17\times10^{21}\ \text{molecules/cm}^3$$

$$= 6.17\times10^{27}\ \text{molecules/cm}^3$$

$$\mathbf{p} = \frac{P}{N} = \frac{\varepsilon_0\chi E}{N} = \frac{\varepsilon_0(\varepsilon_r-1)E}{N} \qquad (\because \varepsilon_r = 1+\chi)$$

$$= \frac{8.85\times10^{-12}\times1.24\times10^{7}}{6.17\times10^{27}} = 1.77\times10^{-32}\ \text{C-m}$$

There are 74 electrons in CCl_4 and if d be average electron displacement, we have

$$p = 74de$$

$$\therefore \quad d = \frac{p}{74e} = \frac{1.77\times10^{-32}}{74\times1.6\times10^{-19}} = 1.5\times10^{-15}\ \text{m}$$

EXAMPLE 4.15: Derive an expression for the induced dipole moment of an atom by assuming a simple classical model of atom. Show that the polarizability is independent of the field.

Solution: According to classical model of the atom, a point positive charge (Ze) called nucleus, is surrounded by a spherically symmetric cloud of negative charge (–Ze) in which the density is uniform upto the atomic radius (r_0) and zero at larger radii.

When an atom is placed in an electric field **E**, the nucleus is displaced in the direction **E** by a certain distance, say d. Obviously, the force, acting on the nucleus in the direction of **E** = Ze **E**. We know that an electrostatic force between the nucleus and the charge cloud will tend to restore the initial configuration. The negative charge attracting the nucleus is, by Gauss's law, the part of the cloud of negative charge within the sphere of radius d and this charge is

$$\frac{4}{3}\pi d^3 \rho = \frac{4}{3}\pi d^3 \frac{Ze}{\frac{4}{3}\pi r_0^3} = \frac{Zed^3}{r_0^3}$$

where ρ is the density of charge.

$$\therefore \quad \frac{(Ze)\left(\frac{Zed^3}{r_0^3}\right)}{4\pi\varepsilon_0 d^2} = ZeE$$

or
$$Ze\mathbf{d} = 4\pi\varepsilon_0 r_0^3 \mathbf{E}$$

$$\therefore \quad \mathbf{p}\text{ (atomic dipole moment)} = Ze\mathbf{d} = 4\pi\varepsilon_0 r_0^3\mathbf{E}$$

or
$$\mathbf{p} = \alpha\varepsilon_0\mathbf{E} \qquad (\text{where } \alpha = 4\pi r_0^3\mathbf{E})$$

Obviously, the polarizability is independent of the field. The dielectrics which satisfy this condition is said to be a *linear dielectric*.

EXAMPLE 4.16: A capacitor of capacity $1\mu f$ is charged to 200 volts. If the charge is leaked through a resistance of 105Ω, calculate the time in which the potential difference will become 100 volts.

Solution: The amount of charge on the condenser after time t during the process of discharging is

$$Q = Q_0 e^{-\frac{t}{RC}}$$

$$\therefore \quad \frac{Q}{Q_0} = \frac{CV}{CV_0} = e^{-\frac{t}{RC}}$$

$$\text{or} \quad \frac{V}{V_0} = e^{-\frac{t}{RC}}$$

$$\text{or} \quad \frac{100}{200} = e^{-t}$$

$$\therefore \quad t = \log e^2 = 0.693\text{s}$$

$V = 100$ Volts;

$V_0 = 200$ Volts

$C = 10^{-6} f$;

$R = 10^6\Omega$

$\therefore \quad RC = 10^{+6} \times 10^{-6} = 1$

EXAMPLE 4.17: A condenser of capacity $0.5\,\mu f$ is discharged through a resistance of 10 mega Ohms. Find the time taken for half the charge on the condenser to escape. Given $\log_e^2 = 0.6931$

Solution: We have

$$\begin{aligned} Q &= Q_0 e^{-\frac{t}{RC}} \\ \text{or} \quad \frac{Q}{Q_0} &= e^{-\frac{t}{RC}} \\ \text{or} \quad \frac{1}{2} &= e^{-\frac{t}{RC}} \\ \therefore \quad e^{\frac{t}{RC}} &= 2 \\ \text{or} \quad \frac{t}{RC} &= \log_e^2 = 0.6931 \end{aligned}$$

$$\begin{aligned} \frac{Q}{Q_0} &= \frac{1}{2}; \\ C &= 0.5\mu f = 0.5 \times 10^{-6} f; \\ R &= 10^7 \Omega \end{aligned}$$

$$\therefore \quad t = 0.6931 \times 10^7 \times 0.5 \times 10^{-6} = 3.4655 \text{ s}$$

EXAMPLE 4.18: After how much time constant, will the energy stored in the condenser in a $R-C$ circuit reach one-half its equilibrium value?

Solution: The energy stored in the condenser at any time t when the charge on it is Q, is given by $U = \frac{1}{2}\frac{Q^2}{C}$. We have

$$Q = Q_0 \left(1 - e^{-\frac{t}{RC}}\right)$$

Here $\frac{Q_0^2}{2C}$ is the equilibrium energy U_∞,

$$\therefore \quad U = U_\infty \left[1 - \exp\left(-\frac{t}{RC}\right)\right]^2$$

From the given data $U = \frac{1}{2} U_\infty$, one obtains

$$\begin{aligned} \frac{1}{2} &= \left[1 - \exp\left(\frac{t}{RC}\right)\right] \\ \therefore \quad t &= 1.22RC = 1.22 \text{ times time constant.} \end{aligned}$$

EXAMPLE 4.19: A condenser is charged to a certain potential through a resistance of 2 megaohms. If it reaches $3/4^{\text{th}}$ of its final potential in 0.5 s, calculate its capacity. Given $\log_e^2 = 0.6391$.

Solution: We have

$$\begin{aligned} Q &= Q_0 \left(1 - e^{-\frac{t}{RC}}\right) \\ \text{or} \qquad CV &= CV_0 \left(e^{-\frac{t}{RC}}\right) \end{aligned}$$

$$\text{or} \qquad \frac{V}{V_0} = 1 - e^{-\frac{t}{RC}}$$

$$\therefore \qquad \frac{3}{4} = 1 - \exp\left(-\frac{0.5}{2 \times 10^6}\right)$$

$$\text{or} \qquad \frac{1}{4 \times 10^6} = 2\log_e^2 = 2 \times 0.6931$$

$$\therefore \qquad C = 0.18\mu f$$

EXAMPLE 4.20: After how many time constants, the charge on a capacitor will be 99% of the value at equilibrium state.

Solution: We have

$$Q = Q_0\left(1 - e^{-\frac{t}{RC}}\right) \quad \Bigg| \quad \frac{Q}{Q_0} = \frac{99}{100}$$

$$\therefore \qquad \frac{99}{100} = 1 - e^{-\frac{t}{RC}}$$

$$\text{or} \quad t = RC\log_e 100$$

$$= 4.6052RC$$

Thus after 4.6052 time constants, the charge will be 99% of the equilibrium state value.

QUESTIONS

1. Define dielectric polarization. What relation it has with electric field?
2. Find relations between **P**, **E** and **D**. Explain the physical significance of these vectors. What is meant by electric susceptibility?
3. What do you mean by the capacitance of a Capacitor? On what factors does it depend? show that the energy density of electric field in the region between the plates of a parallel plate capacitor is given by $U = \frac{1}{2}\varepsilon E^2$.
4. Derive Clausius-Mossotti relation. Discuss its limitations.
5. What do you mean by Polar molecules? How does polarization of such molecules depend upon temperature?
6. What are polar and non-polar dielectrics? What happens when a gaseous dielectric is subjected to an electrostatic field?
7. Derive an expression for the interaction energy of two dipoles.

8. Derive the general form of Gauss's law in the presence of a dielectric.
9. Explain clearly what do you understand by polarization in changing fields.
10. Explain what do you understand by transient behaviour of a $R-C$ circuit. Discuss clearly charging and discharging of a condenser through resistance.
11. What is time constant of a $C-R$ circuit? Show graphically the variation of charge and discharge with time.
12. State Gauss's theorem and prove that the divergence of electrical displacement in a dielectric is equal to the free charge density.
13. A dielectric sphere is placed in a uniform electric field of intensity E_0. Prove that the polarizability (P) produced is given by

$$P = 3E_0 \left(\frac{\varepsilon_r - 1}{\varepsilon_r + 2}\right) \varepsilon_0$$

where ε_r is relative permittivity and ε_0 is the permittivity of free space.

14. Define electrical susceptibility (χ) and atomic polarizability (α). Prove that $\chi = \dfrac{n\alpha}{1 - \dfrac{n\alpha}{3}}$, where n is the number of atoms per unit volume of the medium.
15. Define atomic polarizability. Show that if an empty parallel plate capacitor C_0 is filled with a dielectric its capacity changes to $C = \varepsilon_r C_0$, where ε_r is the dielectric constant of the material filled.
16. If $\mathbf{E}$ is the intensity of electric field applied on a homogeneous and isotropic dielectric, prove that the intensity of electric field acting on an atom of the dielectric is $\mathbf{E}_{\text{eff}} = \mathbf{E} + \dfrac{\mathbf{P}}{3\varepsilon_0}$, where $\mathbf{P}$ is polarizability of the dielectric.
17. Consider a dielectric sphere in a uniform electric field E_0. Prove that the electric field inside the sphere is given $E = \dfrac{3}{2+\varepsilon_r} E_0$, where ε_r is dielectric constant of the sphere.
18. Derive the relations for current during charging and discharging of a capacitor in an $R-C$ circuit.
19. Assuming that the condenser is fully charged in an $R-C$ circuit, now the source of emf is disconnected and condenser allowed to

discharge through the resistance. Derive a relationship for the charge q on the condenser and time t and also plot a graph for q versus t.

20. Write short notes:
    (i) Dielectrics
    (ii) Gauss's law in dielectrics
    (iii) Bound charge current
    (iv) Polar and Non-polar molecules.

PROBLEMS

1. The electric susceptibility of hydrogen at S.T.P. is 5×10^{-4}. Calculate dipole moment of hydrogen when an electric field of 10^5 Volt/m is applied on it. [**Ans.** 1.65×10^{-35} C-m]
2. Using the following given data, calculate the induced dipole moment and atomic polarizability when a uniform field of 6×10^5 is applied on a sample of helium. Given $\chi = 6 \times 10^{-5}$, $\varepsilon_0 = 8.85 \times 10^{-12}$ and Avogadro number $N = 6 \times 10^{23}$ molecules.
[**Ans.** (i) 1.19×10^{-35} C-m (*ii*) 1.98×10^{-41} f - m^2]
3. Calculate the energy of interaction between the electron cloud of a hydrogen atoms and the proton (which is the nucleus of the hydrogen atom). The charge density inside the cloud is $\rho = \frac{e}{\pi a^3} \exp\left(-\frac{2r}{a}\right)$, where a is the Bohr radius.
$$\left[\textbf{Ans.} U = -\frac{e^2}{4\pi\varepsilon_0 a}\right]$$
4. Two large parallel plates each of area A, are separated by a distance d, their potentials being O and V respectively. A third plate currying a charge q is introduced midway between them. Show that the potential of this plate is $\left(V + \frac{qd}{\varepsilon_0 A}\right)/2$.
5. A parallel plate capacitor has a plate separation t. The capacitance with air only between the plates is C. When a slab of thickness t, and relative permittivity ε_r is placed on one of the plates, the capacitance is C_1. Show that
$$C_1 = C\frac{\varepsilon_r t}{[t + \varepsilon_r(t - t_1)]}$$

6. Find the dipole moment induced in each atom when a gas of helium is placed in an electric field, given that the dielectric constant of helium at 0°C and 1 atm. pressure is 1.000074.

7. The electric field strength in a mass of porcelain ($\varepsilon_{r_1} = 6$) in air is 1000 Volt/cm. At the inner surface of the porcelain the field makes an angle of 45° to the normal and emerges into the air. Show that the angle of emergence is 9.5° and the strength of the external field is 4300 Volts/cm.

8. A spherical condenser consists of two concentric spheres of radii a and $r (a < r)$. A spherical shell of permittivity E, bounded by spheres of radii b and c lies between them, with its centre coinciding with that of the spheres. Find the capacity (C) of the condenser.

$$\left[\textbf{Ans.} \frac{1}{C} = \frac{1}{4\varepsilon_0}\left[\frac{1}{a} - \frac{1}{r} + \frac{\varepsilon_r - \varepsilon_0}{\varepsilon}\left(\frac{1}{b} - \frac{1}{c}\right)\right]\right]$$

9. Find the energy expended in polarising a dielectric.

$$\left[\textbf{Ans.} \left(\frac{\varepsilon_r - \varepsilon_0}{2} E^2\right)\right]$$

10. A conducting sphere of radius r and having charge q is placed on the flat boundary between two dielectrics of permittivities ε_1 and ε_2. Show that the potential of this system and charge distribution on the surfaces of the sphere are

$$V = \frac{q}{[2\pi(\varepsilon_1 + \varepsilon_2)r]}, \quad \sigma_1 = \frac{q\varepsilon_1}{[2\pi(\varepsilon_1 + \varepsilon_2)r^2]}$$

and $\sigma_2 = \dfrac{q\varepsilon_2}{[2\pi(\varepsilon_1 + \varepsilon_2)r^2]}$ respectively.

11. A 3 mega ohm resistor and a one micro farad capacitor are connected in series with a battery of 4 volt e.m.f. After one second, what are the rates at which (i) the charge of the capacitor is increasing, (ii) energy is being stored in the capacitor, (iii) heat appearing in the resistor, and (iv) the energy delivered by the battery.

[**Ans.** (i) 9.5×10^{-7} C/s (ii) 1.1×10^{-6} Watt (iii) 2.7×10^{-6} Watt (iv) 3.8×10^{-6} Watt]

12. A condenser is being charged from a *d.c.* source through a resistance of 2 megaohms. If it takes 0.5 sec for the charge to reach $3/4^{\text{th}}$ of its initial value, what is the capacity of the condenser?

[**Ans.** 0.18 F]

13. A dipole moment of 5×10^{-35} *C-m* is developed when an atom of oxygen is polarized. If the centre of negative charge cloud is situated at a distance of 4×10^{-17}m from the nucleus, then calculate atomic polarizability of oxygen. [**Ans.**6.25×10^{-50} f-m^2]
14. A sphere of mass m and radius r floats on a liquid of density ρ and permittivity ε. One quarter of it is immersed in the liquid when it is not charged. Show that the charge on the sphere when it sinks to a depth exactly half of its diameter is given by

$$Q = \left[\frac{8\pi\varepsilon_0 r^2(1+\varepsilon_r)^2}{\varepsilon_r - 1}\right]^{1/2}$$

SHORT ANSWER QUESTIONS

1. What form of energy is stored in the capacitor?

   [**Ans.** Electrostatic energy]
2. What is the formula for energy stored in a capacitor?

   $$\left[\textbf{Ans.} E = \frac{1}{2}CV^2 = \frac{1}{2}QV = \frac{1}{2}\frac{Q^2}{C}\right]$$
3. What is energy density? Write its S.I. Units.

   [**Ans.** Energy density is defined as the energy stored per unit volume of the medium between the plates of the capacitor. Energy density $= \frac{1}{2}\varepsilon_0 E^2$, where E is electric intensity and ε_0 is electric permittivity of free space. The S.I. unit of energy density is J-m^{-3}.]
4. Define the dielectric constant of a dielectric.

   [**Ans.** The dielectric constant of a dielectric is defined as the ratio of external field applied across the dielectric to the net field available in the presence of the dielectric. $K = \frac{E_0}{E - E'}$ where E_0 is the external field applied and E' is the induced (internal) field of the dielectric.]
5. What are the factors on which the capacitance of a parallel plate capacitor depends?

   [**Ans.** (i) Area of the plates, A (ii) separation between the plates, and (iii) the electric permittivity of medium between the plates, ε_0. Thus $C = \frac{A\varepsilon_0}{d}$.]

6. What is the effect of temperature on the dielectric constant of a dielectric?

   [**Ans.** When the temperature rises, K decreases because the induced field decreases.]

7. How will the formula for the capacitance of a parallel plate capacitor modify, when a dielectric slab is inserted between the plates of the capacitor?

   [**Ans.** The capacitance increases, $C' = \dfrac{A\varepsilon_0}{(d-t)+\dfrac{t}{k}}$ where t is thickness of the of the dielectric slab.]

8. How will the formula for the capacitance of a parallel-plate capacitor modify when a conducting slab is inserted between the plates of a capacitor?

   [**Ans.** Capacitance of the capacitor increases, $C' = \dfrac{A\varepsilon_0}{\left(1-\frac{t}{d}\right)}$, where t is the thickness of the conducting slab. when $t = d$, C' is infinite.]

9. What is electric polarization?

   [**Ans.** It is defined as the induced surface charge per unit area.]

10. When a conductor is placed in the electric field, what will be its dielectric constant?

    [**Ans.** Infinite, because the intensity inside the conductor is zero.]

11. What will happen when a dielectric is introduced between the plates of a condenser at a constant potential difference?

    [**Ans.** The capacity of condenser increases. Since the plates are maintained at a constant potential difference, the charge on the plates increases $(Q = CV)$. The extra charge flows to the plates from the source to which the latter are connected.]

12. What is the S.I. unit of atomic polarizability? [**Ans.** m^3.]

13. What are the S.I. units of electric susceptibility?

    [**Ans.** Electric susceptibility has no units.]

14. What is macroscopic polarization density?

    [**Ans.** It is defined as the electric dipole moment per unit volume and it is found by taking the vector sum of the microscopic dipole moments in unit volume.]

15. Give two examples of non-polar molecules.
[**Ans.** H_2O_2 and CO_2]

16. Give two examples of polar molecules [**Ans.** NaCl, H_2O]

17. What are fictitious or bound charges?

[**Ans.** Bound charges are those included in the composition of the atoms and molecules of the dielectric, as well as the charges of ions in chrystalline dielectrics having an ionic lattice. All the other charges are free.]

18. What do you understand by a '*breakdown Voltage*' of a Capacitor?

[**Ans.** Each type of a capacitor has its breakdown voltage which is the potential difference between its plates (or conductors) at which an electric discharge passes through the layer of dielectric. The breakdown voltage depends upon the thickness of the dielectric, its properties and the shape of the conductors in the capacitor.]

19. What are dielectrics?

[**Ans.** These are substances which do not conduct electric current. They have no free electric charges.]

20. On what factors the polarizability of a molecule depends?
[**Ans.** Its volume.]

21. Why the molecules of polar dielectrics (H_2O, NH_3, HCl, CH_3Cl, etc.) possess a permanent dipole moment?

[**Ans.** This is due to the asymmetry in the arrangement of electron clouds and nucleus of these molecules. The "centres of gravity" of the positive and negative charges in such molecules do not coincide.]

22. What do you understand by 'Orientational polarization' of a dielectric with polar molecules?

[**Ans.** This is also called dipole polarization and consists in the aligning of the axes of permanent dipoles along the direction of the intensity vector of the field. It increases with the intensity of the field and decreases with a rise in temperature.]

OBJECTIVE QUESTIONS

1. A point charge in space is attracted towards a dielectric material because of the

   (a) phenomenon of dielectric polarization
   (b) dielectric hysteresis
   (c) fact that dielectric is an insulator
   (d) maximization of the electric flux **[d]**

2. The electric flux and field intensity inside a conducting sphere is

   (a) maximum (b) zero (c) uniform (d) none of the above **[a]**

3. Two parallel plate capacitors of capacitance C and $2C$ are connected in parallel and charged to a potential difference V. The battery is then disconnected and the region between the plates of capacitor C is completely filled with a material of dielectric constant K. The potential difference across the capacitors now becomes

   (a) $\dfrac{3V}{K+2}$ (b) $\dfrac{3V}{K}$ (c) $\dfrac{V}{K+2}$ (d) $\dfrac{V}{3K}$ **[a]**

   [**Hint:** The original capacitance of the system $= 2C + C = 3C$. Total charge $3CV$. Capacitance after inserting dielectric $= KC + 2C = C'$. We know that charge remains the same and potential now will be $\dfrac{3CV}{C'} = \dfrac{3CV}{KC+2C} = \dfrac{3V}{K+2}$]

4. An air-filled parallel plate capacitor has a capacitance of 10^{-12} F. The separation of the plates is doubled and wax is inserted between them, which increases the capacitance to 2×10^{-12} F. The dielectric constant of wax is

   (a) 8 (b) 4 (c) 3 (d) 2 **[b]**

   [**Hint:** $C = \varepsilon_0 \dfrac{A}{d} = 10^{-12}$ F, when dielectric is introduce and d is doubled, the capacitance becomes $C' = \varepsilon_0 \dfrac{kA}{d} = 2 \times 10^{-12}$ F

   $$\therefore \quad K = \frac{d}{\varepsilon_0 A} \times 4 \times 10^{-12}\text{ F} = \frac{4 \times 10^{-12}}{1 \times 10^{-12}} = 4\Big]$$

5. When a dielectric of dielectric constant K is introduced between the plates of a charged parallel plate capacitor, the electric field at

a point between the plates
(a) increases (b) decreases
(c) becomes KE (d) becomes E/K **[b,d]**

[**Hint:** The induced charges produce an electric field in a direction opposite to the field produced by the charged plates.]

6. If a dielectric is placed in an electric field, the field strength
(a) decreases (b) increases
(c) becomes zero (d) remain unchanged **[a]**

7. A parallel plate capacitor of plate area A and plate separation d is charged to potential difference V and then the battery is disconnected. A slab of dielectric constant K is then inserted between the plates so as to fill the space between the plates. If Q, E and W denote respectively, the magnitude of charge on each plate, the electric field between the plates (after the slab is inserted) and the work done on the system is question, in the process of inserting the slab, then
(a) $Q = \varepsilon_0 \frac{AV}{d}$ (b) $Q = \varepsilon_0 KA \frac{V}{d}$
(c) $C = \frac{V}{Kd}$ (d) $W = \varepsilon_0 \frac{AV^2}{2d}\left(1 - \frac{1}{K}\right)$ **[a,c,d]**

[**Hint:** Charge remains the same, $Q = CV = \varepsilon_0 \frac{AV}{d}$, Potential and electric field is reduced by a factor K, $\therefore\ E = \frac{V}{Kd}$. Energy of the system before the dielectric is inserted: $u_1 = \frac{1}{2}\frac{Q^2}{C} = \frac{\varepsilon_0^2 A^2 V^2}{2d^2}\frac{d}{\varepsilon_0 A} = \frac{\varepsilon_0 AV^2}{2d}$. After the dielectric is inserted: $u_2 = \frac{1}{2}\frac{Q^2}{C'}$, where $C' = \varepsilon_0 \frac{KA}{d} = \frac{1}{2}\frac{\varepsilon_0^2 A^2 V^2}{2d^2}\frac{d}{\varepsilon_0 KA} = \frac{\varepsilon_0 AV^2}{2Kd}$

$$\therefore\ u_1 - u_2 = \frac{\varepsilon_0 AV^2}{2d}\left(1 - \frac{1}{K}\right).]$$

8. The dielectric of a charged capacitor experience
(a) tensile force (b) compressive force
(c) lateral force (d) no force **[b]**

9. The relation between the permittivity of the artificial dielectric (ε_r), number of spheres per unit volume (N) and the size of the sphere (v) is

(a) $\varepsilon_r = 1 + 3\nu N$ (b) $\varepsilon_r = \nu N$

(c) $\varepsilon_r = 1 + \frac{N}{\nu}$ (d) $\varepsilon_r = \frac{N}{1+\nu}$ **[a]**

10. If α is atomic polarizability, N is the number of atoms per unit volume and χ is the electric susceptibility, then the Clausius-Mossotti relation can be expressed as

(a) $\alpha = \frac{3\varepsilon_0}{N}\frac{\varepsilon_r - 1}{\varepsilon_r + 2}$ (b) $\frac{Nd}{3\varepsilon_0} = \frac{\chi}{\chi+3}$

(c) $\alpha = \frac{\chi}{\chi+3}$ (d) $d = \varepsilon_0\frac{\chi}{\chi+3}$ **[b]**

11. The Gauss's law in a dielectric medium is

(a) $\nabla \cdot \mathbf{D} = 0$ (b) $\nabla \cdot \mathbf{D} = \rho_{\text{free}}$

(c) $\nabla \cdot \mathbf{D} = \chi$ (d) $\nabla \cdot \mathbf{D} = \varepsilon_r$ **[a]**

12. The bound charge current density is given by

(a) $\mathbf{J}_p = \frac{d\mathbf{P}}{dt}$ (b) $\mathbf{J}_p = \frac{d\mathbf{E}}{dt}$

(c) $\mathbf{J}_p = \frac{\varepsilon_r}{\mathbf{E}}$ (d) $\mathbf{J}_p = \frac{\mathbf{E}}{\varepsilon_0}$ **[a]**

13. A charged capacitor of 0.1μF is discharged through a resistance 10^7 ohms. The charge will reduce to 36.8% of the original value in time

(a) 4 sec. (b) 3 sec. (c) 2 sec. (d) 1 sec. **[d]**

[**Hint:** $\tau = RC = 10^7 \times 0.1 \times 10^{-6} = 1$ sec.]

14. A dielectric sphere is placed in a uniform electric field of intensity E_0. The polarizability produced is given by

(a) $P = 3E_0\left(\frac{\varepsilon_r - 1}{\varepsilon_r + 2}\right)\varepsilon_0$ (b) $P = 3\varepsilon_0 E_0$

(c) $P = \frac{3E_0}{\varepsilon_r + 2}$ (d) none of the above **[a]**

Chapter 5

Solution of Boundary Value Problems in Electrostatic Fields

5.1 INTRODUCTION

The methods discussed in the previous chapters for the calculation of electrostatic fields are useful only in special cases. A more general method, however will involve the solution of Possion's equation, $\nabla^2\phi = -\rho/\varepsilon_0$ or if the space charge density $\rho = 0$, the solution of Laplace's equation $\nabla^2\phi = 0$. Since the solutions of these equations are carried out with the help of boundary conditions, these problems are usually termed as *boundary-value problems.*

5.2 POISSON'S AND LAPLACE'S EQUATIONS

Remembering that the components of the electric field **E** are expressed in terms of the electric potential ϕ by $E_x = -\dfrac{\partial\phi}{\partial x}$ and similar expressions for E_y and E_z, one may write

$$\frac{\partial E_x}{\partial x} = \frac{\partial}{\partial x}\left(-\frac{\partial\phi}{\partial x}\right) = -\frac{\partial^2\phi}{\partial x^2}$$

with similar results for E_y and E_z. Making the substitution in Gauss's theorem in differential form,

$$\nabla\cdot\mathbf{E} = \frac{\rho}{\varepsilon_0} \quad \text{(in free space)} \tag{5.1}$$

or

$$\frac{\partial E_x}{\partial x} + \frac{\partial E_y}{\partial y} + \frac{\partial E_z}{\partial z} = \frac{\rho}{\varepsilon_0} \quad \text{(in free space)}$$

one obtains

$$\frac{\partial^2 \phi}{\partial x^2} + \frac{\partial^2 \phi}{\partial y^2} + \frac{\partial^2 \phi}{\partial z^2} = -\frac{\rho}{\varepsilon_0}$$

or

$$\nabla \cdot \nabla \phi = -\frac{\rho}{\varepsilon_0}$$

or

$$\nabla^2 \phi = -\frac{\rho}{\varepsilon_0} \tag{5.2}$$

The operator $\nabla \cdot \nabla = \nabla^2$ is known as the Laplacian and the Eq. (5.1) is known as *Poisson's equation.* We can use Eq. (5.2) to obtain the electric potential when we know the charge distribution and conversely so long as the charge distribution is time independent. In free space, where there are no charges, $\rho = 0$ and Eq. (5.1) becomes $\nabla \cdot \mathbf{E} = 0$ and Eq. (5.2) gives us

$$\frac{\partial^2 \phi}{\partial x^2} + \frac{\partial^2 \phi}{\partial y^2} + \frac{\partial^2 \phi}{\partial z^2} = 0$$

or

$$\nabla^2 \phi = 0 \tag{5.3}$$

This equation is called *Laplace's equation.* It is one of the most important equations in mathematical physics, and appears in many problems outside the theory of the electromagnetic field, such as fluid motion and elasticity.

∇^2 (The Laplacian) is a scalar operator. This has the following form in the coordinate systems generally used

(i) Cartesian System:

$$\nabla^2 = \frac{\partial^2}{\partial x^2} + \frac{\partial^2}{\partial y^2} + \frac{\partial^2}{\partial z^2} \tag{5.4}$$

(ii) Spherical Polar Coordinate System:

$$\nabla^2 = \frac{1}{r^2}\frac{\partial}{\partial r}\left(r^2 \frac{\partial}{\partial r}\right) + \frac{1}{r^2 \sin\theta}\frac{\partial}{\partial \theta}\left(\sin\theta \frac{\partial}{\partial \theta}\right) + \frac{1}{r^2}\frac{\partial}{\sin^2\theta}\frac{\partial^2}{\partial \phi^2} \tag{5.5}$$

(iii) Cylindrical Polar Coordinates:

$$\nabla^2 = \frac{1}{r}\frac{\partial}{\partial r}\left(r \frac{\partial}{\partial r}\right) + \frac{1}{r^2}\frac{\partial^2}{\partial \phi^2} + \frac{\partial^2}{\partial z^2} \tag{5.6}$$

The Solution of a boundary value problems is usually simplified by a proper selection of the co-ordinate system, for instance, a problem involving a rectangular object may be most readily handled with rectangular coordinates, a cylindrical object by cylindrical coordinates, an elliptical object by elliptical-hyperbolic coordinates etc. In other words, the problem can be solved directly by solving Poisson or Laplace equation, subject to the condition that potential (ϕ), thus obtained, satisfies certain boundary conditions appropriate to the configuration of electrodes. Such problems are termed as boundary value problems. In cases, the boundaries cannot be simply expressed in any coordinate system, efforts must be made to other methods such as graphical, point to point, numerical techniques with the help of computer or experimental methods. There are a few problems, however which because of the high degree of symmetry involved, cannot be solved exactly. These boundary value problems may be handled by the use of series of known functions. Present chapter is devoted to study of different boundary value problems.

5.3 EARNSHOW'S THEOREM

According to this theorem, "*a charged body cannot rest in stable equilibrium under the influence of electrostatic forces alone*". This theorem says that electrostatic containment is also out of questions. However, a hot plasma ($T \approx 10^7 K$) can be confined magnetically.. Such type of confinement of plasma is widely used in controlled thermonuclear reactions. We know that for any extreme value of potential ϕ it must satisfy the condition $\nabla^2\phi = 0$. For ϕ to be maximum $\nabla^2\phi$ must be negative where as for ϕ to be minimum $\nabla^2\phi$ must be positive. Obviously, in the region where there are no charges $\nabla^2\phi = 0$ and hence, the potential cannot have a maximum or minimum value. This shows that for a positive charge to be stable, it must be at a point of minimum potential, i.e. $\nabla^2\phi$ must be positive. However, Poisson's eq. $\nabla^2\phi = -\rho/\varepsilon_0$ tells that $\nabla^2\phi$ is negative indicating that ϕ is maximum, and hence the charge is not stable. In the case of a negative charge, the condition for stability requires that $\nabla^2\phi$ is negative, while Poisson's eq. gives $\nabla^2\phi = \rho/\varepsilon_0$. This means that there can be no points of stable equilibrium in an electrostatic field. This does not mean that it is not possible to balance a charge by electrostatic forces.

5.4 UNIQUENESS THEOREM

Laplace's equation does not itself determine ϕ; in addition, a suitable set of boundary conditions must be known. This means that the solution of Laplace equation or Poisson's equation must satisfy certain conditions at the boundaries, i.e. at some preselected places in the space, the potential must have unique values at the boundaries, since these potentials only are responsible for charge, electric intensity etc. which are unique. In other words, if a function 'ϕ' satisfies Laplace equation, $\nabla^2\phi = 0$, and also the boundary condition, then it is a unique (the only) solution. No other function will satisfy these requirements. Thus, if 'ϕ' is real with its first and second derivative continuous in a region and

(i) the value of 'ϕ' on the closed surface is specified (Dirichlet condition) or

(ii) the normal derivative of potential ϕ, $\frac{\partial\phi}{\partial n}$, i.e., the field vector is specified everywhere on the closed surface (Neumann condition), then the solution of Laplace's equation or Poisson's equation is unique. The proof that a proposed set of boundary conditions will suffice is usually presented in the form of a *uniqueness theorem*. This is a fundamental theorem of potential theory. There are many such theorems for electrostatics, all sharing the same basis format.

To prove the uniqness property, imagine that the solution is not unique and there are on the contrary two such possible solutions ϕ_1, and ϕ_2 of Laplace's equation within the boundary region but satisfying the boundary condition $\phi_1 = \phi_2$ and

$$\frac{\partial\phi_1}{\partial n} = \frac{\partial\phi_2}{\partial n} \tag{5.6}$$

at the boundary. Let $\phi = \phi_1 - \phi_2$. Thus

$$\phi = 0 \quad \text{and} \quad \frac{\partial\phi}{\partial n} = 0 \text{ at the boundary} \tag{5.7}$$

Since ϕ_1 and ϕ_2 are both solutions of Laplace's equation,

$$\nabla^2\phi_1 = 0$$

and

$$\nabla^2\phi_2 = 0$$

or

$$\nabla^2(\phi_1 - \phi_2) = \nabla^2\phi = 0 \tag{5.8}$$

throughout the entire region. Obviously, ϕ is also a solution of Laplace's equation. Making use of Green's theorem, viz.

$$\int_S \psi \frac{\partial \phi}{\partial n} ds = \int_v [\psi \nabla^2 \phi + \nabla \phi \cdot \nabla \psi] dv \tag{5.9}$$

where ψ and ϕ are arbitrary scalar functions. $\frac{\partial \phi}{\partial n}$ is the normal derivative of ϕ on the surface and ds and dv are the surface and volume elements respectively. Now, putting $\psi = \phi = \Phi$, one obtains

$$\int_s \phi \frac{\partial \phi}{\partial n} ds = \int_v [\phi \nabla^2 \phi + |\nabla \phi|^2] dv \tag{5.10}$$

We see that LHS of (5.10) vanishes for any of the boundary conditions (5.7). Moreover $\nabla^2 \phi = 0$ by (5.8). Thus we have

$$\int_v |\nabla \phi|^2 dv = 0 \tag{5.11}$$

The integrand is a positive definite quantity and hence we have

$$\nabla \phi = 0 \tag{5.12}$$

or $\phi = \phi_1 - \phi_2 =$ constant (say c) throughout the volume. This constant must apply even to the boundary, where we know Dirichelt condition is satisfied. This means $c = 0$ on the surface (by eq. 5.6). This shows that it is zero throughout the region, i.e. ϕ_1 and ϕ_2 are identical potential distributions, i.e. $\phi_1 = \phi_2 = 0$, i.e. the **solution is unique**. If on the otherhand, Neumann condition is satisfied on the surface.

$$\frac{\partial \phi}{\partial n} = 0, \; i.e. \; \phi = \phi_1 - \phi_2 = a \text{ constant.}$$

Again the constant is arbitrary and hence one can take it to be zero. Obviously, the solution is again unique and hence

$$\phi_1 = \phi_2$$

We must note that if Dirichelt condition is satisfied over a part of the surface S and Neumann condition over the remaining part, the LHS of Eq. (5.10) still vanishes. The unique theorem remains valid for linear dielectrics. However, if the system involves non-linear dielectrics, then the region may be divided into sub regions each having uniform polarization density and for each sub-region uniqueness theorem holds separately.

5.5 GREEN'S RECIPROCITY THEOREM

With the help of this theorem, one can transform the solution of a known problem into the solution of desired unknown problem.

Let us consider a set of n point charges q_i placed at points, where the potentials due to the other charges are given by a set of numbers ϕ_j. One finds that the potential at j^{th} point due to charges q_i at other points are given by

$$\phi_j = \frac{1}{4\pi\varepsilon_0} \sum_{i=1}^{n}{}' \frac{q_i}{r_{ij}} \tag{5.13}$$

The prime on the summation is used to indicate that $i \neq j$.

Let us now consider a different set of charges q'_i placed at the same points giving rise to the corresponding potential ϕ'_j, then

$$\phi_{j'} = \frac{1}{4\pi\varepsilon_0} \sum_{i=1}^{n}{}' \frac{q'_i}{r_{ij}} \tag{5.14}$$

Multiplying (5.13) by q_j, and (5.14) by q_j, and summing each expression over the index j, one obtains

$$\sum_{j}^{n} \phi_j q'_j = \sum_{j=1}^{n}\sum_{i=1}^{n}{}' \frac{q_i q_{j'}}{r_{ij}} \tag{5.15}$$

and

$$\sum_{j=1}^{n} \phi_{j'} q_j = \sum_{j=1}^{n}\sum_{i=1}^{n}{}' \frac{q'_i q_j}{r_{ij}} \tag{5.16}$$

We know that i and j are summation indices and hence one may interchange them in one product of $q's$, one obtains

$$\sum_{j=1}^{n} \phi_j q'_j = \sum_{j=1}^{n} \phi_{j'} q_j \tag{5.17}$$

This is *Green's reciprocity theorem.* One may generalized it from a set of n point charges to set of n conductors at potential ϕ_j and carrying charges q_j.

5.6 SOLUTION OF POTENTIAL PROBLEMS BY GREEN'S FUNCTION

One can solve large number of potential problems by Green's function. The Green's function is simply the solution of potential problem for a given geometrical arrangement of grounded conducting boundaries when the only charge present is a unit positive charge located at the point r'. According to this theorem the Green's function for a particular geometrical arrangement is a symmetrical function of coordinates of a unit charge located at the point r' and the coordinates of the point of observation r, we have

$$G(r,r') = G(r',r) \tag{5.18}$$

This symmetry property merely represents the physical interchangeability of the positions of the point charge and observation point.

One can solve following two types of problems by the use of Green's function method: (i) When the potential distribution ϕ_s over a certain boundary is known, and (ii) when the charge distribution is known in a region within a conducting boundary.

The solution of above both types of problems can be obtained by means of usual Green's theorem

$$\int (\phi\nabla^2\psi - \psi\nabla^2\phi)dv = \int (\phi\nabla\psi - \psi\nabla\phi)ds \tag{5.19}$$

where ψ and ϕ are arbitrary functions of position which are required to be non-singular throughout the volume v.

Let ϕ be the desired solution of the problems and $\psi = G$ be the Green's function for the geometry of problem, i.e. the solution of the problem of a unit positive charge with grounded surface S. The form of the Green's function $G(r,r')$ is

$$G(r,r') = \frac{1}{4\pi\varepsilon_0}\frac{1}{|r-r'|} + F(r,r') \tag{5.20}$$

The function $F(r,r')$ represents the potential due to the induced charges on S and satisfies Laplace's equation inside the volume V, i.e.

$$\nabla^2 F(r,r') = 0 \tag{5.21}$$

From Eq. (5.20), one gets

$$\nabla^2 G(r,r') = \frac{1}{4\pi\varepsilon_0}\nabla^2\left[\frac{1}{|r-r'|}\right] + \nabla^2 F(r,r')$$

or
$$\nabla^2 G(r,r') = \frac{1}{4\pi\varepsilon_0}\left[\frac{1}{|r-r'|}\right] \quad [\because \ \nabla^2 F(r,r') = 0] \tag{5.22}$$

We must remember that

$$\nabla^2\left[\frac{1}{|r-r'|}\right] = 4\pi\delta(r-r') \tag{5.23}$$

From (5.22) and (5.23), one obtains

$$\nabla^2 G(r,r') = -\frac{\delta(r-r')}{\varepsilon_0} \tag{5.24}$$

Substituting $\psi = G$ in (5.19), one obtains

$$\int (\phi\nabla^2 G - G\nabla^2\phi)dv = \int (\phi\nabla G - G\nabla\phi)ds$$

or

$$\int [\phi(r')\nabla^2 G(r,r') \; - G(r,r')\nabla^2\phi(r')]dv$$
$$= \int \phi(r')\nabla G(r,r') - G(r,r')\nabla\phi(r')]ds$$

Using (5.24), one obtains

$$\int \left[-\phi(r')\frac{\delta(r-r')}{\varepsilon_0} G(r,r')\nabla^2\phi(r') \right] dv$$
$$= \int [\phi(r')\nabla G(r,r') - G(r,r')\nabla\phi(r')]ds$$

or
$$\int \left[\phi(r')\frac{\delta(r-r')}{\varepsilon_0} + G(r,r')\nabla^2\phi(r') \right] dv$$
$$= \int [G(r,r')\nabla\phi(r') - \phi(r') - \phi(r')\nabla G(r,r')]ds \tag{5.25}$$

Since $\int \delta(r-r')dv = 1$ and $G(r,r') = 0$ for r' on surface S, We have

$$\frac{\phi(r)}{\varepsilon_0} + \int G(r,r')\nabla^2\phi(r')dv = \int -\phi(r')\nabla G(r,r')ds$$

or

$$\phi(r) = -\varepsilon_0 \int G(r,r')\nabla^2\phi(r')dv - \varepsilon_0 \int \phi(r')\nabla G(r,r')ds \tag{5.26}$$

Now we shall consider the two cases mentioned earlier.

(i) *The surface surrounding the point r' is grounded,* making $\phi(r') = 0$ and $\nabla^2\phi(r') = -\rho(r)/\varepsilon_0$ due to the charge distribution $\rho(r')$ throughout v. Now, Eq. (5.26) reduces to

$$\phi(r) = -\varepsilon_0 \int G(r,r')\nabla^2\phi dv = \int G(r,r')\rho(r')dv \tag{5.27}$$

This equation merely represents the principle of superposition applied to density of point source within the volume V.

(ii) *When there are no sources of ϕ throughout the volume v.* So that $\nabla^2\phi(r') = 0$, we have

$$\phi(r) = -\varepsilon_0 \int_s \phi(r')\nabla G(r,r')ds \tag{5.28}$$

Eq. (5.28) gives the potential within a given region enclosed by a boundary such that its different parts are raised to a given set of potentials.

5.7 THE METHOD OF ELECTROSTATIC IMAGES

Some typical problems in electrostatics can be solved without specifically solving a differential equation. This method is known as the method of images. This method, originated by Lord Kelvin in 1448, however, does not require a knowledge of the actual charge distribution but involves the conversion of an electrostatic field into another equivalent field which is simpler to calculate. The general principle underlying the method of images is as follows:

(i) For a system of n point charges $q_1, q_2, \ldots q_n$, the potential at any point is given by

$$\phi = \frac{1}{4\pi\varepsilon_0}\sum_{i=1}^{n}\frac{q_i}{r_i} \tag{5.29}$$

(ii) The surface of zero potential, i.e. grounded conductor is the locus of points for which the potential is given by

$$\phi = \frac{1}{4\pi\varepsilon_0}\sum_{i=1}^{n}\frac{q_i}{r_i} = 0 \tag{5.30}$$

(iii) In a problem involving the system of charges $q_1, q_2, \ldots q_j$ and the grounded conductors, the latter may be replaced by the system of charges $q_{j+1}, q_{j+2}, \ldots q_j$ constituting the image if the image charge and the system of real charges combine to produce zero potential over the surface occupied by the grounded conductor. Eq. (5.30) can be rewritten as

$$\phi = \frac{1}{4\pi\varepsilon_0}\sum_{i=1}^{j}\frac{q_i}{r_i} + \frac{1}{4\pi\varepsilon_0}\sum_{i=j+1}^{n}\frac{q_i}{r_i} = 0 \tag{5.31}$$

Eq. (5.31) provides the condition of potential function over the surface occupied by the grounded conductor replaced by electrical image constituted by the system of charges $q_1, q_2, \ldots q_j$.

(iv) If the conductor under consideration is at a constant potential $\phi = 0$, the system of charges and its image must combine to produce the actual potential ϕ. One can write Eq. (5.31) as

$$\phi = \frac{1}{4\pi\varepsilon_0}\sum_{i=1}^{n}\frac{q_i}{r_i} + \frac{1}{4\pi\varepsilon_0}\sum_{i=j+1}^{n}\frac{q_i}{r_i} \tag{5.32}$$

Eq. (5.32) represents the condition of potential function over the surface occupied by the conductor replaced by electrical image constituted by the system of charges $q_{j+1}, q_{j+2}, \cdots q_n$.

Now we shall illustrate the method of images by some typical examples.

(i) A point charge and an infinite conducting plane

Let a point charge $+q$ be placed at a perpendicular distance d from a conducting plate of infinite extent (Fig 5.1(a). Obviously, the charge q will induce charges in the plate. The plate is conducting and hence potential at all points on it must be constant. Let us assume that this potential to be zero, i.e. the induced charges produce at the plate a potential equal and opposite to that of the inducing charge. We are interested in finding the field at a point in the region on the right hand side of the plate and also in the induced charge distribution on conducing plate.

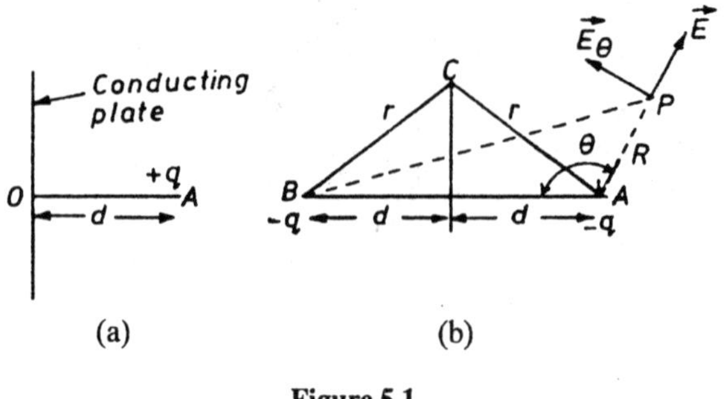

Figure 5.1.

We now consider another situation. Let us consider that a charge $-q$ is placed exactly at a distance d behind the plane of the conducting plate as shown in Fig 5.1(b). The potential at any point on the plane, say at $C = \dfrac{q}{4\pi\varepsilon_0 r} - \dfrac{q}{4\pi\varepsilon_0 r} = 0$. This means, if the conducting plate were removed and the charge $-q$ is placed at B, every point on the plane previously occupied by the conducting plate would still be at zero potential. Obviously, the introduction of the charge $-q$ at B and the removal of the conducting plate does not affect the flux on the right hand side of the plane. This clearly reveals that the conditions of original problem, then remain unaltered by making this change. We must remember that the introduction of charge $-q$ does not describe the situation compatible with the actual situation, i.e. the field on the LHS of the plane is no longer zero. However, one should not worry about it because we are interested only in the region on the RHS of the plane. Now, knowing that the problem remain unaltered by placing a charge $-q$ at B, we can see by the uniqueness theorem, that the field produced by the charge $+q$ at A and $-q$ at B is identical with that produced by the charge $+q$ at A and the induced charge on the plate. Obviously, one can solve the problem by ignoring the conducting plate and assuming a charge $-q$ at B. The charge $-q$ at B is called the 'image' of the charge $+q$ at A in analogy with optical images. We must note that the charge $-q$ has no real existence and it is only a 'virtual' image. The electric potential at P (Fig 5.1(b)) due to charge $+q$ at A and the induced charges on the conductor is

$$\phi_p = \frac{q}{4\pi\varepsilon_0 R} - \frac{q}{4\pi\varepsilon_0\sqrt{R^2+4d^2+4dR\cos\theta}} \tag{5.33}$$

where θ is the angle between BA and AP and $AP = R$. If $R = \infty$, we get $\phi = 0$. Obviously, the boundary condition $\phi = 0$ at $R = \infty$ is also satisfied if the conducting plate (with zero potential) is replaced by a charge $-q$ at B.

The electric field components E_R, E_θ at point P can be obtained by differentiating Eq. (5.33) partially with respect to R, θ respectively, i.e.

$$E_R = -\frac{\partial\phi_p}{\partial R} = \frac{q}{4\pi\varepsilon_0 R^2} - \frac{q(R+2d\cos\theta)}{4\pi\varepsilon_0(R^2+4d^2+4Rd\cos\theta)^{3/2}} \tag{5.34}$$

$$E_\theta = -\frac{1}{R}\frac{\partial V_p}{\partial\theta} = \frac{2qd\sin\theta}{4\pi\varepsilon_0(R^2+4d^2+4Rd\cos\theta)^{3/2}} \tag{5.35}$$

Now, we calculate the field normal to the plate, E at C. We have

$$\begin{aligned} E \text{ at } P &= E_r\cos\theta - E_\theta\sin\theta \\ &= \frac{q}{4\pi\varepsilon_0}\left[\frac{\cos\theta}{R^2} - \frac{R\cos\theta+2d}{(R^2+4d^2+4Rd\cos\theta)^{3/2}}\right] \\ \therefore\quad E \text{ at } C &= -\frac{q}{4\pi\varepsilon_0}\frac{2d}{R^3} \end{aligned}$$

Surface Charge Density

If σ is the surface density of the induced charge on the conducting plane at C, we have from Gauss's law

$$E = -\frac{\sigma}{\varepsilon_0}$$

$$\therefore \quad \sigma = \frac{qd}{2\pi R^3} \tag{5.36}$$

Obviously, the surface density of induced charge varies inversely as the cube of the distance of the position on the plate from the charge $+q$.

The force F exerted on the charge '$+q$' by the induced charge on the conducting plate is the same as that exerted by the image on $+q$ i.e.

$$F = -\frac{1}{4\pi\varepsilon_0}\frac{(q)(-q)}{(2d)^2} = -\frac{q^2}{16\pi\varepsilon_0 d^2} \tag{5.37}$$

The negative sign indicates that the force is *attractive*.

We must note that with the simple geometry as above, it was possible to determine the potential function by introducing one image charge. Thus the problem of charge distribution on the surface and the potential and field produced by the system can be easily solved by introducing an image or a set of images suitably placed in a region external to the region under consideration and ignoring the actual surface of electrification, i.e. replacing boundary conditions by "images". However, we must remember that the success of the method depends upon one's ability to guess a simple equivalent image distribution, i.e. the distribution that keeps the surface still an equipotential surface with requisite potential.

(ii) A point charge in the vicinity of a grounded conducting sphere

The sphere-gap voltmeter and the breakdown tests of liquid and solid insulation make use of spherical electrodes, where in a knowledge of the electric field intensity may sometimes required. The image method if employed to evaluate this field, a group of image charges together with the originally specified charges have to be considered to produce equipotential surfaces at the conductors. The problem is straight forward only in cases where the geometry is simple. Such is the case, however, for a point charge Q in the vicinity of a conducting sphere.

Let us consider a conducting sphere of radius a maintained at zero potential, i.e. grounded and a charge $+q$ at the point A at a distance d

from the centre of the sphere, such that $d > a$ (Fig 5.2(a)). Since the sphere is grounded and hence the boundary conditions are:

(i) The electric potential at ϕ at every point on the surface of the sphere is zero, i.e. $\phi = 0$.
(ii) The electric potential at infinitely far point is zero, i.e. $\phi = 0$.

Let us suppose that the centre of the sphere coincides with the origin. Now, the charge $+q$ is outside the sphere and one is interested in the region outside the sphere and hence the image must be within the sphere. By symmetry we observe that the image charge must be placed on the line joining A to the centre of the sphere. Fig 5.2(b) represents this problem with the boundary conditions

$$\phi(R, \theta)\big|_{|R|=a} = 0 \tag{5.38}$$

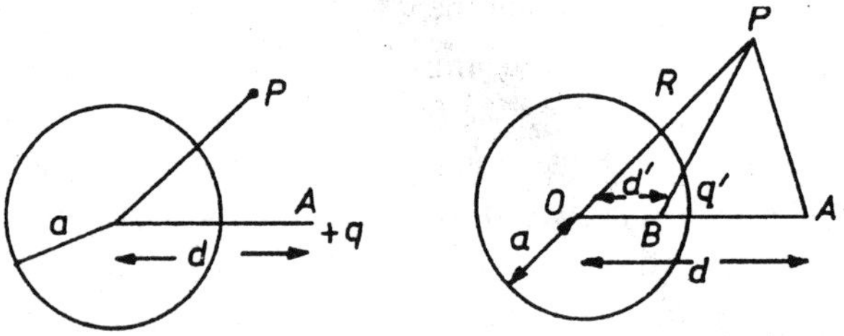

Figure 5.2.

As shown in the Fig 5.2(b), B represents the position of the image charge. We have $OB = d'$. The potential the point P is obtained as

$$\begin{aligned}\phi_P &= \frac{q}{4\pi\varepsilon_0|\mathbf{R}-\mathbf{d}|} + \frac{q'}{4\pi\varepsilon_0|\mathbf{R}-\mathbf{d}'|} \\ &= \frac{q}{4\pi\varepsilon_0|R\hat{a}_r - d\hat{a}_d|} + \frac{q'}{4\pi\varepsilon_0|R\hat{a}_R - d'\hat{a}_d|}\end{aligned} \tag{5.39}$$

Where $\hat{a}_R$ and $\hat{a}_d$ are the unit vectors in the directions OP and OA respectively

$$\therefore \quad \phi_P = \frac{q}{4\pi\varepsilon_0 R|\hat{a}_R - \frac{d}{R}\hat{a}_d|} + \frac{q'}{4\pi\varepsilon_0 d'|\hat{a}_d - \frac{R}{d'}\hat{a}_d|} \tag{5.40}$$

we have on the surface of the sphere $R = a$

$$\therefore \quad V_P\big|_{|R|=a} = \frac{q}{4\pi\varepsilon_0 a|\hat{a}_R - \frac{d}{a}\hat{a}_d|} + \frac{q'}{4\pi\varepsilon_0 d'|\hat{a}_d - \frac{a}{d'}\hat{a}_R|} = 0$$

The above holds true if

$$\frac{q}{a} = -\frac{q'}{d'} \text{ and } \frac{d}{a} = \frac{a}{d'}, \quad \therefore \quad d' = \frac{a^2}{d}$$

and
$$q' = -\frac{qd'}{a} = -\frac{qa}{d} \tag{5.41}$$

We must remember that A and B points are inverse points with respect to the sphere. Alternatively, at $|R| = a$ one obtains

$$\begin{aligned} |\mathbf{R} - \mathbf{d}|^{-1}_{|R|+=a} &= (R^2 + d^2 - 2Rd\cos\theta)^{-1/2}_{|R|=a} \\ &= (a^2 + a^2 - 2ad\cos\theta)^{-1/2} \\ &= d^{-1}(1 - 2h\cos\theta + h^2)^{-1/2} \end{aligned} \tag{5.42}$$

similarly, one obtains

$$|\mathbf{R} - \mathbf{d}'|^{-1}_{|R|=a} = a^{-1}(1 - 2h'\cos\theta + h'^2)^{-1/2} \tag{5.43}$$

where $h' = \dfrac{d'}{a}$ and $h = \dfrac{d}{a}$.

We have

$$(1 - 2hx + h^2)^{-1/2} = \sum_{n=0}^{\infty} h^n P_n(x)$$

where $|h| < 1$ and $|x| \leq 1$. Thus we have

$$|\mathbf{R} - \mathbf{d}|^{-1}_{|R|=a} = \sum_{n=0}^{\infty} a^n d^{-(n+1)} P_n(\cos\theta) \tag{5.44}$$

and
$$|\mathbf{R} - \mathbf{d}'|^{-1}_{|R|=a} = \sum_{n=0}^{\infty} d'^n a^{-(n+1)} P_n(\cos\theta) \tag{5.45}$$

$$\begin{aligned} \therefore \quad \phi_{(a,\theta)} &= \frac{1}{4\pi\varepsilon_0} \sum_{n=0}^{\infty} qa^n d^{-(n+1)} \\ &\quad + q'd'^n a^{-(n+1)} P_n(\cos\theta) = 0 \end{aligned} \tag{5.46}$$

We know that Legend polynominals are linearly independent and hence the coefficient of each Legendre polynomial in the expansion vanishes. Now, for $n = 0$, we obtain

$$qd^{-1} + q'a^{-1} = 0 \quad \text{or} \quad q' = -\left(\frac{a}{d}\right)q \tag{5.47}$$

and for $n = 1$, one obtains

$$qad^{-2} + q'd'a^{-2} = 0$$

or $$qad^{-2} - \left(\frac{a}{d}\right) qd'a^{-2} = 0 \quad (\text{using}(5.47))$$

$$\therefore \quad d' = \frac{a^2}{d} \tag{5.48}$$

We can easily see that for $n = 1$, the conditions (5.47) and (5.48) reduce all succeeding coefficients to zero.

Now. substituting the values of q' and d' (5.47) and (5.48) in eq (5.40), one obtains the desired expression for the potential, i.e.

$$\begin{aligned}
\phi_P &= \frac{q}{4\pi\varepsilon_0 R|\hat{a}_R - \frac{d}{R}\hat{a}_d|} + \frac{\left(-\frac{a}{d}\right)q}{4\pi\varepsilon_0 \frac{a^2}{d}|\hat{a}_d - \frac{dR}{a^2}\hat{a}_R|} \\
&= \frac{q}{4\pi\varepsilon_0 R|\hat{a}_R - \frac{d}{R}\hat{a}_d|} - \frac{q}{4\pi\varepsilon_0\, a|\hat{a}_d - \frac{dR}{a^2}\hat{a}_R|}
\end{aligned} \tag{5.49}$$

We must note that as the charge q is brought nearer and nearer to the surface q' becomes larger and larger and moves away from the centre.

One can obtain the charge density from the relation

$$E_x = -\frac{\partial \phi}{\partial x} = -\frac{\sigma}{\varepsilon_0}$$

Integrating this density over the whole surface of the shere, one obtains the total induced charge

$$\begin{aligned}
q' &= \int_0^{\pi} \sigma(2\pi a^2) \sin\theta d\theta \\
&= -\frac{a}{d}q
\end{aligned} \tag{5.50}$$

This shows that the total induced charge on the real conducting sphere is thus the same as the magnitude of the image charge which replaced the sphere. This must be true from Gauss's law.

(iii) A charge 'q' placed in the vicinity of a charged and insulated sphere

Let us consider that the conducting sphere is insulated and given a charge Q. Let the charge q is at a distance d from the centre as in the case (ii). One can find the potential at a point due to this system by the method of linear superposition. Let us assume that the sphere is grounded, i.e. its potential is zero and the induced charge q' equal to the magnitude of the image charge of q is distributed over it.

Now insulate the sphere by disconnecting it from the ground, and give it an additional amount of charge $Q-q'$. obviously the total charge on the sphere now is Q. Moreover, the forces due to the point charge q' are balanced by the charge q' and hence the added charge $Q-q'$ will be uniformly distributed over the surface of the sphere. The potential at a point outside the sphere is now given by the sum of potentials due to q, q' and added charge $Q-q'$. The contribution to the potential due to $Q-q'$ will be the same as that of a point charge of magnitude $Q-q'$ at the centre. Finally, one obtains

$$\phi_p = \frac{q}{4\pi\varepsilon_0|\mathbf{R}-\mathbf{d}|} - \frac{qa}{4\pi\varepsilon_0 d|\mathbf{R}-a^2/d\,\hat{a}_d|} + \frac{Q+qa/d}{4\pi\varepsilon_0|\mathbf{R}|} \tag{5.51}$$

(iv) When the sphere is maintained at a fixed potential

Let us consider that the sphere is maintained at a potential ϕ. Obviously the charge on the sphere = potential × capacity = ϕa, where a is the capacity of the sphere. The potential at a point p due to this sphere

$$\phi_p = \frac{q}{4\pi\varepsilon_0|\mathbf{R}-\mathbf{d}|} - \frac{qa}{4\pi\varepsilon_0 d|\mathbf{R}-a^2/d\,\hat{a}_d|} + \frac{\phi a}{4\pi\varepsilon_0|R|} \tag{5.52}$$

(v) Conducting sphere placed in a uniform field

Let us consider that a conducting sphere of radius a is placed in a uniform electric field E_0 with centre at the origin O (Fig. 5.3). Due to induced charges on the surface of the sphere the field immediately around the sphere will become distorted as shown in Fig. 5.3.

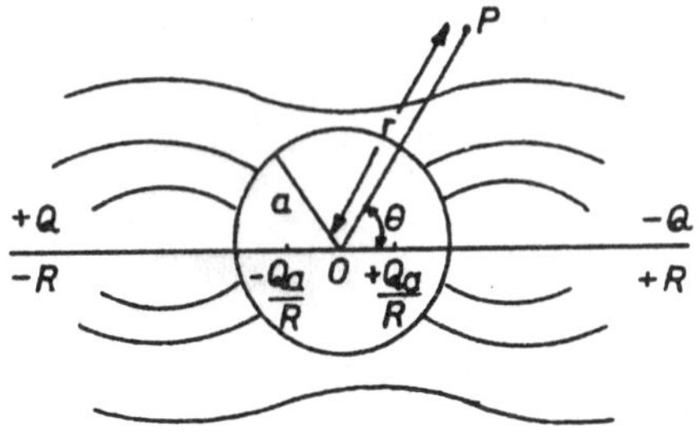

Figure 5.3.

One can regard a uniform field as the field produced by the appropriate positive and negative charges at infinity For example, let us consider a charge $-Q$ at $z=+R$ and a charge $+Q$ at $Z=-R$. Obviously the field produced at a point near the origin O is approximately $\frac{2Q}{4\pi\varepsilon_0 R^2}$ and this will be parallel to z-axis. In the limiting care when $Q \to \infty$, keeping $\frac{Q}{R^2}$ constant, the field is uniform and equal to $E_0 = \frac{2Q}{4\pi\varepsilon R^2}$.

One can easily see that the potential at a point P distant r from O due to the system consisting of charges $\pm Q$ at $z = \mp R$ and the induced charges on the sphere is the same as that produced by the charge $+Q$ at $-R$, $-Q$ at $+R$, the image charge $-\frac{Qa}{R}$ at $-\frac{a^2}{R}$ and the image charge $\frac{Qa}{R}$ at $+\frac{a^2}{R}$. We have

$$\phi_p = \frac{Q}{4\pi\varepsilon_0(r^2+R^2+2rR\cos\theta)^{1/2}} - \frac{Q}{4\pi\varepsilon_0(r^2+R^2-2rR\cos\theta)^{1/2}}$$

$$+ \text{ potentials due to image charges.}$$

Since $r \ll R$ and hence one can easily see that the first two terms give the contribution (expanding and neglecting higher terms) $-\frac{2Q}{4\pi\varepsilon_0 R^2} r\cos\theta$.

Let us regard the image charges $-\frac{Qa}{R}$ and $+\frac{Qa}{R}$ are separated by a distance $\frac{2a^2}{R}$ and forming a dipole, one finds their contribution to the potential at P is $\frac{p\cos\theta}{4\pi\varepsilon_0 r^2}$, where $p = \frac{2Qa^3}{R^2}$ is the dipole moment.

$$\begin{aligned} \therefore \quad \phi_p &= -\frac{2Q}{4\pi\varepsilon_0 R^2} r\cos\theta + \frac{(2Qa^3/R^2)\cos\theta}{4\pi\varepsilon_0 r^2} \\ &= -E_0 r\cos\theta + \frac{E_0 a^3\cos\theta}{r^2} \\ &= -E_0\left(r - \frac{a^3}{r^2}\right)\cos\theta \end{aligned} \tag{5.53}$$

The induced charge density is obtained as

$$\sigma = \varepsilon_0 E = -\varepsilon_0 \frac{\partial\phi_p}{\partial r}\Big|_{r=a} = 3\varepsilon_0 E_0\cos\theta \tag{5.54}$$

From Eq (5.53) it is evident that the resultant potential at any point is due to the uniform field intensity E_0 (first term) plus the potential of a dipole moment (second term). So if we replace the sphere by a dipole of this moment (potential due to a dipole $= \dfrac{p\cos\theta}{4\pi\varepsilon_0 r^2}$, comparing it with the second term in (5.53), one obtains $p = 4\pi\varepsilon_0 E_0 a^3$) located at the centre, the field outside the surface previously occupied by the sphere will remain unchanged. We further note that the surface integral of the charge density in (5.54) vanishes so that there is no difference between a grounded and an insulated sphere.

5.8 IMAGES IN DIELECTRICS

The method of images can also be used to determine potentials and fields in the presence of dielectrics. We shall discuss few typical examples.

(i) A point charge 'q' placed at a perpendicular distance 'd' from a semi infinite dielectric material bounded by a plane surface

Let us consider that the surface coincide with the plane $z = 0$ (Fig 5.4). Let 0 be the origin and A be the point on the z-axis where the charge q is placed. The field of charge q polarizes the dielectric which in turn affects the field in vacuum on the right hand side. Let us assume that the field produced by the polarization charges at a point on the $r.h.s.$ is the same as that produced by charge q' at A' (the image of q). Now we are interested in finding the value of q' which must satisfy our assumption.

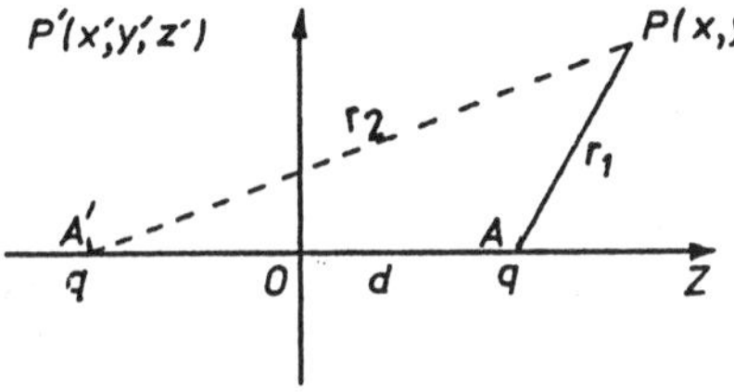

Figure 5.4.

One can write the potential at a point P in vacuum as

$$\phi_P = \frac{q}{4\pi\varepsilon_0 r_1} + \frac{q'}{4\pi\varepsilon_0 r_2}$$

Writing the above in cartesian coordinates, we have

$$\phi_P = \frac{1}{4\pi\varepsilon_0}\left[\frac{q}{[x^2+y^2+(z-d)^2]^{1/2}} + \frac{q'}{[x^2+y^2+(z-d)^2]^{1/2}}\right] \tag{5.55}$$

Now, the question arises, what would be the potential at a point P' in the dielectric medium due point charge q? Obviously it would not the same as q would produce at P' in vacuum. Let us assume that this potential is same as that produced by a point charge q'' at A if the space was vacuum, i.e. there is no dielectric, obviously, one finds this potential at $P'(x', y', z')$ as

$$\phi'_P = \frac{q''}{4\pi\varepsilon_0[x'^2 + y'^2 + (z' - d)^2]^{1/2}} \tag{5.56}$$

The potential in vacuum at P when this point is in the plane $z = 0$, is as

$$\phi_P|_{z=0} = \frac{1}{4\pi\varepsilon_0} \cdot \left[\frac{q + q'}{(x^2 + y^2 + d^2)^{1/2}}\right] \tag{5.57}$$

While that P' when this is at the same point but in the dielectric medium

$$\phi_{P'}|_{z=0} = \frac{q''}{4\pi\varepsilon_0(x'^2 + y'^2 + d^2)^{1/2}} \tag{5.58}$$

On the boundary, we have $\phi_P = \phi'_P$. This satisfies the first boundary condition that the tangential component of field E must be same on either side of boundary. This yields

$$q + q' = q'' \tag{5.59}$$

The second boundary condition requires that the normal component of **D** (displacement vector) must be continuous at the boundary. In accordance with this condition, we have

$$\frac{\partial\phi_P}{\partial z} = \varepsilon\frac{\partial\phi_{P'}}{\partial z} \quad \text{at} \quad z = 0 \tag{5.60}$$

Now, we have

$$\begin{aligned}\frac{\partial\phi_P}{\partial z} &= \frac{1}{4\pi\varepsilon_0}\left[-\frac{q(z-d)}{[x^2 + y^2 + (z-d)^2]^{3/2}} - \frac{q'(z+d)}{[x^2 + y^2 + (z+d)^2]^{3/2}}\right] \\ &= \frac{1}{4\pi\varepsilon_0}\frac{d(q - q')}{(x^2 + y^2 + d^2)^{3/2}} \quad \text{at } z = 0\end{aligned} \tag{5.61}$$

Similarly, one obtains

$$\frac{\partial\phi_{P'}}{\partial z} = \frac{1}{4\pi\varepsilon_0}\frac{q''d}{(x^2 + y^2 + d^2)^{3/2}} \quad \text{at } z' = 0 \tag{5.62}$$

Making use of (5.61) and (5.62), Eq (5.60) yields

$$\frac{d(q - q')}{(x^2 + y^2 + d^2)^{3/2}} = \frac{\varepsilon q''d}{(x^2 + y^2 + d^2)^{3/2}}$$

or
$$q - q' = \varepsilon q'' \tag{5.63}$$

Solving (5.59) and (5.63) one obtains

$$q' = -\frac{q(\varepsilon - 1)}{\varepsilon + 1} \quad \text{and} \quad q'' = \frac{2q}{\varepsilon + 1} \tag{5.64}$$

These combination of charges produce fields which satisfy the boundary conditions. The force of attraction between the dielectric and the point charge is obtained as

$$F = \frac{qq'}{4\pi\varepsilon_0(2d)^2} = -\frac{q^2(\varepsilon - 1)}{16\pi\varepsilon_0 d^2(\varepsilon + 1)} \tag{5.65}$$

(ii) A Dielectric Sphere in a Uniform Field

Let us consider a dielectric sphere of radius a and dielectric constant ε_1 placed in a uniform dielectric field E_0 existing in a medium with dielectric constant ε_2 as shown in Fig 5.5. There are no free charges inside and outside the surface and hence the potential ϕ (outside) and ϕ (inside) the sphere will satisfy Laplace's equation, i.e.

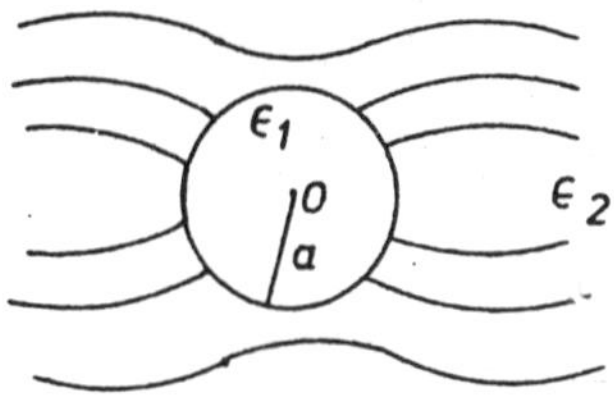

Figure 5.5.

$$\nabla^2\phi_1 = 0 \quad \text{and} \quad \nabla^2\phi_2 = 0 \tag{5.66}$$

Also, the potentials must satisfy the boundary conditions, i.e.

(i) $\phi(r, \theta)$ must be continuous at $r = a$ for all values of θ, i.e. $\phi_1 = \phi_2$ for $r = a$.

(ii) The normal component of **D** must be continuous on $r = a$ for all values of θ. We can express this as

$$\varepsilon_1 \frac{\partial \phi_1}{\partial r} = \varepsilon_2 \frac{\partial \phi_2}{\partial r} \quad \text{on } r = a$$

(iii) At $r = 0$, $\phi_1(r, \theta)$ must be finite and

(iv) $\phi_2(r, \theta) = \phi_0 = -E_0 z = -E_0 r \cos\theta$ as $r \to \infty$.

One can easily see that the potential which is the solution of Laplace's equation can be expressed as

$$\phi = \sum_{i=1}^{\infty} [A_l \phi^l + B_l r^{-(l+1)}] P_l \cos\theta \tag{5.67}$$

In accordance with the condition (iii), the potential at the origin be finite. This means that no term of the form $Br^{-(l+1)}$ should be included in the expression for ϕ_1, as it gives infinite contribution to the boundary at $r = 0$ position. This means

$$\phi_1 = \sum_{l=0}^{\infty} A_l r^l P_l(\cos\theta) \tag{5.68}$$

$$\phi_2 = \sum_{l=0}^{\infty} [C_l r^l + D_l r^{-(l+1)}] P_l \cos\theta \tag{5.69}$$

The constants appearing in the above equations can be determined with the help of boundary conditions. The potential ϕ_2 at infinity is

$$\phi_2 = C_0 P_0(\cos\theta) + C_1 r P_1(\cos\theta) + C_2 r^2 P_2(\cos\theta) + \cdots \tag{5.70}$$

Condition (iv) implies that the above must be equal to $-E_0 r P_1(\cos\theta)$. obviously all terms in (5.70) vanish except the second term and hence

$$l = 1 \quad \text{and} \quad C_1 = -E_0 \tag{5.71}$$

One obtains from boundary condition (i) $\phi_1 = \phi_2$ on $r = a$ for all values of θ

$$\therefore \quad A_1 a = -E_0 a + \frac{D_1}{a^2} \tag{5.72}$$

condition (ii) yields

$$\varepsilon_1 \frac{\partial\phi_1}{\partial r} = \varepsilon_2 \frac{\partial\phi_2}{\partial r} \quad \text{or} \quad r = a \text{ for all } \theta$$

$$\therefore \quad \varepsilon_1 A_1 = -\varepsilon_2\left(E_0 + \frac{2D_1}{a^2}\right) \tag{5.73}$$

Solving (5.72) and (5.73), one obtains

$$D_1 = \frac{\varepsilon_1 - \varepsilon_2}{\varepsilon_1 + 2\varepsilon_2} E_0 a^2 \text{ and } A_1 = \frac{-3\varepsilon_0}{\varepsilon_1 + 2\varepsilon_2} E_0 \tag{5.74}$$

$$\therefore \quad \phi_1 = -\frac{3\varepsilon_2}{\varepsilon_1 + 2\varepsilon_2} E_0 r\cos\theta \text{ and } = \frac{3\varepsilon_0}{\varepsilon_1 + 2\varepsilon_2} E_{0z} \tag{5.75}$$

and

$$\phi_2 = -E_0 r\cos\theta + \frac{\varepsilon_1 - \varepsilon_2}{\varepsilon_1 + 2\varepsilon_2} \frac{E_0 a^3 \cos\theta}{r^2} \tag{5.76}$$

one obtains the field inside the sphere as

$$E_{in} = -\frac{\partial\phi_1}{\partial z} = \frac{3\varepsilon_2}{\varepsilon_1 + 2\varepsilon_2} E_0 \tag{5.77}$$

This field, i.e. E_{in} is constant, uniform and parallel to the field applied. If

$$\varepsilon_1 > \varepsilon_2, \quad E_{in} < E_0 \tag{5.78}$$

This is due to the induced surface polarization charges which give rise to an opposing field.

One can obtain the external field from (5.76). E_{ext} is equal to E_0 plus a dipole field of moment given by

$$p = \frac{\varepsilon_1 - \varepsilon_2}{\varepsilon_1 + 2\varepsilon_2} a^2 E_0 \tag{5.79}$$

This shows that a dielectric sphere in a uniform field acts as a simple dipole.

(iii) A spherical cavity in a dielectric medium

Fig 5.6 exhibit a spherical cavity in a dielectric medium ABCD with dielectric constant ε. One can solve, the problem in the same manner as in the case of the dielectric sphere, i.e. case (ii). Now putting $\varepsilon_1 = 1$ and $\varepsilon_2 = \varepsilon$ in Eq. (5.77), one obtains

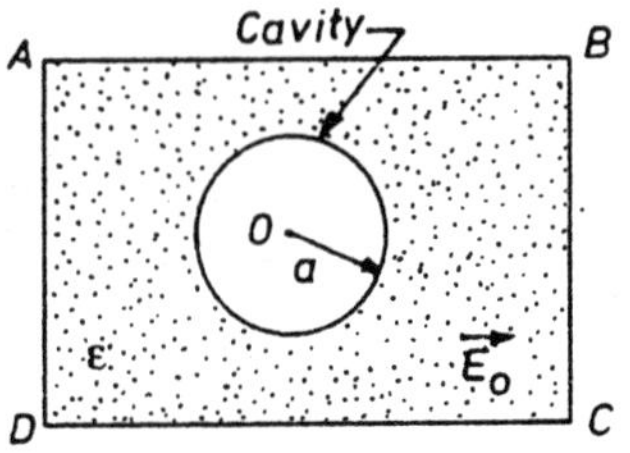

Figure 5.6. A spherical cavity in a dielectric medium.

$$E_{in} = \frac{3\varepsilon}{1 + 2\varepsilon} E_0 \tag{5.80}$$

If $\varepsilon > 1$ and $E_{in} > E_0$.

The outside field is equal to the applied field E_0 plus the field due to the dipole at the origin of dipole moment given by

$$p = \frac{1 - \varepsilon}{1 + 2\varepsilon} a^2 E \quad \text{(from eq. (5.79), using } \varepsilon_1 = 1, \text{ and } \varepsilon_2 = 2)$$

The outside field, i.e. $E_0 + p$ is oriented oppositely to the applied field.

5.9 SOLUTION OF LAPLACE'S EQUATION IN CARTESIAN COORDINATES

We have the Laplace's equation

$$\nabla^2 \phi = 0 \tag{5.81}$$

In rectangular coordinates, we can express (5.81) as

$$\nabla^2\phi = \frac{\partial^2\phi}{\partial x^2} + \frac{\partial^2\phi}{\partial y^2} + \frac{\partial^2\phi}{\partial z^2} = 0 \tag{5.82}$$

One can solve this equation by the method of separation of variables. Let us assume that solution of (5.82), $\phi(x,y,z)$ can be represented as the product of the three functions, viz. $X(x)$, $Y(y)$ and $Z(z)$, each one of which depends on one coordinate only. Thus, we have

$$\phi(x,y,z) = X(x)Y(y)Z(z) \tag{5.83}$$

Substituting (5.83) in (5.82) and dividing throughout by $X(x)Y(y)Z(z)$ one obtains

$$\frac{1}{X(x)}\frac{d^2X}{dx^2} + \frac{1}{Y(y)}\frac{d^2Y}{dy^2} + \frac{1}{Z(z)}\frac{d^2Z}{dz^2} = 0 \tag{5.84}$$

We must note that we have replaced partial derivatives by total derivaties. This can be done in the present case as each term involves a function of one coordinate only.

For Eq. (5.84) to be valid for arbitrary values of independent coordinates, one finds that each term on the left must be separately equal to a constant. One gets

$$\frac{1}{X(x)}\frac{d^2X}{dx^2} = \alpha^2; \quad \frac{1}{Yy}\frac{d^2Y}{dy^2} = \beta^2 \quad \text{and} \quad \frac{1}{Z(z)}\frac{d^2Z}{dz^2} = \gamma^2 \tag{5.85}$$

subject to the condition that

$$\alpha^2 + \beta^2 + \gamma^2 = 0 \tag{5.86}$$

One can easily see that not all these constants can be real nor all imaginary. One finds that atleast one of them must be real and one imaginary. The third constant may be real or imaginary. The real or imaginary nature of constants is decided by the nature of the problem.

Writing the three equations

$$\frac{1}{X(x)}\frac{d^2X}{dx^2} = -\alpha^2, \quad \frac{1}{Y(y)}\frac{d^2Y}{dy^2} = -\beta^2 \text{ and } \frac{1}{Z(z)}\frac{d^2Z}{dz^2} = \gamma^2 \tag{5.87}$$

(Note that we have taken constant α^2 and β^2 as $-\alpha^2$ and $-\beta^2$).

One obtains the solutions of three differential equation (5.87) as

$$\left.\begin{aligned} X(x) &= A_1e^{i\alpha x} + A_2e^{-i\alpha x} \\ Y(y) &= B_1e^{i\beta y} + B_2e^{-i\beta y} \\ \text{and} \quad Z(z) &= C_1e^{rz} + C_2e^{-rz} \end{aligned}\right\} \tag{5.88}$$

It is evident that first two solutions are oscillatory while the last solution in (5.88) is exponential. The constants involved in (5.88) can be found from boundary conditions to be satisfied by the potential. The potential ϕ is the product of three functions as shown by (5.83).

We must note that (5.88) is one particular solution of (5.84). One can also find other values of α, β and γ which could give valid solutions of (5.84). One can write the general solution as

$$\phi(x,y,z) = \sum_{mn,p}^{\infty}(A_1^m e^{i\alpha_m x} + A_2^m e^{-i\alpha_m x}) \\ (B_1^n e^{i\beta_n y} + B_2^n e^{-i\beta_n y})(C_1^p e^{r_p x} + C_2^p e^{-r_p x}) \tag{5.89}$$

5.10 LAPLACE'S EQUATION IN SPHERICAL POLAR COORDINATES

When one is dealing with a problem having axial symmetry, it is generally convenient to use spherical polar coordinates r, θ, ϕ and may chose the axis of symmetry as the polar axis $\theta = 0$.

Laplace's equation for the potential function V is

$$\nabla^2 V = 0 \tag{5.90}$$

In the spherical polar coordinates (r,θ,ϕ), Laplace's equation is written in the form

$$\frac{1}{r^2}\frac{\partial}{\partial r}\left(r^2\frac{\partial V}{\partial r}\right) + \frac{1}{r^2 \sin\theta}\frac{\partial}{\partial\theta}\left(\sin\theta\frac{\partial V}{\partial r}\right) + \frac{1}{r^2\sin^2\theta}\frac{\partial^2 V}{\partial\phi^2} = 0 \tag{5.91}$$

The potential has been denoted by V to avoid confusion with the coordinate ϕ.

It is obvious from Eq. (5.91) that the potential function V is a function of (r,θ,ϕ); hence using method of separation of variables, we may write

$$V = RS \tag{5.92}$$

where R is a function of r only and S is a function of θ and ϕ, i.e. both the angular coordinates. Using (5.92), (5.91) becomes

$$\frac{S}{r^2}\frac{\partial}{\partial r}\left(r^2\frac{\partial R}{\partial r}\right) + \frac{R}{r^2 \sin\theta}\frac{\partial}{\partial\theta}\left(\sin\theta\frac{\partial S}{\partial \theta}\right) + \frac{R}{r^2\sin^2\theta}\frac{\partial^2 S}{\partial\phi^2} = 0 \tag{5.93}$$

Dividing (5.93) throughout by $\frac{RS}{r^2}$ one obtains

$$\frac{1}{R}\frac{\partial}{\partial r}\left(r^2\frac{\partial R}{\partial r}\right) + \frac{1}{S \sin\theta}\frac{\partial}{\partial\theta}\left(\sin\theta\frac{\partial S}{\partial \theta}\right) + \frac{1}{S\sin^2\theta}\frac{\partial^2 S}{\partial\phi^2} = 0 \tag{5.94}$$

One can see that the first term is a function of r only while the remaining two terms are independent of r. The above equation is satisfied if we take

$$\frac{1}{R}\frac{\partial}{\partial r}\left(r^2\frac{\partial R}{\partial r}\right) = m \tag{5.95}$$

and

$$\frac{1}{S\sin\theta}\frac{\partial}{\partial\theta}\left(\sin\theta\frac{\partial S}{\partial\theta}\right) + \frac{1}{S\sin^2\theta}\frac{\partial^2 S}{\partial\phi^2} = -m \tag{5.96}$$

where m is a constant. The solutions of eqs. (5.95) and (5.96) take simpler forms, if one takes the constant m as $l(l+1)$ where the constant l is still arbitrary. We get

$$\frac{1}{R}\frac{\partial}{\partial r}\left(r^2\frac{\partial R}{\partial r}\right) = l(l+1) \tag{5.97}$$

$$\frac{1}{\sin\theta}\frac{\partial}{\partial\theta}\left(\sin\theta\frac{\partial S}{\partial\theta}\right) + \frac{1}{\sin^2\theta}\frac{\partial^2 S}{\partial\phi^2} = -l(l+1)S \tag{5.98}$$

First we may find the solution of radial equation (5.97). This may be expressed as

$$\frac{\partial}{\partial r}\left(r^2\frac{\partial R}{\partial r}\right) - l(l+1)R = 0 \tag{5.99}$$

Let us substitute

$$R(r) = \frac{F(r)}{r} \tag{5.100}$$

Then eq. (5.99) becomes

$$\frac{\partial^2 F(r)}{\partial r^2} - \frac{l(l+1)}{r^2}F(r) = 0 \tag{5.101}$$

From the form of (5.101) it is apparent that a single power of r (rather than of power series) will satisfy it. One finds the solution to be

$$F(r) = Ar^{l+1} + Br^{-l} \tag{5.102}$$

Where l is yet undetermined and A and B are arbitrary constants. Using (5.102), we note that

$$R(r) = Ar^l + \frac{\beta}{r^{l+1}} \tag{5.103}$$

Now, we may try for the solution of (5.98). Any solution of (5.98) S_l is a function of θ and ϕ and is called a surface harmonic of degree 'l'. Adopting the same technique, let the solution of (5.98) be

$$S = P(\theta)Q(\phi) \tag{5.104}$$

Obviously, $P(\theta)$ is a function of θ only and $Q(\phi)$ is a function of ϕ only. Substituting (5.104) in (5.98) one obtains

$$\frac{Q}{\sin\theta}\frac{d}{d\theta}\left(\sin\theta\frac{dP}{d\theta}\right)+\frac{P}{\sin^2\theta}\frac{d^2Q}{d\phi^2}+l(l+1)PQ=0 \qquad (5.105)$$

Dividing (5.105) throughout by $\frac{PQ}{\sin^2\theta}$, one obtains

$$\frac{\sin\theta}{P}\frac{d}{d\theta}\left(\sin\theta\frac{dP}{d\theta}\right)+l(l+1)\sin^2\theta+\frac{1}{Q}\frac{d^2Q}{d\phi^2}=0 \qquad (5.106)$$

One finds that the variables are again separable. The first two terms in (5.106) are functions of θ only and the last term is a function of ϕ only. Let us take

$$\frac{1}{Q}\frac{d^2Q}{d\phi^2}=-m^2 \qquad (5.107)$$

Where m^2 is constant. The solution of (5.107) is

$$Q_m=Ce^{\pm im\phi} \qquad (5.108)$$

where C is a constant. In order that potential V be single valued, it is essential that

$$e^{\pm im\phi}=e^{\pm im(\phi+2\pi)} \qquad (5.109)$$

This is possible only if m is an integer. One can normalize the function Q_m by choosing the constant in such way that $\int_0^{2\pi}Q_m^*Q_m d\phi=1$. For this to be satisfied, constant C must be equal to $\frac{1}{\sqrt{2\pi}}$. We must note that the functions Q_m are also orthogonal, i.e.

$$\int_0^{2\pi}Q_m^*Q_n d\phi = 0 \quad \text{if} \quad m\neq n$$

$$\therefore \quad \int_0^{2\pi}Q_m^*Q_n d\phi = \delta_{mn} \qquad (5.110)$$

The solution of Laplace's equation (5.91) is

$$V=\left(Arl+\frac{B}{r^{l+1}}\right)S_l. \qquad (5.111)$$

5.11 LEGENDRE'S EQUATION

Using (5.107), one obtains the θ part of the Eq. (5.106) as

$$\frac{\sin\theta}{P}\frac{d}{d\theta}\left(\sin\theta\frac{dP}{d\theta}\right)+l(l+1)\sin^2\theta=m^2$$

Dividing throughout by $\frac{\sin^2\theta}{P}$, one obtains

$$\frac{1}{\sin\theta}\frac{d}{d\theta}\left(\sin\theta\frac{dP}{d\theta}\right)+l(l+1)P-\frac{m^2}{\sin^2\theta}P=0 \tag{5.112}$$

Put $x=\cos\theta$, one obtains

$$\frac{d}{dx}\left[(1-x^2)\frac{dP}{dx}\right]+\left[l(l+1)-\frac{m^2}{1-x^2}\right]P=0$$
$$\left(\because\quad \frac{d}{d\theta}=\frac{dx}{d\theta}\cdot\frac{d}{dx}=-\sin\theta\frac{d}{dx}\right) \tag{5.113}$$

Eq. (5.113) is known as the *generalized Legendre's equation.* Putting $m=0$ one obtains the ordinary Legendre's equation, i.e.

$$\frac{d}{dx}\left[(1-x^2)\frac{dP}{dx}\right]+l(l+1)P = 0 \tag{5.114}$$

or

$$(1-x^2)P''-2xP'+l(l+1)P = 0 \tag{5.115}$$

Let us try for the solution of (5.115). One can try its solution by series integration. Let us assume that the solution of (5.115) is

$$P(x)=a_0x^\lambda+a_1x^{1+\lambda}+a_2x^{2+\lambda}+\cdots=\sum_{k=0}^{\infty}a_kx^{k+\lambda} \tag{5.116}$$

The constants a_k and λ are to be determined substituting (5.116) in (5.115), one obtains

$$\sum_{k=0}^{\infty}a_k(k+\lambda)(k+\lambda-1)x^{k+\lambda-2}$$
$$-\sum_{k=0}^{\infty}a_k[(k+\lambda)(k+\lambda-1)+2(k+\lambda)-l(l+1)]x^{k+\lambda}=0 \tag{5.117}$$

Eq. (5.117) must hold good for every value of x. Obviously, the coefficient of every power of x must vanish. One can find the lowest power of x by putting $k=0$ in (5.117). One obtains this to be $\lambda-2$. This means the coefficient of $x^{\lambda-2}$ must be zero, i.e.

$$a_0\lambda(\lambda-1)=0 \tag{5.118}$$

Eq. (5.118) is known as *indicial equation.* Since $a_0\neq 0$, because it is assumed to be the coefficient of lowest power and hence, either

$$\lambda=0 \quad\text{or}\quad \lambda=1$$

one can find the coefficient of $x^{j+\lambda}$ form (5.117) where j is any value of k, by putting $k = j+2$ in the first term and $k = j$ in the second. One obtains

$$a_{j+2}(j+\lambda+2)(j+\lambda+1) - a_j[(j+\lambda)(j+\lambda+1) - l(l+1)]$$

and this must be zero. Thus

$$a_{j+2} = \frac{(j+\lambda)(j+\lambda+1) - l(l+1)}{(j+\lambda+2)(j+\lambda+1)} a_j \tag{5.119}$$

Obviously, if a_j is known, one can find a_{j+2}. Starting from an arbitrary value of a_0, one can easily compute $a_2, a_4, \ldots$ etc. By assigning any arbitrary value to a_1 one can also compute $a_3, a_5, a_7, \ldots$ etc.

For this computation, let us take $\lambda = 0$, since this provides all the necessary independent solutions. one finds

$$a_{j+2} = \frac{j(j+1) - l(l+1)}{(j+2)(j+1)} a_j \tag{5.120}$$

From this, one obtains

$$\left.\begin{aligned}
a_2 &= -\frac{l(l+1)}{\lfloor 2} a_0 \\
a_4 &= \frac{6 - l(l+1)}{4-3} a_2 = -\frac{6 - l(l+1)l(l+1)}{4\cdot 3 \lfloor 2} a_0 \\
&= \frac{(l-2)l(l+1)(l+3)}{\lfloor 4} a_0 \\
a_6 &= -\frac{(l-4)(l-2)l(l+1)(l+3)(l+5)}{\lfloor 6} a_0 \\
\ldots & \ldots \ldots\ldots\ldots\ldots\ldots\ldots \\
\ldots & \ldots \ldots\ldots\ldots\ldots\ldots\ldots \\
a_3 &= \frac{2 - l(l+1)}{3\cdot 2} a_1 = -\frac{(l-1)(l+2)}{\lfloor 3} a_1 \\
a_5 &= \frac{(l-3)(l-1)(l+2)(l+4)}{\lfloor 5} a_1 \\
\ldots & \ldots \ldots\ldots\ldots\ldots\ldots\ldots
\end{aligned}\right] \tag{5.121}$$

Thus, one obtains

$$P_l = a_0 \left\{ 1 - \frac{l(l+1)}{\lfloor 2} x^2 + \frac{(l-2)l(l+1)(l+3)}{\lfloor 4} x^4 \right. \tag{5.122}$$

$$P_2 = a_1\left\{x - \frac{(R-1)(l+2)}{\lfloor 3}x^3 + \frac{(l-3)(l-1)(l+2)(l+4)}{\lfloor 5}x^5\right. \tag{5.123}$$

Eqs. (5.122) and (5.123) are both solutions of Legendre's Eq. (5.115). One can write the general solution as

$$U = AP_1 + BP_2 \tag{5.124}$$

Of course, we have found the solutions of the Legendre's equation, but whether the solutions are of any interest depends on their converging properties. We must note that a series converges if the ratio of two successive terms. i.e. $\frac{a_{j+2}}{a_j}x^2$ is smaller than unity for large j. One can see from (5.119) that

$$\frac{a_{j+2}}{a_j} = 1 \quad \text{as} \quad j \to \infty$$

Obviously, the series will converge if $x^2 < 1$, i.e. if the values of x lie in the range $-1 < x < 1$. We must remember that x in our equation stands for $\cos\theta$, its values range from $+1$ to -1. It is worthwhile to mention that for $x = \pm 1$, the series diverges and the solution becomes unacceptable unless the series terminates and becomes a polynomial. One can see from (5.122) and (5.123) that P_1 terminates for even values of l and P_2 for odd values of l_2. The polynomials in each case have x^l as their highest power of x, the next highest being x^{l-2}, and so on, down to $x^{\circ}(x)$ for l even (odd). By convention these polynomials are normalized to have the value unity at $x = \pm 1$ and are called the *Legendre polynomials* of order l, $P_l(x)$. One can brough normalization by taking

$$a_0 = \frac{1}{1 - \frac{l(l+1)}{\lfloor 2} + \cdots \text{up the coefficients of the highest power of} x} \tag{5.125}$$

$$a_1 = \frac{1}{1 - \frac{(l-1)(l+2)}{3} + \cdots \text{up to the coefficients of the highest power of} x} \tag{5.126}$$

The first few Legendre polynomials are (see (5.122) and (5.123))

$$\text{for } l = 0, \; a_0 = 1 \text{ and } P_0(x) = 1$$

$$l = 1, \; a_1 = 1 \text{ and } p_1(x) = x$$

$$l = 2,\ a_0 = -\frac{1}{2} \text{ and } P_2(x) = \frac{1}{2}(3x^2 - 1)$$

$$l = 3,\ a_1 = -\frac{3}{2} \text{ and } P_3(x) = \frac{1}{2}(5x^3 - 3x)$$

$$l = 4,\ a_0 = -\frac{1}{9} \text{ and } P_4(x) = \frac{1}{8}(35x^4 - 30x^2 + 3)$$

and so on. The general formula for $P_l(x)$ is

$$P_l(x) = \sum_{r=0}^{N} \frac{(-1)r\lfloor(2l-2r)}{2^l \lfloor r \lfloor(l-r) \lfloor(l-2r)} x^{l-2r} \tag{5.127}$$

with $N = \frac{l}{2}$ if l is even and $N = \frac{l-1}{2}$ if l is odd. By manipulation of power series solutions (5.116) and (5.120) one can obtain a compact representation of the Legendre polynomials known as *Rodrigues formula*:

$$P_l(x) = \frac{1}{2^l \lfloor l} \frac{d^l}{dx^l}(x^2 - 1)^l \tag{5.128}$$

One can easily see that (5.128) yields the same expression for $P_l(x)$ as the one obtained from (5.127) for a given value of l.

One can express the Legendre's polynomials in a different way also. If one expands $(1 - 2xs + s^2)^{-1/2}$ by Maclaurin's theorem, one finds that the coefficients of various powers of s are Legendre polynomials, i.e.

$$(1 - 2xs + s^2)^{-1/2} = \sum_{l=0}^{\infty} P_l(x) s^l \tag{5.129}$$

The function $(1 - 2xs + s^2)^{-1/2}$ is called the *generating function* of Legendre polynomials.

The Legendre polynomials form a complete orthogonal set of functions on the interval $-1 \leq x \leq 1$. Let us prove it. To prove the orthogonality, we write the Legendre equation as

$$\frac{d}{dx}[(1 - x^2)P_l'(x)] + l(l+1)P_l(x) = 0 \tag{5.130}$$

Multiplying by $P_q(x)$ and integrating over the interval $-1 \leq x \leq 1$, one obtains

$$\int_{-1}^{1} P_q(x) \frac{d}{dx}[(1 - x^2)P_l'(x)]dx + l(l+1) \int_{-1}^{1} P_q(x)P_l(x)dx = 0$$

$$\therefore \quad [P_q(x)\{(1-x^2)P_l'(x)\}]_{-1}^{1} - \int_{-1}^{1}(1-x^2)P_l'(x)P_q'(x)P_l(x)dx$$
$$+l(l+1)\int_{-1}^{1}P_q(x)P_l(x)dx = 0$$

The first term of the above vanishes at both limits. One obtains

$$-\int_{-1}^{1}(1-x^2)P_l'(x)P_q'(x)dx + l(l+1)\int_{-1}^{1}P_q(x)P_l(x) = 0 \quad (5.131)$$

Interchanging l and q, one obtains

$$-\int_{-1}^{1}(1-x^2)P_q'(x)P_l'(x)dx + q(q+1)\int_{-1}^{1}P_l(x)P_q(x)dx = 0 \quad (5.132)$$

Subtracting (5.132) from (5.131) one obtains

$$[l(l+1) - q(q+1)]\int_{-1}^{1}P_l(x)P_q(x)dx = 0 \quad (5.133)$$

Case (i) when $l \neq q$, one obtains

$$\int_{-1}^{1}P_l(x)\,P_q(x) = 0 \quad (5.134)$$

This shows that the Legendre polynomials of different order are orthogonal.

Case (ii) when $l = q$, the integral is finite. The value of integral $\int_{-1}^{1}[P_l(x)]^2dx$ can be obtained easily by using the generating function $(1-2xs+s^2)^{-1/2}$. We have

$$(1-2xs+s^2)^{-1} = \left[\sum_{l=0}^{\infty}P_l(x)s^l\right]^2 \quad (5.135)$$

$$\therefore \quad \int_{-1}^{1}(1-2xs+s^2)^{-1}dx = \int_{-1}^{1}\left[\sum_{l=0}^{\infty}P_l(x)s^l\right]^2 dx$$

Integrating LHS, one obtains

$$\frac{1}{s}\log\frac{1+s}{1-s} = \int_{-1}^{1}\left[\sum_{l=0}^{\infty}s^lP_l(x)\right]^2 dx$$

The product term in the summation on the RHS side vanish due to orthogonality condition, one obtains

$$\frac{1}{s}\log\frac{1+s}{1-s} = \sum_{l=0}^{\infty}s^{2l}\int_{-1}^{l}[P_l(x)]^2dx \quad (5.136)$$

Now, expanding LHS and equating the coefficients of the power s^{2l}, one obtains

$$\int_{-1}^{1} [P_l(x)]^2 dx = \frac{2}{2l+1}, \quad l = 0, 1, 2, \ldots \tag{5.137}$$

One can express the orthogonality condition as

$$\int_{-1}^{1} P_l(x) P_q(x) dx = \frac{2}{2l+1} \delta_{ql} \tag{5.138}$$

and the orthonormal functions are

$$U_l(x) = \sqrt{\frac{2l+1}{2}} P_l(x) \tag{5.139}$$

Obviously, when $m = 0$, i.e. in the problems in which there is azimuthal symmetry, one can write the solution of Laplace's equation as

$$V = R(r)S(\theta) = \sum_{l=0}^{\infty} (A_l r^l + B_l r^{-l+1)}) \sqrt{\frac{2l+1}{2}} P_l(\cos\theta) \tag{5.140}$$

Since the Legendre polynomials form a complete set of orthogonal functions, any function $f(x)$ on the interval $-1 \le x \le 1$ can be expanded in terms of them, i.e.

$$f(x) = C_0 + C_1 P_1(x) + C_2 P_2(x) + \cdots = \sum_{l=0}^{\infty} C_l P_l(x) \tag{5.141}$$

In order to determine the coefficients C_l, one should multiply both sides by $P_m(x)$ and then integrate. Thus

$$\int_{-1}^{1} f(x) P_m(x) dx = \sum_{l=0}^{\infty} C_l \int_{-1}^{1} P_l(x) P_m(x) dx = \frac{2C_m}{2m+1}$$

$$\therefore \quad C_m = \frac{2m+1}{2} \int_{-1}^{1} f(x) P_m(x) dx \tag{5.142}$$

5.12 ASSOCIATED LEGENDRE FUNCTIONS

So far we have dealt with problems possessing azimuthal symmetry. Now, we discuss the solution of Laplace's equation when the potential problem has azimuthal variation, i.e. $m \neq 0$. We have seen that $y = P_l(x)$ is a solution of the equation

$$(1 - x^2) \frac{d^2 y}{dx^2} - 2x \frac{dy}{dx} + l(l+1) y = 0 \tag{5.143}$$

Differentiating (5.143) m times, one obtains

$$(1-x^2)\frac{d^{m+2}y}{dx^{m+2}} - 2x(m+1)\frac{d^{m+1}y}{dx^{m+1}} + (l-m)(l+m+1)\frac{d^m y}{dx^m} = 0 \tag{5.144}$$

Let us put

$$v = \frac{d^m y}{dx^m} = \frac{d^m P_l(x)}{dx^m} \tag{5.145}$$

One obtains (5.144) as

$$(1-x^2)\frac{d^2 v}{dx^2} - 2x(m+1)\frac{dv}{dx} + (l-m)(l+m+1)v = 0 \tag{5.146}$$

Eq. (5.146) is satisfied by (5.145). Now we make a further substitution

$$y = v(1-x^2)^{m/2} \tag{5.147}$$

One obtains

$$(1-x^2)\frac{d^2 y}{dx^2} - 2x\frac{dy}{dx} + \left[l(l+1) - \frac{m^2}{1-x^2}\right]y = 0 \tag{5.148}$$

One can see that (5.148) is the same as generalized Legendre equation (5.113) and its solution is

$$Y = v(1-x^2)^{m/2} = (1-x^2)^{m/2}\frac{d^m P_l(x)}{dx^m} = P_l^m(x) \tag{5.149}$$

The functions $P_l^m(x)$ are known as *associated Legendre polynomials.*

The orthogonality condition for associated Legendre polynomials can be obtained as

$$\int_{-1}^{1} P_{l'}^m(x)P_l^m(x)dx = \frac{2}{2l+1}\frac{\lfloor l+m}{\lfloor l-m}\delta_{l'l} \tag{5.150}$$

From the normalized condition (5.150) it is clear that the suitably normalized functions, denoted by $Y_{lm}(\theta,\phi)$ are

$$Y_{lm}(\theta,\phi) = \sqrt{\frac{(2l+1)\lfloor l-m}{4\pi\lfloor l+m}}P_l^m(\cos\theta)\,e^{im\phi} \tag{5.151}$$

One can write the general solution of Laplace's equation as

$$V(r,\theta,\phi) = \sum_{l=0}^{\infty}\sum_{m=-l}^{l}[A_{lm}r^l + B_{lm}r^{-(l+1)}]Y_{lm}(\theta,\phi) \tag{5.152}$$

For a few small l values and $m \geq 0$ the list below shows the explicit form of the $Y_{lm}(\theta,\phi)$.

$$l=0 \quad Y_{00} = \frac{1}{\sqrt{4\pi}}$$

$$l=1 \quad \begin{cases} Y_{11} &= \sqrt{\dfrac{3}{8\pi}} sin\theta\, e^{i\phi} \\ Y_{10} &= \sqrt{\dfrac{3}{4\pi}} \cos\theta \end{cases}$$

$$l=2 \quad \begin{cases} Y_{22} &= \dfrac{1}{4}\dfrac{15}{2\pi} sin^2\theta\, e^{2i\phi} \\ Y_{21} &= -\sqrt{\dfrac{15}{8\pi}} \sin\theta\cos\theta\, e^{i\phi} \\ Y_{20} &= \sqrt{\dfrac{5}{4\pi}}\left(\dfrac{3}{2}\cos^2\theta - \dfrac{1}{2}\right) \end{cases}$$

$$l=3 \quad \begin{cases} Y_{33} &= -\dfrac{1}{4}\sqrt{\dfrac{35}{4\pi}} \sin^3\theta\, e^{3i\phi} \\ Y_{32} &= \dfrac{1}{4}\sqrt{\dfrac{105}{2\pi}} \sin^2\theta\cos\theta\, e^{2i\phi} \\ Y_{31} &= -\dfrac{1}{4}\sqrt{\dfrac{21}{4\pi}} \sin\theta(5\cos^2\theta - 1)e^{i\phi} \\ Y_{30} &= \sqrt{\dfrac{7}{4\pi}}\left(\dfrac{5}{2}\cos^3\theta - \dfrac{3}{2}\cos\theta\right) \end{cases}$$

For $m = 0$, we have

$$Y_{l0}(\theta,\phi) = \sqrt{\frac{2l+1}{4\pi}} P_l(\cos\theta) \tag{5.153}$$

EXAMPLE 5.1: Two uniformly and oppositely charged wires run through a metal cylinder of radius R. With the help of method of images show that there will be no force on the wires if they have a separation of $2R = \sqrt{\sqrt{5}-2}$.

Solution: One can obtain the field internal to the cylinder by having to images charges at a distance $d = \dfrac{R^2}{e}$ from the axis of the cylinder as shown in Fig 5.7. Here e is the eccentricity. The field of the line charge $+Q$ is obtained as

$$E = \frac{Q}{2\pi\varepsilon_0}\left(\frac{1}{\frac{R^2}{e}+e} - \frac{1}{2e} + \frac{1}{\frac{R^2}{e}-e}\right)$$

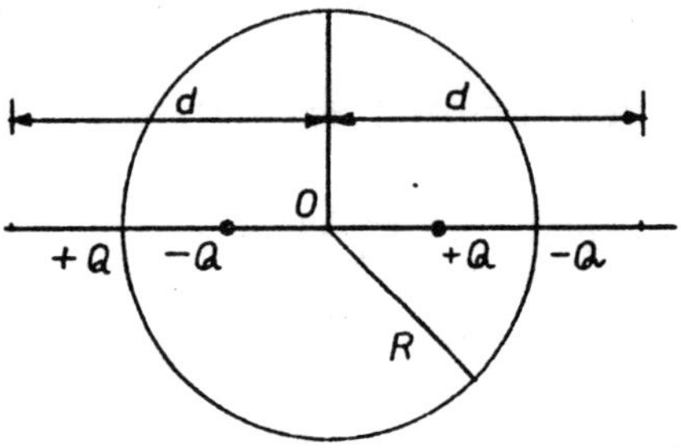

Figure 5.7.

when there is no force on the wire. We have $E = 0$. This yields

$$e^4 + 4e^2R^2 - R^4 = 0, \quad e > 0$$

or

$$e = 2R\sqrt{\sqrt{5}-2}.$$

EXAMPLE 5.2: Consider a rectangular box, located as shown in Fig 5.8 with dimensions $(a, b, c,)$ in the x, y and z directions respectively. If all the surfaces of the box are at zero potential except the surface $z = c$ which is at potential $\phi_0(x, y)$, then determine the potential every where inside the box.

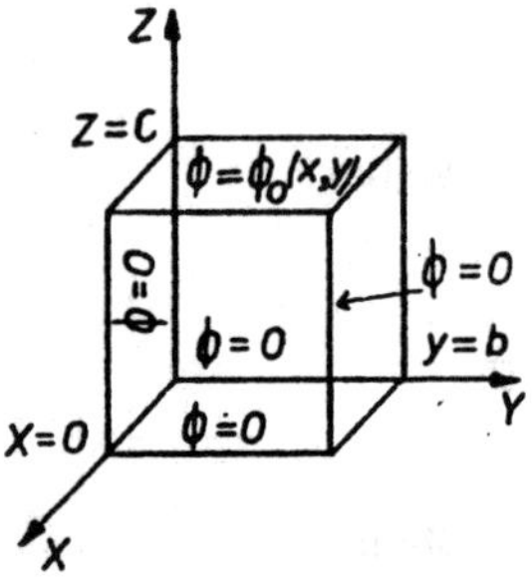

Figure 5.8.

Solution: We have the Laplace's equation

$$\nabla^2\phi = 0 \tag{i}$$

The general solution of Laplace's equation is

$$\phi = X(x)Y(y), Z(z) \tag{ii}$$

Making use of boundary condition that $\phi = 0$ for $x = 0$, $y = 0$, $z = 0$, one finds

$$X(x) = \sin\alpha x$$

$$Y(y) = \sin\beta y$$

and
$$Z(z) = \sinh\sqrt{\{(\alpha^2+\beta^2)z\}} \tag{iii}$$

In order that $\phi = 0$ at $x = a$ and $y = b$, it is essential that

$$\left.\begin{aligned} \alpha_n a &= n\pi \quad \text{and} \quad \beta_m b = m\pi \\ \text{or} \quad \alpha_n &= \frac{n\pi}{a} \quad \text{and} \quad \beta_m = \frac{m\pi}{b} \\ \text{and} \quad r_{nm} &= \sqrt{\alpha^2+\beta^2} = \pi\sqrt{\frac{n^2}{a^2}+\frac{m^2}{b^2}} \end{aligned}\right\} \tag{iv}$$

Now, one can write the partial potential ϕ_{nm}, satisfying all boundary conditions, except one

$$\phi_{nm} = \sin(\alpha_n x)\sin(\beta_m y)\sin h(r_{nm}z) \tag{v}$$

Now, the potential can be expanded in terms of these ϕ_{nm} with initial arbitrary coefficients to be chosen to satisfy the final boundary condition,

$$\begin{aligned} \phi(x,y,z) &= \sum_{nm} A_{nm}\phi_{nm} \\ &= \sum_{n,m-1}^{\infty} A_{nm}\sin(\alpha_n x)\sin(\beta_m y)\sin h(\gamma_{nm}z) \end{aligned} \tag{vi}$$

We have the final boundary condition,

$$\phi = \phi_0(x,y) \quad \text{at} \quad z = c;$$

$$\therefore \quad \phi_0(x,y) = \sum_{n,m-1}^{\infty} A_{nm}\sin(\alpha_n x)\sin(\beta_m y)\sin h(\gamma_{nm}c) \tag{vii}$$

Obviously, this is just a double Fourier series for the function $\phi_0(x,y)$. One obtains the coefficients A_{nm} as

$$A_{nm} = \frac{4}{ab\sin h(\gamma_{nm}c)}\int_0^a dx\int_0^b dy\phi_0(x,y)\sin(\alpha_m x)\sin(\beta_m y) \tag{viii}$$

EXAMPLE 5.3: Two infinitely long grounded metal plates at $y = 0$ and $y = b$ are connected at $x = \pm 1$ by metal strips maintained at potential ϕ_0 (Fig 5.9). A thin silver of insulation at each corner is used to prevent them from shorting out. Show that the potential inside the resulting rectangular pipe is

$$\phi(x,y) = \frac{4\phi_0}{\pi} \sum_{k=1,3,5}^{\infty} \frac{1}{k} \frac{\cos h\, kx}{\cos hk} \sin ky$$

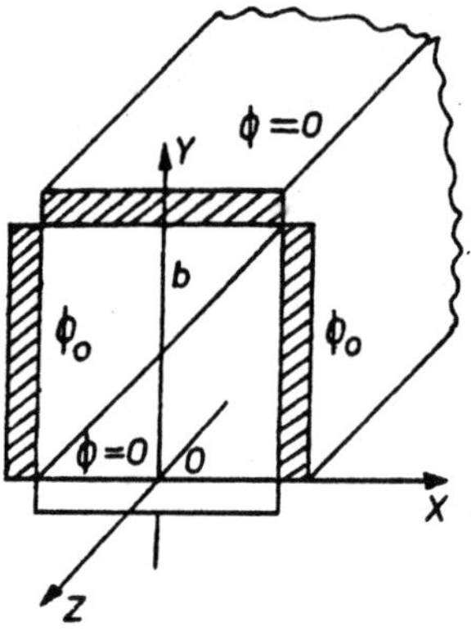

Figure 5.9.

Solution: We have the Laplace's equation

$$\frac{\partial^2 \phi}{\partial x^2} + \frac{\partial^2 \phi}{\partial y^2} = 0 \tag{1}$$

The boundary conditions are $\phi = 0$ when $y = 0$ and $\phi = \phi_0$ when $x = \pm 1$. Thus, we have

$$X = B_1 e^{kx} + B_2 e^{-kx} \tag{2}$$

and

$$Y = A_1 \sin ky + A_2 \sin ky$$

Applying the boundary conditions, one obtains, $A_2 = 0$ and $k = \frac{n\pi}{b}$, where $n = 1, 2, 3, \ldots$.

Obviously here we cannot set $B_1 = 0$, as the region in question does not extend to $x = \infty$ and hence e^{kx} is a perfectly acceptable contribution to the potential. On the other hand, the situation is obviously even with respect to x, i.e. $\phi(-x,y) = \phi(x,y)$. It gives $B_1 = B_2$.

$$\therefore \quad B_1 e^{kx} + B_1 e^{-kx} = 2B_1 \cos h\, kx$$

Absorbing $2B_1$, into A_1 and A_2 one obtains

$$\phi(x,y) = \cos h\, kx (A_1 \sin ky + A_2 \cos h\, ky) \tag{4}$$

Applying the boundary condition, one obtains $A_2 = 0$.

$$\therefore \quad \phi(x,y) = C\cos hkx \sin ky \tag{5}$$

The general solution is

$$\phi(x,y) = \sum_{k=1}^{\infty} C_k \cos hkx \sin ky \tag{6}$$

Again, one finds

$$\phi(1,y) = \sum_{k=1}^{\infty} C_k \cos hk \sin ky = \phi_0 \tag{7}$$

Making use of Fourier analysis, one obtains $C_n \cos hn = \dfrac{4\phi_0}{n\pi}$ when n is odd and zero if n is even

$$\therefore \quad \phi(x,y) = \frac{4\phi_0}{\pi} \sum_{k=1,3,5}^{\infty} \frac{1}{k} \frac{\cos hkx}{\cos hk} \sin ky \tag{8}$$

EXAMPLE 5.4: A set of wires parallel to the Y-axis in the plane $z = 0$ are arranged as shown in Fig 5.10. The wires are equally spaced at $x = \pm n(\frac{a}{2})$, where $n = 0, 1, 2, 3, 4, \ldots$. The wires at odd positions (i.e. for n odd) are at potential ϕ_0 while those at even positions are at potential ϕ_0 (i.e. for n even). Show that the potential at all points in space is given by

$$\phi(x,z) = \sum_k A_k e^{\frac{-2\pi k}{a}} \cos \frac{2\pi kx}{a} \quad \text{for } z > 0$$

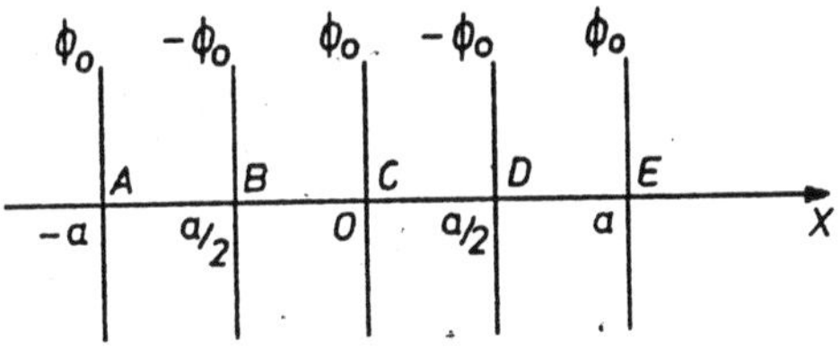

Figure 5.10.

Solution: We can see that the potential ϕ is independent of y and varies periodically with x. Potential is an even function of x, one can represent it by a cosine series $\cos\left(\dfrac{2\pi kx}{a}\right)$. Although the functional dependence of ϕ on z is not known; but from the above one can expect it to be exponential. Obviously, one expect the general form of ϕ as

$$\phi = \sum_{k=-\infty}^{\infty} f_k(z) \cos \frac{2\pi kx}{a} \tag{1}$$

we have to determine the form of $f(z)$. If ϕ is a valid potential then it will have to satisfy Laplace's equation $\frac{\partial^2 \phi}{\partial x^2} + \frac{\partial^2 \phi}{\partial y^2} = 0$ in the region above the wires where there are no charges. Obviously,

$$\sum_{k=-\infty}^{\infty} \left[-\frac{4\pi^2 k^2}{a^2} f_k(z) \cos \frac{2\pi kx}{a} + \frac{\partial^2 f_k(z)}{\partial z^2} \cos \frac{2\pi kx}{a} \right] = 0$$

The above must be true for every value of k. Thus, one obtains

$$-\frac{4\pi^2 k^2}{a^2} f_k(z) + \frac{\partial^2 f_k(z)}{\partial z^2} = 0$$

Solving the above, one obtains

$$f_k(z) = A_k e^{\pm \frac{2\pi k}{a} z}$$

where A_k is a constant.

Nature of the problem demands that for $z > 0$, one should retain only the negative sign of the exponent. Thus, one obtains

$$f_k(z) = A_k e^{-\frac{2\pi k}{a} z}$$

Making use of the above, one obtains from (1)

$$\phi(x, z) = \sum_k A_k e^{-\frac{2\pi k}{a}} \cos \frac{2\pi kx}{a} \quad \text{for} \quad z > 0 \tag{2}$$

At $z = 0$, ϕ (give by (2)) must coincide with the prescribed potential. One can determine A_k from this requirement.

EXAMPLE 5.5: Consider the situation shown in the left of Fig 5.11. The potential at the plane $z = 0$ is constant, so this is equivalent to the situation shown at the right of the Fig 5.11. Keeping in mind that one can determine the electric field for all $z > 0$ and uniqueness allows only one admissible solution. Show that the surface charge density is

$$\rho = \frac{2\varepsilon_0 q l}{(s^2 - l^2)^{3/2}} \quad \text{for} \quad l = z.$$

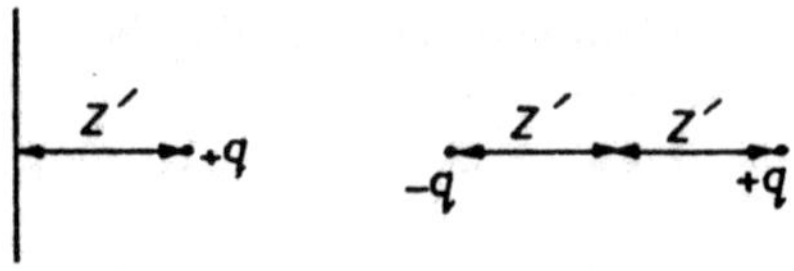

Figure 5.11. Image charges used to simulate a conducting plane.

Solution: The induced surface charge density on the conductor is found from $E = -\left(\frac{\partial\phi}{\partial z}\right)_{z=0}$. One can easily show that

$$\rho = \varepsilon_0 E = \frac{2\varepsilon_0 ql}{(s^2 - l^2)^{3/2}} \quad \text{for} \quad l = z.$$

EXAMPLE 5.6: A particle with mass m and charge q is suspended by a string of length L. At a distance d below the point of suspension there is an infinite conductor. (i) Determine the frequency of the pendulum for small oscillation about its equilibrium point (neglect gravity and radiation damping); and (ii) calculate the average energy loss per unit time by radiation.

Solution: The equivalent of the conducting plane is an image charge $-q$ placed a distance d below as shown in Fig 5.12.

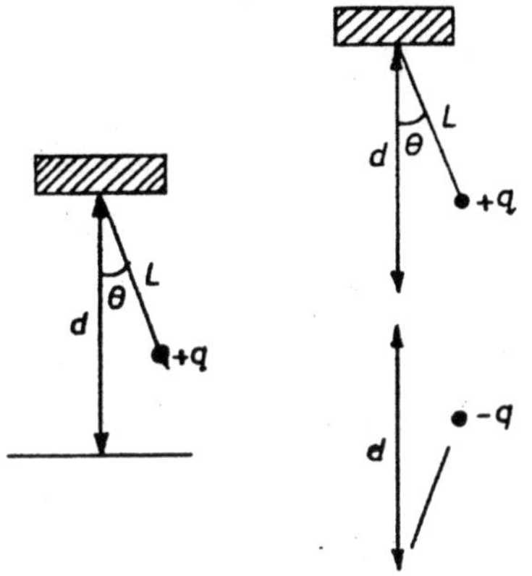

Figure 5.12. Comparison of an oscillating charge above a conductor, left; with an oscillating charge above an image charge, right.

The kinetic energy of the configuration, which is due to both charges, is

$$T = \frac{1}{2}(2m)v^2 = ML^2\dot{\theta}^2 \tag{1}$$

Similarly, the potential energy of the first charge is due to the electric field associated with the image charge, or

$$V = -\frac{q}{4\pi\varepsilon_0 r} = -\frac{q}{4\pi\varepsilon_0 2(d - L\cos\theta)} \tag{2}$$

The Lagrangian is

$$L = T - V = mL^2\dot{\theta}^2 + \frac{q}{4\pi\varepsilon_0 2(d - L\cos\theta)} \tag{3}$$

From Lagrange's equation, one obtains the equation of motion as

$$2mL^2\ddot{\theta} - \left\{-\frac{qL\sin\theta}{4\pi\varepsilon_0\, 2(d - L\cos\theta)^2}\right\} = 0 \tag{4}$$

or

$$\ddot{\theta} + \left\{\frac{q\sin\theta}{4\pi\varepsilon_0 4mL(d - L\cos\theta)^2}\right\} = 0 \tag{5}$$

For small oscillations, $\sin\theta \sim \theta$, $\cos\theta \sim 1$, and

$$\ddot{\theta} + \left\{\frac{q}{4\pi\varepsilon_0 4mL(d - L)^2}\right\}\theta = 0 \tag{6}$$

So that the frequency of oscillation is

$$\omega = \left\{\frac{q}{4\pi\varepsilon_0 4m(d - L)^2}\right\}^{1/2} \tag{7}$$

(ii) The power loss is given by the Larmor radiation formula one obtains

$$\dot{v} = L\ddot{\theta} = L\omega^2\theta \tag{8}$$

and

$$P = \frac{1}{4\pi\varepsilon_0}\frac{2}{3}\frac{q^2}{c^3}\cdot\frac{q}{4mL(d - L)^2}\theta \tag{9}$$

EXAMPLE 5.7: Using Laplace equation in cylindrical coordinates obtain potential between co-axial conducting cylinders.

Solution: Let us consider a cylinder of radius 'a' surrounded by another co-axial cylinder of radius 'b'. Let outer cylinder be grounded and inner cylinder be charged to a potential V_a. Obviously in this problem infinity is not one of the boundaries. Instead the space under consideration is bounded by two given co-axial cylinders. One can easily conclude that the arrangement is similar to that of the cylindrical capacitor. We have the boundary conditions.

$$\left.\begin{array}{rl} V = 0 & \text{when} \quad r=b \\ \text{and} \quad V = V_0 & \text{when} \quad r=a \end{array}\right\} \tag{1}$$

where r is the distance from the axis. Laplace's equation in cylindrical coordinates is

$$\nabla^2 V = \frac{1}{r}\frac{\partial}{\partial r}\left(r\frac{\partial V}{\partial r}\right) + \frac{1}{r^2}\frac{\partial^2 V}{\partial \theta^2} + \frac{\partial^2 V}{\partial z^2} = 0 \tag{2}$$

Assuming that cylinders are long, one can neglect the end effects. This means potential is independent of both θ and z. Thus Eq. (5.2) takes the form

$$\frac{1}{r}\frac{\partial}{\partial r}\left(r\frac{\partial V}{\partial r}\right) = 0 \tag{3}$$

Integration of (3), yields

$$r\frac{\partial V}{\partial r} = \text{Constant}, C_1(\text{say})$$

i.e.

$$\frac{\partial V}{\partial r} = \frac{C_1}{r}$$

Again integrating, one obtains

$$V = C_1 \log_e r + C_2 \tag{4}$$

where C_2 is another constant of integration. One can see from (4) that the potential V varies logarithmically. One can evaluate the constants C_1 and C_2 by applying boundary conditions (1).

Making use of first boundary condition, i.e. $V = 0$ at $r = b$ in (4), one obtains

$$0 = C_1 \log_e a + C_2 \tag{5}$$

Now, using second boundary condition i.e. $V = V_a$ at $r = a$ in (4), one obtains

$$V_a = C_1 \log_e a + C_2 \tag{6}$$

Solving (5) and (6) one obtains

$$C_1 = \frac{V_a}{\log_e(a/b)} \quad \text{and} \quad C_2 = -\frac{V_a \log_{e^b}}{\log_e(a/b)} \tag{7}$$

Using (7), Eq. (5.4) yields .

$$V = \frac{V_a}{\log_e(a/b)} \log e^r - \frac{V_a \log e^b}{\log_e(a/b)}$$
$$= \frac{V_a}{\log_e(a/b)} \log_e(r/b) \quad (8)$$

EXAMPLE 5.8: Find the potential at any point inside the cylinder as shown in Fig 5.13.

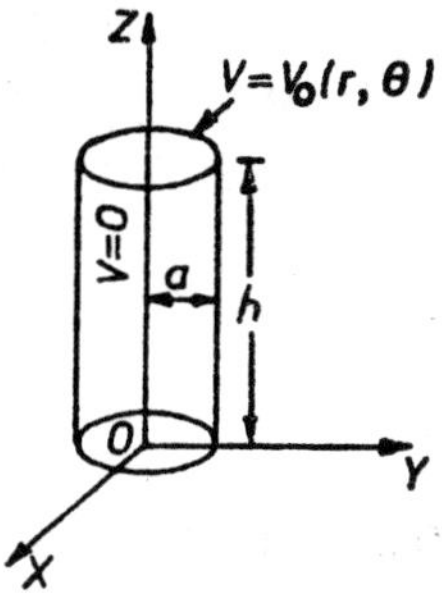

Figure 5.13.

Solution: The radius of cylinder has radius 'a' and height 'h', the top and bottom surfaces being at $z = h$ and $z = 0$. The boundary conditions are (i) $V = 0$ on the side and the bottom of the cylinder (ii) $V = V_0(r, \theta)$ at the top.

The Laplace's equation in cylindrical coordinates is

$$\nabla^2 V = \frac{1}{r}\frac{\partial}{\partial r}\left(r\frac{\partial V}{\partial r}\right) + \frac{1}{r^2}\frac{\partial^2 V}{\partial \theta^2} + \frac{\partial^2 V}{\partial Z^2} = 0 \quad (1)$$

The solution of (1) by the method of separation of variables is

$$V(r, \theta, z) = R(r)\Theta(\theta)Z(z) \quad (2)$$

where

$$R(r) = AJ_m(kr) + BN_m(kr) \quad (3)$$
$$\Theta(\theta) = Ce^{\pm im\theta} \quad (4)$$

and

$$Z(z) = de^{\pm kz} \quad (5)$$

In order that the potential V be signal valued and vanish at $z = 0$. one can express (4) as

$$\Theta(\theta) = C\sin m\theta + D\cos m\theta \quad (6)$$
$$Z(z) = \sinh kz \quad (7)$$

where m is an integer and k is a constant to be determined.

If the potential V is finite at $r = 0$, then $B = 0$ in (3). obviously, the radial solution takes the form

$$R(r) = A J_m(kr) \quad (8)$$

Now, the requirement that the potential at $r = a$ means that k can take only special values, i.e.

$$k_{mn} = \frac{x_{mn}}{a}, \quad n = 1,2,3 \tag{9}$$

where x_{mn} are the roots of $J_{mn}(x_{mn}) = 0$

Combining (2), (6), (7), (8) and (9), one obtains that the general form of the solution is

$$V(r,\theta,z) = \sum_{m=0}^{\infty} \sum_{n=1}^{\infty} J_n(k_{mnr}) \sin h(k_{mnz})[C_{mn} \sin m\theta + D_{mn} \cos m\theta] \tag{10}$$

At $z = h$, $V(r,\theta,z) = V_0(r,\theta)$, we have

$$V_0(r,\theta) = \sum_{mn} \sin h(k_{mn}h) J_m(k_{mn}r)[C_{mn} \sin m\theta + D_{mn} \cos m\theta] \tag{11}$$

One can easily see that this is a Fourier series in θ and a Bessel Fourier series in r. One obtains the coefficients C_{mn} and D_{mn} as

$$C_{mn} = \frac{2 \operatorname{cosec} h(k_{mn}h)}{\pi a^2 J_{m+1}^2 (K_{mn}a)} \int_0^{2\pi} d\theta \int_a^a V_0(r,\theta) J_m(k_{mn}r) \sin m\theta \cdot r dr$$

and

$$D_{mn} = \frac{2 \operatorname{cosec} h(k_{mn}h)}{\pi a^2 J_{m+1}^2 (k_{mn}a)} \int_0^{2\pi} d\theta \int_0^a V_0(r,\theta) J_{mn}(K_{mn}r) \cos m\theta \, r dr \tag{12}$$

EXAMPLE 5.9: Consider a conducting sphere of radius a composed of two hemispheres separated by a small insulated ring. The hemispheres are kept at potentials $\pm V_0$. Calculate the electrostatic potential both inside and outside the sphere using spherical harmonics.

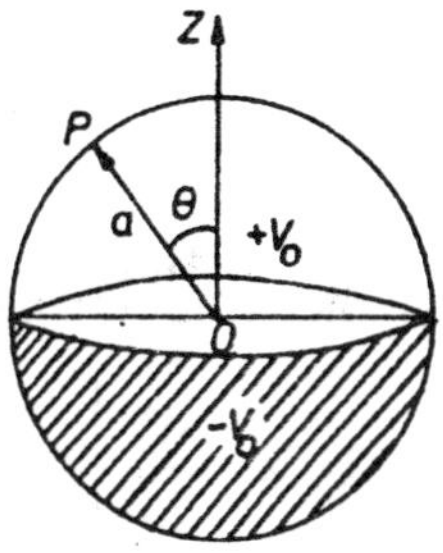

Figure 5.14.

Solution: Fig 5.14 shows a conducting sphere of radius a made up of two hemispheres separated by a small insulated ring. The two hemispheres are at potential $+V_0$ and $-V_0$ (Fig 5.14).

There is azimuthal symmetry and hence we have

$$V = \sum [A_n r^n + B_n r^{-(n+1)}] P_n(\cos\theta) \tag{1}$$

Let us consider the following two situations:

(i) $r < a$, *i.e. point inside the sphere*: Since there are no charges at the origin and hence V is finite at the origin. This is possible only when $\Sigma B_n = 0$. One can express the inside potential V_{in} as

$$V_{\text{in}} = \sum A_n r^n P_n(\cos\theta) \tag{2}$$

At the surface where $r = a$, the potential $(V_{\text{in}})_s$ is obtained as

$$(V_{\text{in}})_s = \sum A_n a^n P_n(\cos\theta) \tag{3}$$

To determine the values of constants A_n, let us multiply eq (3) on both sides by $P_m(\cos\theta)$ and integrating between the limits -1 to $+1$, one obtains

$$\int_{-1}^{+1} (V_{\text{in}})_s P_m(\cos\theta) d(\cos\theta) = \int_{-1}^{+1} \sum A_n a^n P_n(\cos\theta) P_m(\cos\theta) d(\cos\theta)$$

or

$$-\int_{\pi}^{0} (V_{\text{in}})_s P_m(\cos\theta) \sin\theta d\theta = A_n a^n \frac{2}{2n+1} \quad (\text{for } n = m)$$

$$\because \int_{-1}^{1} P_n(\cos\theta) P_m(\cos\theta) d(\cos\theta) = \frac{2}{2n+1}$$

$$\therefore A_n = \frac{(2n+1)}{2a^n} \int_0^{\pi} (V_{\text{in}})_s P_n(\cos\theta) \sin\theta d\theta$$

$$= \frac{(2n+1)}{2a^n} \left[\int_0^{\pi/2} (+V_0) P_n(\cos\theta) \sin\theta d\theta + \int_{\pi/2}^{\pi} (-V_0) P_n(\cos\theta) \sin\theta d\theta \right] \tag{4}$$

Now

$$A_0 = \frac{V_0}{2} \left[(-\cos\theta)_0^{\pi/2} - (-\cos\theta)_{-\pi/2}^{\pi} \right] = 0$$

and

$$A_1 = \frac{3V_0}{2a} \left[\left(-\frac{\cos^2\theta}{2} \right)_0^{\pi/2} + \left(\frac{\cos^2\theta}{2} \right)_{\pi/2}^{\pi} \right] = \frac{3V_0}{2a}$$

Similarly, one obtains

$$A_2 = 0,\ A_3 = -\frac{7}{8}\frac{V_0}{a^3},\ A_4 = 0,\ A_5 = \frac{11}{16}\frac{V_0}{a^5} \text{etc.}$$

Substituting these values in Eq. (5.2) one obtains

$$V_{in} = V_0\left[\frac{3}{2}\left(\frac{r}{a}\right)P_1(\cos\theta) - \frac{7}{8}\left(\frac{r}{a}\right)^3 P_3(\cos\theta) + \frac{11}{16}\left(\frac{r}{a}\right)^5 P_5(\cos\theta) + \cdots\right] \quad (5)$$

(ii) $r > a$, *i.e. point lies outside the sphere*: In this situation where $r \to \infty$, $V = 0$. This is only possible when $\Sigma A_n = 0$.Thus

$$V_{out} = \sum B_n r^{n-1} P_n(\cos\theta) \quad (6)$$

At the surface

$$(V_{out})_s = \sum B_n a^{-(n+1)} P_n(\cos\theta) \quad (7)$$

Multiplying on both sides by $P_m(\cos\theta)$ and integrating between the limits -1 to $+1$, one obtains

$$\int_{-1}^{1}(V_{out})_s P_m(\cos\theta)d(\cos\theta) = \int_{-1}^{1}\sum B_n a^{-(n+1)}P_n(\cos\theta)\, P_m(\cos\theta)d(\cos\theta)$$

or $$-\int_{\pi}^{0}(V_{out})_s P_m(\cos\theta)\sin\theta d\theta = B_n a^{-(n+1)}\left[\frac{2}{2n+1}\right]$$

$$\therefore \quad B_n = \frac{2n+1}{2}a^{(n+1)}\left[\int_0^{\pi/2}(+V_0)P_m(\cos\theta)\sin\theta d\theta + \int_{\pi/2}^{\pi}(-V_0)P_m(\cos\theta)\sin\theta d\theta\right] \quad (8)$$

Putting $n = 0$, one obtains

$$B_0 = \frac{V_0}{2}a\left[(-\cos\theta)_0^{\pi/2} + (\cos\theta)_{\pi/2}^{\pi}\right] = 0$$

Similarly, one obtains $B_1 = \frac{3}{2}V_0 a^2, B_2 = 0,$

$$B_3 = -\frac{7}{8}V_0 a^4,\ B_4 = 0,\ B_5 = \frac{11}{16}V_0 a^6 \text{ etc.}$$

Thus, we have

$$V_{out} = V_0\left[\frac{3}{2}\left(\frac{a}{r}\right)^2 P_1(\cos\theta) - \frac{7}{8}\left(\frac{a}{r}\right)^4 P_3(\cos\theta) + \frac{11}{16}\left(\frac{a}{r}\right)^6 P_5(\cos\theta) + \cdots\right] \quad (9)$$

EXAMPLE 5.10: Show that the potential of a spherical cap of angle α charged to a uniform surface density σ_0 as shown in Fig 5.15 is given by

$$V_p = \frac{a\sigma_0}{2\varepsilon_0}\left[(1-\cos\alpha)\left(\frac{a}{r}\right) + \sum_{n=1}^{\infty} \frac{P_{n-1}(\cos\alpha) - P_{n+1}(\cos\alpha)}{2n+1} \times \left(\frac{a}{r}\right)^{n+1} P_n(\cos\theta)\right]$$

Solution: One can consider that the spherical cap is the surface cut from a sphere by a right circular cone of semi vertical angle α.

To obtain the solution of the problem, one can regard the entire sphere as charged to a surface density σ, where

$$\begin{aligned} \sigma &= \sigma_0 \quad \text{from } \theta=0 \text{ to } \theta=\alpha \\ &= 0 \quad \text{from } \theta=\alpha \text{ to } \theta=\pi \end{aligned} \qquad (1)$$

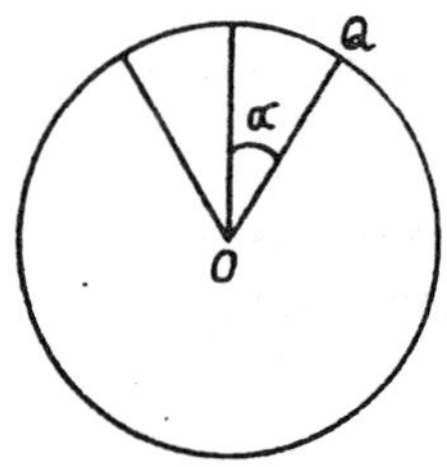

Figure 5.15.

The problem has azimuthal symmetry and therefore, one can express σ in terms of harmonics

$$\sigma = a_0 + a_1P_1(\cos\theta) + a_2P_2(\cos\theta) + \cdots \qquad (2)$$

where

$$\begin{aligned} a_l &= \frac{2l+1}{2}\int_{\theta=\pi}^{\theta=0} \sigma P_l(\cos\theta)d(\cos\theta) \qquad (3) \\ &= \frac{2l+1}{2}\int_{\theta=\pi}^{\theta=\alpha} \sigma P_l(\cos\theta)d(\cos\theta) \\ &+ \frac{2l+1}{2}\int_{\theta=\alpha}^{\theta=0} \sigma P_l(\cos\theta) \times d(\cos\theta) \\ &= \frac{2l+1}{2}\sigma_0\int_{\theta=\alpha}^{\theta=0} P_l(\cos\theta)(d\cos\theta) \\ &(\because \quad \sigma = 0 \text{from} \theta = \pi \text{ to } \theta = \alpha) \end{aligned}$$

$$\therefore \quad a_0 = \frac{1}{2}\sigma_0\int_{\theta=\alpha}^{\theta=0} d(\cos\theta) = \frac{1}{2}\sigma_0(1-\cos\alpha)$$

One can easily show that

$$\int_{\theta=\alpha}^{\theta=0} P_l(\cos\theta)d(\cos\theta) = \left[\frac{P_{l+1}(\cos\theta) - P_{l-1}(\cos\theta)}{2l+1}\right]_{\theta=\alpha}^{\theta=0}$$

$$\therefore \quad \sigma = \frac{1}{2}\left[(1-\cos\alpha) + \sum_{l=1}^{\infty}\{P_{l-1}(\cos\alpha) - P_{l+1}(\cos\alpha)\}P_l(\cos\theta)\right] \tag{4}$$

The potential at a Point P on the polar axis, distance r from the centre is obtained as

$$V_p = \frac{1}{4\pi\varepsilon_0}\iint \frac{\sigma ds}{PQ} = \frac{1}{4\pi\varepsilon_0}\iint \frac{\sigma ds}{(a^2+r^2-2ar\cos\theta)^{1/2}} \tag{5}$$

where ds is an element of area at Q on the surface of the sphere and a is the radius of the sphere. If $r < a$, we have

$$\begin{aligned}
V_p &= \frac{1}{4\pi\varepsilon_0}\iint \frac{\sigma}{a}\left(1+\frac{r^2}{a^2}-\frac{2r}{a}\cos\theta\right)^{-1/2} ds \\
&= \frac{1}{4\pi\varepsilon_0}\iint \frac{\sigma}{a}\sum_{n=0}^{\infty} P_n(\cos\theta)\left(\frac{r}{a}\right)^n ds \\
&= \frac{1}{4\pi\varepsilon_0}\int \frac{\sigma_0}{2a}\left[(1-\cos\alpha) + \sum_{l=1}^{\infty}\{P_{l-1}(\cos\alpha) - P_{l+1}(\cos\alpha)\}P_l(\cos\theta)\right] \\
&\quad \times\left[1 + P_1\cos\theta\left(\frac{r}{a}\right) + P_2(\cos\theta)\left(\frac{r}{a}\right)^2 + \cdots\right] 2\pi a^2 \sin\theta d\theta \\
&= \frac{a\sigma_0}{4\varepsilon_0}\int_{-1}^{1}\left[(1-\cos\alpha) + \sum_{l=1}^{\infty}\{P_{l-1}(\cos\alpha) - P_{l+1}(\cos\alpha)\}P_l(\cos\theta)\right] \\
&\quad \times\left[1 + P_1(\cos\theta)\left(\frac{r}{a}\right) + P_2(\cos\theta)\left(\frac{r}{a}\right)^2 + \cdots\right] d(\cos\theta) \\
&= \frac{a\sigma_0}{2\varepsilon_0}\left[(1-\cos\alpha) + \sum_{n=1}^{\infty}\frac{P_{n-1}(\cos\alpha) - P_{n+1}(\cos\alpha)}{2n+1}\left(\frac{r}{a}\right)^n\right]
\end{aligned}$$

Thus the potential at a point (r,θ) is obtained as

$$\begin{aligned}
V(r,\theta) = \frac{a\sigma_0}{2\varepsilon_0}\Bigg[&(1-\cos\alpha) \\
&+ \sum_{n=1}^{\infty}\frac{P_{n-1}(\cos\alpha) - P_{n+1}(\cos\alpha)}{2n+1}\left(\frac{r}{a}\right)^n P_n(\cos\theta)\Bigg]
\end{aligned} \tag{6}$$

For $r > a$, we have

$$V_p = \frac{a\sigma_0}{2\sigma_0}\left[(1-\cos\alpha)\left(\frac{a}{r}\right) + \sum_{n=1}^{\infty}\frac{P_{n-1}(\cos\alpha)-P_{n+1}(\cos\alpha)}{2n+1}\left(\frac{a}{r}\right)^{n+1}P_n(\cos\theta)\right] \quad (7)$$

EXAMPLE 5.11: Determine the field intensity and capacitance per metre length of a single core cable.

Solution: We have

$$E = -\nabla\phi = -\hat{a}_r\frac{\partial\phi}{\partial r}$$

$$= -\hat{a}_r\frac{\partial}{\partial r}\left[\frac{\phi_1}{\log_e(a/b)}\left\{\log_e\left(\frac{r}{b}\right)\right\}\right]$$

$$= \hat{a}_r\frac{\phi_1}{\log_e(a/b)}\left[\frac{b}{r}\cdot\frac{1}{b}\right] = a_r\frac{\phi_1}{\log_e(b/a)}\left(\frac{1}{r}\right) \quad (1)$$

$$\frac{\text{Capacitance}}{m} = \frac{\sigma A}{\phi_1} = \frac{\varepsilon EA}{\phi_1}$$

$$= \frac{\varepsilon}{\phi_1}\left[\frac{\phi_1}{r\log_e(b/a)}\right]2\pi r\times 1$$

$$= \frac{2\pi\varepsilon}{\log_e(b/a)}\ \frac{\text{farad}}{m} \quad (2)$$

EXAMPLE 5.12: An infinitely long conducting cylinder is placed in an uniform field of intensity E_0 in such a way that its z axis is perpendicular to the field and the centre coincides with the origin (Fig 5.16). Determine the potential.

Solution: We have

$$V = A_0 + B_0\log r + \sum_{n=1}^{\infty}(A_n r^n + B_n r^{-n})\cos n\theta + \sum_{n=1}^{\infty}(C_n r^n + D_n r^{-n})\sin n\theta \quad (1)$$

The boundary condition are (i) $V = 0$ at $r = a$ and $V = -E_0 r\cos\theta$ at $r \gg a$ or ∞.

As desired by the problem, the potential is finite at $r = \infty$ and zero at $r \to 0$. This means $B_0 = 0$. Now, when $r \gg a$, the potential is only a function of $\cos\theta$, i.e. C_n and $D_n = 0$. Now substituting these values in (1) one obtains

$$V = A_0 + \sum_{n=1}^{\infty} (A_n r^n + B_n r^{-n}) \cos n\theta \quad (2)$$

For $r \gg a$, Eq. (5.2) can be expressed as

$$-E_0 r\cos\theta = A_0 + \sum_{n=1}^{\infty} (A_n r^n) \cos n\theta \quad (3)$$

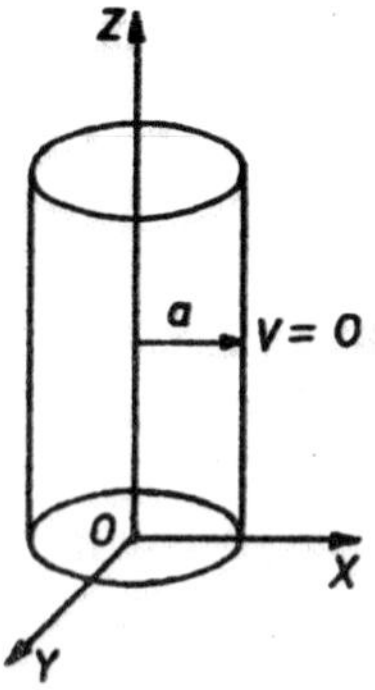

Figure 5.16.

Eq. (3) is valid for any value of r, hence comparing the coefficient of different powers of r on both sides, one obtains

$$A_0 = A_2 = A_3 = 0 \quad \text{and} \quad A_1 = -E_0$$

For $n = 1$, Eq. (5.2) reduces to

$$V = \left[-E_0 r + \frac{B_1}{r}\right] \cos\theta \quad (4)$$

Applying the boundary condition $V = 0$ at $r = a$ in Eq. (5.14), one obtains

$$0 = \left[-E_0 a + \frac{B_1}{a}\right] \cos\theta \quad \text{or} \quad B_1 = E_0 a^2$$

$$\therefore \quad V = \left[E_0 r + \frac{E_0 a^2}{r}\right] \cos\theta \quad (5)$$

EXAMPLE **5.13**: Show that the potential at all points in space exterior to a conducting sphere of radius a placed in a uniform field E_0 is given by

$$V(r,\theta) = -E_0 r\cos\theta \left(1 - \frac{a^3}{r^3}\right)$$

Solution: Let us choose the axis in such way that the polar axis coincides with the direction of the field. The problem has azimuthal symmetry, the symmetry axis being, say z-axis, with $\theta = 0$. We have the potential on the symmetry axis is given by

$$V(r,\theta) = \sum_{l=0}^{\infty}\left[A_l r^l + \frac{B_l}{r^{l+1}}\right] \quad [\because \quad P_l(\cos\theta) = 1] \tag{1}$$

with $r = z$. If one evaluate this potential by some means, at an arbitrary point and the potential function is expanded in power series, then one can obtain the potential at any point in space by multiplying each power of r^l and $\frac{1}{r^{l+1}}$ by $P_l(\cos\theta)$. We have

$$E_0 = -\nabla(\phi_0) \tag{2}$$

$$\therefore \quad \phi_0 = -E_0 z = -E_0 r\cos\theta = -E_0 r P_1(\cos\theta) \tag{3}$$

However, this does not satisfy the condition that $\phi = 0$ as $r \to \infty$. However, we assume that the uniform field of infinite extent, the source of which must lie at infinity and, hence, ϕ will not vanish as $r \to \infty$.

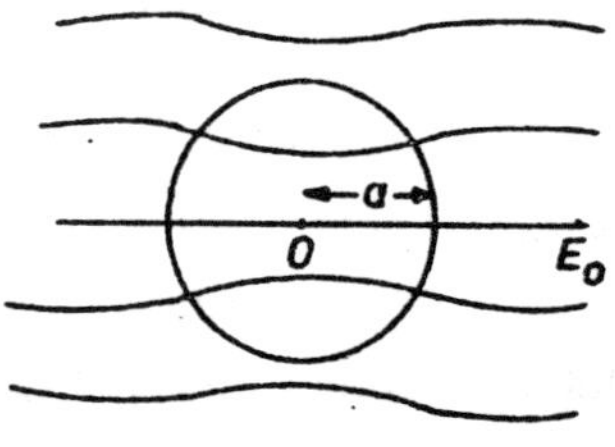

Figure 5.17.

One can easily see that the field immediately around the sphere will be distorted owing due to induced charges on the surface; but at large distances it will be equal to E_0 (Fig. 17). The total potential at a point $V(r,\theta)$ is

$$V(r,\theta) = \phi_0 + \text{ the potential due to the charges induced in the conducting sphere} \tag{3}$$

we know that at very large values of r, however, the potential due to the induced charges may be insignificant and

$$V(r,\theta) = \phi_0$$

Now,

$$V(r,\theta) = \sum_{l=0}^{\infty}\left(A_l r^l + \frac{B_l}{r^{l+1}}\right) P_l(\cos\theta) \tag{4}$$

One can neglect the term involving B_l in (4) at large values of r, i.e.

$$\begin{aligned} V(r,\theta) &= \sum_{l=0}^{\infty} A_l r^l P_l(\cos\theta) \\ &= A_0 P_0(\cos\theta) + A_1 r P_1(\cos\theta) + A_2 r^2 P_2(\cos\theta) + \cdots \end{aligned} \tag{5}$$

This must be identical with ϕ_0 as given by (3)

$$A_1 r P_1(\cos\theta) = -E_0 r P_1(\cos\theta)$$

Hence, $A_1 = E_0$ and all other A_l vanish. Obviously, from (4)

$$\begin{aligned} V(r,\theta) &= A_1 r P_1(\cos\theta) + \sum_{l=0}^{\infty} \frac{B_l}{r^{l+1}} P_l(\cos\theta) \\ &= -E_0 r P_1(\cos\theta) + \sum_{l=0}^{\infty} \frac{B_l}{r^{l+1}} P_l(\cos\theta) \qquad (6) \end{aligned}$$

The sphere under question placed in the uniform field being a conductor, the induced charges generate an electric field within it. This cancels the external field giving zero field within the space and, hence, the potential on the surface ($r = a$) is zero.

$$\therefore \quad V(a,\theta) = -E_0 a P_1(\cos\theta) + \sum_{l=0}^{\infty} B_l a^{-(l+1)} P_l(\cos\theta) = 0$$

$$\text{i.e. } E_0 a P_1(\cos\theta) = \sum_{l=0}^{\infty} B_l a^{-(l+1)} P_l(\cos\theta) \qquad (7)$$

One can obtain the coefficients B_l as

$$\begin{aligned} B_l a^{-(l+1)} &= \frac{2l+1}{2} \int_{-1}^{1} E_0 a P_1(x) P_l(x) dx \\ \therefore \quad B_l &= \left(\frac{2l+1}{2}\right) a^{l+1} E_0 \int_{-1}^{1} P_l(x) P_1(x) dx \\ &= \frac{(2l+1)}{2} a^{l+2} E_0 \frac{2}{2l+1} \delta_{1l} \\ &\therefore B_1 = E_0 a^3, B_2 = B_3 = 0 \qquad (8) \end{aligned}$$

Hence

$$\begin{aligned} V(r,\theta) &= -E_0 r P_1(\cos\theta) + \frac{E_0 a^3}{r^2} P_1(\cos\theta) \\ &= -E_0 r \cos\theta \left(1 - \frac{a^3}{r^3}\right) \qquad (9) \end{aligned}$$

One can also obtain the field components using $E_r = -\frac{\partial V}{\partial r}$, and $E_Q = -\frac{1}{r}\frac{\partial V}{\partial r}$.

QUESTIONS

1. State and prove uniqueness theorem of electrostatics regarding electric potential. With the help of this theorem prove that the electric field inside the hollow conductor is zero.
2. State and prove Green's theorem. How this theorem is applied on the solution of potential problem?
3. Define the Green's function with reference to electrostatic potential problems. Prove that this function satisfies the symmetry
$$G(x,x') = G(x',x).$$
4. State Poisson's equation. Show that the potential function is
$$\phi(r) = \frac{1}{4\pi\varepsilon_0}\int \frac{\rho(\mathbf{r}')}{|\mathbf{r}-\mathbf{r}'|}dv'$$
represents the solution of Poisson's equation.
5. Explain the method of images for the solution of electrostatic problems. A point charge is situated near an infinite plane earthed conductor. Applying the method of images to calculate: (i) surface charge density induced on the plane (ii) the force between the plane and the charge.
6. A point charge is placed outside a conducting sphere. Find by the method of electrical images the intensity of electrical field outside the sphere and also the surface density of induced charge on the sphere when the sphere is at zero potential.
7. Explain the method of images for the solution of electrostatic problems. Obtain the image system for a point charge situated at a distance C form the centre of a conducting sphere of radius $b(< c)$ at zero potential.
8. A point Q is placed at a distance d from the centre of an earthed conducting sphere of radius a. Find the position of the image and show that the induced charge on the sphere is given by $Q' = -\dfrac{Qa}{d}$.
9. Write down Laplace's equation in spherical polar coordinates and obtain its solution. What will be the form of solution for spherically symmetric problems.
10. Deduce the solution of Laplace's equation for a problem having azimuthal symmetry. Use it to obtain potentials inside and outside a dielectric sphere in a uniform field.

PROBLEMS

1. A charge q is placed at a distanced from the centre of an insulated, uncharged conducting sphere of radius a, find the rise in potential due to the presence of the charge and show that there is an attractive force between the sphere and the charge equal to $\frac{q^2}{4\pi\varepsilon_0}\frac{a^3(2d^2-a^2)}{(d^2-a^2)^2}$.
2. A field is produced by a point charge q in the presence of an uninsulated spherical conductor of radius a whose centre is at a distance d from the charge. Find the force with which the charge is attracted to the sphere and show that the force is changed to a repulsion by insulating the sphere and connecting it to a large distance conductor of potential V_0 provided that
$$V_0 > \frac{1}{4\pi\varepsilon_0}\frac{qd^3}{(d^2-a^2)^2}$$
3. Show that the following relations holds good for Legendre's polynomials
$$\int P_l(\cos\theta)d(\cos\theta) = \frac{P_{l+1}(\cos\theta) - P_{l-1}(\cos\theta)}{2l+1}$$
Where the symbols have usual meanings.
4. A rectangular parallelopiped is bounded by the planes $x=0$ and $x=a$; $y=0$ and $y=b$ and $z=0$ and $z=c$. The side at $z=c$ is kept at a constant potential ϕ_0 while the other sides are kept at zero potential. Solve Laplace's equation for the problem and show that the potential inside the enclosure is given by
$$\phi(x,y,z) = \sum_{m,n=1,3,5,\ldots\text{etc}} \frac{16\phi_0}{mn\pi^2}\sin\left(\frac{m\pi x}{a}\right)\sin\left(\frac{n\pi y}{b}\right) \times\frac{\sin h(\gamma_{mn}z)}{\sin h(\gamma_{mn}C)}$$
where
$$\gamma_{mn}^2 = \left[\frac{m^2\pi^2}{a^2}+\frac{n^2\pi^2}{b^2}\right].$$
5. A dielectric sphere is kept in a uniform field. Solve the appropriate Laplace's equation when the potential and electric fields are continuous at the surface of the dielectric sphere. Show that the

potential at a radial distance r outside the sphere is

$$V = -\left[1 - \frac{\varepsilon_r - 1}{\varepsilon_r + 2}\frac{a^3}{r^3}\right] E_0 r \cos\theta$$

where θ is the angle, which the line joining the outside point with the centre of the sphere makes with the electric field.

6. A hemisphere of radius a is charged uniformly with a charge Q. Assuming that the centre of the base of the hemisphere is at the origin, determine the potential at a point $r \gg a$.

$$\left[\textbf{Ans. } V \sim \frac{Q}{4\pi\varepsilon_0 r}\left(1 + \frac{3a}{2r}\cos\theta\right)\right]$$

7. An electron is at distance 10 Åfrom a plane conductor. Calculate the force experienced by the electron and the work done in moving it to infinite distance from the conductor.

[**Ans.** $F = -5.75 \times 10^{-11}$ newton, $W = 5.75 \times 10^{-20}$ J.]

8. Find the capacitance between two long coaxial cones placed tip to tip with infinite gap. [**Ans.** Cap./m $= \dfrac{\pi\varepsilon_0}{\log\cot(\theta/2)}$ f/m]

9. A thunder cloud is stationary above ground level. Considering the earth as a perfect conductor and the thunder cloud as an electric dipole with its axis vertical, show that the electric field at a point on the ground is proportional to $3\sin 5\alpha - \sin^3\alpha$, where α is the elevation of the cloud from the point.

10. Determine the potential at any point in space due to a charge q uniformly distributed around a circular ring of radius a, its axis being the z-axis and its centres at $z = b$.

[**Ans.** $V(r,\theta) = q\sum_{l=0}^{\infty} \frac{c^l}{r^{l+1}} P_l(\cos\alpha)P_l(\cos\theta)$ and if

$r < c$, $V(r,\theta) = q\sum_{l=0}^{\infty} \frac{r^l}{C^{l+1}} P_l(\cos\alpha)P_l(\cos\theta)$]

11. Find the potential and field intensity between anode and cathode (earthed) of a working vacuum diode, assuming uniform space charge

[**Ans.** $\phi_x = \left(\frac{V}{l} + \frac{\rho l}{2\varepsilon_0}\right)x - \frac{\rho x^2}{2\varepsilon_0}$ and $E_x = -i\left(\frac{V}{l} + \frac{\rho l}{2\varepsilon_0}\right) + i\frac{\rho x}{\varepsilon_0}$]

12. An infinite long cylinder of radius 'a' with its axis parallel to z-axis is cut in an infinite dielectric medium of permittivity ε. If

a uniform electric field E_0 is applied in the $+x$ direction, find a solution for the resultant electric field both within and outside the cylindrical cavity.

[**Ans.**
$$V_i(\rho,\phi) = -\frac{2}{\varepsilon_r+1}\rho\cos\phi \text{ and}$$
$$V_0(\rho,\phi) = -\left[1+\frac{(\varepsilon_r-1}{\varepsilon_r+1}\frac{a^2}{\rho^3}\right]\frac{E_0}{\varepsilon_r}\rho\cos\phi$$

and the components of electric field E just outside the cavity are
$$E_\rho = -\frac{\partial V_0}{\partial\rho} = \left[1-\frac{(\varepsilon_r-1}{\varepsilon_r+1}\frac{a^2}{\rho^2}\right]\frac{E_0}{\varepsilon_0}\cos\phi$$
$$E_\phi = +\frac{1}{\rho}\frac{\partial V_0}{\partial\phi} = \left[1-\frac{(\varepsilon_r-1}{\varepsilon_r+1}\frac{a^2}{p^2}\right]\frac{E_0}{\varepsilon_r}\cos\phi]$$

13. A grounded conducting sphere of radius a is placed in a potential field $V(r,\theta,\phi) = \frac{r^2\sin 2\theta}{4\pi\varepsilon_0}e^{i\phi}$ with its centre at the origin. Show that the potential at a point outside the sphere is given by
$$V = \frac{1}{2\pi\varepsilon_0}\left(r^2-\frac{a^5}{r^3}\right)\sin\theta\cos\theta\, e^{i\phi}$$

14. A spherical cavity of radius a is cut out of a conducting block of metal maintained at zero potential. A charge q is placed at a distance d from the centre of gravity. Calculate the force on it.

[**Ans.** $F = \frac{q^2ad}{[4\pi\varepsilon_0(a^2-d^2)^2]}$]

15. A conducting sphere of radius a is kept at zero potential. An electric dipole of moment p pointing away from the sphere is at a distance d from the centre of the sphere. Calculate the dipole moment and charge of the image at the inverse point.

[**Ans.** Dipole moment $= \frac{pa^3}{d^3}$ and Charge $= \frac{pa}{d^3}$]

16. A conducting circular cylinder of radius a and infinite length is placed in a uniform field with its axis at right angles to the field. If the cylinder is maintained at zero potential, then show that the potential at a point outside the cylinder is given by
$$V = -E_0\left(r-\frac{a^2}{r}\right)\cos\theta$$

17. An electron is placed at a height h above an infinite dielectric slab of relative permeability ε. Show that the force on the electron is given by:

$$F = \frac{e^2}{4\pi\varepsilon_0}\left(\frac{\varepsilon-1}{\varepsilon+1}\right)\frac{1}{(2h)^2}$$

SHORT ANSWER QUESTIONS

1. Give a simple example of method of images.
[*Ans.* A point charge located in front of an infinite plane conductor at zero potential.]
2. Write Laplace's equation in spherical polar coordinates.
[**Ans.** $\frac{1}{r^2}\frac{\partial}{\partial r}\left(r^2\frac{\partial V}{\partial r}\right) + \frac{1}{r^2\sin\theta}\frac{\partial}{\partial\theta}\left(\sin\theta\frac{\partial V}{\partial\theta}\right) + \frac{1}{r^2\sin^2\theta}\frac{\partial^2 V}{\partial\phi^2} = 0$]
3. A function (say Q_m) is normalized and also orthogonal. How will you express it mathematically? [**Ans.** $\int_0^{2\pi} Q_m^* Q_n d\phi = \delta_{mn}$]
4. Write Laplace's equation in cylindrical Coordinates
[**Ans.** $\frac{1}{r}\frac{\partial}{\partial r}\left(r\frac{\partial V}{\partial r}\right) + \frac{1}{r^2}\frac{\partial^2 V}{\partial\phi^2} + \frac{\partial^2 V}{\partial z^2} = 0$]
5. Is the Green's function with reference to the electrostatic potential problems satisfy symmetry problem? If yes, How will you express it mathematically? [**Ans.** Yes, $G(x,x') = G(x',x)$.]
6. What do you understand by method of images and image charges?
[**Ans.** The method of images concerns itself with the problem of one or more point charges in the presence of boundary surfaces, e.g., conductors either grounded or held at fixed potentials. Under favourable conditions it is possible to infer from the geometry of the situation that a small number of suitably placed charges of appropriate magnitudes, external to the region of interest, can stimulate the required boundary conditions. These charges are called image charges, and the replacement of the actual problem with boundaries by an enlarged region with images charges but not boundaries is called the method of images.]

OBJECTIVE QUESTIONS

1. A point charge is kept somewhat midway between two infinitely extended earthed metallic planes, then
   (a) there will be an infinite number of images of alternate polarity and the charge will tend to occupy a position exactly midway between the planes.
   (b) there will be no image at all.
   (c) there will be an infinite number of images of alternate polarity and the charge will be in equilibrium.
   (d) there will be an infinite number of images of the same polarity and the charge will be in equilibrium. **[a]**

2. A point charge is placed near an earthed sphere. The induced image charge depends on position and magnitude of point charge in the following manner:
   (a) increases with magnitude and decreases with distance.
   (b) increases with both distance and magnitude.
   (c) decreases with both distance and magnitude.
   (d) increases with distance and decreases with magnitude.
   [a]

3. A long cylindrical dielectric is introduced in a uniform electric field in a direction perpendicular to the longitudional axis. The potential inside the cylinder and that outside (at a long distance) compare as
   (a) the inside potential is greater.
   (b) the inside potential is smaller.
   (c) both-equal.
   (d) none of the above is correct. **[b]**

4. A grounded conducting sphere of radius a is placed in a uniform electric field E_0 with its centre at the origin. The induced charge density is
   (a) $3\varepsilon_0 E_0$; (b) $3\varepsilon_0 E_0 \cos\theta$; (c) Zero; (d) $3\varepsilon_0 E_0 \sin\theta$. **[b]**

Chapter 6

Magnetostatics

6.1 INTRODUCTION

So far we have dealt with the electric charges that are stationary. Now, we shall consider the motion of charges and the forces associated with them. The concept of electric field has been developed starting from the experimental observation that a charge brought into the vicinity of other charges experiences a force. One can also experimentally determine that a current element (e.g. a small loop carrying current) will be acted on by force if it is brought in the vicinity of another current or system of currents. The region in which such forces exist is often called as a region of magnetic field. This concept may appear to exclude the well-known magnetic effects arising from permanent magnets, but these effects may be included conceptually if one thinks of them as arising from groups of atomic currents in the ferromagnetic material. This aspect will be discussed later in this chapter. obviously, similar to a static electric charge which has an electric field, an electric current, on the other hand will possess a magnetic field.

6.2 ELECTRIC CURRENT

An electric current consists of a stream of charged particles or ions. This applies to the ions in an accelerator of any kind, to those in an electrostatic solution, to those in an ionized gas, or plasma, or to the electrons in metallic conductors. We have already studied that when a conductor is placed in an electric field, the electrons acquire an average velocity in the direction opposite to the field. This results into a net drift of charges and one can say that there is an electric current flowing in the conductor. Obviously, in order for an electric current to be produced, an electric field must be applied to move the charged particles in a well-defined direction. We must remember that the conductor carrying the current is electrically neutral since the number of conducting electrons is balanced by the number of positively charged atoms.

Let us consider that there are N charges per unit volume at a point P. each carries a charge q and moving with mean velocity $\mathbf{u}$. The amount of charge crossing unit area perpendicular to mean drift velocity, per unit time, is

$$\mathbf{J} = Nq\mathbf{u} \tag{6.1}$$

where **J** is called the current density and it is defined as the electric current (I) held normal to the local velocity of the current carriers. It is vector quantity and the direction of **J** and **u** being identical, i.e. the direction of **J** is the direction in which the positive charges move. If ds is an element of area in the conductor (or say material), the amount of charge flowing across that area in unit time is $\mathbf{J} \cdot \hat{a}_n ds$, where $\hat{a}_n$ is the unit vector normal to ds. Obviously, this is the flux of **J** across ds.

As stated above, the electric charge in motion makes an electric current. The net charge that passes through the surface per unit time is called the electric current and is usually denoted by I. Thus the intensity of electric current is

$$I = \frac{Nq}{t} = \frac{Q}{t} \tag{6.2}$$

Actually, (2) gives the average current in time t; the instantaneous current is

$$I = \frac{dQ}{dt} = \int \mathbf{J} \cdot \hat{a}_n ds \tag{6.3}$$

The electric current is expressed in Coulomb per second or C/s, a unit called the ampere (abbreviated A). An ampere is the intensity of an electric current corresponding to a charge of one Coulomb passing through a section of material.

Let us consider a closed surface S enclosing a volume V. If ρ is the volume density of charge, then the total charge within the volume is $\int \rho dV$. Since the charge is conserved, i.e. charge can neither be created nor destroyed, it follows that the net flow of the charge, if any, out of this volume must be equal to the rate of decrease of the total charge inside the volume. This can be expressed as

$$\int \mathbf{J}\cdot \hat{a}_n ds = -\frac{d}{dt}\int \rho dv \tag{6.4}$$

Since the surface S is fixed in space and hence the rate of change of the charge within the volume is solely due to the time variation of ρ. Obviously, we have

$$\int_s \mathbf{J}\cdot \hat{a}_n ds = -\int \frac{\partial \rho}{\partial t} dv \tag{6.5}$$

making use of divergence theorem, we transform the surface integral into a volume integral,

$$\int_s \mathbf{J}\cdot \hat{a}_n ds = \int_v \nabla\cdot \mathbf{J} dV = -\int \frac{\partial \rho}{\partial t} dV$$

$$\therefore \quad \int_v \left(\nabla\cdot \mathbf{J} + \frac{\partial \rho}{\partial t}\right) dV = 0 \tag{6.6}$$

Since the above equation is true for any volume V, we can equate the integrands on both sides and obtain the so called *equation of continuity*, i.e.

$$\nabla\cdot \mathbf{J} + \frac{\partial \rho}{\partial t} = 0 \tag{6.7}$$

We must remember that the above equation is based on the law of conservation of charge. In most practical situations of interest, the charge density is constant in time at each point. This means $\frac{\partial \rho}{\partial t}$ is zero at each point. This is called a *steady state*. Thus for steady state conditions, the equation of continuity gives.

$$\nabla\cdot \mathbf{J} = 0 \tag{6.8}$$

Eq.(8) is valid in the region which does not contain a source or a sink of current.

6.3 SURFACE CURRENT DENSITY

In electrostatics, we have seen that a very thin slab of thickness t containing a volume charge density ρ may be considered to be a surface charge density of value $\sigma = \rho t \frac{C}{m^2}$. Similarly, in current electricity one can define a surface current density **K** as a vector field on the surface. The direction of **K** is the direction of velocity of the current carriers (assumed positively charged) at that point and the magnitude of **K** is the amount of charge crossing a unit length held normal to the local direction of **K**. The total rate at which the charge is crossing the curve of length L from A to B (Fig. 6.1) can be expressed as

$$\frac{dQ}{dt} = \int_A^B K dl \sin\theta = \int_A^B |\mathbf{k} \times d\mathbf{l}| \qquad (6.9)$$

one can see from Fig. 6.1 that $dl \sin\theta$ is the projection of $d\mathbf{l}$ normal to **K**. Obviously, the integral of $kdl \sin\theta$ over appropriate limits gives the current crossing the given curve.

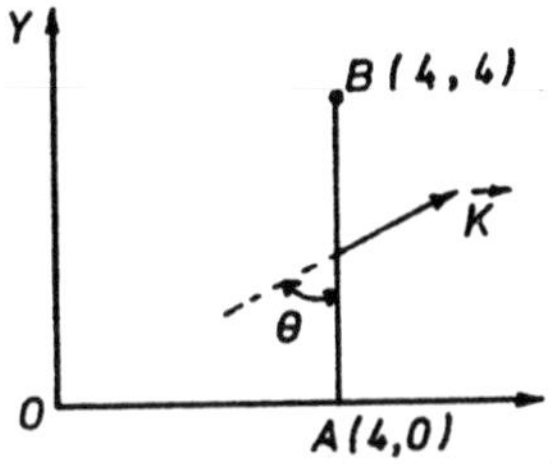

Figure 6.1. Surface electric current in XOY plane.

6.4 ELECTRICAL CONDUCTIVITY: OHM'S LAW

We have read that when an electric field is applied to a dielectric, a polarization of the dielectric results. But if the field is applied in a region when free charges exist, the charges are set in motion and an electric current instead of a polarization of the medium results. The charges are accelerated by the field and therefore gain energy.

When free charges are present within a body, such as electrons-in a metal, their motion is, hindered by the interaction with the positive ions that form the crystal lattice of the metal. Let us consider, for example, a metal with the positive ions regularly arranged in three dimensions, as shown in Fig. 6.2. The free electrons move in a electric field that exhibits the same periodicity as the lattice, and during their motion they are very frequently scattered by the field. One can describe this type of electronic motion with the help of methods of quantum mechanics. Because the electrons are moving in all directions, no net charge transport or electric current results. However, if an external electric field is applied, a drift motion is superposed on the natural random motion of the electrons and an electric current results. Experimentally, it is found that for many materials, the current density in the steady state is linearly proportional to the applied field **E**, i.e.

$$\mathbf{J} \propto \mathbf{E} \quad \text{or} \quad \mathbf{J} = \sigma\mathbf{E} \qquad (6.10)$$

where σ is a constant called the *electrical conductivity* of the material. It is expressed in $\Omega^{-1} m^{-1}$ or $m^{-3} kg^{-1} s C^2$. σ is a constant for a given material at a given temperature. The reciprocal of the conductiv-

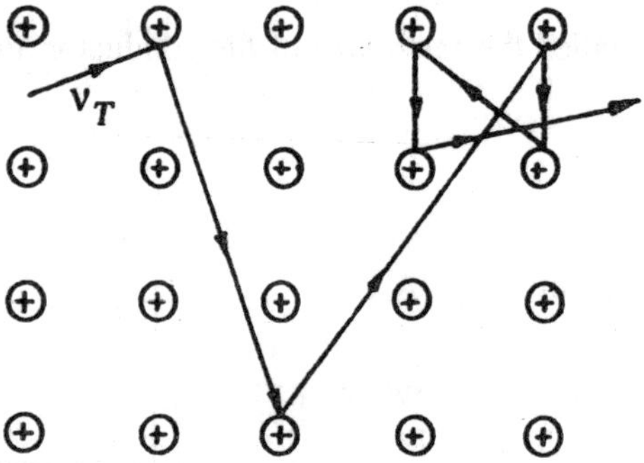

Figure 6.2. Electron motion through the crystal lattice of a metal. V_T represent the thermal velocity of the electrons.

ity of a material is called its *resistivity* (ρ). It is also called the *specific resistance* of the material. The resistivity ρ and therefore σ is a characteristic property of a given material at a given temperature. In terms of ρ, Eq. (6.10) can be written as

$$\mathbf{E} = \rho \mathbf{J} \tag{6.11}$$

Eq. (6.10) states that the proportionality constant σ between **J** and **E** is the same for different values of **E** and and that the two vectors are parallel in an isotropic, homogeneous material. This is found to be true for metallic conductors over a wide range of experimental conditions. It forms the basis of Ohm's law.

One obtains the steady current I as

$$I = \int_s \mathbf{J} \cdot \hat{a}_n dS = \sigma \int_s \mathbf{E} \cdot \hat{a}_n ds \tag{6.12}$$

If the conductor has uniform cross sectional area A, we have

$$I = \sigma A E \tag{6.13}$$

Let V be the potential difference maintained across the ends of a conductor of length l then the electric field within the conductor will be uniform and given by

$$E = \frac{V}{l} \tag{6.14}$$

$$\therefore \quad I = \frac{\sigma A V}{l}$$

This leads to

$$\frac{V}{I} = \frac{l}{\sigma A} = \frac{\rho l}{A} \tag{6.15}$$

The ratio $\frac{V}{I}$ is called the *resistance* of the conductor and Eq. (6.15) is often written in the form

$$V = IR \tag{6.16}$$

which is known as *Ohm's Law*. The unit of resistance, $R = \frac{\rho l}{A}$ is an Ohm(symbol Ω) and it is defined as

$$1 \text{ Ohm} = \frac{1 \text{ volt}}{1 \text{ ampere}} \tag{6.17}$$

Experimentally, it is found that the resistivity ρ of a material generally increases with temperature. The fractional change in ρ for 1°C rise in temperature compared to the value of ρ at 0°C is called the *temperature coefficient of resistivity*. It is generally denoted by α. Thus we have

$$\rho_t = \rho_0(1 + \alpha t) \tag{6.18}$$

where t is the temperature in degrees centigrade. One can easily see that that the resistance $R = \frac{\rho L}{A}$ also has the same temperature coefficient α and one can write

$$R_t = R_0(1 + \alpha t) \tag{6.19}$$

α is about $0.004°C^{-1}$ for many metals. The materials obeying Ohm's law (6.16) are said to be *ohmic*. Table 1 gives the value of σ, ρ and α for some common materials.

Table 6.1. Conductivity (σ), Resistivity (ρ) and Temperature Coefficient of Resistivity (α) for some substances

Material	σ (at 20°C) in mh/m	ρ (at 20°C) in $\Omega - m$	$\alpha °C^{-1}$
Aluminium	3.54×10^7	0.282×10^{-7}	0.004
Copper	5.8×10^7	0.172×10^{-7}	0.004
Silver	6.14×10^7	0.162×10^{-7}	0.0038
Mercury	0.01×10^7	100×10^{-7}	0.0009
Tungsten	1.81×10^7	0.552×10^{-7}	0.0045
Brass (66% cu + 33% Zn)	1.39×10^7	0.719×10^{-7}	0.002
Germanium (300K)	0.022	45	Semiconductor
Silicon (300K)	4.35×10^{-6}	2.3×10^5	Semiconductor
Glass	1.1×10^{-12}	9×10^{11}	
Paraffin Oil	1×10^{-14}	1×10^{14}	

Ohm's law essentially states that the resistance R is independent of the current density in the material. However, there are many instances when Ohm's law fails: As an example one may cite the example of a space charge limited diode. I–V characteristics for a vacuum diode has the form shown in Fig. 6.3. When the anode is positive with respect to the cathode, the diode is said to be forward biased. Obviously, there cannot be a current in the reversed biased condition because the electrons are repelled by the cathode. We know that the diode is not a metallic conductor of the type for which Eq. (6.10) is normally used. Still one can represent its behaviour by its I–V characteristics, which indicates that in the reversed biased condition, this device offers an *infinite resistance* to the passage of current. We all know that this property of the diode is utilized in rectifiers to convert alternating current into direct current. One can easily see that the potential at a point in the space charge limited region between the electrodes is given by

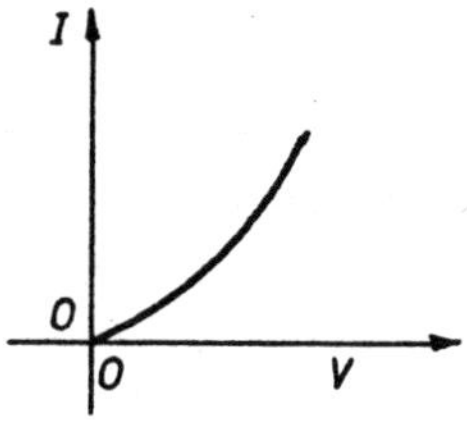

Figure 6.3.

$$V = \left(\frac{3xm^{1/4}}{2^{5/4}\, e^{1/4}\, \varepsilon_0^{1/2}}\right)^{4/3} J^{2/3} \tag{6.20}$$

The current density J is given by

$$J = \frac{I}{A}$$

$$\therefore V = \left(\frac{3xm^{1/4}}{2^{5/4}\, e^{1/4}\, \varepsilon_0^{1/2}}\right) \frac{I^{2/3}}{A^{2/3}}$$

or

$$I = \frac{4}{9}\varepsilon_0 A \left(\frac{2e}{m}\right)^{1/2} \frac{V^{3/2}}{x^2} \tag{6.21}$$

i.e.

$$I \propto V^{3/2} \tag{6.22}$$

Obviously, (6.22) is not in conformity with Ohm's law. One can also find such non-ohmic behaviour at semiconductor junctions.

There are also few elements and alloys which exhibit zero resistivity, i.e. infinite conductivity at very low temperatures. The temperature at which these materials exhibit the properly of zero resistivity is called critical temperature (T_c) and the phenomenon is known as *superconductivity.* This phenomenon was first observed in the laboratory in 1911 by Kamerlingh onnes. Today the highest known critical temperature (T_c) is 134K of $HgBa_2Ca_2Cu_3O_8$. Table 6.2 lists some superconducting metals and alloys along with their transition temperatures.

Table 6.2. Some Superconducting materials and their transition temperatures

Material	T_c (K)	Material	T_c (K)
Nb	9.3	$La_{2-x}M_xCuO_{4-y}$ (M = Ba, Sr, Ca) $x =\simeq 0.15$, $y \to$ small	38
Pb	7.2		
Ta	4.5	$HgBa_2Ca_2Cu_3O_{1+x}$	133
Sn	3.7	Y_9Co_7	3
Zr	0.8	UBe_{13}	0.85
Nb_3Sn	18	UPd_2Al_3	2.0
K_3C_{60}	19.3	$BaPb_xBi_{1-x}O_{3-y}$ $(0.05 \leq x \leq 0.3)$	13
Rb_3C_{60}	29.6		

It is now believed that many more high temperature superconductors must exist and await discovery. Superconductors find several applications. Some important fields are: Magnets, Electronics and small devices, computers and information processing, Energy related fields, e.g. magnetic energy storage, production by magnetic fusion and MHD, electric power transmission and transportation, e.g. high speed trains (MAGLEV), ship drive system etc.

6.5 MAGNETIC INTERACTION

Another type of interaction observed in nature is one called magnetic. Magnetic phenomena were known since ancient times. Before the means to produce steady electric currents became available, the study of magnetism was confined to naturally occurring magnetic objects such as load-stones. Lode stone have the property of attracting small pieces of iron. The property is exhibited in the natural state by iron, cobalt and manganese, and by many compounds of these metals. The regions of a body where the magnetism appears to be concentrated are called *magnetic poles.* A magnetized body is called a *magnet.*

The earth itself is a huge magnet. When we suspend a magnetized rod at any point on the earth's surface and allow it to rotate freely about the vertical, the rod orients itself so that the same end always points toward the north geographic pole. This result shows that the earth is exerting an additional force on magnetized rod which it does not exert on unmagnetized rods. This experiment also suggests that there are two kinds of magnetic poles, e.g. north (N) and south (S). The interaction between like magnetic poles is repulsive and the interaction between unlike magnetic poles is attractive.

When we attempt to measure the strength of a magnetic pole by defining a *magnetic mass* or charge, and investigate the dependence of the magnetic interaction on the distance between the poles, a fundamental difficulty arises. Although we have been able to isolate positive and negative electric charges and associate a definite amount of electric charge with the fundamental particles constituting all atoms, we have not been able to isolate a magnetic pole or identify a fundamental particle having only one kind of magnetism, either N or S. Magnetised bodies always exhibit poles in pairs, equal and opposite. On the other hand, the notions of magnetic pole and magnetic mass have been found unnecessary for the description of magnetism. Electric and magnetic interactions are very closely related, and in fact are only two different aspects of one property of matter, its electric charge; *magnetism is a manifestation of electric charges in motion.* Electric and magnetic interactions are considered together under the general name of electromagnetic interaction. A magnetic field is one of the forms of an electromagnetic field. Its distinguishing feature is that it acts only on moving particles and bodies having an electric charge, as well as on magnetized bodies regardless of their state of motion. A magnetic field is produced by current-carrying conductors, by moving electrically charged particles and bodies by magnetized bodies, or by a variable electric field (by displacement currents). Experiments by Biot and Savart, Ampere and others helped to establish the laws of magnetism as we know them today.

6.6 MAGNETIC FORCE ON A MOVING CHARGE

Since we observe interactions between magnetized bodies, we may say, in analogy with the gravitational and electrical cases, that a magnetized body produces a *magnetic field* around it. When we place an electric charge at rest in a magnetic field, no special force or interaction is observed on the charge. But when an electric charges moves in a region where there is a magnetic field, a new force is observed on the charge in addition to those due to its gravitational and electric interactions.

By measuring, at the same point in a magnetic field, the force experienced by different charges moving in different ways, one may obtain a relation between the force, the charge and its velocity. In this way we may obtain that '*the force exerted by a magnetic field on a moving charge is proportional to the electric charge and to its velocity, and the direction of the force is perpendicular to the velocity of the charge*'.

The force characteristic of a magnetic field is the vector of magnetic induction **B**. It is defined through its action on a moving charge q moving with velocity **V** in a region where **B** exists. Such a charged particle experiences a force given by

$$\mathbf{F} = q\mathbf{V} \times \mathbf{B} \tag{6.23}$$

Obviously, **B** describes a property that is characteristic of the magnetic field, and one may call it the magnetic field strength. Another name, imposed by usage, is magnetic induction.

When F is measured in newtons, q in Coulombs and V in m/s, the dimensions of B are $Ns/C - m$. This unit of B in the MKSA system of units is called 1 Weber/m^2 = 1 tesla = 1 Ns/C-m. The more familiar unit in the cgs system, viz 1 gauss, equals 10^{-4} tesla. The earth produces a magnetic induction of about 0.36×10^{-4} tesla on its surface.

When the particle moves in a region where there are electric and a magnetic field, the total force is the sum of the electric force $q\mathbf{E}$ and the magnetic force $q\mathbf{V} \times \mathbf{B}$. That is

$$\mathbf{F} = q(\mathbf{E} + \mathbf{V} \times \mathbf{B}) \tag{6.24}$$

This expression is called the *Lorentz force*.

Because of the property of vector product, Eq. (6.23) gives a force perpendicular to the velocity **V**, but also perpendicular to the magnetic field **B**. Eq. (6.23) also implies that when **V** is parallel to **B**, the force **F** is zero. In fact it is observed' that at each point in every magnetic field there is a certain direction of motion in which no force is exerted on the moving charge. This direction s defined as the direction of the magnetic field at the point. In Fig 6.4 the relation between the three vectors **V**, **B** and **F** is illustrated for both a positive and a negative charge. The direction of the force can be-determined with right hand rule as shown in Fig. 6.4.

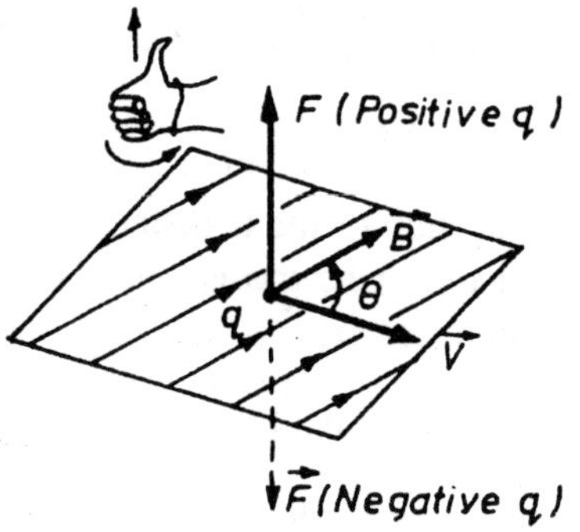

Figure 6.4. Vector relation between magnetic force, magnetic field, and charge velocity. The force (**F** is perpendicular to the plane containing **B** and **V**).

If θ is the angle between **V** and **B**, the magnitude of **F** is

$$F = qVB\sin\theta \tag{6.25}$$

The maximum force occurs when $\theta = \frac{\pi}{2}$ or **V** is perpendicular to **B**, resulting in

$$F = qVB \tag{6.26}$$

The minimum force F, zero, occurs when $\theta = 0$ or when **V** is parallel to **B**.

Because the magnetic force $\mathbf{F} = q\mathbf{V} \times \mathbf{B}$ to the velocity, its work is zero, and therefore, it does not produce any change in the kinetic energy of the particle. Although the magnetic force is not conservative but when a particle moves in Combined electric and magnetic fields, its total energy (i.e. Kinetic + Potential due to its different interactions) remains constant.

The Lorentz force law (6.24) is found to be true even for particles moving at speeds close to the speed of light. This means that the charge on a body is independent of its speed. Obviously, the charge, like rest mass, is relativistically invariant.

6.7 FORCE ON A CURRENT

Let us now consider the situation, when we have a number of moving charges as, for example, conduction electrons in metals. Let the number electrons per unit volume is N, then the force per unit volume will be

$$\mathbf{F} = Nq\mathbf{V} \times \mathbf{B} = \mathbf{J} \times \mathbf{B} \tag{6.27}$$

(if electric field E is assumed to be zero). The total force

$$\mathbf{F}_{tot} = \int_v \mathbf{J} \times \mathbf{B} d\tau = \int_s \int_l (\mathbf{J} \times \mathbf{B})(d\mathbf{l} \cdot \hat{a}_n ds) \tag{6.28}$$

where ds is the cross section of the conductor, and dl an elements of its length, with its direction the same as that of the current. We have $\mathbf{J}$ and $d\mathbf{l}$ are parallel, and hence

$$\mathbf{F}_{\text{tot}} = \int_s \int_l (d\mathbf{l} \times \mathbf{B})(\mathbf{J} \cdot \hat{a}_n ds) = \int I d\mathbf{l} \times \mathbf{B} \tag{6.29}$$

If the conducting wire carrying the current is in the form of a closed loop, we have

$$\mathbf{F} = \oint I d\mathbf{l} \times \mathbf{B} = I \oint d\mathbf{l} \times \mathbf{B} \tag{6.30}$$

6.8 BIOT SAVART LAW

Biot and Savart analysed the various experiments and formulated appropriated laws relating the magnetic flux density $\mathbf{B}$ to the current and also the law of force between one current and another. These are:

(i) Let us consider that dl, be an element of length of a wire with its sense taken in the direction of the current I_1 in the wire as shown in Fig. 6.5. One obtains the magnetic flux density $d\mathbf{B}$ due to this element at a point P specified by the location vector $\mathbf{r}$ as,

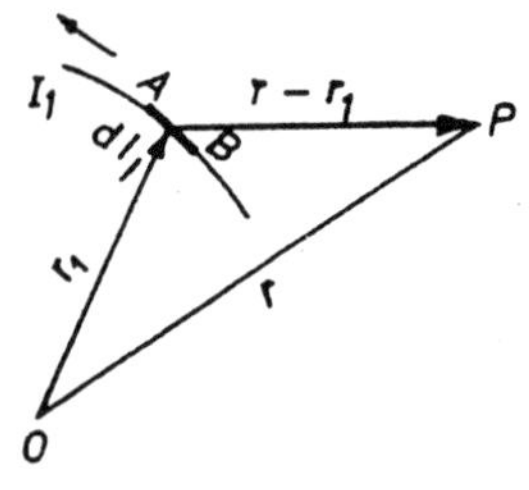

Figure 6.5.

$$d\mathbf{B} = \frac{\mu_0}{4\pi} \frac{I_1 \mathbf{dl}_1 \times (\mathbf{r} - \mathbf{r}_1)}{|\mathbf{r} - \mathbf{r}_1|3} \tag{6.31}$$

where $\mathbf{r}_1$ is the position or location vector of the element dl. Obviously, the flux density is directly proportional to the current flowing in the circuit and to the length of the element of the wire and inversely proportional to the square of the distance of the point from the element of the wire Thus we have

$$\mathbf{B} = \frac{\mu_0 I_1}{4\pi} \int \frac{d\mathbf{l}_1 \times (\mathbf{r} - \mathbf{r}_1)}{|\mathbf{r} - \mathbf{r}_1|3} \tag{6.32}$$

The integration extends for the whole length of the current element. The constant μ_0 is termed the *Permeability of free space* and its value is exactly $4\pi \times 10^{-7}$ NA^{-2} or $4\pi \times 10^{-7}$ henry/m ($= 1.3566 \times 10^{-6}$ mkg C^{-2}). Eqs. (6.31) and (6.32) are statements of the Bio-Savart Law.

For a closed loop of current as shown in Fig. 6.6, the resultant magnetic induction at P is

$$B = \frac{\mu_0}{4\pi} \oint \frac{I_1 d\mathbf{l}_1 \times \mathbf{r}_{12}}{r_{12}^3} \qquad (6.33)$$

where the integration is taken over the entire closed loop circuit.

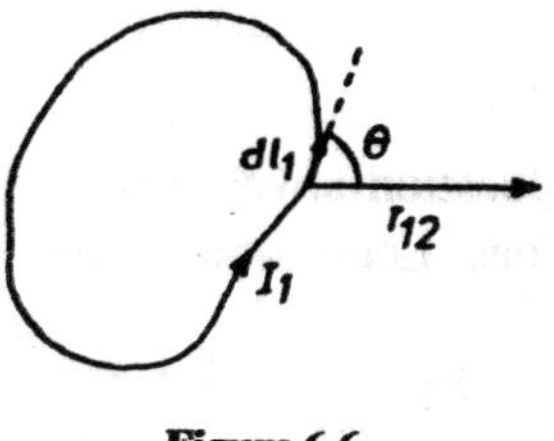

Figure 6.6.

(ii) If there are two closed circuits as shown in Fig. 6.7, the force on circuit 2 due to circuit 1 is

$$\mathbf{F}_2 = \oint I_2 d\mathbf{l}_2 \times \mathbf{B}_2 \qquad (6.34)$$

where B_2 is the magnetic flux density due to circuit at the position where circuit 2 is located as shown in Fig 6.7. Let us assume that B_2 be constant over the region occupied by the circuit. Thus we have

$$\begin{aligned} \mathbf{F}_2 &= \oint_2 I_2 dl_2 \times \frac{\mu_0}{4\pi} \oint_1 \frac{I_1 d\mathbf{l}_1 \times (\mathbf{r}_2 - \mathbf{r}_1)}{|\mathbf{r}_2 - \mathbf{r}_1|^3} \\ &= \frac{\mu_0}{4\pi} I_1 I_2 \oint_2 \oint_1 \frac{d\mathbf{l}_2 \times [d\mathbf{l}_1 \times (\mathbf{r}_2 - \mathbf{r}_1)]}{|\mathbf{r}_2 - \mathbf{r}_1|^3} \end{aligned} \qquad (6.35)$$

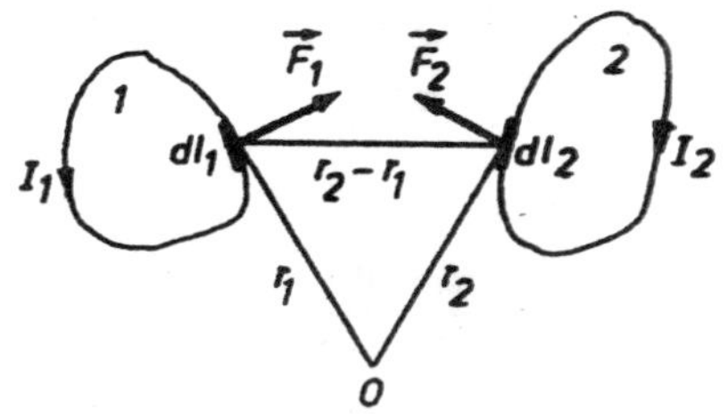

Figure 6.7.

Similarly, one obtains the force exerted by the circuit 2 at the position where circuit 1 is located,

$$\mathbf{F}_1 = \frac{\mu_0}{4\pi} I_1 I_2 \oint_1 \oint_2 \frac{d\mathbf{l}_1 \times [d\mathbf{l}_2 \times (\mathbf{r}_1 - \mathbf{r}_2)]}{|\mathbf{r}_1 - \mathbf{r}_2|^3} \qquad (6.36)$$

However, Eqs. (6.35) and (6.36) seem to give rise to an unacceptable solution. We must expect that the forces $\mathbf{F}_2$ and $\mathbf{F}_1$ be equal and opposite. However, we have

$$\frac{d\mathbf{l}_2 \times [d\mathbf{l}_1 \times (\mathbf{r}_2 - \mathbf{r}_1)]}{|\mathbf{r}_2 - \mathbf{r}_1|^3} \neq \frac{d\mathbf{l}_1 \times [d\mathbf{l}_2 \times (\mathbf{r}_1 - \mathbf{r}_2)]}{|\mathbf{r}_1 - \mathbf{r}_2|^3}$$

i.e.

$$\mathbf{F}_2 \neq \mathbf{F}_1$$

Question arises, where this violates Newton's law? It seems that it does. Let us now expand the integrand in (6.35)

$$\mathbf{F}_2 = \frac{\mu_0}{4\pi} I_1 I_2 \oint_2 \oint_1 \left[\frac{\{d\mathbf{l}_2 \cdot (\mathbf{r}_2 - \mathbf{r}_1)\} d\mathbf{l}_1}{|\mathbf{r}_2 - \mathbf{r}_1|^3} - \frac{(d\mathbf{l}_1 \cdot d\mathbf{l}_2)(\mathbf{r}_2 - \mathbf{r}_1)}{|\mathbf{r}_2 - \mathbf{r}_1|^3} \right]$$

The first integral vanishes,

$$\oint_2 \frac{d\mathbf{l}_2 (\mathbf{r}_2 - \mathbf{r}_1)}{|\mathbf{r}_2 - \mathbf{r}_1|^3} = -\oint \nabla_2 \left(\frac{1}{|\mathbf{r}_2 - \mathbf{r}_1|} \right) \cdot d\mathbf{l}_2 = 0$$

(We must remember that ∇_2 operates on the terms involving r_2). One obtains

$$\mathbf{F}_2 = -\frac{\mu_0}{4\pi} I_1 I_2 \oint_1 \oint_2 \frac{(d\mathbf{l}_1 \cdot d\mathbf{l}_2)(\mathbf{r}_2 - \mathbf{r}_1)}{|\mathbf{r}_2 - \mathbf{r}_1|^3} = -\mathbf{F}_1 \tag{6.37}$$

when we compare with the expression obtained from (6.36) by similar transformations we find that it is in conformity with Newton's law.

Now we apply Biot and Savart's law to calculate the *force between parallel wires carrying currents.*

Fig. 6.8 shows two parallel wires distant a apart and carrying current I_1, I_2 respectively. Let us assume that the first wire is along the z-axis and second wire is parallel to the first wire and passing through the point $(a, 0, 0)$ as shown in Fig. 6.8. Let us take a point $P(a, 0, z)$ and calculate the field at this point due to the current I_1, in the first wire. Let r be the distance of the point P from the origin.

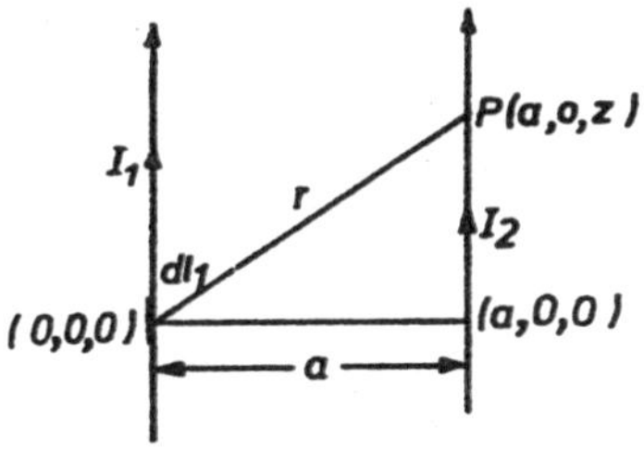

Figure 6.8.

Let us consider an element dl_1 of the first wire at the origin (Fig 6.8). One can easily see that its components are: (o, o, dz) and those of r are: (a, o, z). Obviously, the components of $d\mathbf{l}_1 \times \mathbf{r}$ are (o, adz, o). This clearly reveals that the magnetic field dB at P due to the element dl_1 is parallel to y-axis. Obviously, B also will be parallel to y-axis with components $B_x = 0$, $B_z = 0$ and

$$B_y = \frac{\mu_0}{4\pi} I_1 \int_{-\infty}^{\infty} \frac{|d\mathbf{l} \times \mathbf{r}|}{|\mathbf{r}|^3} = \frac{\mu_0}{4\pi} I_1 \int_{-\infty}^{\infty} \frac{adz}{(a^2+z^2)^{3/2}}$$

Now, putting $Z = a\tan\theta$, one obtains

$$B_y = \frac{\mu_0}{4\pi} I_1 \int_{-\pi/2}^{\pi/2} \frac{a \times a\sec^2\theta}{a\sec^3\theta} d\theta = \frac{2\mu_0 I_1}{4\pi a} = \frac{\mu_0 I_1}{2\pi a} \qquad (6.38)$$

From (6.38) it is evident that **B** is directly proportional to I_1 and inversely proportional to a. Moreover, **B** will always be normal to the plane containing the wire and the radius vector **r** and hence the lines of constant **B** form closed circles with their centres on the wire in planes perpendicular to the current.

Let us now find out the force dF acting on an element dl_2 of the second wire due to the current in the first wire. Force is given by (6.29). Now , we see that B is in positive y direction and dl_2 in the z-direction, the force acting on dl_2 will be in the negative x-direction, its component in this direction being

$$dF_x = -\frac{\mu_0 I_1 I_2 dl_2}{2\pi a} \qquad (6.39)$$

This result indicates that if the currents are in the same direction, the force is one of attraction. If the currents are in the opposite direction, the force is one of repulsion. This result confirms the Ampere's observations.

The above result can be extended to currents of any configuration. These interactions among currents are of great practical importance for electric motors and other engineering applications.

6.9 MAGNETIC FIELD INDUCTION ON THE AXIS OF A CIRCULAR-LOOP

Let us consider a circular loop of radius a, carrying current I ampere with its perpendicular to the plane of the paper (Fig 6.9). Let P be a point on the axis of this circular loop at a distance r from an element dl. Let us calculate the magnetic flux density at P. Since dB, the magnetic field produced by the element dl, is normal to dl as well as r, it will act in the direction shown in Fig 6.9. We must remember that the resultant magnetic field **B** at P produced by the current is the sum of a large number of very small or elementary contributions $d\mathbf{B}$ by each of the segments or length elements dl composing the circuit. On integrating round the coil, the sum of the components of dB, normal to the axis, is zero. One obtains the component parallel to the axis as

$$dB = \frac{\mu_0}{4\pi}\frac{Idl}{r^2}\cos\phi$$

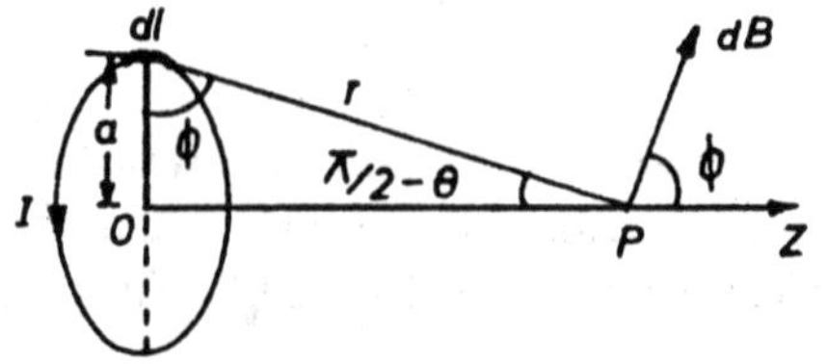

Figure 6.9.

where ϕ is the angle between the magnetic field dB and the axis, Now integrating the above, we obtain

$$B = \frac{\mu_0 I}{4\pi r^2}\oint \cos\phi dl = \frac{\mu_0 I\cos\phi}{4\pi r^2}2\pi a \quad \left(\because \oint dl = 2\pi a\right)$$

$$\text{or} \quad B = \frac{\mu_0 I a}{2r^2}\frac{a}{r} \quad \left(\because \cos\phi = \frac{a}{r}\right)$$

$$= \frac{\mu_0 I a^2}{2r^3} \tag{6.40}$$

Noting that $r = (a^2 + z^2)^{1/2}$, we can write the magnetic field for points on the axis of a circular current as

$$B = \frac{\mu_0 I a^2}{2(a^2+z^2)^{3/2}} \tag{6.41}$$

The magnetic field of a circular current has been represented in Fig 6.10. If the coil contains n turns, then

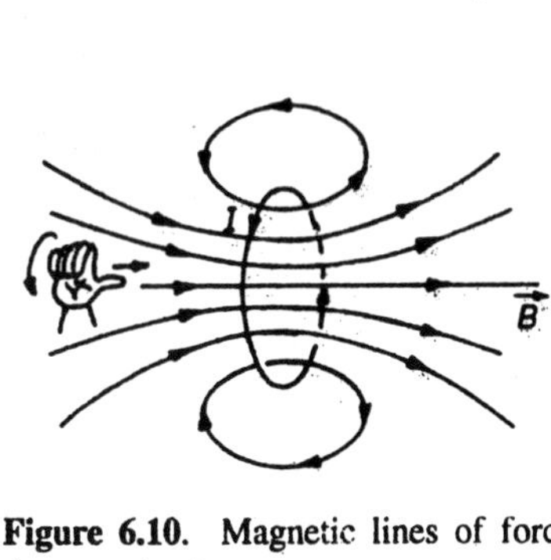

Figure 6.10. Magnetic lines of force due to a circular current.

$$B = \frac{\mu_0 n I a^2}{2(a^2+z^2)^{3/2}} \tag{6.41(a)}$$

Different Cases

(i) *The magnetic field due to induction at the centre of the circular coil:* At the centre of the circular coil, $Z = 0$, hence

$$B = \frac{\mu_0 I}{2a} \frac{W}{m^2} \tag{6.42}$$

If the coil contains n turns, then

$$B = \frac{\mu_0 nI}{2a} \tag{6.42(a)}$$

(ii) *The magnetic field due to a small circular coil:* An interesting case occurs when the circuit is very small, so that the radius a can be neglected in comparison with the distance Z. Then Eq. (6.41) reduces to

$$B = \frac{\mu_0 I a^2}{2Z^3} \tag{6.43}$$

The magnetic dipole moment of the circuit is $M = I(\pi a^2)$. Then

$$B = \frac{\mu_0 M}{2\pi Z^3} = \frac{\mu_0 (2M)}{4\pi Z^3} \tag{6.44}$$

When we compare (6.44) with the dipole expression, with $\theta = 0$, i.e., $E_r = \frac{1}{4\pi\varepsilon_0}\left(\frac{2p}{r^3}\right)$, we see that the magnetic field along the axis of the small current is identical to the electric field of an electric dipole along its axis if we make $\left(\frac{\mu_0}{4\pi}\right) M$ correspond to $\frac{p}{4\pi\varepsilon_0}$. For that reason the circuit is called a *magnetic dipole*. Therefore we can apply electric dipole results for a magnetic dipole so that the magnetic field off-axis may be computed (Fig 6.11). One obtains

$$B_r = \frac{\mu_0}{4\pi} \frac{2M\cos\theta}{Z^3}$$

and

$$B_\theta = \frac{\mu_0}{4\pi} \frac{M\sin\theta}{Z^3} \tag{6.45}$$

We have seen that the lines of force of an electric field go from the negative to positive charges or perhaps, in some cases, from or to infinity. However, in the case of magnetism (Fig 6.10), we may see that the lines of force of a magnetic field are closed lines, linked with the current. The reason for this is that the magnetic field is not originated

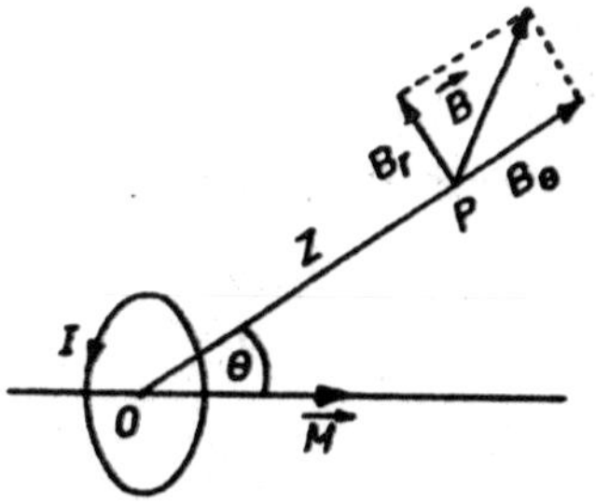

Figure 6.11. Magnetic field at point p due to a magnetic dipole current.

by magnetic poles. A field of this kind, which does not have point sources, is called *solenoidal.*

6.10 MAGNETIC FIELD DUE TO SOLENOID

Solenoid is a cylindrically shaped coil consisting of a long wire wound usually on a non-magnetic frame in a closed-packed helix and carrying an electric current. A Solenoidal current is a current composed of several coaxial circular loops of the same radius, all carrying the same current (Fig 6.12). Such an arrangement may be used in electromagnetic relay and in some of the electrical instruments.

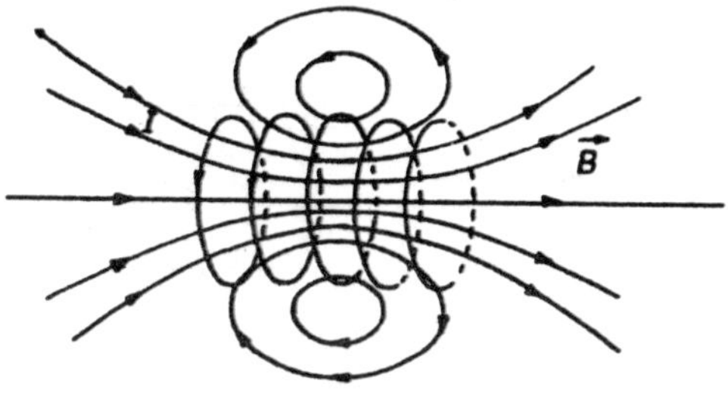

Figure 6.12. Magnetic lines of force due to a solenoidal current.

Fig 6.13 shows a solenoid of length L having N turns uniformly wound round a cylinder of radius a and carrying a current I. One can obtain its magnetic field by adding the magnetic fields of each of the component circular currents. The field is indicated by lines of magnetic force in the Figure 6.12, where some fluctuations in the space between loops have been smoothed out. Let us calculate the field at a point P on the axis of the solenoid. The point P is at a distance z_0 from O. Since the turns in the solenoid are closely wound, one may regard the current flowing uniformly round the cylinder. Let us consider that the length of the solenoid is divided into elements dz, such as the one shown at a distance z from O. The current in this elemental section dz of the coil is $\frac{N}{L}dzI$ and the field at axial point P due to this element is

$$dB = \frac{1}{2}\frac{\mu_0 NI}{L}\frac{a^2dz}{[(z_0-z)^2+a^2]^{3/2}} \text{ (See Eq. (6.41))} \quad (6.46)$$

$$\therefore \quad B = \frac{1}{2}\frac{\mu_0 NI}{L}\int_0^L \frac{a^2dz}{[(z_0-z)^2+a^2]^{3/2}}$$

$$= \frac{1}{2}\frac{\mu_0 NI}{L}\left[\frac{z_0-z}{\sqrt{(z_0-z)^2+a^2}}\right]_0^L$$

$$= \frac{\mu_0 NI}{2L}\left[\frac{z_0}{\sqrt{z_0^2+a^2}}+\frac{L-z_0}{\sqrt{(L-z_0)^2+a^2}}\right]$$

$$= \frac{\mu_0 NI}{2L}[\cos\alpha+\cos\beta] \quad (6.47)$$

where α and β are the angles subtended by the ends E' and F' of the solenoid at point P with axis (Fig 6.13).

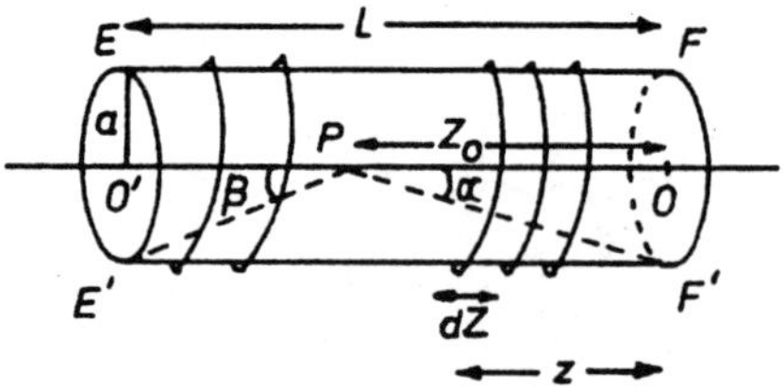

Figure 6.13. Computation of the magnetic field at a point P locate along the axis of a solenoidal current.

If the solenoid is *very long*, i.e. for an *infinite* solenoid $\alpha = \beta = 0$ and

$$B = \frac{\mu_0 NI}{L} \quad (6.48)$$

Obviously, B does not depend upon the position of the point P and, hence a long solenoid is used to produce fairly uniform fields in limited regions around its centre.

For a point at one end, i.e. point P is at one end of the long solenoid, (say F') of the long solenoid, then $\alpha = \frac{\pi}{2}$, $\beta = 0$ or π, so the magnetic field at an end point of long solenoid is obtained as

$$B = \frac{\mu_0 NI}{2L} \tag{6.49}$$

or one-half the value at the centre. A long solenoid is used to produce fairly uniform magnetic fields in limited regions around its centre. The variation in magnitude of B from point to point over the complete length of a long solenoid is shown in Fig 6.14.

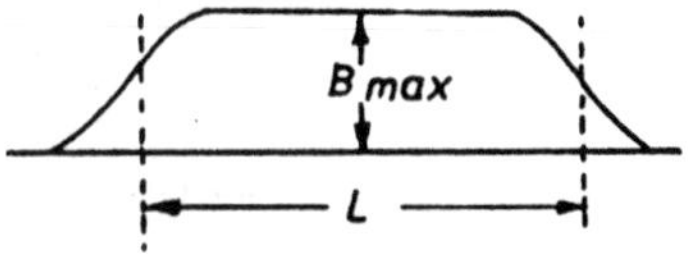

Figure 6.14.

6.11 LAWS OF MAGNETOSTATICS

(i) $\nabla \cdot B = 0$

One can express the expression for the magnetic flux density (6.32) in terms of current density. Thus we have

$$\mathbf{B} = \frac{\mu_0 I}{4\pi}\int \frac{d\mathbf{l}_1 \times (\mathbf{r}-\mathbf{r}_1)}{|\mathbf{r}-\mathbf{r}_1|^3} = \frac{\mu_0}{4\pi}\int \frac{\mathbf{J}(r_1)\times(\mathbf{r}-\mathbf{r}_1)}{|\mathbf{r}-\mathbf{r}_1|^3}dV$$

Here we have used $Id\mathbf{l}_1 = \mathbf{J}(r_1)dl_1 \cdot \hat{a}_n ds = \mathbf{J}(r_1)dv$

$$\therefore \quad \mathbf{B} = -\frac{\mu_0}{4\pi}\int_v \left[\mathbf{J}(r_1)\times\nabla\left(\frac{1}{|\mathbf{r}-\mathbf{r}_1|}\right)\right]dV \tag{6.50}$$

Taking divergence, one obtained

$$\nabla\cdot\mathbf{B} = -\frac{\mu_0}{4\pi}\int_v \nabla\cdot\left[\mathbf{J}(r_1)\times\nabla\left(\frac{1}{|\mathbf{r}-\mathbf{r}_1|}\right)\right]dV$$

Using the vector identity

$$\nabla\cdot(\mathbf{A}\times\mathbf{C}) = \mathbf{C}\cdot(\nabla\times\mathbf{A}) - \mathbf{A}\cdot(\nabla\times\mathbf{C}),$$

one obtains

$$\begin{aligned}\nabla\cdot\mathbf{B} = & -\frac{\mu_0}{4\pi}\int_v \nabla\left(\frac{1}{|\mathbf{r}-\mathbf{r}_1|}\right)\cdot\{\nabla\times\mathbf{J}(r_1)\}dV \\ & +\frac{\mu_0}{4\pi}\int \mathbf{J}(r_1)\cdot\left\{\nabla\times\nabla\left(\frac{1}{|\mathbf{r}-\mathbf{r}_1|}\right)\right\}dV\end{aligned} \tag{6.51}$$

we can see that ∇ operates on r and hence the first integral in (6.51) is zero. The second term in (6.51) contains a factor curl grad $\left(\frac{1}{|\mathbf{r}-\mathbf{r}_1|}\right)$.

Which is identically zero. Obviously

$$\nabla \cdot \mathbf{B} = 0 \tag{6.51}$$

The divergence-less character of **B** is an important property. Eq. (6.51), is one of the four equations of Maxwell that form the basis of electromagnetic theory; it reflects the absence of magnetic monopoles. One can understood this by noting that the divergence of a vector field is a measure of its source density. We have already seen this result in Poisson's equation in electrostatics, i .e. $\nabla \cdot \mathbf{E} = \frac{\rho}{\varepsilon_0}$. Similarly $\nabla \cdot \mathbf{B}$ would give us the local density of magnetic charges or monopoles. Since **B** is always divergence less, it implies that no volume element contains magnetic charges of one sign only. We must note that the mathematical description of **B** is arranged in such a manner that $\nabla \cdot \mathbf{B}$ is zero. This is necessary to confirm with the fact that magnetic monopoles have not been observed so far. Eq. (6.51) is the first law in magnetistatics corresponding to the electrostatic relation $\nabla \times \mathbf{E} = 0$. This relation shows that the magnetic field is **solenoidal** in contrast to the electric field which is **irrotational**.

(ii) Ampere's Law for the Magnetic field (Ampere's Circuital law)

Now, we determine the value of $\nabla \times \mathbf{B}$. This will help us to complete the analongy with electrostatics. We have

$$\mathbf{B} = \frac{\mu_0}{4\pi} \int_v \frac{\mathbf{J}(r_1) \times (\mathbf{r} - \mathbf{r}_1)}{|\mathbf{r} - \mathbf{r}_1|^3} dV = -\frac{\mu_0}{4\pi} \int_v \left[\mathbf{J}(r_1) \times \nabla \left(\frac{1}{|\mathbf{r} - \mathbf{r}_1|} \right) \right] dV$$

$$= \frac{\mu_0}{4\pi} \nabla \times \int \frac{\mathbf{J}(r_1)}{|\mathbf{r} - \mathbf{r}_1|} dV \tag{6.52}$$

Taking curl of the above, one obtains

$$\nabla \times \mathbf{B} = \frac{\mu_0}{4\pi} \nabla \times \left(\nabla \times \int \frac{\mathbf{J}(r_1)}{|\mathbf{r} - \mathbf{r}_1|} dV \right)$$

Using the vector identity $\nabla \times (\nabla \times \mathbf{A}) = \nabla(\nabla \cdot \mathbf{A}) - \nabla^2 \mathbf{A}$, one obtains

$$\nabla \times \mathbf{B} = \frac{\mu_0}{4\pi} \nabla \left(\nabla \cdot \int \frac{\mathbf{J}(\mathbf{r}_1)}{|\mathbf{r} - \mathbf{r}_1|} dV \right) - \frac{\mu_0}{4\pi} \nabla^2 \int \frac{\mathbf{J}(r_1)}{|\mathbf{r} - \mathbf{r}_1|} dV$$

$$= \frac{\mu_0}{4\pi} \nabla \int \mathbf{J}(r_1) \cdot \nabla \left(\frac{1}{|\mathbf{r} - \mathbf{r}_1|} \right) dV$$

$$- \frac{\mu_0}{4\pi} \nabla \int \mathbf{J}(r_1) \nabla^2 \left(\frac{1}{|\mathbf{r} - \mathbf{r}_1|} \right) dV \tag{6.53}$$

We have

$$\nabla^2\left(\frac{1}{|\mathbf{r}-\mathbf{r}_1|}\right) = -4\pi\delta(\mathbf{r}-\mathbf{r}_1) \tag{6.54}$$

and

$$\nabla\left(\frac{1}{|\mathbf{r}-\mathbf{r}_1|}\right) = -\nabla_1\left(\frac{1}{|\mathbf{r}-\mathbf{r}_1|}\right) \tag{6.55}$$

where ∇_1 operates only on r_1.

Using (6.54) and (6.55), (6.53) reduces to

$$\begin{aligned}
\nabla\times\mathbf{B} &= -\frac{\mu_0}{4\pi}\nabla\int \mathbf{J}(r_1)\cdot\nabla_1\left(\frac{1}{|\mathbf{r}-\mathbf{r}_1|}\right)dv \\
&\quad +\frac{\mu_0}{4\pi}\int \mathbf{J}(r_1)4\pi\delta(\mathbf{r}-\mathbf{r}_1)dv \\
&= -\frac{\mu_0}{4\pi}\nabla\int \mathbf{J}(r_1)\cdot\nabla_1\left(\frac{1}{|r-r_1|}\right)dv+\mu_0\mathbf{J}(r) \\
&= -\frac{\mu_0}{4\pi}\nabla\int\left[\nabla_1\cdot\frac{\mathbf{J}(r_1)}{|\mathbf{r}-\mathbf{r}_1|}-\frac{\nabla_1\cdot\mathbf{J}(r_1)}{|\mathbf{r}-\mathbf{r}_1|}\right]dv+\mu_0\mathbf{J}(r)
\end{aligned} \tag{6.56}$$

We must remember that $\nabla(\phi A) = \mathbf{A}\cdot\nabla\phi+\phi\nabla\mathbf{A}$.

One can easily show with the help of divergence theorem by taking a large enough surface outside the current carrying region that the first term in (6.56) vanishes, and second is identically zero since $\nabla\cdot\mathbf{J} = 0$ for steady state. Thus one obtains

$$\nabla\times\mathbf{B} = \mu_0\mathbf{J} \tag{6.57}$$

This is *second law of magnetostatics*. This corresponds to $\nabla\cdot\mathbf{E} = \dfrac{\rho}{\varepsilon_0}$ in electrostatics. One can easily found the integral form the above. We have

$$\int_s \nabla\times\mathbf{B}\cdot\hat{a}_n ds = \mu_0\int \mathbf{J}(r)\cdot\hat{a}_n ds \tag{6.58}$$

From Stoke's theorem

$$\int_s \nabla\times\mathbf{B}\cdot\hat{a}_n ds = \oint \mathbf{B}\cdot d\mathbf{l} \tag{6.59}$$

Using (6.59), (6.58) reduces to

$$\oint \mathbf{B}\cdot d\mathbf{l} = \mu_0\int \mathbf{J}(r)\cdot\hat{a}_n ds = \mu_0 I \tag{6.60}$$

where S is any surface bounded by l and I stands for the total current linked by the path l.

This is Ampere's law for the magnetic field, i.e. *Ampere's circuital law*. The sign of integral in (6.60) depends upon the direction in which the current is encircled. The sign is positive, if the path followed for line integral is parallel to **B** and negative, if the path followed is anti-parallel. According to Ampere's circuital law the line integral of flux density round any closed path is equal to μ_0 times the current flowing through the area enclosed by the path. The fact that the circulation of the magnetic field **B** generally is not zero indicates that the magnetic field does not have a magnetic potential in the same sense that the electric field has an electric potential. The law further indicates that the magnetic field of a current is *non-conservative*. Ampere's law is particularly useful when we wish to compute the magnetic field produced by current distributions having certain geometrical symmetries.

We must note that in the integral statement of Ampere's law and also in the differential statement, the magnetic field **B** need not be only that field which is produced by the encircled current I (or the local current density **J**). It may include the field produced by any other neighbouring currents. The closed line integral for the total **B** gives μ_0 times the current linked with the path of integration. One may compare this with Gauss's law in electrostatics. We have read that the surface integral $\oint_S$ of **E** over a closed surface gives $\frac{1}{\varepsilon_0}$ times q, the charged enclosed within that surface. The **E** that figures in the calculation of flux ϕ_E can be due to all charges, q included. The charges outside the closed surface make zero contribution to the integral that gives ϕ_E. One can summarize the laws of electrostatics and magnetostatics as follows:

$$\left.\begin{array}{ll} \text{Electrostatics} & \text{Magnetostatics} \\ \nabla \times \mathbf{E} = 0 & \nabla \cdot \mathbf{B} = 0 \\ \nabla \cdot \mathbf{E} = \dfrac{\rho}{\varepsilon_0} & \nabla \times \mathbf{B} = \mu_0 \mathbf{J} \end{array}\right\} \tag{6.61}$$

6.12 MAGNETIC POTENTIALS

(i) Magnetic Scalar Potential

Ampere's law states that the line integral of **B** for a closed path is equal to μ_0 times the current I encircled by the path. Obviously, when the current linkage is monzero, the magnetic field along such a path must be considered as non conservative. However, there can be situations where the current encircled is zero, e.g. when one is confined to a region outside of the current sources. In such a region there is no possibility of any closed path being linked with a current. Obviously, this makes $\oint \mathbf{B}\cdot d\mathbf{l}=0$ for all closed paths. But this is the situation of a conservative field. This means that the magnetic field in such a region can be described as a conservative field. Obviously, one can therefore express the magnetic field for such situations as the gradient of a scalar potential field.

In electrostatics we have seen that once the electrostatic potential V is found, one can easily found $\mathbf{E}$ using the relation $E=-\nabla V$. (Since $\nabla\times\mathbf{E}=0$ and hence one can express $\mathbf{E}$ as the negative gradient scalar quantity V, i.e. electrostatic potential). In case of magnetism, if $\mathbf{J}=0$, one finds an analogous situation, i.e. $\nabla\times\mathbf{B}=0$. We may, therefore, express $\mathbf{B}$ as a gradient of scalar quantity, ϕ_m (say). Thus, we have

$$\mathbf{B}=-\nabla\phi_m \tag{6.62}$$

Let us consider ϕ_m as *magnetic scalar potential.* Using the relation $\nabla\cdot\mathbf{B}=0$, Eq. (6.62) yields,

$$\nabla\cdot\mathbf{B}=-\nabla\cdot\nabla\phi_m=-\nabla^2\phi_m=0$$

This shows that ϕ_m satisfies Laplace's equation. Let us consider a situation shown in Fig. 6.15. There is a circuit carrying a current I. One obtains the magnetic flux density at the point due the current I flowing the circuit as

$$\mathbf{B}=\frac{\mu_0}{4\pi}I\int\frac{d\mathbf{l}\times\mathbf{r}}{|\mathbf{r}|^3}$$

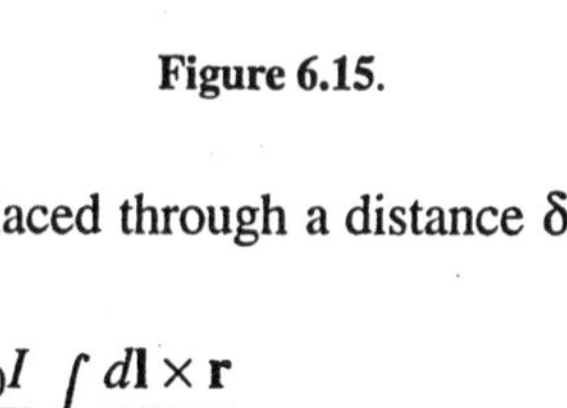

Figure 6.15.

Let us consider that the point P is displaced through a distance δS, then the change in the potential will be

$$\begin{aligned}\delta\phi_m=-\mathbf{B}\cdot\delta\mathbf{S} &= -\delta\mathbf{S}\cdot\frac{\mu_0 I}{4\pi}\int\frac{d\mathbf{l}\times\mathbf{r}}{|\mathbf{r}|^3}\\ &= -\frac{\mu_0 I}{4\pi}\int\frac{\delta\mathbf{S}\cdot(d\mathbf{l}\times\mathbf{r})}{|\mathbf{r}|^3}\end{aligned} \tag{6.63}$$

we have taken $\delta\mathbf{S}$ inside the sign of integration as it is constant during the integration.

Let us now consider that the point P is fixed and the circuit is displaced through a small amount $-\delta\mathbf{S}$. Obviously, the change in the potential should be the same as above. In such a small displacement, one can easily see that an element dl of the circuit sweeps an area $dl\delta S\sin\theta$, where θ is the angle between dl and δS. One can easily express the area swept out as $-\delta\mathbf{S}\times d\mathbf{l}$. One finds the angle subtended at P by this area as

$$d\Omega = -\frac{\mathbf{r}\cdot(\delta\mathbf{S}\times d\mathbf{l})}{|\mathbf{r}|^3} = -\frac{\delta\mathbf{S}\cdot(d\mathbf{l}\times\mathbf{r})}{|\mathbf{r}|^3} \tag{6.64}$$

Using (6.64), one can express (6.63) as

$$\delta\phi_m = \frac{\mu_0 I}{4\pi}d\Omega$$

where $d\Omega$ is the solid angle subtended by the area element at P. Obviously, the potential at P is obtained as

$$\phi_m = \frac{\mu_0 I}{4\pi}\Omega \tag{6.65}$$

Here Ω is the solid angle subtended at P by the boundary of the given current loop.

Let us compare (6.64) with the potential of an electrostatic double layer, i.e. $\phi(r) = \frac{D}{4\pi\varepsilon_0}d\Omega$. This reveals that the magnetic scalar potential has mathematically the same properties as the electrostatic double layer potential.

We must note that the concept of magnetic scalar potential is of practical utility only to derive magnetic field in the absence of continuous current distributions. The scalar potential ϕ_m can be used if line integrals encircling any currents are considered or if the fields within current carrying media are desired. Now

$$\int \mathbf{B}\cdot d\mathbf{l} = -\int \nabla\phi_m\cdot d\mathbf{l} = -\phi_m \tag{6.66}$$

The quantity $-\phi_m$ is sometimes called as *magnetomotive force.*

(ii) Vector Potentials

We have seen that the electrostatic field can be expressed as the negative gradient of a scalar potential ϕ, i.e. $\mathbf{E} = -\nabla\phi$. Since the curl of a gradient is always zero the above formulation immediately leads to $\nabla\times\mathbf{E} = 0$, which is an expression of the conservative nature of the electrostatic field.

The question arises: can one define the magnetic induction **B** in terms of some potential function? One can easily see that the answer is: yes, one can. The guiding equation for this purpose is

$$\nabla \cdot \mathbf{B} = 0.$$

We know that

$$\text{div curl}\mathbf{A} = \nabla \cdot \nabla \times \mathbf{A} = 0$$

Obviously, one can express **B** as curl of **A**, i.e.

$$\mathbf{B} = \nabla \times \mathbf{A} \tag{6.67}$$

The vector **A** which satisfies (6.67) is known as the *vector potential.*

We find that **A** is not uniquely defined by (6.67). Let us add to **A** any function whose curl is zero, say gradient of a scalar ψ and still **B** is same.

$$\nabla \times (\mathbf{A} + \nabla\psi) = \nabla \times \mathbf{A} + \nabla \times (\nabla\psi) = \nabla \times \mathbf{A} = \mathbf{B}$$

We must remember that in electrostatics the scalar potential V was not completely specified by the definition $E = -\nabla V$. If V is the potential for some problem, one would need a different potential $V' = V + b$, where b is a constant, gives the same field. Thus

$$-\nabla V' = -\nabla(V + b) = -\nabla V - \nabla b = -\nabla V = \mathbf{E}$$

One can have **A** more specific by imposing an additional restriction to it, without affecting **B**. One can easily take,

$$\nabla \cdot \mathbf{A} = 0 \tag{6.68}$$

This is possible because it makes calculations easier that with any other choice in magnetostatics. In electrodynamics, one have to choose it differently.

Now, we have $\mathbf{B} = \nabla \times \mathbf{A}$ and $\nabla \cdot \mathbf{A} = 0$. These two together define the vector potential **A** which satisfies the fundamental equation $\nabla \cdot \mathbf{B} = 0$.

It is more appropriate to drop the word 'magnetic' from 'magnetic vector potential', because in the case of time dependent phenomena, the potential **A** appears in the description of the electric field **E** also.

Now, we try to find a relation in magnetostatics corresponding to Poisson's equation $\nabla^2 V = \frac{\rho}{\varepsilon_0}$ in electrostatics. We have

$$\nabla \times \mathbf{B} = \mu_0 \mathbf{J} \quad \text{(Ampere's Law)}$$

Also

$$\nabla\times\mathbf{B} = \nabla\times\nabla\times\mathbf{A}=\nabla(\nabla\cdot\mathbf{A})-\nabla^2\mathbf{A}=-\nabla^2\mathbf{A}$$

$$\therefore\quad \nabla^2\mathbf{A} = -\mu_0\mathbf{J} \tag{6.69}$$

We can see that (6.69) is similar to Poisson's equation except that **A** is a vector. However, this should not cause any difficulty. Since each component of **A** must satisfy the differential equation. Thus

$$\nabla^2 A_x=-\mu_0 J_x, \nabla^2 A_y=-\mu_0 J_y \quad \text{and} \quad \nabla^2 A_z=-\mu_0 J_z \tag{6.70}$$

One finds the solution as

$$A_x=\frac{\mu_0}{4\pi}\int\frac{J_x}{|\mathbf{r}|}dV \text{ etc.} \tag{6.71}$$

One can write the general solution as

$$\mathbf{A}=\frac{\mu_0}{4\pi}\int\frac{\mathbf{J}}{|\mathbf{r}|}dV \tag{6.72}$$

Obviously, one can compute the field produced by a current by first determining **A** and then substituting this value in $\mathbf{B}=\nabla\times\mathbf{A}$.

6.13 MAGNETIC DIPOLE

While discussing the field of an electric dipole **p**, we have considered it as a combination of two point charges, $+q$ and $-q$, separated by a distance a such that $p = qa$. However, this approach is ruled out for a magnetic dipole m unless one is willing to assume the existence of magnetic charges. This means m cannot be broken down into simpler objects.

One can easily show the equivalence of a current loop and a magnetic dipole by showing that the induction produced by a loop has the same form as the electric field produced by an electric dipole. In the case of the electric dipole we have seen that the dumb-bell model, i.e. $+q$ and $-q$ separated by a distance a, produces a dipole field only for distance $r \gg a$. In the case of magnetic dipole, the corresponding condition is that r, the distance from the loop should be large compared with the dimensions of the loop. The actual shape of the loop—assumed to be in one plane—does not come in the picture.

Let us consider a simple loop of wire in rectangular form carrying current I and suspended in vertical plane in an horizontal magnetic field B in such a way that the normal to the plane to the coil makes an angle θ with the direction of the field as shown in Fig 6.16.

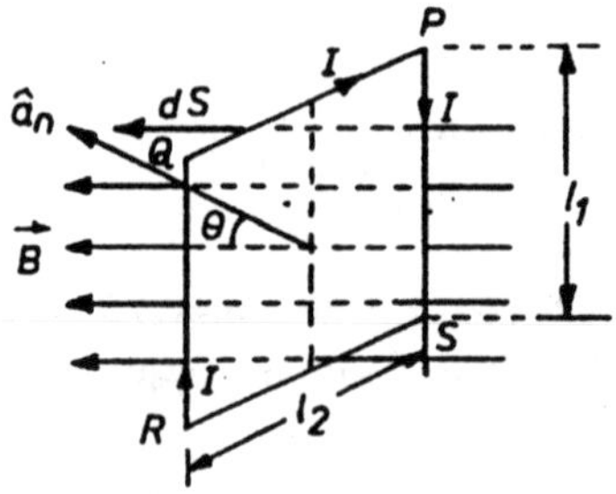

Figure 6.16.

Let the cross-section of the wire be a. If N be the number of free electrons per unit volume, e the charge on the electrons and u be the velocity of electrons, then the current flowing in the wire,

$$I = Neua \tag{6.73}$$

The force acting on a small element dl, of the side PS of the loop (Fig 6.16) is obtained as

$$d\mathbf{F}_1 = Neaud\mathbf{l}_1 \times \mathbf{B} = Id\mathbf{l}_1 \times \mathbf{B} \tag{6.74}$$

we can easily see that $d\mathbf{l}_1$, is in the direction of the current. Thus the total force on the side PS is

$$\mathbf{F}_1 = I\mathbf{l}_1 \times \mathbf{B} = Il_1 B\hat{a}_y \quad (\because \quad PS \text{ is perpendicular to } \mathbf{B}). \tag{6.75}$$

Let us assume that current I is parallel to z-axis and field $\mathbf{B}$ is parallel to x-axis. The force $\mathbf{F}$ acts in the positive y-direction (Fig 6.17).

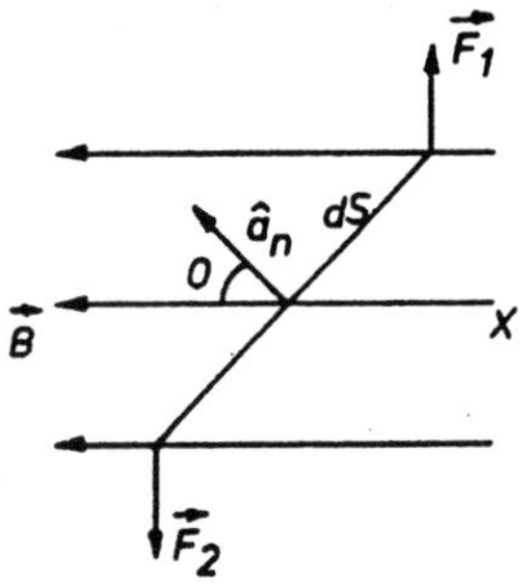

Figure 6.17.

One can see that the figure presents a view looking down on the coil. From Fig. 6.17 it is clear that force $\mathbf{F}_2$, equal and opposite to $\mathbf{F}_1$, acts on the side QR of the loop. We can easily say that in the same way there will be forces acting on the other two sides of the loop, i.e. PQ and SR which are equal and opposite. Obviously the net force on the loop, is zero and hence there will be no translational motion of the loop. However, there will be a torque τ acting on the coil. This torque will tend to turn the coil about the vertical axis. This torque is obtained as

$$|\tau| = Il_1 Bl_2 \sin\theta = IdsB\sin\theta \tag{6.76}$$

where $l_1 l_2 = ds$ is the area enclosed of the loop. Expressing the torque in the vector notation, we have

$$\tau = I\hat{a}_n ds \times \mathbf{B} \tag{6.77}$$

We can see that torque tends to decrease the angle θ. Let U be the potential energy of the loop, then the torque tending to increase θ is $-\frac{\partial U}{\partial \theta}$. This means

$$-\frac{\partial U}{\partial \theta} = -IdsB\sin\theta$$

or

$$U = -IdsB\cos\theta + c$$

where c is constant of integration. When $\theta = \frac{\pi}{2}$ we have $U = 0$. Thus

$$U = -IdsB\cos\theta = I\hat{a}_n ds \cdot \mathbf{B} \tag{6.78}$$

where $\phi = \mathbf{B} \cdot \hat{a}_n ds$ is the magnetic flux through the circuit. Thus one obtains

$$U = -I\phi \tag{6.79}$$

we must remember that ϕ and ϕ_m are different. Comparing (6.78) with potential energy of an electric dipole of dipole moment $\mathbf{p}$ in an electric field $\mathbf{E}$, i.e. $U = -\mathbf{p} \cdot \mathbf{E}$, one finds that the magnetic field produced by a current loop is similar in form to the electrostatic field of an electric dipole. Thus magnetic dipole moment of the loop is given by

$$\mathbf{m} = I\hat{a}_n ds \tag{6.80}$$

$$\therefore U = -\mathbf{m} \cdot \mathbf{B} \tag{6.81}$$

and the torque is given by

$$\tau = \mathbf{m} \times \mathbf{B} \tag{6.82}$$

we must note that the above result is true for a small loop of any shape.

Dipole in a Non-Uniform Magnetic field

When a current loop, which is equivalent to a magnetic dipole is placed in a non-uniform magnetic field, then in addition to the torque given by (6.82), the dipole also experiences a translatory force.

We have already obtained the expression for the magnetic potential energy of a magnetic dipole $\mathbf{m}$ in a magnetic field $\mathbf{B}$ as

$$U = -\mathbf{m} \cdot \mathbf{B}$$

Now, the force **F** is related to the potential energy U by the relation

$$\begin{aligned} \mathbf{F} &= -\nabla U \\ \therefore \quad \mathbf{F} &= \nabla(\mathbf{m}\cdot\mathbf{B}) \\ &= (\mathbf{B}\cdot\nabla)\mathbf{m}+(\mathbf{m}\cdot\nabla)\mathbf{B}+\mathbf{B}\times(\nabla\times\mathbf{m})+\mathbf{m}\times(\nabla\times\mathbf{B}) \end{aligned} \tag{6.83}$$

we know that **m** is not a function of the space coordinates, and hence one gets

$$\begin{aligned} (\mathbf{B}\cdot\nabla)\mathbf{m} &= 0, \nabla\times\mathbf{m}=0 \\ \therefore \quad \mathbf{F} &= \mathbf{m}\times(\nabla\times\mathbf{B})+(\mathbf{m}\cdot\nabla)\mathbf{B} \end{aligned}$$

But $\nabla\times\mathbf{B}=0$ and hence

$$\mathbf{F}=(\mathbf{m}\cdot\nabla)\mathbf{B} \tag{6.84}$$

(6.84) is the general formula for force **F** on a magnetic dipole **m** in a non-uniform field **B**.

6.14 APPLICATIONS OF VECTOR POTENTIAL

(i) Vector Potential due to a small current loop

Let us consider a small current loop of radia a and carrying a current I (Fig 6.18).

Let us assume that the origin of the frame of reference coincides with the centre of the coil O and its z-axis is perpendicular to the coil. Let us consider an element dl of the circuit. Let r be the distance of the point P from the element dl. ϕ is the azimuthal angle of the element dl. One can easily see that the components of the element $dl = ad\phi$ are $-a\sin\phi d\phi$, $a\cos\phi d\phi$ and O. We have

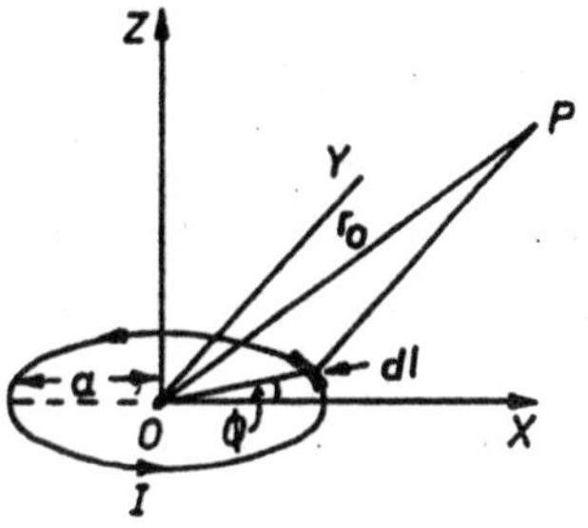

Figure 6.18. Magnetic Vector potential due to a small current loop.

$$\mathbf{A}=\frac{\mu_0}{4\pi}\int\frac{\mathbf{J}}{|\mathbf{r}|}dv=\frac{\mu_0 I}{4\pi}\int\frac{dl}{|\mathbf{r}|}$$

Now, the x component is obtained as

$$A_x=\frac{\mu_0 I}{4\pi}\int\frac{-a\sin\phi d\phi}{|\mathbf{r}|}$$

we have $r^2 = (x - a\cos\phi)^2 + (y - a\sin\phi)^2 + z^2$ Since $a \ll r_0$, we have

$$\frac{1}{r} = \frac{1}{r_0} + \frac{ax\cos\phi + ay\sin\phi}{r_0^2} + \cdots$$

where $r_0 = OP$. Therefore,

$$\begin{aligned} A_x &= -\frac{\mu_0 Ia}{4\pi}\int_0^{2\pi}\left(\frac{1}{r_0} + \frac{ax\cos\phi + ay\sin\phi}{r_0^3}\right)\sin\phi d\phi \\ &= -\frac{\mu_0 Ia^2\pi y}{4\pi r_0^3} \end{aligned} \tag{6.85}$$

Similarly, one obtains

$$A_y = \frac{\mu_0 Ia^2\pi x}{4\pi r_0^3} \quad \text{and} \quad A_z = 0 \tag{6.86}$$

we have the magnitude of the dipole moment equivalent to the circuit (small current loop) is as

$$m = \text{area} \times \text{current} = \pi a^2 I \tag{6.87}$$

The direction of dipole moment **m** is normal to the plane of the coil, i.e. along the z-axis. Obviously, $m_x = 0$, $m_y = 0$ and $m_z = \pi a^2 I$. One can easily see that the components of vector potential **A** given by Eqs. (6.85) and (6.86) are proportional to those of the vector $\mathbf{m} \times \mathbf{r}$. Dropping the subscripts, one can write,

$$\mathbf{A} = \frac{\mu_0}{4\pi r^3}(\mathbf{m} \times \mathbf{r}) = -\frac{\mu_0}{4\pi}\left[\mathbf{m} \times \nabla\left(\frac{1}{r}\right)\right] \tag{6.88}$$

Magnetic induction **B** can be determined by taking the curl of (6.88). We have

$$\begin{aligned} B(\mathbf{r}) &= \nabla \times \mathbf{A} = \nabla \times \left[\frac{\mu_0}{4\pi r^3}(\mathbf{m} \times \mathbf{r})\right] \\ &= \frac{\mu_0}{4\pi}\left[-(\mathbf{m} \cdot \nabla)\frac{\mathbf{r}}{r^3} + m\nabla \cdot \left(\frac{\mathbf{r}}{r^3}\right)\right] \end{aligned}$$

(for constant **m**)

But

$$\nabla \cdot \left[\frac{\mathbf{r}}{r^3}\right] = 0 \tag{6.89}$$

$$\text{and } (\mathbf{m} \cdot \nabla)\frac{\mathbf{r}}{r^3} = \frac{m}{r^3} - 3\frac{(\mathbf{m} \cdot \mathbf{r})r}{r^5}$$

$$\therefore \quad \mathbf{B}(r) = \frac{\mu_0}{4\pi}\left[\frac{3(\mathbf{m} \cdot \mathbf{r})\mathbf{r}}{r^5} - \frac{\mathbf{m}}{r^3}\right] \tag{6.90}$$

One can also determine the scalar magnetic potential ϕ_m by the relation

$$\mathbf{B} = -\nabla\phi_m \tag{6.91}$$

Eq. (6.90) can be written as

$$\mathbf{B} = -\nabla\left(\frac{\mu_0}{4\pi}\frac{\mathbf{m}\cdot\mathbf{r}}{r^3}\right) \tag{6.92}$$

Comparing (6.91) and (6.92), one obtains

$$\phi_m = \frac{\mu_0}{4\pi}\left(\frac{\mathbf{m}\cdot\mathbf{r}}{r^3}\right) \tag{6.93}$$

Alternate Method

Let us consider a filament circuit C carrying a current I as shown in Fig 6.19. We have the vector potential A at à point $P(x, y, z,)$ due to the current loop is

$$\mathbf{A} = \frac{\mu_0 I}{4\pi}\oint\frac{d\mathbf{l}'}{r'} \tag{6.94}$$

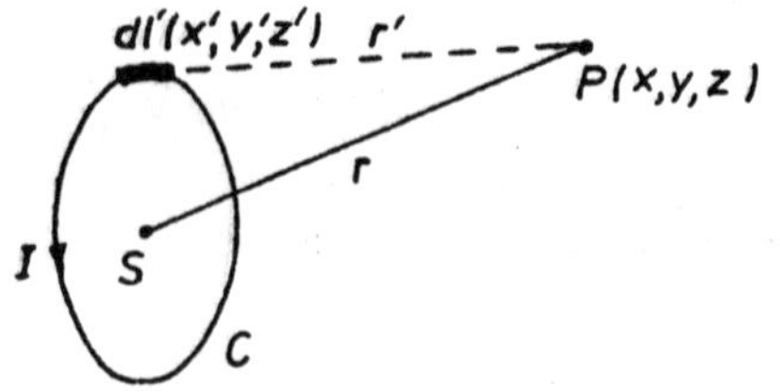

Figure 6.19.

where $dl'(x', y', z')$ is an element of the circuit C. Making use of Stokes' theorem.

$$\oint\frac{d\mathbf{l}'}{r'} = \int_s \hat{a}_n \times \nabla'\left(\frac{1}{r'}\right) ds$$

one obtains

$$\mathbf{A} = \frac{\mu_0 I}{4\pi}\int_s \hat{a}_n \times \nabla'\left(\frac{1}{r'}\right) ds = \frac{\mu_0 I}{4\pi}\int_s\left(\hat{a}_n \times \frac{\hat{a}_{r'}}{r'^2}\right) ds \tag{6.95}$$

where $\hat{a}_n$ and $\hat{a}_r'$, are the unit vectors perpendicular to the surface enclosed by the circuit C and along r' respectively. Since the dimensions of the current loop are very small compared with the distance to the point one can consider the factor $\frac{\hat{a}_r'}{r^{2\prime}}$ very nearly constant during integration. Now taking $\frac{\hat{a}_r'}{r'^2}$ as $\frac{\hat{a}_r}{r^2}$, one obtains

$$\begin{aligned}
\mathbf{A} &= -\frac{\mu_0 I}{4\pi}\frac{\hat{a}_r}{r^2} \times \int_s \hat{a}_n ds = -\frac{\mu_0 I}{4\pi r^2}\hat{a}_r \times \hat{a}_n s \\
&= \frac{\mu_0}{4\pi}\left(\mathbf{m} \times \frac{\hat{a}_r}{r^2}\right) = \frac{\mu_0}{4\pi}\left(\frac{\mathbf{m}\times\mathbf{r}}{|\mathbf{r}|^3}\right) \quad (\because \quad \mathbf{m} = \hat{a}_n I s)
\end{aligned} \tag{6.96}$$

Now, one can obtains the magnetic flux density **B**. We have

$$\mathbf{B} = \nabla \times \mathbf{A} = \frac{\mu_0}{4\pi}\nabla \times \left(\frac{\mathbf{m}\times\mathbf{r}}{|\mathbf{r}^3|}\right)$$

$$= \frac{\mu_0}{4\pi}\left[m\nabla\cdot\left(\frac{\mathbf{r}}{|\mathbf{r}|^3}\right) - (\mathbf{m}\cdot\nabla)\left(\frac{\mathbf{r}}{|\mathbf{r}|^3}\right)\right]$$

$$\because \quad \nabla\cdot\left(\frac{\mathbf{r}}{|r|^3}\right) = 0 \quad \text{and hence}$$

$$\mathbf{B} = -\frac{\mu_0}{4\pi}(\mathbf{m}\cdot\nabla)\left(\frac{\mathbf{r}}{|\mathbf{r}|^3}\right)$$

Also

$$\nabla\left(m\cdot\frac{\mathbf{r}}{|\mathbf{r}|^3}\right) = (m\cdot\nabla)\frac{\mathbf{r}}{|\mathbf{r}|^3} + m\times\left(\nabla\times\left|\frac{\mathbf{r}}{|\mathbf{r}|^3}\right|\right)$$

The last term vanishes and hence one obtains

$$(\mathbf{m}\cdot\nabla)\frac{\mathbf{r}}{|\mathbf{r}|^3} = \nabla\left(\mathbf{m}\cdot\frac{\mathbf{r}}{|\mathbf{r}|^3}\right) = \left[\frac{\mathbf{m}}{|\mathbf{r}|^3} - \frac{3(\mathbf{m}\cdot\mathbf{r})\mathbf{r}}{|\mathbf{r}|^3}\right]$$

and magnetic induction **B** as

$$\mathbf{B} = \frac{\mu_0}{4\pi r^3}\left[\frac{3(\mathbf{m}\cdot\mathbf{r})\mathbf{r}}{r^2} - \mathbf{m}\right] \tag{6.97}$$

One can easily verify that (6.97) is similarly the relation for the field of an electric dipole.

(ii) Magnetic-Induction due to an infinite current carrying wire

Let us consider a long wire whose axis coincides with z-axis and carrying a current. Obviously, the only component of current density and hence magnetic vector potential is z component.

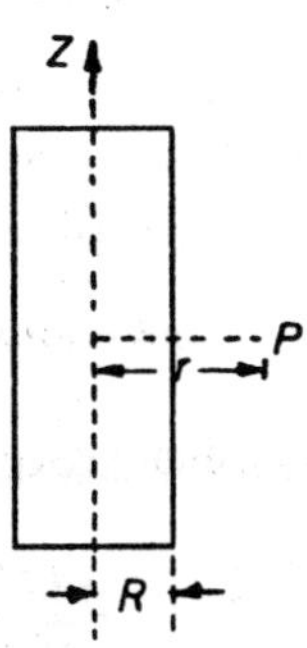

Figure 6.20.

(a) *Inside the wire*: In cylindrical coordinates (r, θ, z), the Poisson's equation is

$$\nabla^2 A_z = -\mu J_z \tag{6.98}$$

where μ is the permeability inside the wire. Let R be the radius of the wire, then

$$J_z = \frac{I}{\pi R^2}$$

we have

$$\nabla^2\phi = \frac{1}{r}\frac{\partial}{\partial r}\left(r\frac{\partial\phi}{\partial r}\right) + \frac{1}{r^2}\frac{\partial}{\partial\theta}\left(\sin\theta\frac{\partial\phi}{\partial\theta}\right) + \frac{\partial^2\phi}{\partial z^2}$$

Since A_z is independent of θ and z, Eq. (6.98) can be expressed as

$$\frac{1}{r}\frac{\partial}{\partial r}\left(r\frac{\partial A_z}{\partial r}\right) = -\frac{\mu I}{\pi R^2} \tag{6.99}$$

Integrating, one obtains

$$r\frac{\partial A_z}{\partial r} = -\frac{\mu I r^2}{2\pi R^2} + b$$

where b is constant of integration. We have $r = 0$, $r\frac{\partial A_z}{\partial r} = 0$ and therefore $b = 0$. Thus

$$r\frac{\partial A_z}{\partial r} = -\frac{\mu I r^2}{2\pi R^2}$$

or

$$\frac{\partial A_z}{\partial r} = -\frac{\mu_0 I r}{2\pi r^2} \tag{6.100}$$

Integrating again, one obtains

$$A_z = -\frac{\mu_0 I r^2}{4\pi R^2} + b_1 \tag{6.101}$$

If one assumes that $A_z = 0$ at $r = R$, then

$$b_1 = \frac{\mu_0 I R^2}{4\pi R^2} = \frac{\mu I}{4\pi}$$

Therefore Eq. (6.101) becomes

$$A_z = \frac{\mu_0 I}{4\pi}\left[1 - \frac{r^2}{R^2}\right] \tag{6.102}$$

We have $\mathbf{B} = \nabla\times\mathbf{A}$, one can express the cylindrical components of $\nabla\times A$ as

$$\left.\begin{aligned}
(\nabla\times\mathbf{A})_r &= \frac{1}{r}\frac{\partial A_z}{\partial\theta} - \frac{\partial A_z}{\partial z}\\
(\nabla\times\mathbf{A})_\theta &= \frac{\partial A_r}{\partial z} - \frac{\partial A_z}{\partial r}\\
(\nabla\times\mathbf{A})_z &= \frac{\partial A_\theta}{\partial r} + \frac{A_\theta}{r} - \frac{1}{r}\frac{\partial A_r}{\partial\theta}
\end{aligned}\right\} \tag{6.103}$$

One obtains,

$$B_\theta = \frac{\mu I r}{2\pi R^2} \tag{6.104}$$

(b) *Outside the wire*: Poisson's equation takes the form as

$$\frac{1}{r}\frac{\partial}{\partial r}\left(r\frac{\partial A_z}{\partial r}\right)=0 \tag{6.105}$$

Integrating, one obtains

$$r\frac{\partial A_z}{\partial r} = C(\text{constant})$$

or

$$\frac{\partial A_z}{\partial r} = \frac{C}{r} \tag{6.106}$$

Integrating again, one obtains

$$A_z = C\log_e r + C' \quad (\text{constant})$$

we have $r = R$, $A_z = 0$

$$\therefore \quad C' = -C\log_e R$$

$$\therefore A_z = C\log_e r - C\log_e R = C\log_e\left(\frac{r}{R}\right) \tag{6.107}$$

Magnetic field induction

$$\mathbf{B} = \nabla\times\mathbf{A}$$

$$\therefore \quad B_\theta = (\nabla\times\mathbf{A})_\theta = \frac{\partial A_r}{\partial z} - \frac{\partial A_z}{\partial r} = -\frac{\partial A_z}{\partial r}$$

As B_θ must be continuous at $r = R$, we have

$$\frac{\partial A_z}{\partial r} = \frac{\mu I}{4\pi}\left(0-\frac{2r}{R}\right)_{r=R} = -\frac{\mu I}{2\pi R} \quad (\text{from eq.(6.12)})$$

Also from Eq. (6.106), $\left(\frac{\partial A_z}{\partial r}\right)_{r=R} = \frac{C}{R}$

$$\therefore \quad \frac{C}{R} = -\frac{\mu I}{2\pi R} \quad \text{or} \quad C = -\frac{\mu I}{2\pi}$$

Substituting this value in Eq. (6.107), one obtains

$$A_z = -\frac{\mu I}{2\pi}\log_e\left(\frac{r}{R}\right) \tag{6.108}$$

$$\therefore \quad B_\theta = -\frac{\partial A_z}{\partial r} = \frac{\mu I}{2\pi R} \tag{6.109}$$

6.15 MAGNETIC FIELD IN A SUBSTANCE

We assumed in the preceding chapter that conductors carrying a current are in vacuum. If the conductors carrying a current are in a medium, the magnetic field changes. The explanation is that any substance is a *magnetic*, i.e. is capable of acquiring a magnetic moment under the action of a magnetic field (of becoming magnetized). The magnetized substance sets up the magnetic field $\mathbf{B}'$, that is superposed onto the field $\mathbf{B}$ produced by the currents. Both field produce the resultant field

$$\mathbf{B} = \mathbf{B}_0 + \mathbf{B}' \tag{6.110}$$

The true (microscopic) field in a magnetic medium varies greatly within the limits of intermolecular distances. Here $\mathbf{B}$ denotes the averaged (macroscopic) field. This clearly reveals that a material substance acquires a '*magnetic polarization*' when placed in a magnetic field, just as a dielectric medium acquires an electric polarization.

The response of the materials to an applied magnetic field depends on the properties of the individual atoms and molecules, and on their interactions. The orbital motion of electrons in atoms and molecules provide currents which give rise to the magnetic dipoles. There are several materials, e.g. water, Cu, Pb, NaCl, S, graphite etc. in which the small electric currents associated with the orbital motion and spin of the electrons average to zero. When such substances are placed in a magnetic field, minute electron currents are generated by induction in the clouds of electrons, the direction of which is such that the magnetic field associated them opposes the inducing field $\mathbf{B}$. Substances which exhibit such a behaviour are called *diamagnetic substances*. Diamagnetism is found only in substances whose atoms have no magnetic moment (the vector sum of the orbital and spin magnetic moments of the atom electrons is zero). We must remember that diamagnetic properties are present in all atoms, however, they are observed only when other effects are not present to mask the comparatively weak diamagnetic effects.

There are materials which may possess an intrinsic magnetic dipole moment on account of the fact that the currents from the orbital motions and spins of the electrons do not average to zero. Although electron spins tend to pair and cancel one another, there are atoms in which pairing is incomplete. When such materials are placed in a external magnetic field, the magnetic moments of the atoms and induced magnetism enhances the external field. Such substances are called *paramagnetic materials*. Examples of paramagnetic substances are: Al, Na, $CuCl_2$, $NiSO_4$, etc.

Paramagnetic effect is more pronounced at low temperatures. This effect arise due to the alignment of permanent dipoles due to an external magnetic field. When the external field is removed, thermal agitations disturbs the atomic dipoles and produce random orientations. This means that even in the presence of an external field, lowering of the temperature of the substance should enhance the paramagnetic effects that are observed. This is observed to be true, because the average thermal kinetic energy of an atom is proportional to the absolute temperature of the substance.

There are a third and important kind of substances capable of having magnetization in the absence of an external field. According to the name of their most wide spread representative-ferrum (iron)—they have been called *ferromagnetic*. These are counterparts of electrets and do not shed their magnetic properties even at high temperatures. In addition to iron, they include nickel, cobalt, gadolinium, their alloys and compounds, and also certain alloys and compounds of maganese and chromium with non-ferromagnetic elements. All these substances display *ferromagnetism* only in the crystalline state.

Ferromagnetics are strongly magnetic substances. Their magnetization exceeds that of dia- and paramagnetics which belong to the category of weakly magnetized substances an enormous number of times (up to 10^{10}). For these substances, the atomic dipoles get aligned by the application of an external magnetic field and they remain locked in the aligned position, even after the external field is removed. Obviously, these substances retain a memory of their history.

It follows from experiments involving the studying of gyromagnetic phenomena that the intrinsic (spin) magnetic moments of electrons are responsible for the magnetic properties of ferromagnetics. In definite conditions, exchange forces may appear in crystals that make the magnetic moments of the electrons become lined up parallel to one another. The result is the setting up of regions of *spontaneous magnetization*, also called *domains*. Within the confines of each domain, a ferromagnetic is spontaneously magnetized to saturation and has a definite magnetic moment. The directions of

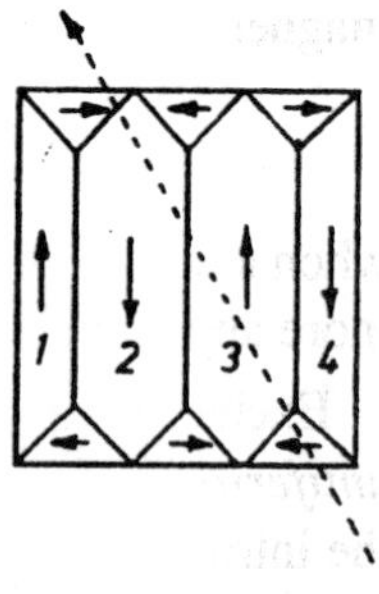

Figure 6.21. Domain Structure

these moments are different for different domains (Fig 6.21), so that in the absence of an external field the total moment of an entire body is zero. Domains have dimensions of the order of 1 to 10 μm.

The action of a field on domains at different stages of the magnetization process is different. The resulting net macroscopic field in the vicinity of ferromagnetic material can be expected to depend on the relative sizes of these domains and on the orientation of the molecular dipoles within each domain. First, with weak fields displacement of the domain boundaries is observed. There is a net alignment of the domains in the field direction resulting in a very large paramagnetic effect. With an increase in the field strength, these domains which are already aligned become enlarged at the expense of the others. This process goes on further and further until the domains which have a smaller energy in magnetic field completely absorb the domains that are less advantageous from the energy point of view. In the next stage, the magnetic moments of the domains turn in the direction of the field. The moments of the electrons within the confines of a domain turn simultaneously without violating their strict parallelism to one another. These processes (excluding slight displacements of the boundaries between the domains in very weak fields) are irreversible, and this is exactly what causes *hysteresis*.

There is a definite temperature too for every ferromagnetic at which the regions of spontaneous magnetization (domains) break up and the substance loses it ferromagnetic properties. This temperature is called the *Curie point*. It is 768°C for iron and 365°C for nickel. At a temperature above the Curie point, a ferromagnetic becomes on ordinary paramagnetic whose magnetic susceptibility obeys the *Curie-Weiss Law*.

$$\chi = \frac{c}{T - T_c} \tag{6.111}$$

when a ferromagnetic is cooled to below its Curie point, domains once more appear in it.

Exchange forces sometimes result in the appearance of so-called *antiferromagnetics* (Chromium, manganese etc.). In these substances, the intrinsic magnetic moments of the electrons are spontaneously oriented antiparallel to one another. Such an orientation involves adjacent atoms in pairs. The result is that antiferromagnetics have an extremely low magnetic susceptibility and behave like very weak paramagnetics. There is also a temperature T_n for antiferromagnetics at which the antiparallel orientation of the spins vanishes. This temperature is known as the *Neel point* or *Neel temperature*. Some antiferromagnetics for e.g., erbium, dysprosium, alloys of manganese and copper have two such points (an upper and a lower Neel point), the antiferromagnetic properties being observed only at the intermediate temperatures. Above the upper point, the substance behaves like a paramagnetic, and at temperatures below the lower Neel point it becomes a ferromagnetic.

Another type of magnetization is called *ferrimagnetism*. It is similar to antiferromagnetism, but the atomic or ionic magnetic moments in one direction are different from these oriented in the opposite direction resulting in a net magnetization. These substances are called *ferrites*, and can be generally represented by the chemical formula $MoFe_2O_3$, where M stands for Co, Ni, Cu, Mg, Zn, Cd etc. Note that if M is Fe, the compound Fe_3O_4, or magnetic, results.

6.16 MAGNETIZATION (M)

We have indicated that every atom or molecule may be regarded as a tiny magnetic dipole with the magnetic moment

$$Q_m\mathbf{l} = -\hat{a}_n I ds \tag{6.112}$$

where Q_m is the magnetic pole strength, l is the pole separation, I is the current and ds is the area of the loop. Atoms may or may not exhibit a net magnetic dipole moment, depending on their symmetry or on the relative orientation of their electronic orbits. Since most molecules are not spherically symmetric, they may exhibit a permanent magnetic dipole moment because of special orientation of the electronic orbits. For example, diatomic molecules have axial symmetry, and may passes a magnetic dipole moment parallel to the molecule axis. Even so, matter in bulk, with the exception of ferromagnetic materials, does not exhibit a net magnetic moment, because of random orientation of the molecules, a situation similar to that found in the electric polarization of matter. However, the presence of an external magnetic field distorts the electronic motion, giving rise to a net magnetic polarization or *magnetization* of the material.

The *magnetization vector* **M** of a material is defined as the magnetic dipole moment of the medium per unit volume. If **m** is the magnetic dipole moment contributed by each atom or molecule and n is the number of atoms or molecules per unit volume, the magnetization is $\mathbf{M} = n\mathbf{m}$. The magnetic moment of an elementary current is expressed in $A - m^2$, and therefore the magnetization $\mathbf{M}$ is expressed in $A - m^2/m^3 = Am^{-1}$ or $m^{-1}s^{-1}C$, and is equivalent to current per unit length. We must note that the magnetization $\mathbf{M}$ is not necessarily uniform throughout the substance. Thus

$$M = \frac{dm}{dv} \tag{6.113}$$

Let us consider an elementary volume with sides δx, δy and δz at a point (x, y, z) within a magnetized substance (Fig 6.22). One obtains the component of the magnetic moment associated with this volume in the z-direction is

$$M_z \, \delta x \, \delta y \, \delta z$$

Using Eq.(6.112), one obtains

$$M_z \, \delta x \, \delta y \, \delta z = I \, \delta x \, \delta y \tag{6.114}$$

where I is the current flowing in the loop in a plane parallel to XY plane as shown in Fig 6.22.

$$\therefore \quad I = M_z \, \delta z \tag{114}$$

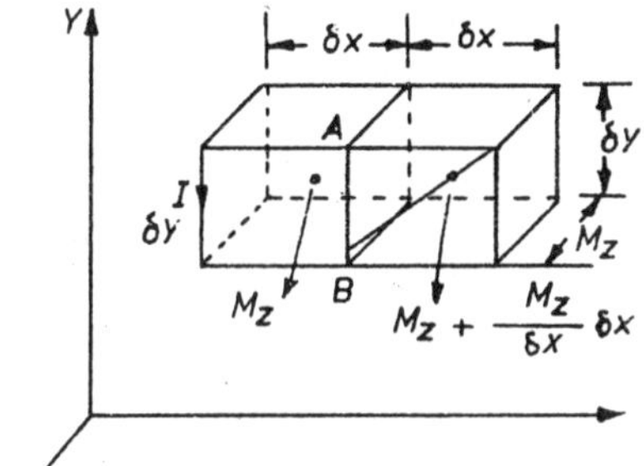

Figure 6.22.

One can obtain the current in the adjacent loop at $(x + \delta x, y, z)$ as

$$I' = I + \frac{\partial I}{\partial x}\delta x = M_z \delta z + \frac{\partial M_z}{\partial x}\delta x \, \delta z$$

obviously, the current in the interface AB of the two elements has a component in the y-direction and it is given by

$$I - I' = -\frac{\partial M_z}{\partial x}\delta x \, \delta z$$

Similarly, by considering a current loop in the YZ plane, one can find another component of current in the y-direction. Its magnitude-is $\frac{\partial M_z}{\partial z}\delta x \delta z$. Thus, one obtains the total current in the y-direction as

$$\left(\frac{\partial M_x}{\partial z} - \frac{\partial M_z}{\partial x}\right)\delta x\,\delta z$$

one can express the above in terms of the y-component of the magnetization current (J_M) as

$$J_{M_y}\,\delta x\,\delta z = \left(\frac{\partial M_x}{\partial z} - \frac{\partial M_z}{\partial x}\right)\delta x\,\delta z$$

$$\therefore \quad J_{M_y} = \frac{\partial M_x}{\partial z} - \frac{\partial M_z}{\partial x} \tag{6.114}$$

Similarly, one obtains the other components

$$J_{M_x} = \frac{\partial M_z}{\partial y} - \frac{\partial M_y}{\partial z} \tag{6.115}$$

and

$$J_{M_z} = \frac{\partial M_y}{\partial x} - \frac{\partial M_x}{\partial y} \tag{6.116}$$

One can easily see that the expressions appearing on the right hand side of eqs.(6.114) to (6.116) are the components of $\nabla \times \mathbf{M}$. Therefore

$$\mathbf{J}_M = \nabla \times \mathbf{M} = \begin{vmatrix} \hat{a}_x & \hat{a}_y & \hat{a}_z \\ \frac{\partial}{\partial x} & \frac{\partial}{\partial y} & \frac{\partial}{\partial z} \\ M_x & M_y & M_z \end{vmatrix} \tag{6.117}$$

obviously, the magnetization current density is the curl of the magnetization. Moreover, a current density equal to that given by (6.117) can produce the same magnetic field as the field produced by the magnetization $\mathbf{M}$. We must note that magnetization current flow exists only if $\mathbf{M}$ is varying. If $\mathbf{M}$ is uniform, we have $\mathbf{J}_M = 0$. The magnetization current I_M is a bound current in the sense that one cannot tap the I_M of a bar magnet to set up a perpetual source of current at no cost. One can compare this with the polarization charge (Q_p) on an electret. One cannot wipe out the Q_p from the surface of an electret to make an inexhaustible supply of electric charge.

Alternative Method

One can obtain the above result using vector potential. We have

$$\mathbf{A} = \frac{\mu_0}{4\pi}\int \left\{\mathbf{M}(R) \times \nabla\left(\frac{1}{|\mathbf{r}|}\right)\right\} dV \tag{6.118}$$

Making use of the vector identity

$$\nabla \times (\phi \mathbf{B}) = \phi \nabla \times \mathbf{B} - \mathbf{B} \times \nabla \phi$$

One can express (6.118) in the form

$$\mathbf{A} = \frac{\mu_0}{4\pi}\int \frac{\nabla \times \mathbf{M}}{|\mathbf{r}|} dV - \frac{\mu_0}{4\pi}\int \left\{\nabla \times \frac{\mathbf{M}}{|\mathbf{r}|}\right\} dV \tag{6.119}$$

We can also see that the following vector relation holds good

$$\int_s \mathbf{B} \times \hat{a}_n ds = -\int_v \nabla \times \mathbf{B}\, dV \tag{6.120}$$

[$\int (\nabla \times \mathbf{B}) dV = -\int \mathbf{B} \times \hat{a}_n ds$, If **B** be a constant vector, then

$$\begin{aligned} \mathbf{C} \cdot \int_s \mathbf{B} \times \hat{a}_n ds &= \int_s (\mathbf{C} \times \mathbf{B}) \cdot \hat{a}_n ds \\ &= \int_v \nabla \cdot (\mathbf{C} \times \mathbf{B}) dV \quad \text{(divergence theorem)} \\ &= -\mathbf{C} \cdot \int_v \nabla \times \mathbf{B}\, dV \end{aligned}$$

where **C** is an arbitrary vector]

Making use of (6.120), eq.(6.119) takes the form

$$\mathbf{A} = \frac{\mu_0}{4\pi}\int \frac{\nabla \times \mathbf{M}}{|\mathbf{r}|} dV + \frac{\mu_0}{4\pi}\int_s \frac{\mathbf{M} \times \hat{a}_n ds}{|\mathbf{r}|}$$

We know that **M** is localized. This means that the surface integral taken over a surface outside the region in which the current flows, vanishes. Thus

$$\mathbf{A} = \frac{\mu_0}{4\pi}\int \frac{\nabla \times \mathbf{M}}{|\mathbf{r}|} dV \tag{6.121}$$

We have already obtained that

$$\mathbf{A} = \frac{\mu_0}{4\pi}\int \frac{J(\mathbf{r})}{|\mathbf{r}|} dV \tag{6.122}$$

Comparing (121) and (122), one obtains

$$\mathbf{J}_M = \nabla \times \mathbf{M} \tag{6.123}$$

6.17 MAGNETIC FIELD INTENSITY (H)

When a medium is electrically conducting and also magnetizable, then there will be present both a real current density (**J**) and the magnetization current density ($\mathbf{J}_M$). Obviously, one will be have take into account of both these, Thus

$$\nabla \times \mathbf{B} = \mu_0(\mathbf{J}+\mathbf{J}_M) = \mu_0(\mathbf{J}+\nabla\times\mathbf{M}) \quad (6.124)$$

$$\therefore \quad \nabla \times (\mathbf{B}-\mu_0\mathbf{M}) = \mu_0\mathbf{J} \quad (6.125)$$

The vector within the parenthesis is such that its curl depends only on the conduction current density **J** at the point, and not on the equivalent current. Thus, analogous to the displacement vector **D** which in an electrostatic system is defined in such a way as to make **D** depend only on the free charges, it is convenient to define a vector **H**, usually called the magnetic field intensity, which depends only on total external currents. Whence it follows that

$$\mathbf{H} = \frac{\mathbf{B}}{\mu_0} - \mathbf{M} \quad (6.126)$$

or

$$\mathbf{B} = \mu_0(\mathbf{M}+\mathbf{H}) \quad (6.127)$$

In accordance with (6.126), eq.(6.125) can be expressed as

$$\nabla \times \mathbf{H} = \mathbf{J} \quad (6.128)$$

obviously, the curl of vector **H** equals the vector of the density of the macroscopic currents. H is expressed in Am^{-1} or $m^{-1}s^{-1}C$, which are the units of the two terms that appear on the right hand side of Eq. (6.126). Eq. (6.128) is a new form a Ampere's law which holds both in vacuum and in a medium and, hence, is more general than obtained earlier. In vacuum, we have

$$\mathbf{B} = \mu_0\mathbf{H}$$

$$\therefore \quad \nabla \times \mathbf{B} = \mu_0\mathbf{J}$$

which is a particular case of (6.128).

The expression in integral form corresponding to (128) is

$$\int_s \nabla \times \mathbf{H}\cdot\hat{a}_n ds = \oint \mathbf{H}\cdot d\mathbf{l} = \int \mathbf{J}\cdot\hat{a}_n ds$$

or

$$\oint \mathbf{H}\cdot d\mathbf{l} = I \quad (6.129)$$

The line integral $\oint \mathbf{H}d\mathbf{l}$ is called the *magnetomotance* and Eq. (6.129) is completely a general expression, except that I must be interpreted as total current if both conduction and displacement currents exist. Eq. (6.129) express the *theorem on the circulation* of the *vector* **H**, i.e. the circulation of the magnetic field strength vector **H** around a loop equals the algebraic sum of the macroscopic currents enclosed by this loop.

6.18 MAGNETIC SUSCEPTIBILITY AND PERMEABILITY

Eq. (6.126) may be written in a somewhat different form for certain classes of materials. This follows from the fact that experimentally for most non ferromagnetic materials the magnetization is proportional to and in the same direction as the magnetic field. Thus

$$\mathbf{M} = \chi_m \mathbf{H} \tag{6.130}$$

where the quantity χ_m is called the magnetic susceptibility of the material, and is a pure number (dimensionless) independent of the units chosen for **M** and **H**. χ_m is a function of temperature. Paramagnetic substances have small positive values of the susceptibility (χ_m), while diamagnetic substances have negative values of χ_m. Table 6.3 gives the susceptibility values of some common substances.

Table 6.3. χ_m of some Materials

Paramagnetic		Diamagnetic	
Material	χ_m	Material	χ_m
Aluminium	2.3×10^{-5}	Antimony	-7×10^{-5}
Magnesium	1.2×10^{-5}	Gold	-3.6×10^{-5}
Neodymium	3.0×10^{-3}	Copper	-9.8×10^{-6}
Tungsten	6.8×10^{-5}	Germanium	-1.5×10^{-6}
Oxygen (NTP)	19.2×10^{-7}	Water	-9.1×10^{-6}

Substituting (6.130) into Eq. (6.127), one obtains

$$\mathbf{B} = \mu_0(\mathbf{H} + \chi_m \mathbf{H}) = \mu_0(1+\chi_m)\mathbf{H} = \mu\mathbf{H} \tag{6.131}$$

where
$$\mu = \frac{\mathbf{B}}{\mathbf{H}} = \mu_0(1+\chi_m) \tag{6.132}$$

is called the *permeability* of the medium, and is expressed in the same units as μ_0; that is, m kg^{-2}C. The ratio

$$\frac{\mu}{\mu_0} = \mu_r \tag{6.133}$$

is called the *relative permeability* of the medium and is a pure number independent of the system of units. μ differs from unity only by few parts for non-ferromagnetic substances. For paramagnetic substances $\mu > 1$ and for diamagnetic $\mu < 1$. For ferromagnetic substances μ has a large value around 1000.

We must note that the relation (6.132), i.e. $\mathbf{B} = \mu\mathbf{H}$ is not valid for certain materials, such as crystals, solid under strain, iron or steel which has been subjected to rolling or hammering or materials which have a definite structure. Moreover, an anisotropic magnetic material must be described in terms of a tensor permeability:

$$\left.\begin{aligned} B_x &= \mu_{xx}H_x + \mu_{xy}H_y + \mu_{xz}H_z \\ B_y &= \mu_{yx}H_x + \mu_{yy}H_y + \mu_{yz}H_z \\ B_z &= \mu_{zx}H_x + \mu_{zy}H_y + \mu_{zz}H_z \end{aligned}\right\} \tag{6.134}$$

Obviously, μ is a tensor for anisotropic materials in the relationship $\mathbf{B} = \mu\mathbf{H}$, but $\mathbf{B} = \mu_0(\mathbf{H} + \mathbf{M})$ remains valid, although $\mathbf{B}$, $\mathbf{H}$ and $\mathbf{M}$ are no longer parallel in general. The most common anisotropic magnetic material is a single ferro magnetic crystal. Thin magnetic films also exhibit anisotropy. Most applications of ferromagnetic materials, however, involve polycrystalline arrays that are much easier to make.

One can easily show that the magnetization is equal to

$$\mathbf{M} = \frac{Nm^2}{3kT}\mathbf{B} = \frac{Nm^2\mu_0\mathbf{H}}{3kT} \tag{6.135}$$

where m is the permanent magnetic dipole moment associated with each molecule and N is the number of molecules per unit volume. One obtains the magnetic susceptibility

$$\chi_m = \frac{Nm^2\mu_0}{3kT} \tag{6.136}$$

6.19 MAGNETIC BOUNDARY CONDITIONS

In a single medium the magnetic field is continuous. That is, the field, if not constant, changes only by an infinitesimal amount in an infinitesimal distance. However, at the boundary between two different media, the magnetic field may change abruptly both in magnitude and direction. Obviously, it is important to know the relations for magnetic fields at the interface between two magnetic media.

We have divergence theorem

$$\int_v \nabla \cdot \mathbf{B} dv = \int_s \mathbf{B} \cdot \hat{a}_n ds \tag{6.137}$$

we must note that the condition $\nabla \cdot \mathbf{B} = 0$ remains unaltered by the presence of magnetic materials. Therefore

$$\int \mathbf{B} \cdot \hat{a}_n ds = 0 \tag{6.138}$$

This is *Gauss's theorem.* This shows that the flux of field **B** out of any closed surface is zero. Consider a small disc of height that sits astride the boundary between two media with permeabilities μ_1 and μ_2 as shown in Fig 6.23. If h is very small, i.e. $h \to 0$, the integral $\int \mathbf{B} \cdot \hat{a}_n ds = 0$ has contribution from the top and bottom ends only, i.e.

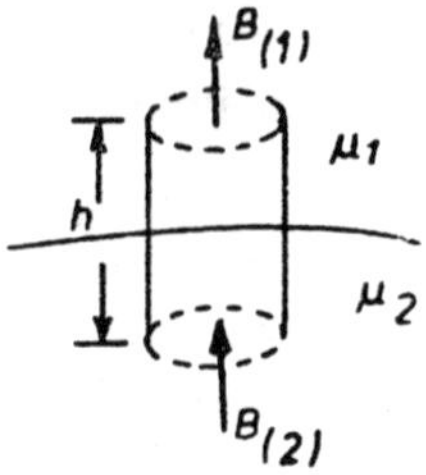

Figure 6.23.

$$\int_1 \mathbf{B} \cdot \hat{a}_n ds - \int_2 \mathbf{B} \cdot \hat{a}_n ds = 0 \tag{6.139}$$

we have $\mathbf{B} \cdot \hat{a}_n ds = B_\perp ds$, where $B_\perp$ is the component of **B** normal to ds. Therefore

$$\int_1 \mathbf{B} \cdot \hat{a}_n ds = \int \mathbf{B}_{(1)\perp} ds$$

and

$$\int_2 \mathbf{B} \cdot \hat{a}_n ds = -\int B_{(2)\perp} ds$$

This means

$$\int B_{(1)\perp} ds = \int B_{(2)\perp} ds \tag{6.140}$$

We must note that Eq. (6.140) is true whatever the size of ds. Therefore

$$\mathbf{B}_{(1)\perp} = B_{(2)\perp} \tag{6.141}$$

According to (6.141) *the normal component of the flux density* **B** *is continuous across the boundary between two medium.* This relation applies at the boundary of any two media for both static and time changing fields.

One can obtain the condition for **H** by means of Ampere's law. Let us consider a small circuit *ABCDA* as shown in Fig. 6.24.

$AB = CD = dl$ and the sides *BC*, *DA* are extremely small. Now,

$$\oint \mathbf{H} \cdot \mathbf{dl} = I \qquad (6.142)$$

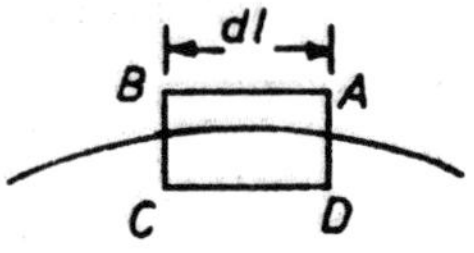

Figure 6.24.

where I is the current. Since *BC* and *DA* are extremely small and hence tend to zero. We have

$$H_{(1)||}dl - H_{(2)||}dl = I \qquad (6.143)$$

where $H_{(1)||}$ and $H_{(2)||}$ are the tangential components of **H** in the two media. Obviously, there is a discontinuity in the component of the magnetic field **H** equal to the surface current. When there is no surface current. One obtains

$$H_{(1)||} = H_{(2)||} \qquad (6.144)$$

According to (6.144) the tangential component of **H** are continuous across the boundary between two media provided the boundary has no current sheet of infinitesimal thickness. Let us compare the boundary conditions in magnetostatics and electrostatics. One finds,

(i) The normal component of **D** is continuous only if there is no charge change, where as the normal component of **B** is strictly continuous.

(ii) The tangential component of **E** is strictly continuous, while that of **H** is continuous across the boundary only if there is no surrent current.

6.20 UNIFORMLY MAGNETIZED SPHERE IN AN EXTERNAL FIELD

This problem is analogous to that of a dielectric sphere in an external field. Let us consider a sphere of radius R with a uniform magnetization **M** along z-axis without an external magnetic field.

Now, suppose that the sphere is placed in a uniform field H_0, the direction of which is parallel to z-axis (Fig 6.25)

Let μ_1 and μ_2 be the relative permeabilities inside and outside the sphere respectively. Let the potential inside and outside the sphere to be

$$\phi_1 = -H_1 r \cos\theta \quad (r < a) \qquad (6.145)$$

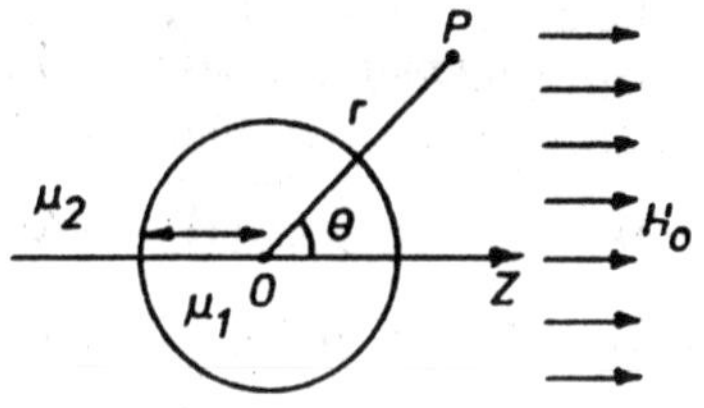

Figure 6.25.

and $$\phi_2 = -H_0 r\cos\theta + Ar^{-2}\cos\theta \quad (r > a) \tag{6.146}$$

where θ is the angle which the radius vector r makes with the field direction, i.e. z-axis and $\mathbf{H}_1$ is the field inside the sphere. We can see that the term in r^{-2} is not included in ϕ_1, as the inclusion of such a term in ϕ_1 would make it infinite at $r = 0$.

We have from boundary conditions that the tangential component of **H** be continuous at the boundary. This makes $\phi_1 = \phi_2$ at $r = a$. Hence

$$-H_a a\cos\theta = -H_0 a\cos\theta + Aa^{-2}\cos\theta$$

or $$H_1 = H_0 - Aa^{-3} \tag{6.147}$$

Let us assume that the magnetization $\mathbf{M}_1$ inside the sphere is parallel to the field $\mathbf{H}_0$ and made up of two components: (i) a permanent component M_0 and a component induced by the field $\mathbf{M}_1$. This is given by

$$\mathbf{M}' = \chi_m \mathbf{H}_1 = (\mu_1 - 1)\mathbf{H}_1$$

$$\therefore \quad \mathbf{M}_1 = (\mu_1 - 1)\mathbf{H}_1 + \mathbf{M}_0$$

But we have

$$\mathbf{B}_1 = \mu_0(\mathbf{H}_1 + \mathbf{M}_1) = \mu_0[\mathbf{H}_1 + (\mu_1 - 1)\mathbf{H}_1 + \mathbf{M}_0]$$

$$= \mu_0\mu_1\mathbf{H}_1 + \mu_0\mathbf{M}_0 \tag{6.148}$$

and $$\mathbf{B}_2 = \mu_0\mu_2\mathbf{H}_0 \tag{6.149}$$

One obtains the radial components of **B** as

$$-\mu_0\mu_1\left(\frac{\partial\phi_1}{\partial r}\right) + \mu_0 M_0\cos\theta = \mu_0(\mu_1 H_1 + M_0)\cos\theta \tag{6.150}$$

(inside the sphere)

and $$\mu_0\mu_2\left(\frac{\partial\phi_2}{\partial r}\right) = \mu_0\mu_2(H_0 + 2Ar^{-3})\cos\theta \tag{6.151}$$

(outside the sphere)

In accordance with the boundary conditions, these two components must be equal to one another at $r = a$. Thus we obtain

$$\mu_0(\mu_1 \mathbf{H}_1 + \mathbf{M}_0)\cos\theta = \mu_0\mu_2(\mathbf{H}_0 + 2\mathbf{A}a^{-3})\cos\theta$$

$$\therefore \quad \mu_1\mathbf{H}_1 + \mathbf{M}_0 = \mu_2\mathbf{H}_0 + 2\mu_2\mathbf{A}a^{-3}$$

But
$$\mathbf{A} = \frac{\mathbf{H}_0 - \mathbf{H}_1}{a^{-3}}$$

$$\therefore \quad \mu_1\mathbf{H}_1 + \mathbf{M}_0 = \mu_2\mathbf{H}_0 + 2\mu_2(\mathbf{H}_0 - \mathbf{H}_1)$$

or
$$(\mu_1 + 2\mu_2)\mathbf{H}_1 = 3\mu_2\mathbf{H}_0 - \mathbf{M}_0$$

$$\therefore \quad \mathbf{H}_1 = \frac{3\mu_2}{\mu_1 + 2\mu_2}\mu_0 - \frac{1}{\mu_1 + 2\mu_2}\mathbf{M}_0 \qquad (6.152)$$

The difference of the fields inside and outside the sphere, i.e.

$$\mathbf{H}_d = \mathbf{H}_1 - \mathbf{H}_0 \qquad (6.153)$$

is called the *demagnetizing field* i.e. magnetization produces a reverse field inside the sphere. For a paramagnetic substance H_1 is slightly less than H_0 and the induction is greater than the free space value $\mu_0 H_0$. This means the lines of induction are crowded together in a paramagnetic sphere. In a diamagnetic substance the reverse is the case.

Table 6.4 provides a comparison of static electric and magnetic field equations.

Table 6.4.

	Property	Electric field	Magnetic field
(i)	Force	$\mathbf{F} = q\mathbf{E}$	$d\mathbf{F} = id\mathbf{l} \times \mathbf{B}$
(ii)	Basic equation for the field	$\nabla \times \mathbf{E} = 0$	$\nabla \cdot \mathbf{B} = 0$
(iii)	Field and potential relation	$\left.\begin{matrix} E = -\nabla\phi \\ \phi = \frac{1}{4\pi\varepsilon_0}\int \frac{\rho dv}{r} \end{matrix}\right\}$	$\left.\begin{matrix} \mathbf{B} = \nabla \times \mathbf{A} \\ \mathbf{A} = \frac{\mu_0}{4\pi}\int \frac{\mathbf{J}}{\lvert\mathbf{r}\rvert}dv \end{matrix}\right\}$
(iv)	Constitutive relations for field	$\mathbf{D} = \varepsilon\mathbf{E}$	$\mathbf{B} = \mu\mathbf{H}$
(v)	Sources of the fields	$\nabla \cdot \mathbf{D} = \rho$	$\nabla \times \mathbf{H} = \mathbf{J}$
(vi)	Derivation from scalar potentials	$E = -\nabla\phi$	$\mathbf{H} = -\nabla U$ (in current free region)

EXAMPLE 6.1: Show that the magnetic field at the focus of a carrying current I in the form of the parabola is given by

$$\mathbf{B} = \frac{\mu_0 \mathbf{I}}{4d}$$

where d is the distance from focus to the apex.

Solution: We have magnetic field for a wire at a,

$$\mathbf{B} = \frac{\mu_0 I}{4\pi} \int \frac{\mathbf{r} \times d\mathbf{r}}{|\mathbf{r}|^3} \tag{1}$$

The wire is in the form of a parabola Fig (6.26), we have

$$\frac{\mathbf{r} \times d\mathbf{r}}{|\mathbf{r}|^3} = \frac{r d\theta}{r^2} = \frac{d\theta}{r} \tag{2}$$

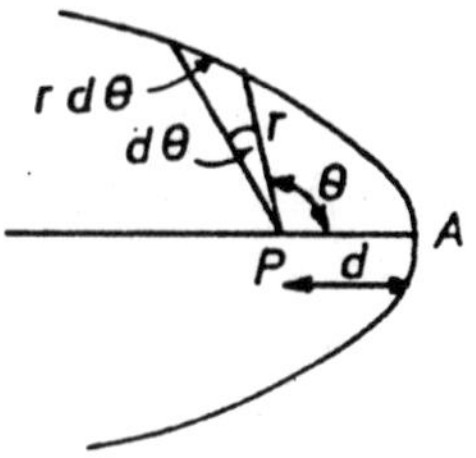

Figure 6.26.

The field at the focus P of the parabola

$$B = \frac{\mu_0 I}{4\pi} \int_{-\pi}^{\pi} \frac{d\theta}{r}$$

In the case of parabola, we have $r(1 + \cos\theta) = 2d$

$$\therefore \quad B \text{ at } P = \frac{\mu_0 I}{4\pi} \int_{-\pi}^{\pi} \frac{1 + \cos\theta}{2d} d\theta = \frac{\mu_0 I}{4d} \tag{3}$$

EXAMPLE 6.2: Use Ampere's theorem to obtain an expression for the magnetic field at the centre of a coil in the form of a square of $2d$ carrying a current I.

Solution: $ABCD$ is a square of side $2d$ meters carrying a current I. Let P be the centre of this square. The magnetic field at the centre P due to the side CD carrying current I is given by

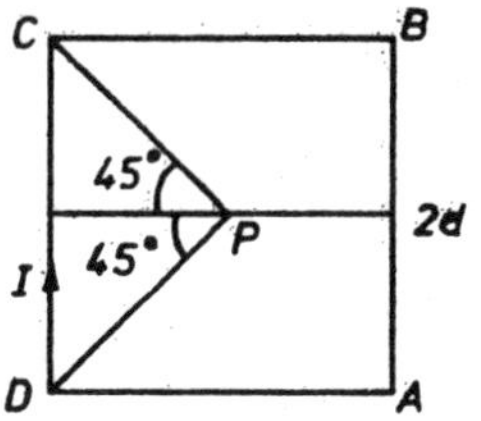

Figure 6.27.

$$\begin{aligned} B &= \frac{\mu_0 I}{4\pi d}(\sin 45° + \sin 45°) \\ &= \frac{\sqrt{2}\mu I}{4Hd} \quad (\because \alpha = \beta = 45° \text{ and } R = d) \end{aligned}$$

By symmetry the magnetic field at P due to each side will be $\dfrac{\sqrt{2}\mu I}{4\pi d} \dfrac{W}{m^2}$.

∴ The magnetic field at the centre of the square, i.e. at P is given by

$$B = 4\sqrt{2}\frac{\mu I}{4\pi d} = \frac{\sqrt{2}\mu I}{\pi d}\frac{W}{m^2}$$

Obviously, the direction of the magnetic field is perpendicular to the plane of the paper.

EXAMPLE 6.3: Show that the magnetic field of hollow circular cylindrical conductor carrying current vanishes everywhere inside the hollow region.

Solution: A section of a hollow circular cylindrical conductor perpendicular to its length is shown in Fig 6.28. The axis of the cylindrical conductor passes through the point O. Let r_1 and r_2 be the internal and external radii of the cylindrical conductor respectively. Let us consider a circular path of radius r around O inside the hollow conductor. By symmetry the magnetic field produced due to the current flowing in the conductor at any point of this path will be tangential and any point of this path will be tangential and equal in magnitude everywhere. Therefore the line integral of **B** along this path is

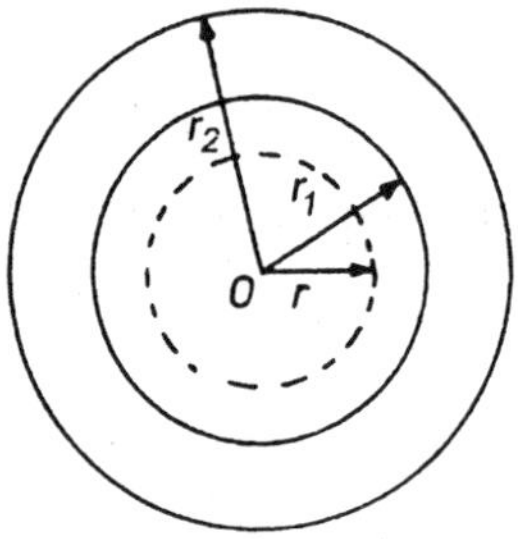

Figure 6.28.

$$\oint \mathbf{B}\cdot d\mathbf{l} = B2\pi r$$

But $\oint \mathbf{B}\cdot d\mathbf{l} = \mu\times$ (current enclosed by the circular path of radius r) i.e.

$$B = \mu\times 0 = 0$$

obviously, the magnetic field of a hollow circular cylindrical conductor carrying current vanishes everywhere inside the hollow region.

EXAMPLE 6.4: A current of 20 amperes is flowing in a long straight wire. An electron is moving with a velocity 10^7 m/s. How much force will act on the electron if wire is 2 cms away from it and the velocity of electron is (i) towards the wire (ii) parallel to the wire and (iii) perpendicular to the directions given by (i) and (ii).

Solution: Magnetic field at a distance r from a current carrying wire is

$$B = \frac{\mu_0 I}{2\pi r} \quad \left| \begin{array}{l} \mu_0 = 4\pi \times 10^{-7} \frac{W}{A-m} \\ I = 20 \text{ amp} \\ r = 2 \times 10^{-2} \text{ m} \end{array} \right.$$

$$= \frac{4\pi \times 10^{-7} \times 20}{2\pi \times 2 \times 10^{-2}} = 2 \times 10^{-4} \text{ W/m}^2$$

The magnetic field **B** is perpendicular to the plane containing wire and **r**. If wire is in y-direction and **r** in x-direction, **B** will be in z-direction (Fig 6.29).

$\because$ Force on electron

$$\mathbf{F} = q(\mathbf{V} \times \mathbf{B})$$

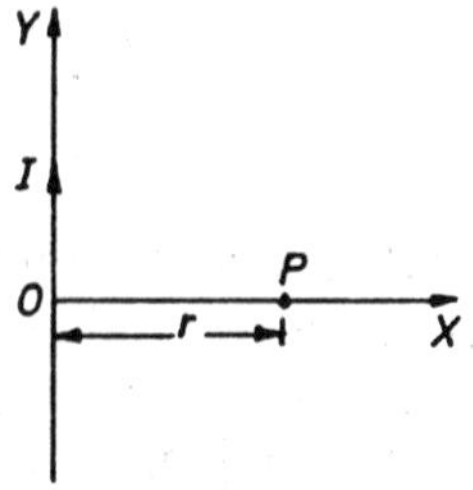

Figure 6.29.

(i) When electron is moving towards the wire, then

$$\begin{aligned} \mathbf{V} &= V(-\hat{i}) \\ \therefore &= q[V(-\hat{x}) + B(-\hat{k})] = -eVB(-\hat{j}) \\ &= eVB(\hat{j}) \\ &= 1.6 \times 10^{-19} \times 10^7 \times 2 \times 10^{-4} \hat{j} \\ &= 3.2 \times 10^{-16} \quad \text{newton in the direction of current.} \end{aligned}$$

(ii) When electron is moving parallel to the wire, then $\mathbf{V} = V(\hat{j})$

$$\therefore \quad F = -eVB[\hat{j} \times (-\hat{k})] = eVB\hat{i} = 3.2 \times 10^{-16}$$
newton $\perp^r$ to wire

(iii) In this case the force will be zero as the motion of the electron, i.e. **V** is in the $\hat{k}$- direction.

EXAMPLE 6.5: The vector potential corresponding to a magnetic field is given by $\mathbf{A} = -\hat{k}\frac{I}{C}\ln(x^2 + y^2)$. Determine the field.

Solution: We have

$$\mathbf{B} = \nabla\times\mathbf{A} = \begin{vmatrix} \hat{i} & \hat{j} & \hat{k} \\ \dfrac{\partial}{\partial x} & \dfrac{\partial}{\partial y} & \dfrac{\partial}{\partial z} \\ 0 & 0 & -\dfrac{I}{C}\ln(x^2+y^2) \end{vmatrix}$$

$$= \hat{i}\frac{\partial}{\partial y}\left[\frac{I}{C}\ln(x^2+y^2)\right] - \hat{j}\frac{\partial}{\partial y}\left[-\frac{I}{C}\ln(x^2+y^2)\right]$$

$$= -\frac{I}{C}\frac{2(y\hat{i}-x\hat{j})}{(x^2+y^2)}$$

EXAMPLE 6.6: A toroid is wound uniformly. It has N turns, and its inner and outer radii are 'a' and 'b' respectively. Find the magnetic induction inside the toroid when the current is I and determine the ratio b/a so that **B** in the ring does not vary by more than 25%.

Solution: Let us consider a closed circular path of radius r concentric with the toroid, $a < r < b$. If **B** is magnetic induction along this path then by Ampere's law

$$2\pi rB = \mu_0 NI$$

or

$$B = \mu_0\frac{NI}{2\pi r}$$

$$\therefore B_{max} = \frac{\mu_0 NI}{2\pi a} \quad \text{and} \quad B_{min} = \frac{\mu_0 NI}{2\pi b}$$

Hence

$$B_{max} - B_{min} = \frac{\mu_0 NI}{2\pi}\left(\frac{1}{a}-\frac{1}{b}\right)$$

$$= \frac{\mu_0 NI}{2\pi}\left(\frac{b-a}{ab}\right) = \frac{1}{4}\left(\frac{\mu_0 NI}{2\pi a}\right)$$

or

$$\frac{b-a}{ab} = \frac{1}{4a} \quad \therefore \frac{b}{a} = \frac{4}{3}.$$

EXAMPLE 6.7: An infinitely long cylindrical conductor of radius a carries a steady current of I amperes, show that the fields inside and outside the conductors are

$$H_{\phi(\text{in})} = \frac{Ir}{2\pi a^2} \quad (\text{for } r < a)$$

and $$H_{\phi(\text{out})} = \frac{I}{2\pi r} \quad (\text{for } r > a)$$

Solution: The problem can be solved with the help of Ampere's circuital law. From consideration of symmetry it is evident that the magnetic intensity has only a ϕ component. Considering field point P at $r(< a)$ inside the cylinder, we obtain

$$H_\phi \times 2\pi r = \text{Current enclosed} = \frac{Ir^2}{a^2}$$

Since the current is uniformly distributed over the cross-section. Therefore, one obtains

$$(H_\phi)_{\text{in}} = \frac{Ir}{2\pi a^2} \quad (r < a)$$

when the field point is outside the conductor, i.e. $r > a$ the current enclosed is I. Thus

$$(H_\phi)_{\text{out}} = \frac{I}{2\pi r} \quad (r > a)$$

EXAMPLE 6.8: Show that the current required to balance the earth's gravitational force of a horizontal wire of mass m and length l, and oriented from the east to the west in the earth's magnetic field is given by $I = \frac{mg}{lB}$, where B is the magnitude of earth's magnetic field directed north-south.

Solution: The earth's magnetic field is directed north-south and has a magnitude of B. The force due to interaction of current and flux $= IlB$. This force is equal to the weight mg. Therefore

$$IlB = mg$$

or $$I = \frac{mg}{Bl}.$$

EXAMPLE 6.9: An infinitely long straight wire carrying current I_1 and a rectangular loop carrying current I_2 are situated in the same plane with two long sides of the rectangle parallel to the infinitely long wire. Show that the force between them is given by $F = \frac{\mu_0 a I_1 I_2 b}{2\pi d(a+b)}$.

Solution: Fig 6.30 is drawn in accordance with the problem. We see that the conductor *AB* experiences a force of attraction towards the infinitely long conductor, and conductor *CD* a force of repulsion, whereas conductor *BC* and *DA* do not experience any force. The flux density acting on *AB* due to the infinitely long conductor $= \dfrac{\mu_0 I}{2\pi d}$ in a direction penetrating the paper. Therefore, one obtains the force on conductor *AD* as

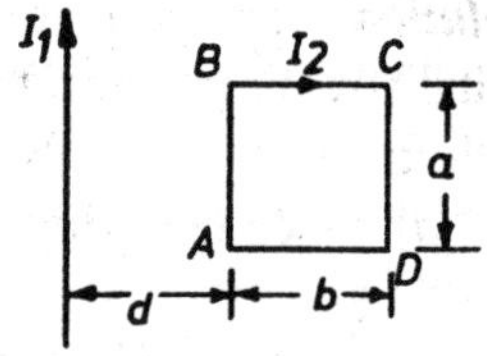

Figure 6.30.

$$F_1 = I_2 \frac{\mu_0 a I_1}{2\pi d} \quad \text{directed towards the infinitely long conductor}$$

Similarly, one can obtain the force on conductor *CD* as

$$F_2 = I_2 \left[\frac{\mu_0 a I}{2\pi(a+b)} \right] \quad \text{away from the conductor.}$$

$$\text{The net force } F = F_1 - F_2 = \frac{\mu_0 I_1 I_2 ab}{2\pi d(a+b)}$$

QUESTIONS

1. State and explain Ampere's circuital law.
2. State Biot-Savart law and deduce Ampere's circuital law from this for the steady current.
3. Show that the divergence of magnetic induction is always zero.
4. Define the magnetic vector potential for steady current and show that it satisfies the Poisson's equation $\nabla^2 \mathbf{A} = -\mu_0 \mathbf{J}$.
5. State Ampere's circuital law. Discuss why and how it was modified to include the displacement current.
6. How would you define the magnetic field at a point? Describe some general properties of the magnetic field.
7. State Ampere's law for magnetic field of a rectilinear current. Using it discuss the magnetic field produced by a current along a circular cylinder of infinite length.
8. Prove that the line integral of magnetic field in a current less closed path is always zero.

9. What do you understand by scalar and vector potentials? What is their significance?
10. Derive expression for the magnetic vector potential and magnetic induction due to a small element of current carrying circuit at a large distance.
11. What do you understand by magnetic permeability and susceptibility of a substance?
12. Distinguish between para, dia and ferromagnetism.
13. What do you understand by the intensity of magnetization **m**. Establish the relation $\mathbf{J} = \nabla \times \mathbf{m}$, where **J** is current density vector in a non-uniformly magnetized material at a point where intensity of magnetization vector is **m**.
14. Obtain the magnetic boundary conditions and compare with electrostatic boundary conditions.
15. Show that the lines of magnetic field strength **H** are refracted at a change of media. Also, show that $\mu_1 \cot\theta_1 = \mu_2 \cot\theta_2$, where μ_1, μ_2 are permeabilities of the media and θ_1, θ_2 are the angles made with the normal.

PROBLEMS

1. A current carrying loop consists of three straight edges, each of length a and a semicircular portion of radius $a/2$, all in one plane as shown in Fig 6.31. The current is I. (i) Calculate the magnetic

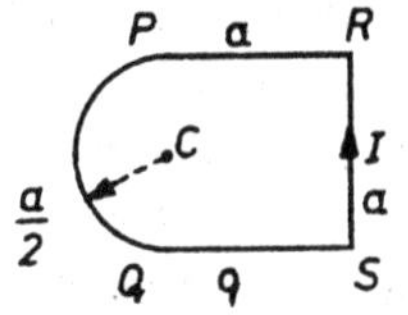

Figure 6.31.

field at the point C which is the mid point of PQ and (ii) the Semi-circular portion is bent about the line PQ so that its plane is at 90° to the plane of the remaining loop. What is the magnetic field at point C for the same current.

[**Ans.** (i) $\dfrac{\mu_0 I}{a}\left(\dfrac{1}{4}+\dfrac{\sqrt{5}}{2\pi}\right)$ (ii) $0.435\mu_0 I/a$ at an angle of 35° to the normal to the square area].

2. The current following into a conductor varies in time in accordance to the relation $I = I_0 \exp(-\alpha t)$, where I_0 and α are constants. Show that the charge Q which accumulated on the conductor after a time t_0 is given by $Q = I_0\alpha^{-1}[1 - \exp(-\alpha t_0)]$.
3. Show that the magnetic vector potential for two long, straight, parallel wires carrying the same current I, in opposite directions is given by $\mathbf{A} = \frac{\mu_0 \mathbf{I}}{2\pi} \log_e\left(\frac{r_2}{r_1}\right) \hat{a}_n$, where r_1 and r_2 are the distances from the field point to the wires and $\hat{a}_n$ is a unit vector parallel to the wires.
4. Show that in a medium in which a uniform electric current of constant density exists, the magnetic field is $\mathbf{B} = \frac{1}{2}\mu_0 \mathbf{J} \times \mathbf{r}$.
5. A closed surface in the form of a hemisphere has radius a. Its base is in the XOY plane and it is kept in a uniform field $\mathbf{B} = B\hat{k}$. Determine the magnetic flux over the surface by direct integration. [Ans. zero]
6. The radius and number of turns of a circular coil are 6 cm and 20 respectively. If a current of 1.5 ampere is passed through the coil, then calculate the magnetic field at a point on the axis of the coil at a distance of 8 cm from its centre. [Ans. 6×10^{-5} tesla]
7. A current of 10 amperes is flowing in a long straight wire. Find the magnetic field at a distance of 2 cms from the wire. [Ans. 10^{-4} tesla]
8. A long, cylindrical conductor of radius a carries a current along its length. The current density at a distance r from the axis of the conductor is given by $J(r) = J_0 r/a$, where J_0 being a constant. Show that the magnetic field for $r \leq a$ and $r \geq a$ are given by (i) $B(r \leq a) = \mu_0 J_0 r^2/3a$ and (ii) $B(r \geq a) = \mu_0 a^2 J_0/3r$ respectively.
9. Using Biot and Savart's law show that the magnetic flux density due to a circular coil of radius a and carrying current I at a point z distance from its centre and lying on its axis is given by
$$B = \frac{\mu_0 I a^2}{2(a^2 + z^2)^{3/2}}.$$
10. Two infinite thin wires [one at (0,0,0) and other at $(R,0,0)$] parallel to z-axis carry current I in opposite directions, where R is very large. Calculate the magnetic potential at (r,ϕ,z).
[Ans. $\phi_m = -\frac{I}{2\pi}(\pi - \phi)$]

11. A circular conductor of radius $r_0 = 1$ cm has an internal field

$$\mathbf{H} = \frac{10^4}{r}\left(\frac{1}{a^2}\sin ar - \frac{r}{a}\cos ar\right)\hat{a}_\phi A/m.$$

If $a = \frac{\pi}{4r_0}$, show that the total current in the conductor is $\frac{8}{\sqrt{2}}\left(\frac{4}{\pi} - 1\right)$ ampere.

12. A vector potential is given by

$$\mathbf{A} = x^2 y\hat{a}_x + y^2 x\hat{a}_y - 4xyz\hat{a}_z$$

Determine the field = **B**. [**Ans.** $-4zx\hat{a}_x - 4yz\hat{a}_y + (y^2 - x^2)\hat{a}_z$]

13. A current is flowing in a long circular conductor of radius a. The current is distributed in the wire in such a way that the current density at a distance r from the axis is given by $J = J_0\left(1 + \frac{r^2}{a^2}\right)$. Show that the total current in the wire is $I = \frac{3}{2}\pi J_0 a^2$.

14. In problem 13, Calculate the magnetic flux density both inside and outside the wire.

[**Ans.** (i)$B_{\text{inside}} = \frac{\mu_0 J_0 r}{2}\left(1 + \frac{r^2}{2a^2}\right)$ and (ii)$B_{\text{outside}} = \frac{3\mu_0 J_0 a^2}{4r}$]

15. A circular coil of radius a carries a current I. Show that the magnetic scalar potential ϕ_m at a point P on its axis at a distance x from of the coil is given by $\phi_m = \frac{I}{2}\left[1 - \frac{x}{\sqrt{x^2 - a^2}}\right]$.

16. A uniformly charged sphere whose radius is a and charge density ρ, rotates with a constant angular velocity ω. calculate the magnetic flux density **B** at the centre of the sphere.

[**Ans.** $B = \frac{\mu_0 \rho \omega a^2}{3}$]

17. The molecular weight of a compound is 400 and its density is 2×10^3 kg/m^3. Its magnetic susceptibility is 293 K at 2.56×10^{-4} Show that the permanent dipole associated with each molecule is $m = 2.87 \times 10^{-23}$ A-m^2. Given Avogadro's number is 6.02×10^{23}.

SHORT ANSWER QUESTIONS

1. What will be the force on a particle carrying a charge q and moving with a velocity $\mathbf{V}$ in a region of magnetic induction $\mathbf{B}$?

   [**Ans.** $\mathbf{F} = q(\mathbf{V} \times \mathbf{B})$]

   [Note: We must note that the eq. $\mathbf{F} = q(\mathbf{V} \times \mathbf{B})$ enables one to measure $\mathbf{B}$ without reference to its sources which are moving charges or currents].
2. What is the unit of $\mathbf{B}$? [**Ans.** $\text{Wb/m}^2 = \text{N/A} - \text{m} = \text{tesla}$]
3. What is the magnetic flux over a closed surface? [**Ans.** zero]
4. How will you define ampere?

   [**Ans.** An ampere is that current which when flowing through two thin, parallel conductors of infinite length, produces a mutual force of 2×10^{-7} N/m between the conductors. 1 amp is the same as a rate of flow of 1 C/s.]
5. What is an ideal solenoid?

   [**Ans.** An ideal solenoid is essentially a circular current loop, where the current is not filamentary but in the form of a broad sheet flowing on the curved surface of a cylinder. In practice, one can consider a closely wound, long solenoid (length $\gg$ radius) with n turns per unit length as an ideal solenoid.]
6. What will the magnetic induction at interior points of an ideal solenoid defined in Q 5? [**Ans.** $B = \mu_0 NI$]
7. How will you define magnetic field $\mathbf{B}$ due to any type of current source?

   [**Ans.** $\mathbf{B} = \nabla \times \mathbf{A}$, where $\mathbf{A}$ is a vector potential given by $\mathbf{A}(r) = \frac{\mu_0 I}{4\pi} \int \frac{\mathbf{dl}}{r}$]
8. When the right hand side of Ampere's law equation $\int \mathbf{B} \cdot d\mathbf{l} = \mu_0 I$ is nonzero, what it signifies?

   [**Ans.** $\mathbf{B}$ is a non-conservative field, i.e. $\mathbf{B} = \nabla \times \mathbf{A}$, where $\mathbf{A}$ is magnetic vector potential]
9. Is the dimensions of magnetic scalar potential ϕ_m is of an ampere?

   [Ans. Yes]
10. What will be the scalar potential for a magnetic dipole?

    [**Ans.** $\phi_m(r, \theta) = \frac{1}{4\pi} \frac{m \cos\theta}{r^2}$]

11. The atoms of certain elemental substances and the molecules of some compounds possess an intrinsic magnetic dipole moment. What are these substances called?

    [**Ans.** Paramagnetic substances]

12. For diamagnetic and paramagnetic substances, what is the relation between magnetic moment density **M** and magnetic intensity **H**?

    [**Ans.** $\mathbf{M} = \chi \mathbf{H}$]

13. What is relative permeability? [**Ans.** $\mu_r = \dfrac{\mu}{\mu_0} = 1 + \chi$]

14. Which one of the following three groups of magnetic materials exhibit the phenomenon of hysteresis? (i) Diamagnetic, (ii) Paramagnetic, (iii) Ferromagnetic [**Ans.** Ferromagnetic]

OBJECTIVE QUESTIONS

1. Biot-Savart law states the relation between magnetic intensity and
   (a) electric intensity, (b) electric current,
   (c) work done, (d) vector potential. **[b]**

2. Which one of the following statement is incorrect in connection with Ampere's circuital law of magnetostatics?
   (a) There is no scalar magnetic field in regions including a current carrying conductor
   (b) The vector field at a point depends on current strength and inverse distance
   (c) The field due to a filament conductor is directed along it
   (d) The work done of a unit north pole in enclosing a current is independent of the shape and size of the path. **[c]**

3. A conducting circular loop of radius r carries a constant current I. It is placed in a uniform magnetic field B such that B is perpendicular to the plane of the loop. The magnetic force acting on the loop is given by (a) IrB, (b) zero, (c) πIrB, (d) $2\pi IrB$. **[b]**

   [**Hint:** magnetic field produced by the current and B are along the same direction].

4. A small circular flexible loop of wire of radius r carries a current I. It is placed in a uniform magnetic field B. The tension in the loop will be doubled if
   (a) I is doubled, (b) Both B and I are doubled,
   (c) r is doubled, (d) Both B and I are doubled. **[(a) and (b)]**

[Hint: $|\mathbf{F}| = |\mathbf{m} \times \mathbf{B}| = mB\sin\theta$ **and** $m = I\pi r^2$ **obviously, tension will be doubled if** I **or** B **is doubled].**

5. Two thin long parallel wires separated by a distance b are carrying a current I amperes each. The magnitude of the force per unit length exerted by one wire on the other is
(a) $\mu_0 I^2/b^2$, (b) $\mu_0 I^2/2\pi b$, (c) $\mu_0 I/2\pi b$, (d) $\mu_0 I/2\pi b^2$. **[b]**

6. A square loop carrying current is introduced in a uniform magnetic field inclined at angle α with the perpendicular to the plane. The coil experiences a torque proportional to
(a) $\sin\alpha$, (b) $\cos\alpha$, (c) $\cot\alpha$, (d) $\sin\alpha\cos\alpha$. **[b]**

7. The magnetic vector potential due to a single conductor carrying current is
(a) infinite, (b) zero, (c) finite and definite,
(d) none of the above. **[a]**

8. The chief characteristic of ferromagnetic substances is
   (a) permanent magnetization
   (b) No hysteresis under the action of a magnetic field.
   (c) atomic dipoles are not aligned by the application of external field,
   (d) none of the above. **[a]**

9. The unit of magnetic permeability is
(a) henry-metre, (b) Weber, (c) Weber per square metre,
(d) henry/metre. **[d]**

10. The magnetic field inside a current carrying toroidal solenoid is B. If its radius is doubled and the current through it is also doubled, the magnetic field inside the solenoid will be
(a) B, (b) $B/2$, (c) $2B$, (d) $4B$. **[c]**

11. A long solenoid has 20 turns/cm. The current necessary to produce a magnetic field of 20 milli tesla inside the solenoid is approximately
(a) 8A, (b) 4A, (c) 2A, (d) 1A. **[a]**

12. Force between two long straight parallel current-carrying wires is F. If the current in one of them is doubled, the force between them will be
(a) $2F$, (b) $\sqrt{2}F$, (c) $4F$, (d) $2\sqrt{2}F$. **[a]**

Chapter 7

Electromagnetic Induction and Maxwell's Equations

7.1 INTRODUCTION

One of the many electromagnetic phenomena familiar to us is *electromagnetic induction*, which was discovered, almost simultaneously, around 1830 by British physicist and chemist Michel Farady and Joseph Henry independently. They discovered that an electric current is produced in a closed conducting loop when the flux of magnetic induction through the surface enclosed by this loop changes. This phenomenon is called *electromagnetic induction*, and the current produced an *induced current*. In this chapter we shall investigate the behaviour of time-dependent electric and magnetic fields. In terms of fields, one may say that a time-varying magnetic field produces an electromotive force which may establish a current in a suitable closed circuit. Obviously, there exists a closer connection between the $\vec{E}$- and $\vec{B}$- fields, which is exhibited in phenomena of electromagnetic induction.

7.2 ELECTROMOTIVE FORCE

We are all familiar that when a study current flows in a single conductor a potential difference exist between its two ends. This potential difference is maintained by a source of *electromotive force* (e.m.f.). There are many devices which provide e.m.f. such as cell, battery, dynamo, etc. We all know that these sources of e.m.f. convert some other kind of energy into electrical energy. We know that a battery uses stored chemical energy to drive a current in the circuit. If the battery maintains a constant potential difference of V volts between its terminals, its e.m.f. is said to be V volts. When the terminals of the cell or battery are connected by a conductor, an electric field is established in the conductor. This electric field is static. The line integral of the electric field over a path between two points (say P and Q) is equal to the potential difference between the two points. If these two points are the terminals of the battery, one obtains the e.m.f. of the battery as

$$V = \int_P^Q \mathbf{E} \cdot d\mathbf{l} \tag{7.1}$$

We know that when the current flows in the conductor, the battery has to do work in order to keep the potential difference between its terminals constant. If a charge q moves from one terminal of the battery to the other terminal through the conductor, the amount of work done by the battery is

$$Vq = \int_P^Q q\mathbf{E} \cdot d\mathbf{l} \tag{7.2}$$

Therefore, the e.m.f. of the battery is

$$V = \frac{1}{q} \int_P^Q q\mathbf{E} \cdot d\mathbf{l} \tag{7.3}$$

V is expressed in $Volt(V)$ or $\mathrm{m^2 kg\ s^{-2} C^{-1}}$. Obviously, V has the dimensions of energy per charge. However, we must remember that the term e.m.f. is not a force.

Let us consider a metal rod moving with constant velocity $\mathbf{u}$ in a direction perpendicular to a uniform field $\mathbf{B}$ as shown in Fig. 7.1.

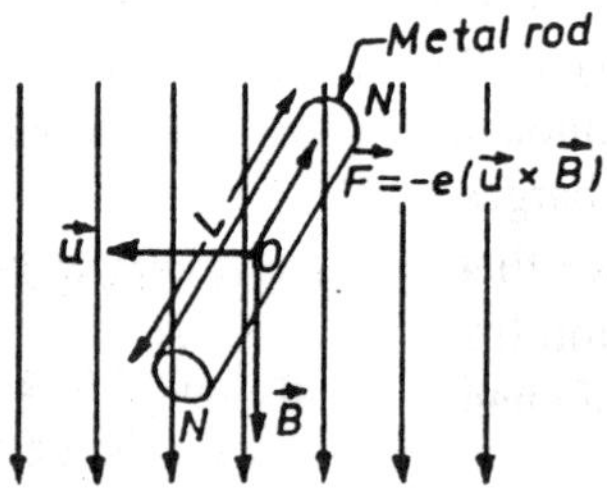

Figure 7.1.

As discussed in chapter 6, there will be a magnetic force on each electron in the rod given by

$$\mathbf{F} = -e(\mathbf{u} \times \mathbf{B})$$

The free electrons will move towards the end of the bar and a field $\vec{E}$ will be set as

$$\mathbf{E} = \mathbf{u} \times \mathbf{B} \tag{7.4}$$

The potential difference between the two ends MN of the bar is

$$V_{MN} = \int_N^M \mathbf{E} \cdot d\mathbf{l} = uBL \tag{7.5}$$

However, this potential difference will not produce any current flow. If however, the metal rod forms a part of a circuit as shown in Fig. 7.2, a current I will flow.

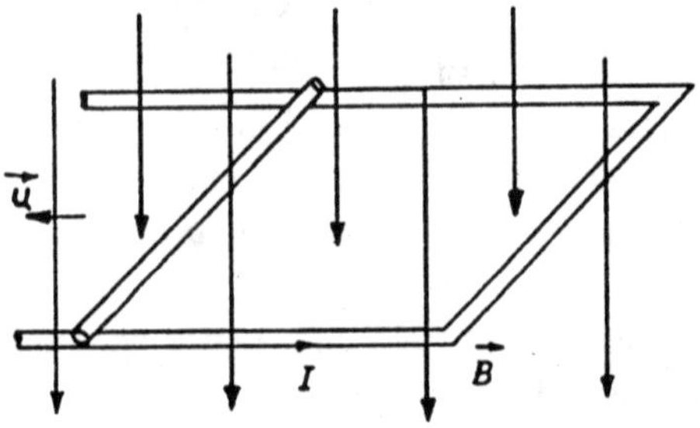

Figure 7.2.

One obtains the line integral of the force on a charge q round the circuit as

$$\mathbf{F} = \oint (q\mathbf{E} \cdot d\mathbf{l}) = q u \mathbf{B} L$$

Thus the induced e.m.f. in the closed circuit, due to the motion of the conductor is

$$V = \frac{1}{q} \oint (q\mathbf{E} \cdot d\mathbf{l}) = uBL \tag{7.6}$$

One may call it a motional e.m.f. because it depends on the velocity of the conductor and not on its position. Obviously, we have two kinds of voltage: (i) electrostatic potential difference due to stationary charges and (ii) electromotive force due to moving charges. We can see that uBL is the magnetic flux through the area swept by the rod in unit time. If ϕ is total magnetic flux through the circuit, uBL is the rate of change of flux through the circuit. One can express

$$|\text{e.m.f.}| = uBL = \frac{d\phi}{dt} \tag{7.7}$$

7.3 THE FARADAY-HENRY LAW

One can summarize the observations of M. Faraday and J. Henry as

(i) When a magnet is moved about in the neighbourhood of a conductor in the form of a closed circuit but with no source of e.m.f., i.e. battery, then a current is produced in the conductor *so long as the movement lasts*. However, the current disappears on cessation of movement of the magnet.

(ii) One observes the same effect if the magnet is kept still and the circuit containing the conductor is moved. Obviously, for a current to be produced in the conductor, there must be a relative moment. We must note that the current can be produced without any mechanical movement.

(iii) A transient current is induced in a loop of conducting wire if the stationary current in an adjacent circuit is switched on and off. This means that the transient current flowed when the flux through the circuit is altered. Obviously, the changing flux induced an electric field and, hence an e.m.f. in the circuit which causes the current to flow.

Faraday and Henry called this phenomena *electromagnetic induction*. One can sum up Faraday's results in the form of a "*flux rule*". We may state this in words as:

When the magnetic flux through a closed circuit is changing, an e.m.f. is induced in the circuit, the magnitude of which is proportional to the time rate of change of the magnetic flux through the circuit, one may write

$$|\varepsilon| \propto \frac{d\phi}{dt} \tag{7.8}$$

where ε is the induced e.m.f. and ϕ, the magnetic flux.

The direction of the induced e.m.f. is determined by *Lenz's Law*. According to Lenz's Law.

The direction of the induced e.m.f. is such that the magnetic flux associated with the current generated by it, counteracts the cause generating it. In otherwords, the induced current creates a magnetic flux which prevents the variation of the magnetic flux generating the induced e.m.f. Therefore, one may write

$$\varepsilon = -\frac{d\phi}{dt} \tag{7.9}$$

ϕ is expressed in Wb or $m^2 kg\, s^{-1} C^{-1}$ and hence $\frac{d\phi}{dt}$ must be expressed in Wb/s or $m^2 kg\, s^{-2}/C^{-1}$. We must note that the induced e.m.f. is often called the *back* e.m.f. Lentz's Law expresses an important physical fact, viz. the tendency of a system to counteract the change its state (electromagnetic inertia). In general, (7.9) can also be expressed in the following way

$$\varepsilon = \oint_L \mathbf{E} \cdot d\mathbf{l} = -\frac{d\phi}{dt} = -\frac{d}{dt} \int_s \mathbf{B} \cdot \hat{a}_n \, ds \tag{7.10}$$

Referring to Fig. 7.3, if we divide the area limited by L into infinitesimal area elements, the magnetic flux through L is $\phi = \int_s \mathbf{B} \cdot \hat{a}_n \, ds$. Also the e.m.f. V implies the existence of an electric field E such that $V = -\oint \mathbf{E} \cdot d\mathbf{l}$. Thus we may write Eq. (7.10).

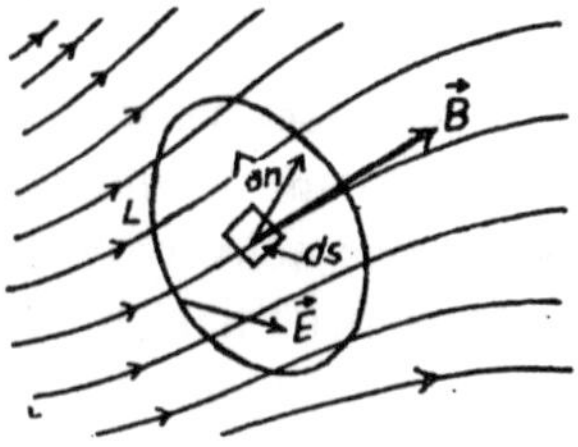

Figure 7.3.

The electric field in the above equation is produced by the changing magnetic flux through the surface. The line integral of **E** is taken around the closed curve that forms the boundary of the surface S. This line integral sometime called the e.m.f. around the circuit (Fig. 7.3(a)).

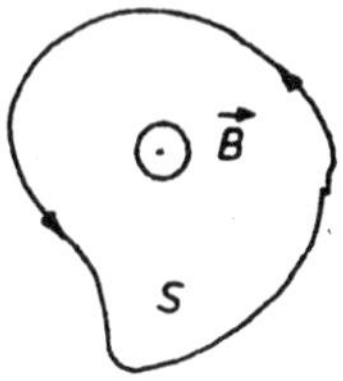

Figure 7.3a. Surface S is an open surface. Magnetic field is changing over this surface.

For convenience, we have taken the loop to be in one plane only. Faraday's law does not require this. Faraday's law is applicable to an open surface of an arbitrary shape. We have seen in chapter 6 that the surface integral of **B** is the same for all open surfaces bounded by a given closed curve. This is a consequence of divergence character of the field **B**. In Fig. 7.3 we have therefore taken the simplest surface, viz. a plane surface bounded by the loop. We must note that there is no loss of generality in this. Then Eq. (7.10) is equivalent to saying:

> *a time dependent magnetic field implies the existence of an electric field such that the circulation of the electric field along an arbitrary closed path is equal to the negative of the time rate of change of the magnetic flux through a surface handed by the path.*

This is another way of stating the Faraday-Henry law of e.m. induction. It gives us a deeper insight into the physical content of the phenomenon of electromagnetic induction, i.e., the fact that an electric field must exist wherever a magnetic field is changing with time, the two fields related by (7.10). The electric field can be determined by measuring the force on a charge at rest in the region where the magnetic field is varying.

7.4 INDUCTION LAW FOR MOVING CIRCUITS

Let us consider a circuit of any arbitrary shape moving with instantaneous velocity **V** in uniform magnetic field **B**. An element $d\mathbf{l}(PQ)$ will move in time dt a distance Vdt to a position $P'Q'$. The displacement of element in time $\delta t = Vdt$.

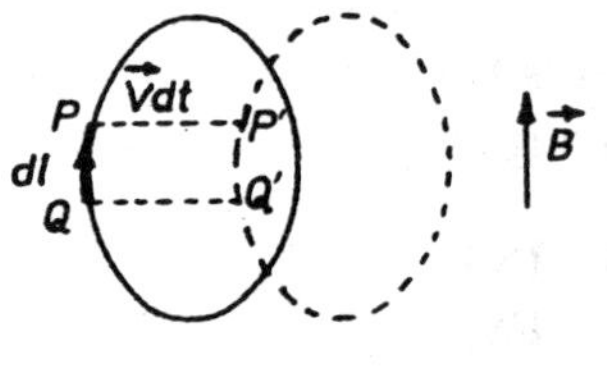

Figure 7.4.

Let the velocity of the conducting electrons of element, relative to wire along the length of wire is **u**, then relative velocity is $\mathbf{v}+\mathbf{u}$.

The magnetic force on each electron of charge $-e$,

$$\mathbf{F}_m = -e(\mathbf{v}+\mathbf{u})\times\mathbf{B} \tag{7.11}$$

The component of this force along the element δl is $\mathbf{F}_{11} = [-e(\mathbf{v}+\mathbf{u})\times\mathbf{B}]\cdot\hat{a}_l$, where $\hat{a}_l$ is unit vector along **PQ**. As **u** and $\hat{a}_l$ are along the same direction, we have

$$\begin{aligned} \mathbf{u}\times\mathbf{B}\cdot\hat{a}_l &= \mathbf{B}\cdot\hat{a}_l\times\mathbf{u} = 0 \\ \therefore \quad \mathbf{F}_{11} &= -e\mathbf{V}\times\mathbf{B}\cdot\hat{a}_l \end{aligned} \tag{7.12}$$

Eq. (7.12) shows that there is an electric field $\mathbf{E} = \mathbf{V}\times\mathbf{B}$ induced in the wire. Its component along the wire is $\mathbf{V}\times\mathbf{B}\cdot\hat{a}_l$.

The induced e.m.f. is the line integral of this field round the circuit, i.e. the induced e.m.f. is given by

$$\varepsilon = \oint(\mathbf{V}\times\mathbf{B})\cdot\hat{a}_l\,dl \tag{7.13}$$

But $\mathbf{V}dt\times\hat{a}_l dl$ = area $PP'Q'Q$ swept by element dl in time dt and hence $(\mathbf{V}\,dt\times\hat{a}_l\,dl)$. **B** represents the flux change across with the element in time dt. Obviously, the flux change across entire area

$$d\phi = \oint (\mathbf{V}\,dt \times \hat{a}_l\,dl)\cdot\mathbf{B}$$

$$\therefore \quad \frac{d\phi}{dt} = \oint (\mathbf{V}\times\hat{a}_l\,dl)\cdot\mathbf{B} = \oint \mathbf{B}\cdot(\mathbf{V}\times\hat{a}_l)\,dl$$

$$= -\oint (\mathbf{V}\times\mathbf{B})\cdot\hat{a}_l\,dl \tag{7.14}$$

Comparing (7.13) and (7.14), one obtains

$$\varepsilon = -\frac{d\phi}{dt} \tag{7.15}$$

This is Faraday-Henry law. Obviously, the law also holds for circuits moving in uniform magnetic field **B**.

7.5 INTEGRAL AND DIFFERENTIAL FORMS OF FARADAY-HENRY LAW

We have read that the induced e.m.f. is equal to the line integral of the induced electric field **E** around the coil. Thus we have

$$\varepsilon = \oint \mathbf{E}\cdot\hat{a}_l\,dl$$

The magnetic flux through the coil is

$$\phi = \int_s \mathbf{B}\cdot\hat{a}_n\,ds$$

where the integral is taken over any area bounded by the circuit. One can express (7.15) as

$$\oint \mathbf{E}\cdot\hat{a}_n\,dl = -\frac{d}{dt}\int \mathbf{B}\cdot\hat{a}_n\,dS$$

Since the surface S does not change its shape and nor its position with time, one can express the above equation as

$$\oint \mathbf{E}\cdot\hat{a}_n\,dl = -\int_S \frac{d\mathbf{B}}{dt}\hat{a}_n\,dS \tag{7.16}$$

Eq. (7.16) is the *integral form* of Faraday-Henry law.

Making use of Stoke's theorem, one obtains

$$\oint \mathbf{E}\cdot\hat{a}_l\,dl = \int_S \nabla\times\mathbf{E}\cdot\hat{a}_n dS = -\int \frac{\partial\mathbf{B}}{\partial t}\cdot\hat{a}_n\,dS$$

$$\therefore \quad \int\left(\nabla\times\mathbf{E}+\frac{\partial\mathbf{B}}{\partial t}\right)\cdot\hat{a}_n dS = 0 \tag{7.17}$$

We have replaced the total derivative by partial derivative as one is only concerned with the changes in field **B** with time at the fixed position of the elemental area dS.

Eq. (7.17) must hold for any arbitrary surface, we have

$$\nabla \times \mathbf{E} = -\frac{\partial \mathbf{B}}{\partial t} \tag{7.18}$$

This is called the *differential form* of Faraday-Henry Law of electromagnetic induction. A knowledge of the time rate of change of **B** at any point thus enables us to find the curl of the electric field at that point. One can compare this situation with the differential form of Gauss's law which relates the charge density at any point with the divergence of electric field at that point. Eq. (7.18) togetherwith $\nabla \cdot \mathbf{E} = \frac{\rho}{\varepsilon_0}$ show that the electric field has a non conservative part due to changing magnetic flux density as well as a conservative part due to electric charge density.

So far we have seen that there are two kinds of sources of electromagnetic field. The first kind of source is associated with a system such as electrostatics in which energy is conserved during a cyclic process within the system. The sources of such a *irrotational* or conservative system are well described by the relation $\nabla \cdot \mathbf{E} = \frac{\rho}{\varepsilon_0}$, where $\mathbf{E} = -\nabla\phi$. We have seen that such a field has no curl because $\nabla \times (\nabla\phi) = 0$.

The second kind of electromagnetic field source is associated with a system in which there is energy transferred in a cyclic process (e.g. magnetic field of a solenoid). One can specify such a field by the curl sources and has no divergence. In general, one finds that electromagnetic fields have both types of sources, and hence, the complete specification of such a vector field should include both types of sources. We must note that such a specification is not only necessary but sufficient. It is important to note that *any vector field is uniquely determined if its divergence and curl sources are given*. This is *Helmholtz theorem*.

Taking divergence of (7.18), one obtains

$$\nabla \cdot \nabla \times \mathbf{E} = -\frac{\partial}{\partial t}(\nabla \cdot \mathbf{B}) = 0$$

Obviously, $\nabla \cdot \mathbf{B}$ is necessarily independent of time at every point in space and this condition can reasonably be satisfied if one assumes

$$\nabla \cdot \mathbf{B} = 0 \tag{7.19}$$

This means, **B** is always *solenoidal*.

One can draw two following important conclusions from Faraday's law (7.18).

(i) The electric field **E** is no longer a conservative field when the magnetic field varies with time. This means that energy can flow between electric and magnetic forms through time-varying fields, and

(ii) No free magnetic monopoles can exist. As stated earlier, all magnetic poles occur in pairs, positive and negative.

We have seen that how Faraday-Henry law of induction shows that the electric and magnetic fields are interrelated. When we consider the time-dependence of these fields, then their independent nature disappears. Obviously, one can look upon these fields as a single field, i.e. *electromagnetic field.*

7.6 SELF-INDUCTANCE

Let us consider a circuit carrying a current I (Fig. 7.5). According to Ampere's law the current produces a magnetic field which, at each point is proportional to I. One may compute the magnetic flux through the circuit due to its own magnetic field, and call it the *self flux*. The flux, designated by ϕ_I, is then proportional to the current I, and we may write

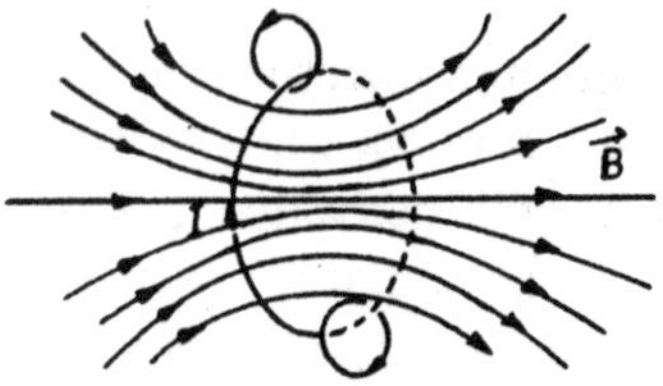

Figure 7.5. Self magnetic flux in a circuit

$$\phi_I = LI \tag{7.20}$$

where L is constant and depends on the geometrical shape of the conductor and is called the *self inductance* of the circuit. It is expressed in $Wb - A^{-1}$, a unit called the *henry*, in honour of Joseph Henry, and abbreviated H, i.e. $H = \dfrac{Wb}{A} = \text{m}^2\text{kg C}^{-2}$.

If the current I changes with time, the magnetic flux ϕ_I through the circuit also changes and, according to the law of electromagnetic induction, an e.m.f. is induced in the circuit. This special case of electromagnetic induction is called *self induction.*

Let us consider a situation in which the circuit is stationary but the magnetic field and, hence, the flux changes with time. Suppose, for example, that the current in the circuit is time dependent. Obviously, the flux in this situation is a function of time, i.e. $\phi = \phi(t)$. Therefore,

$$\frac{d\phi}{dt} = \frac{d\phi}{dI} \cdot \frac{dI}{dT} = L\frac{dI}{dt} \qquad \text{(using 7.20).}$$

$$\therefore \quad L = \frac{d\phi}{dt} \tag{7.21}$$

Using (7.21), Eq. (7.15) can now be written as

$$\varepsilon_L = -L\frac{dI}{dt} \tag{7.22}$$

The minus sign indicates that ε_L is opposed to change in the current. So if the current increases, $\frac{dI}{dt}$ is positive and ε_L acts in the same direction as the current [Fig. 7.6].

Therefore ε_L always acts in a direction that opposes the change in the current. When we wrote Eq. (7.22), we assumed the circuit rigid and therefore we considered L constant while computing the time derivative. If the shape of the circuit is variable, L is not constant, and instead of Eq. (7.22) we must write

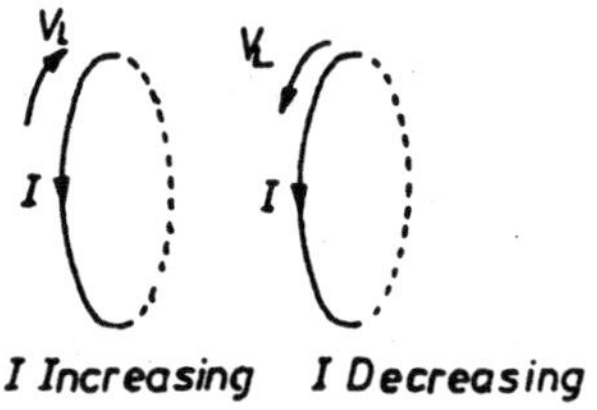

Figure 7.6. Direction of self induced e.m.f. in a circuit

$$\varepsilon_L = -\frac{d}{dt}\,(LI) \tag{7.23}$$

In diagrams drawn to indicate that a conductor has an appreciable inductance, the symbol of Fig. 7.7 is used. However, we must note that the self-inductance of a circuit is not concentrated at a particular point, but is a property of whole circuit.

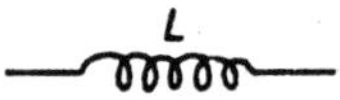

Figure 7.7. Representation self inductance a circuit

Self inductance of the circuit can be defined in two equivalent ways: (i) L can be defined as the magnetic flux linked with the circuit due to unit current flowing in it and (ii) L can also be defined as the induced e.m.f. in the circuit per unit rate of change of current in it.

When the current in the circuit is first switched on, the back e.m.f. opposes the grown of current, so that the current flows against the back e.m.f. and does work against it. Fig. 7.8 shows an electric circuit containing a resistance and a self inductance. When am e.m.f. V is applied to a circuit by closing a switch (Fig. 7.8), the current does not instantaneously attain the value V/R corresponding to ohm's law. This process is due to the self-induced e.m.f. ε_L which opposes the change in the current and is present while the current increases from zero upto its final constant value. The total e m f applied to the circuit is then $V - \varepsilon_L = V - L\left(\frac{dI}{dt}\right)$. Ohm's law is now

$$IR = V - \varepsilon_L \quad \text{or} \quad IR = V - L\left(\frac{dI}{dt}\right) \tag{7.24}$$

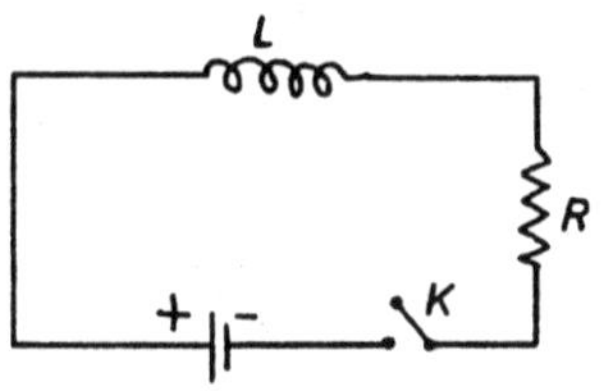

(a)

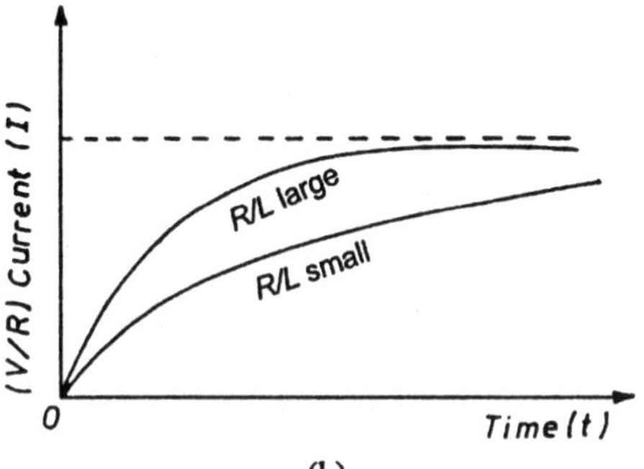

(b)

Figure 7.8. (a) Electric circuit contains a resistance and a self inductance (b) Establishment of a current in a circuit

This may be written as $R(I - V/R) = -L\left(\frac{dI}{dt}\right)$ or, separating the variables I and t,

$$\frac{dI}{I - V/R} = -\frac{R}{L}dt$$

Integrating and taking care that at $t = 0$, the current is also zero $(I = 0)$,

we have

$$\int_0^I \frac{dI}{I - V/R} = -\frac{R}{L}\int_0^t dt$$

or

$$I = \frac{V}{R}\left(1 - e^{-\frac{RL}{t}}\right) = I_0\left(1 - e^{-\frac{RL}{t}}\right) \qquad (7.25)$$

The second term in the parentheses decreases with time and the current asympotically approaches the value V/R which is given by Ohm's Law (Fig. 7.8(b)). If R/L is large, the current reaches this value very fast, but if R/L is small it may take a long time before the current stabilizes.

When switch is made off, then the only e.m.f. that remains is $\varepsilon_L = -L\frac{dI}{dt}$ and Ohm's law becomes

$$RI = -L\frac{dI}{dt} \quad \text{or} \quad \frac{dI}{I} = -\frac{R}{L}dt$$

If we count time ($t = 0$) from the instant that V is removed from the circuit, the initial current is V/R. Integrating, one obtains

$$\int_{V/A}^I \frac{dI}{I} = -\frac{R}{L}\int_0^t dt$$

or

$$I = \left(\frac{V}{R}\right)e^{-\frac{Rt}{L}} = I_0 e^{-\frac{RL}{t}} \qquad (7.26)$$

The current decreases exponentially, as shown in Fig. 7.9. The larger the resistance R, or the smaller the inductance L, the faster is the drop in the current. The time required for the current to drop to $\frac{1}{e}$, or approximately 63% of the initial value, is $\tau = \frac{L}{R}$. This time is called *relaxation time*.

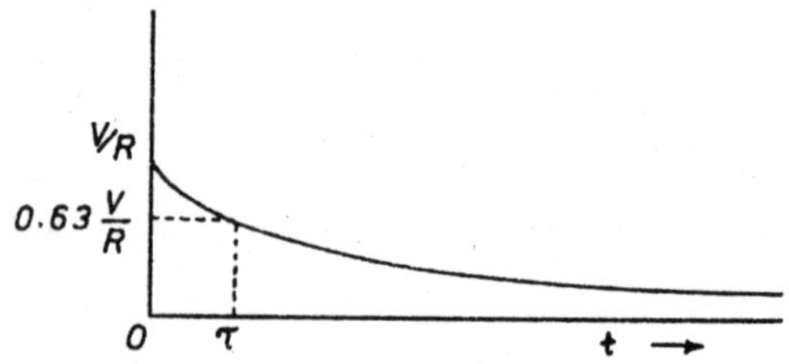

Figure 7.9. Decay of the current in a circuit after the e.m.f. has been removed

Let us now calculate the work done against back e.m.f. in a short time dt when the current is growing in a circuit (Fig. 7.8). We have

$$dW = -\varepsilon I\,dt = L\frac{dI}{dt}\cdot I\,dt$$

Hence the total work done in bringing the current from zero to a steady maximum value I_0 is

$$W. = L\int_0^{I_0} I\frac{dI}{dt} = L\int_0^{I_0} I\,dt = \frac{1}{2}LI_0^2 \qquad (7.27)$$

This work is stored as the energy of the magnetic field. Eq. (7.20), (7.23) and (7.27) enable one to define the self induce (L) of a circuit in the following ways:

(a) We have $\phi = LI \quad \therefore \quad L = \frac{\phi}{I}$. If $I = 1$ ampere, $L = \phi$ Henries. Thus the self inductance of a circuit (in henry) is defined as the magnetic flux (in Webers) linked with the circuit when 1 ampere current flows through it.

(b) We have $\varepsilon = L\frac{dI}{dt} \quad \therefore \quad L = e/-\frac{dI}{dt}$, when $\frac{dI}{dt} = 1$ ampere/s, we have $L = \varepsilon$. Thus the self inductance of a circuit is the e.m.f. induced in the circuit when the rate of decay of current in circuit is unity. We must note that e.m.f. is in volts, current in amperes and L in Henries.

(c) We have $W = \frac{1}{2}LI_0^2 \quad \therefore \quad L = \frac{2W}{I^2}$. Obviously, the self inductance of a circuit is defined as twice the work done against the induced e.m.f. in establishing unit current in the coil. We must note that work is in joules, current in amperes and L in henries.

We note that the above three definitions of self inductance are identical only when the permeability of the medium round which the coil is wound remains constant. The role of self-inductance in an electric circuit is same as that of inertia in mechanical motion of bodies. Obviously, the self inductance of a coil is a measure of its ability to oppose the change in current through it.

7.6.1 Self Inductance of a Solenoid

Consider a long air cored solenoid having length l, cross sectional area A and the number of turns per unit length n. If I ampere is the current flowing through the solenoid, then the magnetic field inside the solenoid is $A = \mu_0 nI \frac{W}{m^2}$.

The magnetic flux linked with the solenoid is, $\phi = \mu_0 nI \times$ effective area of solenoid $= \mu_0 nI(nlA)$ Webers. ($\because$ Effective area of the solenoid = area of each turn × total number of turns in solenoid). Thus

$$\phi = \mu_0 n^2 AlI \tag{7.28}$$

But

$$\phi = LI$$

$$\therefore \quad LI = \mu_0 n^2 AlI \quad \text{or} \quad \mu_0 n^2 Al \text{ Henry.}$$

7.7 MUTUAL INDUCTANCE (COUPLED CIRCUITS)

Let us consider a coil (such as 1 in Fig. 7.10(a) carries a current I. If a second coil (2) (Fig. 7.10(a)) is brough near this coil there will be a magnetic flux ϕ_2 through coil (2) due to the current in coil (1). Since ϕ_2 is linearly proportional to I,

$$\phi_2 = L_{21} I_1 \tag{7.29}$$

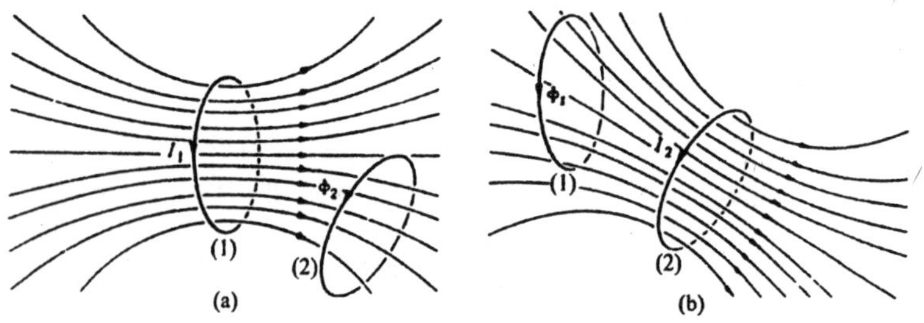

Figure 7.10. Mutual Induction

where L_{21} is a constant and represents the magnetic flux through circuit (2) per unit current in circuit (1). Similarly, if a current I_2 circulates in circuit (2), a magnetic field is produced, and it in turn produces a magnetic flux ϕ_1 through circuit (1) which is proportional to I_2 (Fig. 7.10(b)). Hence we may write

$$\phi_1 = L_{12} I_2 \tag{7.30}$$

where L_{12} is again a constant. Now, question arises what is the relationship between these two constants? The potential energy of the system is given by

$$U = -\phi_2 I_2 = -L_{21} I_1 I_2 = -\phi I_1 = -L_{12} I_2 I_1$$

$$\therefore \quad L_{21} = L_{12} = M \tag{7.31}$$

Calculation show (and experiments confirm) (7.31) in the absence of ferromagnetics, the magnetic flux through circuit (1) due to the unit current in circuit (2) is the same as before. This common coefficient (M) is called the *mutual inductance* of the two circuits and it is same in both cases. This remarkable property of mutual inductance is usually called the *reciprocity theorem.* Obviously, mutual inductance is symmetrical. The coefficient M depends on the shapes of the circuits and their relative orientation. It is also measured in henry, since it correspond to $\frac{Wb}{A}$.

Now

$$U = -\phi_1 I_1 = -I_1 \int B_1 \cdot \hat{a}_n ds$$

$$= I_1 \int \nabla \times A_{12} \cdot \hat{a}_n ds = -I_1 \int A_{12} \cdot \hat{a}_n dl_1$$

(By Stokes' theorem)

$$= -I_1 \int \left(\frac{\mu}{4\pi} \int \frac{I_2 \hat{a}_{12} dl_2}{|\mathbf{r}|} \right) \cdot \hat{a}_{l_1} dl_1$$

$$= -\frac{\mu}{4\pi} I_1 I_2 \iint \frac{\hat{a}_{l_1} dl_1 \cdot \hat{a}_{l_2} dl_2}{\mathbf{r}} = -L_{12} I_1 I_2$$

$$\therefore \quad L_{12} = L_{21} = M = \frac{\mu}{4\pi} \iint \frac{\hat{a}_{l_1} dl_1 \cdot a_{l_2} dl_2}{\mathbf{r}} \tag{7.32}$$

Relation (7.32) is known as *Neumann's formula.*

If the current I_1, is variable, the flux ϕ_2 through circuit (2) changes and an e.m.f. V_{M_2} is induced in this circuit. This e.m.f. is given by

$$V_{M_2} = -M \frac{dI_1}{dt}$$

In writing this equation, we assume that the circuits are rigid and fixed in space, so that M is constant. Similarly, if the current I_2 is variable, an e.m.f. V_{M_1} is induced in circuit (1), given by

$$V_{M_1} = -M \frac{dI_2}{dt} \tag{7.33}$$

This is why M is called "mutual inductance", since it describes the mutual effect or influence between the two circuits. In addition, if the circuits are moved relative to each other, resulting in a change in M, e.m.f.'s are also induced in them.

We see that there will be an exchange of energy between the circuits. Common and practical applications of this process are the *transformer* and the *induction generator*. Another application of mutual induction, in a broader sense, is the transmittal of a single from one place to another producing a variable current in one circuit, called the *transmitter*. This current in turn acts on another circuit coupled to it, called the *receiver*. This is the case for telegraph, radio, television, radar, etc.

The most important and fundamental aspect of mutual induction is that *energy can be exchanged between two circuits via the electromagnetic field*, one can say that the electromagnetic field produced by the currents in the circuits acts as a carrier of energy, transporting the energy through space from one circuit to the other.

We note that mutual induction between two circuits is a macroscopic phenomenon, resulting from elementary interactions between the moving charges which constitute their respective currents. Obviously we may conclude from the phenomenon of mutual induction that the electromagnetic interaction between two charged particles can be described as an exchange of energy via their mutual electromagnetic field.

7.8 ENERGY OF THE MAGNETIC FIELD

We have seen that to maintain a current in a circuit, energy must be supplied. Consider a simple circuit as shown in Fig. 7.11 in which R

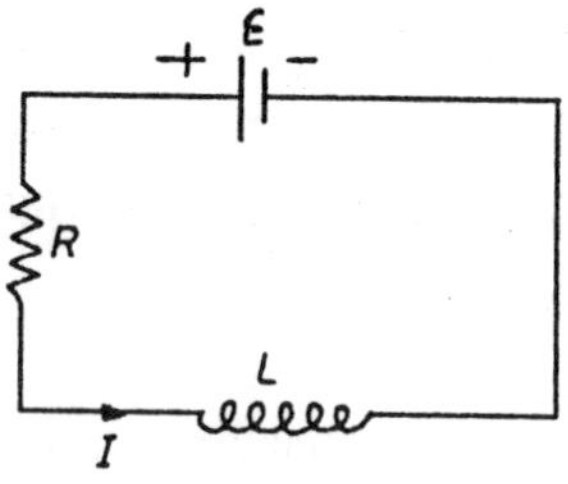

Figure 7.11.

is the resistance, L an inductive coil with self-inductance L and ε is the e.m.f. of the battery. Let us consider that the current flowing in the circuit at any time t be $I(t)$. We have the voltage drop across the coil is $L\frac{dI}{dt}$ and hence the forward e.m.f. is $= -\varepsilon - L\frac{dI}{dt}$. Using Ohm's law, we have

$$\varepsilon - L\frac{dI}{dt} = IR \tag{7.34}$$

Now, we consider the work done by the e.m.f. ε in moving a small amount of charge Q through the circuit

$$\begin{aligned} dW &= \varepsilon dQ = \varepsilon I dt \\ \therefore \quad \frac{dW}{dt} &= I\varepsilon = IL\frac{dI}{dt} + I^2R \end{aligned} \tag{7.35}$$

Therefore total work done by the battery in time interval T in which the current changes from 0 to I_T is obtained as

$$\begin{aligned} W &= \int_0^T I\varepsilon dt = L\int_0^T I\frac{dI}{dt}dt + R\int_0^T I^2 dt \\ &= \frac{1}{2}LI_T^2 + R\int_0^T I^2 dt \end{aligned} \tag{7.36}$$

The first term on the r.h.s. of (7.36) is the energy stored in the inductance in time T and the second term is the energy dissipated as heat in the resistance. Obviously the magnetic energy required to increase a current from zero to the value I is thus $\frac{1}{2}LI^2$.

7.8.1 Magnetic Energy Stored in a Series of Inductances

Let us consider a number of coils connected in series. Assume that the initial current in each coil is zero and that they attain their equilibrium value in all the circuits simultaneously at $t = T$. Obviously, at any instant t, in the interval $0 \le t \le T$, the current in each circuit $I_k(t)$ and the flux through it $\phi_k(t)$, will be some fraction α of their ultimate values, i.e.

$$I_k(t) = \alpha I_k \quad \text{and} \quad \phi_k(t) = \alpha\phi_k$$

The induced e.m.f. in the kth circuit is

$$\varepsilon_k = \frac{d\phi_k(t)}{dt}$$

$\therefore$ Total work done by applied source for k^{th} circuit,

$$W_k = \int_0^t \varepsilon_k I_k(t)dt = I_k\phi_k\int_0^T \alpha\frac{d\alpha}{dt}dt = I_k\phi_k\int_0^1 \alpha d\alpha = \frac{1}{2}I_k\phi_k \tag{7.37}$$

On summing for all circuits, we have

$$W = \frac{1}{2}\sum_k I_k \phi_k \tag{7.37a}$$

We know that the magnetic flux depends on the self as well as mutual inductances, and hence we have

$$\phi_k = L_k I_k + \sum_{j=k} M_{kj} I_j \tag{7.38}$$

$$\therefore \quad W = \frac{1}{2}\sum_k L_k I_k^2 + \frac{1}{2}\sum_k \sum_j M_{kj} I_k I_j \tag{7.39}$$

Thus, we have for a pair of coils

$$W = \frac{1}{2}L_1 I_1^2 + \frac{1}{2}L_2 I_2^2 + M I_1 I_2 \tag{7.40}$$

where $M = M_{12} = M_{21}$

One can express (7.39) in a more general and standard form. For any circuit

$$\phi_k = \int_k \mathbf{B}\cdot\hat{a}_n\, ds = \oint \nabla\times\mathbf{A}\cdot\hat{a}_n\, ds = \oint_k \mathbf{A}\cdot\hat{a}_l\, dl \tag{7.41}$$

Now, Eq. 7.37(a) can be expressed as

$$W = \frac{1}{2}\sum_k I_k \phi_k = \frac{1}{2}\sum_k I_k \oint_k \mathbf{A}\cdot\hat{a}_l\, dl$$

Let us now make a transition from the discrete to the continuous case by using the relation $I\hat{a}_l\, dl = j\, d\tau$ and taking the integral over all space, since contributions arise only from regions where j is finite, i.e. by using the relation

$$\sum_k \oint_k \longrightarrow \int_v$$

$$\therefore \quad W = \frac{1}{2}\int_v (\mathbf{A}\cdot\mathbf{J})\, d\tau \tag{7.42}$$

But $\mathbf{J} = \nabla\times\mathbf{H}$

$$\therefore \quad W = \frac{1}{2}\int_v \{\mathbf{A}\cdot(\nabla\times\mathbf{H})\}\, d\tau \tag{7.43}$$

Using vector identity

$$\nabla\cdot(\mathbf{A}\times\mathbf{H}) = \mathbf{H}\cdot(\nabla\times\mathbf{A}) - \mathbf{A}\cdot(\nabla\times\mathbf{H})$$

One obtains (7.43) as

$$W = \frac{1}{2}\int_v \{\mathbf{H}\cdot(\nabla\times\mathbf{A})\}d\tau - \frac{1}{2}\int_v \{\nabla\cdot(\mathbf{A}\cdot\mathbf{H})\}d\tau \tag{7.44}$$

Using Gauss's divergence theorem to change volume integral in second term on R.H.S of (7.44), one obtains

$$W = \frac{1}{2}\int_v \{\mathbf{H}\cdot\mathbf{B}\}d\tau - \frac{1}{2}\int_v (\mathbf{A}\times\mathbf{H})\cdot\hat{a}_n ds \tag{7.45}$$

As the volume integral is to be taken for all space, the surface integral must be taken over the sphere at infinity. Because **H** and **A** fall off rapidly at large distance ($H \sim r^{-3}, A \sim r^{-2}$ the surface integral vanishes as $r \to \infty$. Hence Eq. (7.45) reduces to

$$W = \frac{1}{2}\int_v (\mathbf{H}\cdot\mathbf{B})d\tau \tag{7.46}$$

Eq. (7.46) shows that the magnetic energy may be regarded as distributed throughout the region occupied by the field with density $\frac{1}{2}(\mathbf{H}\cdot\mathbf{B})$. As $\mathbf{B} = \mu\mathbf{H}$ for linear medium, we have the density $= \frac{1}{2}\mu H^2 = \frac{1}{2}\frac{B^2}{\mu}$. We must note that this relation has been obtained by assuming that medium is linear. If the medium is non-linear, then the analysis will have to be modified suitably.

7.9 HYSTERESIS

We shall now consider the behaviour of ferromagnetic substances such as iron, cobalt, gadolinium and their alloys and compounds under the action of a magnetic field. As pointed out earlier, whether magnetized or not, these substances are made up of *magnetic domains*, each of which has an average volume of the order of 10^{-15}m^3 and are completely magnetized. Ordinarily the magnetic axes of these domains are oriented in various directions. In an unmagnetized iron crystal the domains are parallel to the direction of easy magnetization which in an iron crystal is six. The application of the external field to ferromagnetics causes the walls of the domains to shift in such a way as to increase the sizes of the domains with magnetic axes in the direction of field and decrease the sizes of the domains which have unfavourable direction of the axes. Obviously, the pattern of the domains changes and the specimen exhibits magnetization. Increasing the intensity of the applied field increases the magnetization of the specimen. The magnetization of weakly magnetized substance (*dia* and para magnetic) varies linearly with the field strength. The magnetization of ferromagnetics depends on H in an intricate way, i.e. the relation between **H** and the flux density **B** of the field produced by a ferromagnetic specimen is not a simple direct proportion. In order to understand this relationship let us consider a magnetization experiment and examine its results.

Consider a toroidal coil wound around an iron core in the form of ring as shown in Fig. 7.12. Let us suppose that a second winding

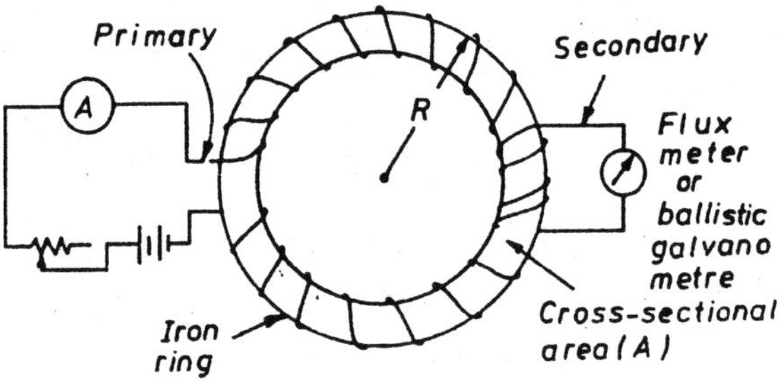

Figure 7.12. Rowland-ring method of obtaining magnetization curve

is placed on the ring with its terminals connected to a fluxmeter or ballistic galvanometer (B.G.) [See Appendix 7.1]. With the iron core specimen unmagnetized at the begining of the experiment, H may be increased in small increments, say $\Delta \mathbf{H}$. One can determine the corresponding increment $\Delta \mathbf{B}$ in **B** from the deflection of the B.G. A typical magnetization curve for a ferromagnetic material is shown in Fig. 7.13. From figure it is evident that as the magnetization current in the winding is steadily increased from zero until the magnetic intensity corresponds to the value corresponding to the point X_1 on the x-axis, the flux density is given by OY_1. We see that the curve is not a straight line and hence it is not possible to speak of a single value of the permeability (μ) of the substance. If now the magnetic intensity is increased further to OX_2 (at which point the magnetization curve levels off, i.e. ferromagnetic specimen (say iron) saturates) and then decreased to OX_1, one should expect the flux density to fall to OY_1. However, the magnetic state of the sample follows the path *OAPC* (Fig. 7.14) and the flux density when H is reduced to OX_1 is given by OY_2 rather than OY_1. Obviously, the flux density is different, although the magnetic intensity is the same. This shows that the magnetization curve is not

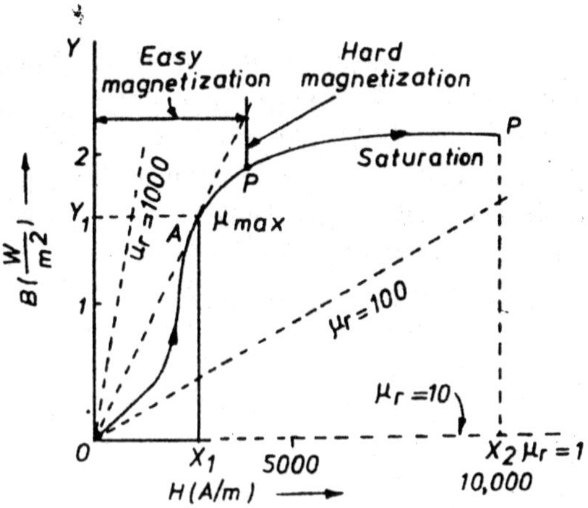

Figure 7.13.

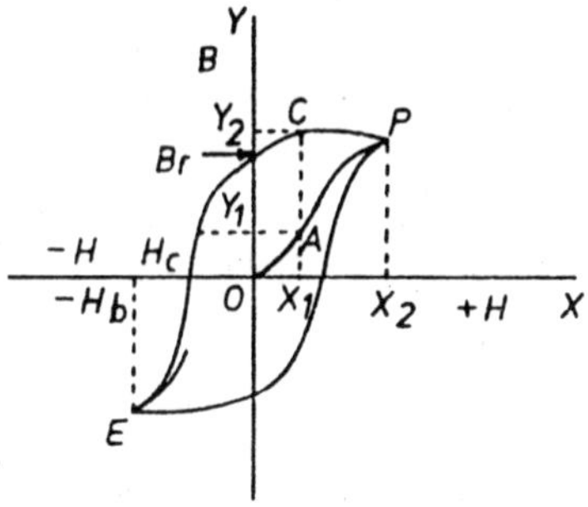

Figure 7.14. Hysteresis loop

reversible, i.e. an increment $-\Delta H$ will not produce an increment $-\Delta B$ of the same magnitude as was associated with $+\Delta H$ at the same point. Obviously, the flux density does not depend on the magnetic intensity alone but on the *magnetic history* of the sample as well. This particular behaviour, i.e. lag of the magnetization of a ferromagnetic substance behind the strength of the applied field is called hysteresis. This shows that once the atomic magnetic dipoles have been aligned in a particular direction, they get locked in that state and it takes a reverse field **H** to go back to the original value.

If after the core is saturated, **H** is reduced to zero, some magnetization still remains in the ferromagnetic specimen and **B** is reduced to a value $\mathbf{B}_r$. The magnitude of $\mathbf{B}_r$ is called the *remanence* or *retentivity* of the sample. Obviously, the specimen has now become a permanent magnet. One can see from the curve that a further change in **H** to $\mathbf{H}_c$ will be required to reduce **B** to zero. H_c is called the *coercive force* or the *corecivity*. A still further change of **H** to $-\mathbf{H}_b$ will cause the iron core to saturate again with **B** in the opposite direction. We can see that reversing **H** again will reverse the process and when iron is again saturated in the original direction, a closed loop is formed. This loop is called the '*hysteresis loop*'. If we alternate the current between large positive and negative values the B–H plot traces very nearly the same curve. It can be shown that the area enclosed by the hysteresis loop gives the work done in taking a unit volume of the material through the cycle represented by the loop. Though the main features of the hysteresis loop are common to all ferromagnetic materials, the actual shape of the loop can show considerable variation. Fig. 7.15(a) shows the B–H curve for magnetically hard iron. One can easily see that this curve is characterized by large values of remanence and of the coercive force. Fig. 7.15(b) shows the hysteresis loop for a specially made alloy, which can be taken almost to saturation level magnetization by a small magnetic intensity of the order of a few ampere-turns. We note that the coercive force for this alloy is small. One can use the pieces of such an alloy for recording a binary code of memory. For example, the two possible directions of magnetization of a pin made of such an alloy can indicate 1 and 0 respectively. A series of such pins can then be magnetized in the required directions to read any number. For e.g. 10111 could be represented by a sequence $\uparrow\downarrow\uparrow\uparrow\uparrow$ indicates the number 23 in the binary code. Erasing a recorded number and writing another would then involve switching on magnetic fields in the appropriate directions to magnetize the pins in the required directions. We must note that the magnetization would remain even when the fields are switched off. Instead of pins, we can use small circular rings of magnetic material, which can be magnetized either clockwise or anticlockwise direction to denote 1 or 0 respectively. With such an arrangement one could express the number 23 as a sequence $\curvearrowright\curvearrowleft\curvearrowright\curvearrowright\curvearrowright$. The possible way of demagnetizing a sample of ferromagnetic material is shown in Fig. 7.15(c). We can see from the B–H curve that switching off the magnetic intensity does not lead to zero magnetization. One can demagnetize a sample by taking it through a series of hysteresis loops of diminishing amplitudes. Starting from a state indicated by Q (Fig. 7.15(c)), the sample goes through cycles of smaller and smaller size until the magnetization is sufficiently small to be considered is zero. One can also demagnetize a sample by heating it to a sufficiently high temperature, where the ferromagnetic properties of the material vanish and in the absence of an external magnetic intensity, the sample magnetization becomes zero.

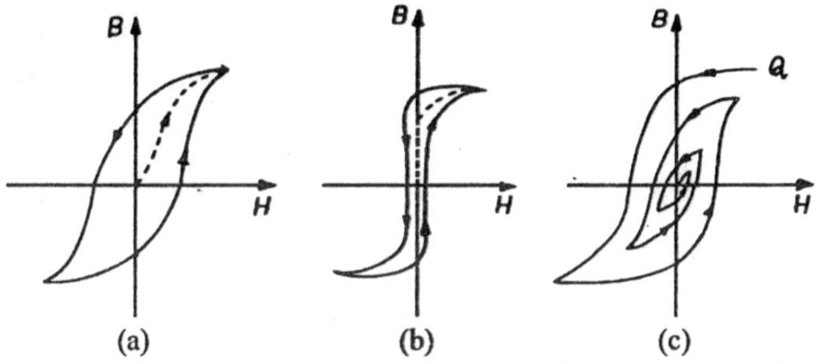

Figure 7.15. (a) B–H loop for magnetically hard iron. (b) B–H loop for a specially made alloy having a small coercive force. (c) Method of demagnetization of a sample of iron

Steinmetz obtained an empirical relationship between the area of a hysteresis loop and the maximum value of B attained in a loop (B_m). The area of a hysteresis loop represents the work done in taking unit volume of the sample through one hysteresis cycle and as such one can express the Steinmetz relationship as

$$\text{Work done per unit volume} = \eta(B_m)^n \tag{7.47}$$

where n has the value 1.6 for a large number of materials and n is the Steinmetz coefficient which is characteristic of a given material. One can estimate the order of magnitude knowing n and η.

When a specimen of a ferromagnetic material is taken through a cycle of magnetization, it results into an expenditure of irrecoverable energy. As the elementary magnets within the iron core try to line up with **H**, first one way and then the other, the tendency to turn around causes to develop mechanical stresses in the iron specimen; those, in turn produce heating. One can easily show that the energy dissipated per unit volume $\frac{dU}{d\tau}$ in each magnetization cycle, is proportional to the area enclosed by the hysteresis loop.

Let us consider the case when the current in the coil is I and the magnetic flux through the circuit is ϕ. The instantaneous induced e.m.f. in the circuit is equal to $-\frac{d\phi}{dt}$. Now, the rate at which work is being done against the induced e.m.f. by the source supplying the current is obtained as

$$\frac{dU}{dt} = I\frac{d\phi}{dt} \tag{7.48}$$

or

$$dU = I\frac{d\phi}{dt}dt$$

Using $\phi = \int_S \mathbf{B}\cdot\mathbf{dS}$, one obtains

$$dU = -\int \frac{\partial \mathbf{B}}{\partial t}\cdot I\mathbf{dS}dt = -\int \nabla\times\mathbf{E}\cdot I\mathbf{dS}dt \tag{7.49}$$

Using Stoke's theorem, we have

$$dU = -\oint \mathbf{E}\cdot I\mathbf{dl}dt$$

Since $I\,dl = Jd\tau$, we have

$$dU = -\int \mathbf{E}\cdot\mathbf{J}d\tau dt = -\int \mathbf{E}\cdot\nabla\times\mathbf{H}d\tau\, dt \tag{7.50}$$

Making use of vector identity

$$\nabla\cdot(\mathbf{A}\times\mathbf{B}) = B\cdot\nabla\times\mathbf{A} - \mathbf{A}\cdot\nabla\times\mathbf{B}$$

One obtains

$$dU = -\int \mathbf{H}\cdot\nabla\times\mathbf{E}d\tau dt + \int \nabla\cdot(\mathbf{E}\times\mathbf{H})d\tau dt \tag{7.51}$$

Using divergence theorem, one can transform the second integral in (7.51). Finally one obtains,

$$dU = -\int \mathbf{H}\cdot\nabla\times E d\tau dt + \int (\mathbf{E}\times\mathbf{H})\cdot dS dt \tag{7.52}$$

The second integral in (7.52) vanishes, as **E** and **H** vanish over a surface very far away from the circuit. Thus we have

$$\begin{aligned} dU &= -\int \mathbf{H}\cdot\nabla\times\mathbf{E}d\tau dt = \int \mathbf{H}\cdot\frac{\partial\mathbf{B}}{\partial t}d\tau dt \\ &= \mathbf{H}\cdot d\mathbf{B}d\tau \end{aligned} \tag{7.53}$$

or

$$W = \frac{dU}{d\tau} = \int \mathbf{H}\cdot d\mathbf{B} \tag{7.54}$$

Obviously, the work done per unit volume of the material per cycle is equal μ_0 times the area of $\mathbf{M}-\mathbf{H}$ loop or the area of the $\mathbf{B}-\mathbf{H}$ loop. The work is measured in J/m^3 per cycle and is dissipated in the form of heat. The above result further reveals that if a ferromagnetic material is to be subjected to a field which is continuously reversing its direction, e.g. transformer, it is desirable that the hysteresis loop of the material shall be narrow to minimize losses.

7.10 MAXWELL'S EQUATIONS

Maxell's theory and field equations are systematic generalizations of the fundamental laws describing electric and electromagnetic phenomena. Until Maxwell's work, the known basic laws of electricity and magnetism were:

(i) Gauss's law for electrostatics

$$\nabla \cdot \mathbf{D} = \rho \tag{7.55}$$

(ii) Gauss's law for magnetic field

$$\nabla \cdot \mathbf{B} = 0 \tag{7.56}$$

(iii) Faraday's law for electromagnetic induction

$$\nabla \times \mathbf{E} = -\frac{\partial \mathbf{B}}{\partial t} \quad \text{and} \tag{7.57}$$

(iv) Ampere's law for magneto motive force

$$\nabla \times \mathbf{H} = \mathbf{J} \tag{7.58}$$

Eqs. (7.55) to (7.57) enable us to calculate the divergence and curl of the electric field in any situation. If the charge density is zero, and there are no time varying magnetic fields, one obtains

$$\nabla \cdot \mathbf{E} = 0 \quad \text{and} \quad \nabla \times \mathbf{E} = 0$$

as special cases of (7.55) and (7.57) respectively.

Mathematically it can be shown that if the divergence and the curl of a vector field are known everywhere, that field can be calculated at any point. Eqs. (7.55) and (7.57) enable us to do this for the electric field. Obviously, Eqs. (7.55) to (7.57) are general equations and are valid for static as well as dynamic field. Eq. (7.58), i.e. Ampere's law was derived from steady state observations and one will have to examine its validity for time varying fields.

Taking the divergence of both sides of Eq. (7.58), one obtains

$$\nabla \cdot (\nabla \times \mathbf{H}) = \nabla \cdot \mathbf{J} = 0 \tag{7.59}$$

However, Eq. (7.59) is true for steady-state phenomena only. However, when the currents are changing with time, the result is incompatible with the principle of conservation of charge as evident in the equation of continuity

$$\nabla \cdot \mathbf{J} + \frac{\partial \rho}{\partial t} = 0 \tag{7.60}$$

Maxwell appreciated this and suggested that Ampere's law must be modified when dealing with time-dependent fields Maxwell realised that the difficulty arose from an incomplete definition of total current density in Eq. (7.58) by putting forward an argument that the current density consists of two parts: (i) first part is $\mathbf{J}_f$, the free current density developed due to the motion of free charge carriers and (ii) $\mathbf{J}_d$ developed due to the time varying electric field. Thus the total current density is

$$\mathbf{J} = \mathbf{J}_f + \mathbf{J}_d \tag{7.61}$$

Using (7.61) and taking divergence of (7.58), one obtains

$$\nabla \cdot \nabla \times \mathbf{H} = \nabla \cdot (\mathbf{J}_f + \mathbf{J}_d) = 0$$

or

$$\nabla \cdot \mathbf{J}_d = -\nabla \cdot \mathbf{J}_f \tag{7.62}$$

We have from the equation of continuity,

$$\nabla . \mathbf{J}_f = \frac{\partial \rho}{\partial t} = 0$$

$$\therefore \quad \nabla \cdot \mathbf{J}_d = \frac{\partial \rho}{\partial t} \tag{7.63}$$

Using (7.55), one obtains from (7.63)

$$\nabla \cdot \mathbf{J}_d = \frac{\partial}{\partial t}(\nabla \cdot \mathbf{D})$$

or

$$\nabla \cdot \mathbf{J}_d = \nabla \cdot \frac{\partial \mathbf{D}}{\partial t} \quad (\because \text{Space and time coordinates are independent})$$

$$\therefore \quad \mathbf{J}_d = \frac{\partial \mathbf{d}}{\partial t} \tag{7.64}$$

Obviously, $\mathbf{J}_d$ is developed due to change in **D** or electric field **E** with respect to time. It is called '*displacement current*'. Maxwell replaced **J** (we have also denoted it by $\mathbf{J}_f$) by $\mathbf{J} + \frac{\partial \mathbf{D}}{\partial t}$ in Ampere's law (7.59) and obtained

$$\nabla \times \mathbf{H} = \mathbf{J} + \frac{\partial \mathbf{D}}{\partial t} \tag{7.65}$$

This is Ampere-Maxwell law. This expresses a relation between the electric current at a point in space and the electric and magnetic fields at the same point in space and magnetic field at the same point. In free space, where there are no currents, $\mathbf{J} = 0$ and Eq. (7.65) become

$$\nabla \times \mathbf{H} = \frac{\partial \mathbf{D}}{\partial t} \tag{7.66}$$

or

$$\nabla \times \mathbf{B} = \mu_0 \varepsilon_0 \frac{\partial \mathbf{E}}{\partial t} \tag{7.66a}$$

This equation shows the relationship between the magnetic field and the time rate of change of the electric field at the same point.

We may observe in Eq. (7.65) that the effect of a time dependent electric field is to add to the current a density a term $\frac{\partial \mathbf{D}}{\partial t}$. Maxwell interpreted this as an additional current and called it the '*displacement current*'. Maxwell's reasoning was the following. In a circuit containing a capacitor C (Fig. 7.16), the current I is interrupted by the capacitor C. In order to "close" the circuit, there must be a current from one plate to the other, and this current is just $\left(\frac{\partial \mathbf{D}}{\partial t}\right) S$, where $\mathbf{E}(\mathbf{D} = \varepsilon_0 \mathbf{E})$ is the electric field within the capacitor and S its surface area. However, the term "*displacement current*" is misleading and Maxwell's picture unnecessary, since there is no such current between the places of the capacitor, and Eq. (7.65) simply expresses a correlation between **D**, **H** and **J** at the same point in space.

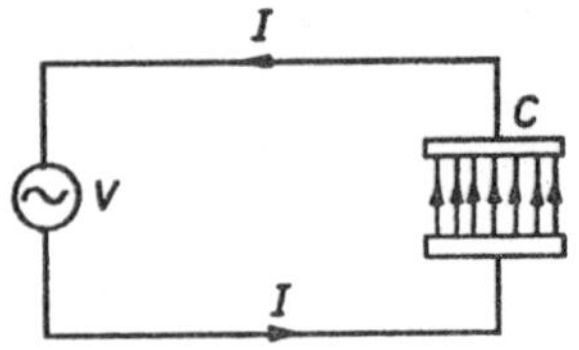

Figure 7.16. Maxwell's electric displacement current in a capacitor

To have an idea of the relative size of the two types of currents in conductors, let us consider a copper wire in which there is an electric field $E = E_0 \exp(-i\omega t)$ we have

$$\mathbf{J} = \sigma \mathbf{E} = \mathbf{E}_0 \exp(-i\omega t)$$

$$\therefore \quad |\mathbf{J}| = \sigma^2 |\mathbf{E}_0|^2$$

and

$$\mathbf{D} = \varepsilon_0 \mathbf{E} = \varepsilon_0 \mathbf{E}_0 \exp(-i\omega t)$$

$$\therefore \quad \frac{\partial \mathbf{D}}{\partial t} = i\omega \varepsilon_0 \mathbf{E}_0 e^{-i\omega t}$$

Hence
$$\left|\frac{\partial \mathbf{D}}{\partial t}\right|^2 = \omega^2 \varepsilon_0^2 |E_0^2|$$

$\therefore$
$$\frac{\mathbf{J}}{\left|\frac{\partial \mathbf{D}}{\partial t}\right|} = \frac{\sigma}{\omega \varepsilon_0}$$

For copper, we have $\sigma = 5.9 \times 10^7$. Therefore $\dfrac{\sigma}{\omega\varepsilon_0} \sim \dfrac{10^{10}}{\omega}$. Obviously, the ratio is very large for all frequencies. This shows that displacement currents are not as significant as the currents due to the motion of free charges in the study of the continuous circuits. However, displacement currents have far reaching consequences in other respects. It is mainly due to the idea of the displacement current that led to uncover a rich variety of new electromagnetic phenomena.

We are now in a position to write down all the four equations of Maxwell in *differential* as well as in *integral form*. There are

(i) *Maxwell's first equation*: (*Gauss's law in electrostatics*)

$$\nabla \cdot \mathbf{E} = \frac{\rho}{\varepsilon_0} \quad \text{or} \quad \nabla \cdot \mathbf{D} = \rho \tag{7.67}$$

The integral form of the above law is

$$\oint \mathbf{E} \cdot d\mathbf{S} = \frac{1}{\varepsilon_0} \quad \text{(net charge enclosed)} \tag{7.67a}$$

In terms of displacement vector **D**, we can write

$$\oint \mathbf{D} \cdot d\mathbf{S} = \text{charge enclosed} \tag{7.67b}$$

(ii) *Maxwell's second equation*: *Solenoidal nature of the magnetic induction* **B**

$$\nabla \cdot \mathbf{B} = 0 \quad \text{or} \quad \oint \mathbf{B} \cdot d\mathbf{S} = 0 \tag{7.68}$$

This indicates that **B** is a solenoidal field. Obviously, the source density of the field **B** is always zero. This does not, of course, mean that the sources of **B** does not exist. In other words lines of **B** are closed loops, i.e. magnetic monopoles cannot be accomodated in this formalism.

(iii) *Maxwell's third equation*: *Faraday's law of electromagnetic Induction*:

$$\nabla \times \mathbf{E} = -\frac{\partial \mathbf{B}}{\partial t} \quad \text{or} \quad \oint \mathbf{E} \cdot d\mathbf{l} = -\frac{\partial}{\partial t}\int \mathbf{B} \cdot d\mathbf{S} \tag{7.69}$$

This equation essentially states that a time varying magnetic field produces an electric field.

(iv) *Maxwell's fourth equation: Ampere's Law*

$$\nabla \times \mathbf{H} = \mathbf{J} + \frac{\partial \mathbf{D}}{\partial t}$$

or

$$\nabla \times \mathbf{B} = \mu_0 \mathbf{J} + \mu_0 \frac{\partial \mathbf{D}}{\partial t}$$

and in integral form

$$\oint \mathbf{B} \cdot d\mathbf{l} = \mu_0 \int \mathbf{J} \cdot d\mathbf{S} + \mu_0 \varepsilon_0 \int \mathbf{E} \cdot d\mathbf{S} \tag{7.70}$$

The importance of the *displacement current density*, $\mathbf{J}_d$ becomes apparent when we consider the above equation for the special case when $\mathbf{J} = 0$, i.e. when there are no true currents present. Eq. (7.70) then takes the form

$$\nabla \times \mathbf{B} = \mu_0 \frac{\partial \mathbf{D}}{\partial t} = \mu_0 \varepsilon_0 \frac{\partial \mathbf{E}}{\partial t} = \frac{1}{c^2} \frac{\partial \mathbf{E}}{\partial t} \quad \left(\because c^2 = \frac{1}{\mu_0 \varepsilon_0} \right) \tag{7.71}$$

Together with Eq. (7.69), we have now the pair of equation

$$\nabla \times \mathbf{E} = -\frac{\partial \mathbf{B}}{\partial t}$$

and

$$\nabla \times \mathbf{B} = \frac{1}{c^2} \frac{\partial \mathbf{E}}{\partial t}$$

we can easily see that except for the negative sign and the factor C^2, the two equations are almost symmetrical in **E** and **B**. These equations show that a time-varying magnetic (electric field) produces an electric (magnetic) field. Now, one can understand that the concept of the displacement current was essential to arrive at the above two equations. In the next chapter it will be shown that the idea of displacement current makes it possible to understand the formation of electromagnetic waves in free space, which is one of the most exciting and beautiful phenomena in nature.

Maxwell's equations are used in integral or in differential form, depending on the problem to be solved. These equations are compatible with the principle of relativity in the sense that they remain invariant under a Lorentz transformation.

The synthesis of electromagnetic interactions is expressed by Maxwell's equation is one of the greatest achievements in physics, and that is what places these interactions in a unique position.

We must recognize, however, that Maxwell's equations, as they have been presented have their limitations. They work very well when dealing with electromagnetic interactions between large aggregates of charges, such as radiating antennas, electric circuits, and even beams of ionized atoms or molecules. But it has been found that the electromagnetic interactions between fundamental particles (especially at high energies) must be treated in a somewhat different way, and according to the laws of quantum mechanics, constituting a technique called '*quantum electrodynamics*'. Even granted these limitations, the results derived from the form of Maxwell's equation are an excellent approximation for describing electromagnetic interactions between elementary particles. This method is called classical electrodynamics.

7.11 DECAY OF FREE CHARGE

We can use Maxwell's equations for one of the important deduction regarding the decay of free charges.

Writing Eq. (7.70) as

$$\nabla \times \mathbf{H} = \mathbf{J} + \frac{\partial \mathbf{D}}{\partial t} = \sigma \mathbf{E} + \varepsilon \frac{\partial \mathbf{E}}{\partial t} \tag{7.72}$$

Taking the divergence on both the sides of Eq. (7.72) and assuming that σ and ε are constants one obtains

$$\begin{aligned} \nabla \cdot \nabla \times \mathbf{H} &= \sigma \nabla \cdot \mathbf{E} + \varepsilon \frac{\partial}{\partial t}(\nabla \cdot \mathbf{E}) = \frac{\sigma \rho}{\varepsilon} + \frac{\partial \rho}{\partial t} \\ &= 0 \qquad (\because \quad \nabla \cdot \mathbf{D} = \rho) \end{aligned}$$

Integrating the above equation, one obtains

$$\rho = \rho_0 \exp\left(-\frac{t}{\tau}\right) \tag{7.73}$$

where

$$\tau = \frac{\varepsilon}{\sigma} \tag{7.74}$$

τ is known as relaxation time. Relation (7.73) shows that any original distribution of charge decays exponentially at a rate which is independent of any other electromagnetic disturbances that may be taking place.

Obviously, any excess of charge density in the interior of a copper-conductor $\left(\sigma = \dfrac{58 \times 10^6}{\Omega m}\right)$ will disappear with a time constant

$$\tau = \frac{\varepsilon_0}{\sigma} = \frac{8.85 \times 10^{-12}}{58 \times 10^6} = 1.5 \times 10^{-19} \text{ s.}$$

7.12 POTENTIALS OF ELECTROMAGNETIC FIELD

One can obtain the complete description of an electromagnetic field by solving Maxwell's four equations. If one writes these equations in a suitable form then the process becomes simpler. For this purpose, we introduce new quantities called "*electromagnetic potentials*". With the help of these quantities one can easily reduce the number of equations. We have seen that the electrostatic fields can be expressed in terms of scalar potential (ϕ) as $\mathbf{E} = -\nabla\phi$ and the magnetic fields in terms of a vector potential (**A**) as $\mathbf{B} = \nabla \times \mathbf{A}$. Let us consider potentials in electromagnetic fields when electric and magnetic fields are time-varying.

One finds that in the time-dependent field, the Maxwell equation

$$\nabla \cdot \mathbf{B} = 0 \tag{7.75}$$

still holds and hence one can express **B** in terms of a vector potential **A**, i.e.

$$\mathbf{B} = \nabla \times \mathbf{A} \tag{7.76}$$

Let us now consider another Maxwell's equation which does not involve any currents or charges, i.e.

$$\nabla \times \mathbf{E} = -\frac{\partial \mathbf{B}}{\partial t} = -\frac{\partial}{\partial t}(\nabla \times \mathbf{A}) = -\nabla \times \frac{\partial \mathbf{A}}{\partial t}$$

$$\therefore \quad \nabla \times \left(\mathbf{E} + \frac{\partial \mathbf{A}}{\partial t}\right) = 0 \tag{7.76}$$

We know that the curl of the gradient of a scalar function vanishes and hence the quantity within the brackets can be expressed as a gradient of a scalar function ϕ, i.e.

$$\mathbf{E} + \frac{\partial \mathbf{A}}{\partial t} = -\nabla\phi$$

or

$$\mathbf{E} = -\nabla\phi - \frac{\partial \mathbf{A}}{\partial t} \tag{7.77}$$

Obviously, we have solved the two Maxwell's homogeneous equations in terms of A and ϕ. Once A and ϕ are determined, one can obtain **B and E.**

We must note that Eq. (7.76) does not completely define **A**. Let us add the gradient of any arbitrary scalar function to the vector potential, i.e. we change **A** to

$$\mathbf{A}' = \mathbf{A} + \nabla\Psi \tag{7.78}$$

We must note that the magnetic field remains unchanged. In order that the addition of $\nabla\Psi$ should not affect the electric field, the scalar potential ϕ must also be simultaneously transformed to ϕ, where

$$\phi' = \phi - \frac{\partial\psi}{\partial t} \tag{7.79}$$

One can easily verify this by substituting $\mathbf{A}'$ and ϕ' in Eq. (7.77). Thus any physical law that can be expressed in terms of the electromagnetic potentials A and ϕ remains unaffected by the transformations of the type (7.78) and (7.79). These transformations are called gauge transformations. Obviously, equations involving these potentials must be gauge invariant.

The condition $\nabla\cdot\mathbf{A} = 0$ in electrostatics together with $\mathbf{B} = \nabla\times\mathbf{A}$ specifies **A**. However, one will have to make a different choice in electromagnetism. In order to specify **A**, one has to impose an additional condition on **A** which does not change the physics, i.e. it must remain consistent with the transformations (7.78) and (7.79). Obviously, **E** and **B** must remain unaffected. For this purpose, let us consider the remaining two inhomogeneous equations of Maxwell viz.(7.67) and (7.70). Now, substituting (7.77) in (7.67), one obtains

$$\nabla\cdot\mathbf{D} = \nabla\cdot\varepsilon_0\mathbf{E} = \varepsilon_0\nabla\cdot\left(-\nabla\phi - \frac{\partial\mathbf{A}}{\partial t}\right) = \rho$$

or

$$-\nabla^2\phi - \frac{\partial}{\partial t}(\nabla\cdot\mathbf{A}) = \frac{\rho}{\varepsilon_0} \tag{7.80}$$

one can express Eq. (7.70) as

$$\nabla\times\mathbf{H} - \frac{\partial\mathbf{D}}{\partial t} = \mathbf{J}$$

or $$\nabla\times\frac{\mathbf{B}}{\mu} - \varepsilon_0\frac{\partial\mathbf{E}}{\partial t} = \mathbf{J}$$

or $$\frac{1}{\mu}(\nabla\times\nabla\times\mathbf{A}) - \varepsilon_0\frac{\partial}{\partial t}\left(-\nabla\phi - \frac{\partial\mathbf{A}}{\partial t}\right) = \mathbf{J}$$

or $$\nabla\times\nabla\times\mathbf{A} - \mu\varepsilon_0\frac{\partial}{\partial t}\left(-\nabla\phi - \frac{\partial\mathbf{A}}{\partial t}\right) = \mu\mathbf{J}$$

or $$-\nabla^2\mathbf{A}+\nabla(\nabla\cdot\mathbf{A})+\mu\varepsilon_0\frac{\partial}{\partial t}\nabla\phi+\mu\varepsilon_0\frac{\partial^2\mathbf{A}}{\partial t^2}=\mu\mathbf{J} \qquad (7.81)$$

In obtaining (7.81), we have used the vector identity

$$\nabla\times\nabla\times\mathbf{A}=\nabla(\nabla\cdot\mathbf{A})-\nabla^2\mathbf{A}$$

Now, we determine the condition for **A**. Let us choose **A** and ϕ such that

$$\nabla\cdot\mathbf{A}=-\mu\varepsilon_0\frac{\partial\phi}{\partial t}=-\frac{1}{c^2}\frac{\partial\phi}{\partial t} \qquad (7.82)$$

One can see that with the substitution of (7.82), the two middle terms of (7.81) cancel and one obtains

$$\nabla^2\mathbf{A}+-\mu\varepsilon_0\frac{\partial^2\mathbf{A}}{\partial t^2}=-\mu\mathbf{J} \qquad (7.83)$$

Using (7.82), Eq. (7.80) becomes

$$-\nabla^2\phi-\frac{\partial}{\partial t}(\nabla\cdot\mathbf{A}) = -\nabla^2\phi+\mu\varepsilon_0\frac{\partial^2\phi}{\partial t^2}=\rho/\varepsilon_0$$

or $$\nabla^2\phi-\mu\varepsilon_0\frac{\partial^2\phi}{\partial t^2} = -\frac{\rho}{\varepsilon_0} \qquad (7.84)$$

We have obtained two independent equations one for **A** (Eq. (7.10)) and other for ϕ (Eq. (7.84)). We note that **A** is connected with **J** and ϕ with scalar ρ. Moreover, both the equations have the same form, i.e. both potentials satisfy the same equations. This shows that the condition introduces complete symmetry between the scalar and vector potentials. For the steady-state, the time derivatives vanish and obtains

$$\nabla^2\mathbf{A}=-\mu\mathbf{J} \quad \text{and} \quad \nabla^2\phi=-\frac{\rho}{\varepsilon_0} \qquad (7.85)$$

The condition (7.85) is called *Lorentz gauge condition*. The gauge used in magnetostatics viz. $\nabla\cdot\mathbf{A}=0$ is known as *Coulomb gauge.*

Transformation Eqs. (7.78) and (7.79) shows that the electric field **E** and the magnetic field **B** are invariant. The potentials thus transformed will have to satisfy the Lorentz condition. This means, the gauge function Ψ which so far remained arbitrary must satisfy a certain condition.

The original and the transformed potentials have to satisfy Lorentz condition. We have

$$\nabla\cdot\mathbf{A} + \mu\varepsilon_0\frac{\partial\phi}{\partial t}=0 \qquad (7.86)$$

and $$\nabla\cdot\mathbf{A}' + \mu\varepsilon_0\frac{\partial\phi'}{\partial t} = 0 \tag{7.87}$$

or $$\nabla\cdot(A + \nabla\Psi) + \mu\varepsilon_0\frac{\partial}{\partial t}\left(\phi - \frac{\partial\Psi}{\partial t}\right) = 0$$

or $$\nabla\cdot\mathbf{A} + \nabla^2\Psi + \mu\varepsilon_0\frac{\partial\phi}{\partial t} - \frac{\partial^2\Psi}{\partial t^2} = 0$$

Hence, we get

$$\nabla^2\Psi - \mu\varepsilon_0\frac{\partial^2\Psi}{\partial t^2} = 0 \tag{7.88}$$

Thus, one obtains the restricted gauge transformations as

$$\mathbf{A}' \to \mathbf{A} + \nabla\phi \tag{7.89}$$

$$\phi' \to \phi - \frac{\partial\Psi}{\partial t} \tag{7.90}$$

where Ψ satisfies the condition

$$\nabla^2\Psi - \mu\varepsilon_0\frac{\partial^2\Psi}{\partial t^2} = -0 \tag{7.91}$$

Obviously, this preserve the Lorentz condition.

We can easily see that the Lorentz gauge condition is not quite arbitrary as it appears at first sight. One can relate it to the first principles of electromagnetism viz. Coulomb's law, Biot-Savart's law and principle of conservation of energy.

We have seen while deriving Ampere's law from Biot-Savart's law for stationary current distribution, the Coulomb gauge condition $\nabla\cdot\mathbf{A} = 0$ appears a natural way. One can easily extend this derivation for quasistationary conditions.

One obtains the following expression using Biot-Savarat's law

$$\mathbf{A}(r_2) = \frac{\mu_0}{4\pi}\int_{v_1}\frac{\mathbf{J}(r_1)}{|\mathbf{r}_2 - \mathbf{r}_1|}d\tau \tag{7.92}$$

where the subscripts 1 and 2 refer to the source and field coordinates. Thus we have

$$\nabla_2\cdot\mathbf{A}(r_2) = \frac{\mu_0}{4\pi}\int_{v_1}\mathbf{J}(r_1)\cdot\nabla_2\left[\frac{1}{|\mathbf{r}_2 - \mathbf{r}_1|}\right]d\tau \tag{7.93}$$

We must note that ∇_2 operates on r_2 only.

We have

$$\nabla_2\left[\frac{1}{|\mathbf{r}_2 - \mathbf{r}_1|}\right] = -\nabla_1\left[\frac{1}{|\mathbf{r}_2 - \mathbf{r}_1|}\right]$$

or
$$\nabla_2 \cdot \mathbf{A}(r_2) = -\frac{\mu_0}{4\pi}\int_{v_1} \mathbf{J}(r_1)\cdot\nabla_1\left[\frac{1}{|\mathbf{r}_2-\mathbf{r}_1|}\right]d\tau$$
$$= \frac{\mu_0}{4\pi}\left[\int_{v_1}\left\{\frac{1}{|\mathbf{r}_2-\mathbf{r}_1|}\right\}\nabla_1\cdot\mathbf{J}(r_1)d\tau_1 - \int_{v_1}\nabla_{v_1}\cdot\left\{\frac{\mathbf{J}(r_1)}{|\mathbf{r}_2-\mathbf{r}_1|}\right\}\right]$$

With the help of divergence theorem, one can write the second integral for bounded source distribution as

$$\int_{v_1}\nabla_1\cdot\left\{\frac{\mathbf{J}(r_1)}{|\mathbf{r}_2-\mathbf{r}_1|}\right\}d\tau_1 = \int\frac{\mathbf{J}(r_1)}{|\mathbf{r}_2-\mathbf{r}_1|}\cdot\hat{a}_n\,dS = 0$$

$$\therefore \quad \nabla_2\cdot\mathbf{A}(r_2) = \frac{\mu_0}{4\pi}\int_{v_1}\left\{\frac{1}{|\mathbf{r}_2-\mathbf{r}_1|}\right\}\nabla_1\cdot\mathbf{J}(r_1)d\tau_1 \qquad (7.94)$$

The conservation of charge relation leads to

$$\nabla\cdot\mathbf{J} = -\frac{\partial\rho}{\partial t}$$
$$\therefore \quad \nabla_2\cdot\mathbf{A}(r_2) = -\frac{\mu_0}{4\pi}\int_{v_1}\left\{\frac{1}{|\mathbf{r}_2-\mathbf{r}_1|}\right\}\frac{\partial\rho(r_1)}{\partial t}d\tau_1 \qquad (7.95)$$

One obtains from Coulomb's law

$$\phi(r_2) = \frac{1}{4\pi\varepsilon_0}\int_{v_1}\frac{\rho(r_1)}{|\mathbf{r}_2-\mathbf{r}_1|}d\tau_1$$
$$\therefore \quad \nabla_2\cdot A(r_2) = -\mu_0\varepsilon_0\frac{\partial}{\partial t}\left\{\frac{1}{4\pi\varepsilon_0}\int_{v_1}\frac{e(r_1)}{|r_2-r_1|}d\tau_1\right\}$$
$$= -\mu_0\varepsilon_0\frac{\partial\phi(r_2)}{\partial t} \qquad (7.96)$$

Generalizing, one obtains

$$\nabla\cdot\mathbf{A} = -\mu_0\varepsilon_0\frac{\partial\phi}{\partial t} \qquad (7.97)$$

Obviously, (7.97) is the Lorentz Gauge Condition. Now, we have

$$\nabla_2\times\mathbf{B}(r_2) = \nabla_2\times\{\nabla_2\times\mathbf{A}(r_2)\} = \nabla_2\{\nabla_2\cdot\mathbf{A}(r_2)\} - \nabla_2^2\mathbf{A}(r_2)$$

One finds that the first term of the above is

$$-\nabla_2\left[\mu_0\varepsilon_0\frac{\partial\phi(r_2)}{\partial t}\right] = \mu_0\frac{\partial}{\partial t}[-\varepsilon_0\nabla_2\,\phi(r_2)]$$
$$= \mu_0\frac{\partial}{\partial t}(\varepsilon_0\mathbf{E}) = \mu_0\frac{\partial\mathbf{D}}{\partial t} \qquad (7.98)$$

$$\text{The second term} = -\nabla_2^2 \mathbf{A}(r_2) = -\frac{\mu_0}{4\pi}\int_{v_1} \mathbf{J}(r_1)\nabla_2^2\left\{\frac{1}{\mathbf{r}_2 - \mathbf{r}_1}\right\}d\tau$$
$$= \mu_0 \mathbf{J}(r_2) \tag{7.99}$$

Thus we have

$$\nabla \times \mathbf{B} = \mu_0\left(\mathbf{J} + \frac{\partial \mathbf{D}}{\partial t}\right) \tag{7.99a}$$

Obviously, this treatment also leads to the displacement.

7.13 ELECTROMAGNETIC FIELD ENERGY AND FIELD MOMENTUM

The general expressions obtained earlier for the electrostatic and magnetostatic field energies are

$$W_E = \frac{1}{2}\int_v (\mathbf{E}\cdot\mathbf{D})d\tau \quad \text{and} \quad W_M = \frac{1}{2}\int_v (\mathbf{B}\cdot\mathbf{H})d\tau \tag{7.100}$$

Now, we can find the expression for the electromagnetic energy in time dependent situations.

The Lorentz force on a moving q is given by

$$\mathbf{F} = q(\mathbf{E} + \mathbf{V} \times \mathbf{B}) \tag{7.101}$$

The rate of doing work on this charge is

$$\mathbf{F}\cdot\mathbf{V} = q(\mathbf{E} + \mathbf{V}\times\mathbf{B})\cdot\mathbf{V} = q(\mathbf{E}\cdot\mathbf{V}) \tag{7.102}$$

The magnetic field does no work as it is perpendicular to the velocity of the charge.

If there exists a continuous distribution of charge, the total rate at which the work is done in a given volume is

$$\int \rho(\mathbf{E}\cdot\mathbf{V})d\tau = \int (\mathbf{E}\cdot J)d\tau \tag{7.103}$$

Substituting for **J** from (7.70), one obtains

$$\int (\mathbf{E}\cdot\mathbf{J})d\tau = \int\left[\mathbf{E}\cdot(\nabla\times\mathbf{H} - \frac{\partial\mathbf{D}}{\partial t}\right]d\tau \tag{7.104}$$

But
$$\mathbf{E}\cdot(\nabla\times\mathbf{H}) = \nabla\cdot(\mathbf{H}\times\mathbf{E}) + \mathbf{H}\cdot(\nabla\times\mathbf{E})$$
$$= \nabla\cdot(\mathbf{H}\times\mathbf{E}) - \mathbf{H}\cdot\frac{\partial\mathbf{B}}{\partial t}$$

Therefore, one obtains

$$\int(\mathbf{E}\cdot\mathbf{J})d\tau = \int\left[\nabla\cdot(\mathbf{H})\times\mathbf{E}-\mathbf{E}\cdot\frac{\partial\mathbf{D}}{\partial t}-\mathbf{H}\cdot\frac{\partial\mathbf{B}}{\partial t}\right]d\tau$$

$$= \int[\nabla\cdot(\mathbf{H}\times\mathbf{E})]d\tau-\int\left(\mathbf{E}\cdot\frac{\partial\mathbf{D}}{\partial t}+\mathbf{H}\cdot\frac{\partial\mathbf{B}}{\partial t}\right)d\tau$$

One can transform the first integral by using divergence theorem. Thus, one obtains

$$\int_v(\mathbf{E}\cdot\mathbf{J})d\tau=\int_S(\mathbf{H}\times\mathbf{E})\cdot\hat{a}_n dS-\int_v\left(\mathbf{E}\cdot\frac{\partial\mathbf{D}}{\partial t}+\mathbf{H}\cdot\frac{\partial\mathbf{B}}{\partial t}\right)d\tau$$

We must note that S is the surface boundary of the volume V. We get

$$-\int_v\left(\mathbf{E}\cdot\frac{\partial\mathbf{D}}{\partial t}+\mathbf{H}\cdot\frac{\partial\mathbf{B}}{\partial t}\right)d\tau=\int_v(\mathbf{E}\cdot\mathbf{J})d\tau+\int_S(\mathbf{E}\times\mathbf{H})\cdot\hat{a}_n dS \quad (7.105)$$

If ε and μ are assumed to be constant, one obtains for the linear media

$$\int_v\left(\mathbf{E}\cdot\frac{\partial\mathbf{D}}{\partial t}+\mathbf{H}\cdot\frac{\partial\mathbf{B}}{\partial t}\right)d\tau = \frac{\partial}{\partial t}\int_v\frac{1}{2}(\mathbf{E}\cdot\mathbf{D}+\mathbf{B}\cdot\mathbf{H})d\tau$$

$$\therefore \quad -\frac{\partial}{\partial t}\int_v\frac{1}{2}[\mathbf{E}\cdot\mathbf{D}+\mathbf{B}\cdot\mathbf{H}]d\tau = \int(\mathbf{E}\cdot\mathbf{J})d\tau + \int(\mathbf{E}\times\mathbf{H})\cdot\hat{a}_n dS \quad (7.106)$$

We know that an electromagnetic field consists of electric and magnetic fields and hence it is reasonable to assume that the sum of the energies given in (7.100) represents the total electromagnetic energy. Obviously, one may write $\varepsilon_M = \frac{1}{2}[\mathbf{E}\cdot\mathbf{D}+\mathbf{B}\cdot\mathbf{H}]$ as the electromagnetic energy density. This means that one can interpret L.H.S. of (106) as the rate at which the energy stored in the electromagnetic field diminishes. The first term on the r.h.s. gives the work done by the field forces on the charges contained in the volume.

The dimensions of the vector $[\mathbf{E}\times\mathbf{H}]$ are of $\frac{\text{energy}}{\text{area}\times\text{time}}$ and hence we can interpret that the last term represent the flow of energy out of the boundary per unit time. Obviously, the vector $[\mathbf{E}\times\mathbf{H}]$ which gives the rate at which the energy flows across unit area of boundary, is called the *Poynting Vector*. Poynting vector is usually represented by the symbol N, i.e.

$$[\mathbf{E}\times\mathbf{H}]=\mathbf{N} \quad (7.107)$$

Obviously, **N** represents the power being transmitted through unit area held, normal to **N**. Writing Eq. (7.106) in differential form, we have

$$\frac{d\varepsilon_M}{dt} + \nabla \cdot \mathbf{N} = -\mathbf{E} \cdot \mathbf{J} \tag{7.108}$$

If the conductivity of the medium is zero, we have $\mathbf{J} = \sigma \mathbf{E} = 0$. Thus

$$\frac{d\varepsilon_M}{dt} + \nabla \cdot \mathbf{N} = 0 \tag{7.109}$$

We can see that Eq. (7.109) has exactly the same form as the equation of continuity (7.60). We note that here ε_M behaves as volume density of conserved quantity where as **N** is "current density". One can easily see that this analogy leads us to same conclusion as the identification of the flux of electromagnetic energy with the Poynting vector $\mathbf{N} = \mathbf{E} \times \mathbf{H}$.

Eq. (7.106) represents the law of conservation of energy, i.e. the decrease of electromagnetic energy per unit time in a certain volume V is equal to the work done by the field forces per unit time plus the flux flow outwords per unit time. This theorem is called as *Poynting theorem.*

We know that energy and momentum are closely related. One can easily show that an electromagnetic field possesses momentum. We have the force on a region containing both charges and currents is

$$\mathbf{F} = \int_V (\rho \mathbf{E} + \mathbf{J} \times \mathbf{B}) d\tau \tag{7.110}$$

Let the sum of the momenta of all the particles be P_{mech}, then

$$\frac{dP_{\text{mech}}}{dt} = \int_V (\rho \mathbf{E} + \mathbf{J} \times \mathbf{B}) d\tau \tag{7.111}$$

We have from Maxwell's equations

$$\rho = \nabla \cdot \mathbf{D} \quad \text{and} \quad \mathbf{J} = \nabla \times \mathbf{H} - \frac{\partial \mathbf{D}}{\partial t}$$

Using the above, Eq. (7.111) becomes

$$\begin{aligned} \frac{dP_{\text{mech}}}{dt} &= \int_V \left\{ \nabla \cdot \mathbf{D})\mathbf{E} + \left(\nabla \times \mathbf{H} - \frac{\partial \mathbf{D}}{\partial t} \right) \times \mathbf{B} \right\} d\tau \\ &= \int_V \left\{ \nabla \cdot \mathbf{D})\mathbf{E} + \mathbf{B} \times \frac{\partial \mathbf{D}}{\partial t} - \mathbf{B} \times (\nabla \times \mathbf{H}) \right\} d\tau \end{aligned}$$

Since $\frac{\partial}{\partial t}(\mathbf{D} \times \mathbf{B}) = \mathbf{D} \times \frac{\partial \mathbf{B}}{\partial t} + \frac{\partial \mathbf{D}}{\partial t} \times \mathbf{B},$ we have

$$\frac{dP_{\text{mech}}}{dt} = \int_v \left[(\nabla \cdot \mathbf{D})\mathbf{E} + \left(\mathbf{D} \times \frac{\partial \mathbf{B}}{\partial t}\right) - \frac{\partial}{\partial t}(\mathbf{D} \times \mathbf{B}) \right.$$
$$\left. - \mathbf{B} \times (\nabla \times \mathbf{H}) \right] d\tau$$

we have $\nabla \cdot \mathbf{B} = 0$. This means addition of $(\nabla \cdot \mathbf{B})\mathbf{H}$ to the square bracket does not alter the result.

$$\frac{dP_{\text{mech}}}{dt} + \frac{d}{dt}\int_v (\mathbf{D} \times \mathbf{B}) d\tau = \int_v [(\nabla \cdot \mathbf{B})\mathbf{E} + (\nabla \cdot \mathbf{B})\mathbf{E} + (\nabla \cdot \mathbf{B})\mathbf{H}$$
$$- \{\mathbf{D} \times (\nabla \times \mathbf{E})\} - \{\mathbf{B} \times (\nabla \times \mathbf{H})\}] d\tau$$

Here we have used $\nabla \times \mathbf{H} = -\frac{\partial \mathbf{B}}{\partial t}$ (7.112)

We can see that the integral in the second term of L.H.S. of (7.112) represents momentum. Further, we note that it is not associated with the mass of particles and consists only of fields, one can identify it as the electromagnetic moment (P_{field}). Writing $[\mathbf{D} \times \mathbf{B}] = \mathbf{g}$, we find that $\mathbf{g}$ is *electromagnetic momentum density.* One can convert R.H.S. of Eq. (7.112) into a surface integral and we find that it represent the momentum flow. One can easily see from (7.112) that the total momentum of the closed system consisting of a field and particles is conserved.

The relation between $\mathbf{g}$ and $\mathbf{N}$ is

$$\mathbf{g} = [\mathbf{D} \times \mathbf{B}] = [\varepsilon \mathbf{E} + \mu \mathbf{H}] = \mu\varepsilon \mathbf{N} \tag{7.113}$$

EXAMPLE 7.1: A plane circuit composed of N turns, each of area S, is placed perpendicular to an alternating uniform magnetic field that varies with time. The equation of the field is $B = B_0 \sin \omega t$. Show that the induced e.m.f. in the circuit is oscillatory and given by $V = -NSB_0\omega \cos \omega t$.

Solution: The flux through one turn of the circuit,

$$\phi = SB = SB_0 \sin \omega t$$

∴ Total flux through N turns

$$\phi' = \phi N = NSB_0 \sin \omega t$$

The induced e.m.f.

$$V = -\frac{d\phi'}{dt} = -NSB_0\omega \cos \omega t$$

Obviously, the induced e.m.f. is oscillatory or alternating with the same frequency as the magnetic field.

EXAMPLE 7.2: A circuit is composed of two coaxial metallic sheets of radii a and b, each carrying a current I, but in the opposite direction (Fig. 7.17). The space between the cylinders is filled with a substance whose permeability is R. Show that the self inductance per unit length of the circuit is

$$L = \frac{\mu}{2\pi} \log\left(\frac{b}{a}\right)$$

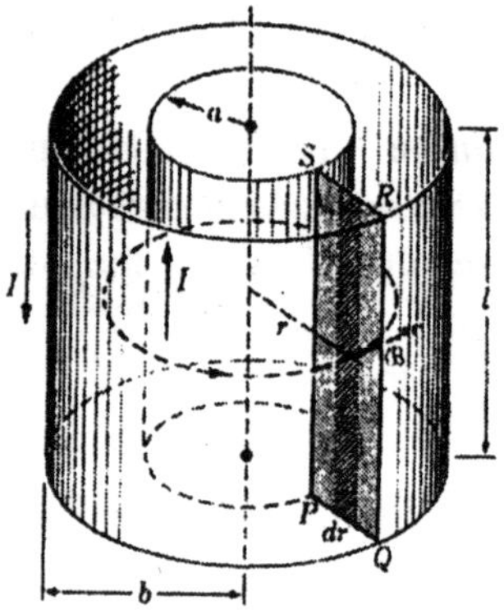

Figure 7.17.

Solution: One can show that magnetic field for this current arrangement is given by $B = \dfrac{\mu I}{2\pi r}$ in the region within two cylinders, and zero elsewhere. Here μ is the permeability of the medium filling the space between the two cylinders. To obtain the self inductance, one will have to compute the magnetic flux through any section of the conductor, such as $PQRS$, having a length l. Let us divide this section into strips of width dr, the area of each strip = ldr. The magnetic field B is perpendicular to $PQRS$. Therefore,

$$\phi = \int_{PQRS} B\,dS = \int_a^b \left(\frac{\mu I}{2\pi r}\right)(l dr)$$

$$= \frac{\mu I l}{2\pi}\int_a^b \frac{dr}{r} = \frac{\mu I l}{2\pi}\log\left(\frac{b}{a}\right)$$

$\therefore$ Self-inductance of a portion of length l is

$$L = \frac{\phi}{I} = \frac{\mu l}{2\pi}\log\left(\frac{b}{a}\right)$$

$\therefore$ Self inductance per unit length $\dfrac{L}{l} = L' = \dfrac{\mu}{2\pi}\log\left(\dfrac{b}{a}\right)$.

EXAMPLE 7.3: A superconducting ring of radius a and inductance L is in uniform magnetic field $\mathbf{B}$. In the initial position, the plane of the ring is parallel to the vector **B**, and the current in the ring is equal to zero. The ring is turned to the position perpendicular to vector **B**. Find the current in the ring in the final position and the magnetic induction at its centre.

Solution: The magnetic flux through the ring does not change upon its rotation and remains equal to zero. This means that the magnetic fluxes of the field of the induced current and of the external current through the ring are equal in magnitude and opposite in sign. Hence $LI = \pi a^2 B$, where

$$I = \pi a^2 B/L$$

This current creates a field $B_I = \pi\mu_0 aB/2L$ at the centre of the ring. The resultant magnetic induction at this point is given by

$$\mathbf{B}_{\text{res}} = \mathbf{B} - \mathbf{B}_1 = \mathbf{B}\left(1 - \frac{\pi\mu_0 a}{2L}\right)$$

EXAMPLE 7.4: Along ferromagnetic cylinder of volume V has two windings (one over the other). One winding contains n_1 turns per unit length, while the other contains n_2 turns. Ignoring the edge effects, find their mutual inductance.

Solution: According to Eq. (7.29), $L_{21} = \dfrac{\phi_2}{I_1}$. This means that we must create current I_1, in winding (1) and calculate the total magnetic flux through all the turns of winding (2). If winding (2) contains N_2 turns, one obtains

$$\phi_2 = N_2 B_1 S$$

where S is the cross-sectional area of the cylinder. Consider that $N_2 = n_2 l$, where l is the cylinder length, and $B_1 = \mu_1\mu_0 n_1 I_1$, where μ_1 is the magnetic permeability for current I_1, we write $\phi_2 = \mu_1\mu_0 n_1 n_2 V I_1$, where $V = lS$. Hence

$$L_{21} = \mu_1\mu_0 n_1 n_2 V$$

Similarly, one finds, $L_{12} = \mu_1\mu_0 n_1 n_2 V$. We must note that values of μ_1 and μ_2 in the last two expressions are generally different (they depend on currents I_1 and I_2 in ferromagnetics), the values of L_{21} and L_{12} do not coincide.

EXAMPLE 7.5: A coil containing N turns is wrapped around the central portion of a toroidal solenoid having n turns per unit length and a cross-section of area S. Calculate the mutual inductance of the system (Fig. 7.18).

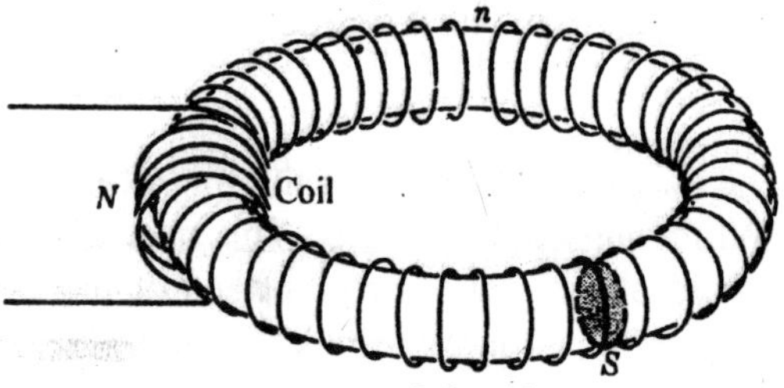

Figure 7.18.

Solution: One can solve the problem either by finding the magnetic flux through the solenoid when a current flows along the coil, or alternatively, by finding the magnetic flux through the coil when a current flows along the solenoid. We will follow the second procedure, which is easier. We have seen that in the case of a toroidal solenoid, the magnetic field is confined to its interior and has a value given by, $B = \mu_0 nI$. The magnetic flux through any cross section of the solenoid is

$$\phi = BS = \mu_0 nIS$$

where S is the cross-sectional area of the solenoid. This is same as the flux through any turn of the coil, even if its cross-section is larger. Therefore the magnetic flux through the coil is

$$\phi' = N\phi = \mu_0 nNSI$$

$$\therefore \qquad \text{Mututal Inductance } M = \mu_0 nNS$$

We must note that this arrangement is widely used in the laboratory when a standard mutual inductance is required.

EXAMPLE 7.6: An all metal aeroplane drives down vertically at 30 km/s at a place where the horizontal component of the earth field is 0.4 Oersted. If the wing span is 30 m, show that the resulting potential difference between the tips is 360 V.

Solution: The vertical distance covered in one sec.

$$= 30 \text{ km} = 30 \times 10^3 \text{ m}$$

The distance between the wing tips = 30 m. The area swept by the wing span in one sec.

$$\frac{dA}{dt} = 30 \times 10^3 \times 30 = 9 \times 10^5 \text{ m}^2/\text{s}.$$

$\therefore$ The potential difference between the wing tips

$$\begin{aligned} V = -\frac{d\phi}{dt} &= -\frac{d}{dt}(\mathbf{B} \cdot \mathbf{A}) = \frac{d}{dt}(HA) = H\frac{dA}{dt} \text{ (numerically)} \\ &= (0.4 \times 10^{-4} \text{W/m}^2) \times (9 \times 10^5 \text{m}^2/\text{s}) \\ &= 360 \text{ V}. \end{aligned}$$

EXAMPLE 7.7: A car ignition coil consists of two insulated coils, one of 16000 turns and the other of 400 turns, wound over each other. The length of each coil is 10 cm and the turns have the radius of 3 cm. A current of 3A is allowed to pass through the primary coil and switched off in 10^{-4}s. Show that the voltage induced in the secondary coil is 6814 V.

Solution: We have the magnetic flux density inside the solenoid

$$B = \mu_0 N_1 I_1$$

where I_1, is the current flowing through the coil and N_1 is the number of turns per unit length since the magnetic flux through each turn of the loop $= \mu_0 N_1 I_1 \pi r^2$ and hence the total flux

$$\phi_2 = \mu_0 N_1 N_2 l \pi r^2 I_1$$

where l is the length of the coil and N_2 is the number of turns per unit length of the second coil. The induced e.m.f. is given by

$$e = -\frac{d\phi_2}{dt} = -\mu_0 N_1 N_2 l \pi r^2 \frac{dI_1}{dt} = -M\frac{dI_1}{dt}$$

we have

$$\begin{aligned} \frac{dI_1}{dt} &= 3 \times 10^4 \text{ A/s} \\ \therefore \quad e &= 4\pi \times 10^{-7} \times 16 \times 10^4 \times 4 \times 10^3 \times 10^{-1}\pi \times (0.03)^2 \\ &\quad \times 3 \times 10^4 \\ &= 6814 \text{ V}. \end{aligned}$$

EXAMPLE 7.8: Show that the self inductance of two long parallel wires is given by

$$L = \frac{\mu}{\pi} \log_e \frac{d-r}{r}$$

where r is the radius of each wire and d is the distance between their axes through a medium of absolute permeability μ.

Solution: Fig. 7.19 shows two infinitely long wires A and B each of radius r. d is the distance between their axes and I ampere is the current flowing in each wire in opposite directions. Let us consider a thin strip of thickness dx between the wires at a distance x from the wire A.

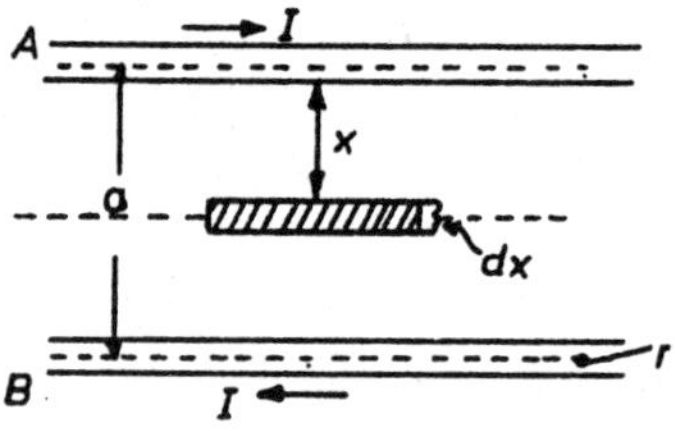

Figure 7.19.

The magnetic field at any point on the strip is given by

$$B = \frac{\mu I}{2\pi x} + \frac{\mu I}{2\pi(d-x)}$$

∴ Magnetic flux linked with the Unit length of the strip is

$$d\phi = B dx \qquad (\because \text{ For length } l \text{ we have } dQ = Blx)$$

$$= \left[\frac{\mu I}{2\pi x} + \frac{\mu I}{2\pi(d-x)}\right] dx$$

∴ Magnetic flux linked with unit length of the system is

$$\phi = \frac{\mu I}{2\pi} \int_r^{d-r} \left(\frac{1}{x} + \frac{1}{x-d}\right) dx$$

$$= \frac{\mu I}{2\pi} [\log_e x - \log_e (d-x)]_r^{d-r}$$

$$= \frac{\mu I}{2\pi} \left[\log_e \frac{d-r}{r} - \log_e \frac{r}{d-r}\right]$$

$$= \frac{\mu I}{\pi} \log_e \frac{d-r}{r} \qquad \text{(i)}$$

we have $\phi = LI$ (ii)

Comparing (i) and (ii)

$$L = \frac{\mu}{\pi} \log_e \frac{d-r}{r}$$

EXAMPLE 7.9: Show that the mutual inductance between two coplanar concentric rings of radii R_1 and R_2 respectively is given by

$$M = \mu_0 n_1 n_2 \pi R_2^2 / 2R_1$$

where n_1 and n_2 are turns of wire in outer and inner rings respectively.

Solution: Fig. 7.20 shows the arrangement of two rings. If the current flowing in the outer ring is I_1 ampere, then the magnetic field at the centre of the ring due to this current is

$$B_1 = \frac{\mu_0 I_1}{2R_1}$$

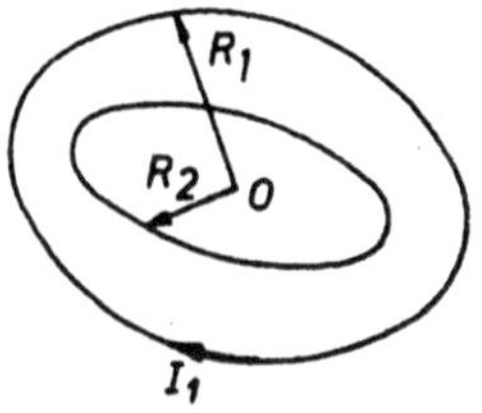

Figure 7.20.

The magnetic flux through the inner ring of radius R_2 is

$$\phi = \frac{\mu_0 I_1}{2R_1} \pi R_2^2 \tag{i}$$

We have assumed that $R_2 \ll R_1$ so that the magnetic field B_1 remains nearly uniform over this area.

We have the flux through the inner ring is

$$\phi = MI_1 \tag{ii}$$

where M is the mutual inductance between the two rings. Comparing (i) and (ii), one obtains

$$M = \frac{\mu_0 \pi R_2^2}{4R_1} \text{ henries}$$

The outer ring has n_1 number of turns of wire and the inner ring has n_2 turns. The magnetic field at the centre of the ring will be $n_1 B_1$ and the flux through the inner ring $n_1 B_1 \pi R_2^2 n_2$. Thus the mutual inductance between two rings $= \mu_0 n_1 n_2 \pi R_2^2 / 2R_1$

EXAMPLE 7.10: The self inductances of the primary and secondary coils of a transformer are 20 and 10 henry respectively. If the magnetic flux leakage linked with transformer coils is negligible then find the rate of change of current in primary to produce an e.m.f. of 200 volts in the secondary.

Solution: When the magnetic flux leakage is negligible the mutual inductance is given by

$$M = \sqrt{L_P L_S}$$

where L_P and L_S are self inductances of primary and secondary coils of the transformer. The e.m.f. induced in the secondary is

$$e_s = -M\frac{dI_P}{dt}, \quad \text{where } I_P \text{ is the current in the primary}$$

$$\therefore \quad M = \sqrt{20\times 10} = \sqrt{200}\ \text{henry} \qquad \left| \begin{array}{l} L_P = 20\ \text{henry} \\ L_S = 10\ \text{henry} \\ e_s = 200\ \text{V} \end{array}\right.$$

$$\therefore \quad \frac{dI_P}{dt} = \frac{e_s}{M} = \frac{200}{\sqrt{200}} = 14.14\ \text{A/s}$$

EXAMPLE 7.11: An antenna fixed vertically on a car is of length 1 m. If the car is travelling from east to west with a velocity of 60 km/hr then calculate the induced e.m.f. between the two ends of the antenna (Horizontal component of Earth's magnetic field is $0.4\times 10^{-4} Wb/m^2$).

Solution: Induced e.m.f.

$$\varepsilon = -\frac{\partial \phi}{\partial t}$$

and

$$\phi = \mathbf{B}\cdot\mathbf{A}$$

$$\therefore \quad e = B\cdot\frac{\partial \mathbf{A}}{\partial t} = BlV$$

If the magnetic field is uniform and constant

$$\begin{aligned} \therefore e &= BlV \\ &= 0.4\times 10^{-4}\times 1\times\frac{50}{3} \\ &= 6.67\times 10^{-4}\ \text{Volt.} \end{aligned} \qquad \left| \begin{array}{l} B = 0.4\times 10^{-4}\ \text{resla} \\ l = 1\ \text{m} \\ V = 60\ \text{km/hour} \\ = 50\sqrt{3}\ \text{m/s} \end{array}\right.$$

EXAMPLE 7.12: For a medium, conductivity σ is = 5 mho/m and dielectric constant is 1. If an electric field $E = 250\sin(10^{10}t)$ is applied then find conducting current and displacement current in the medium. Show that both the currents will be equal at frequency $\omega = 5.6\times 10^{11}$ radian/s.

Solution: Conducting current density $J_f = \sigma E$

$$\begin{aligned} J_f &= 5\times 250\sin(10^{10}t) \\ &= 1250\sin(10^{10}t)\ \text{Amp/m}^2 \end{aligned} \qquad \left| \begin{array}{l} \sigma = 5\ \text{mho/m} \\ E = 250\sin(10^{10}t) \end{array}\right.$$

Displacement current density $J_d = \varepsilon \frac{\partial \mathbf{E}}{\partial t}$

$$
\begin{aligned}
\mathbf{J}_d &= \varepsilon_0 \varepsilon_r \frac{\partial E}{\partial t} \\
&= 8.85 \times 10^{-12} \times \frac{\partial}{\partial t}(250 \sin 10^{10} t) \\
&= 8.85 \times 10^{-12} \times 250 \times 10^{10} \cos 10^{10} t \\
&= 8.85 \times 10^{-12} \times 1 \times 250 \times 10^{10} \cos(10^{10} t) \\
&= 2.21 \cos(10^{10} t) \text{ Amp/m}^2
\end{aligned}
$$

When the magnitudes of both the currents are same them

$$J_f = J_d$$

or
$$\sigma E = \varepsilon_0 \varepsilon_r \omega E$$

$$\therefore \quad \omega = \frac{\sigma}{\varepsilon_0 \varepsilon_r} = \frac{5}{8.85 \times 10^{-12} \times 1} = 5.6 \times 10^{11} \text{ rad/s}$$

EXAMPLE 7.13: If the radius of the Sun is 7×10^8 m and power radiated by it is 3.8×10^{26}, then calculate the magnitude of Poynting vector at the surface of the Sun.

Solution: We know that Poynting vector **S**, represents the power radiated per unit area. Surface area of the sun is $4\pi R^2$, where R is the radius of the Sun. If P is the total power radiated by Sun, then

$$P = S 4\pi R^2$$

or
$$S = \frac{P}{4\pi R^2} = \frac{3.8 \times 10^{26}}{4 \times 3.14 \times (7 \times 10^8)^2} = 6.1 \times 10^7 \text{ W/m}^2$$

EXAMPLE 7.14: A parallel plate capacitor is formed by two discs, the space between which is filled with a homogeneous, poorly conducting medium The capacitor was charged and then disconnected from a power source. ignoring edge effects, show that the magnetic field inside the capacitor is absent.

Solution: Magnetic field will be absent since the total current (conduction current plus displacement current) is equal to zero. We now prove this. Let us consider the current density. Suppose that at a certain instant the density of conduction current is **J**. Obviously, $\mathbf{J} \propto \mathbf{D}$, and

$\mathbf{D} = \sigma n$, where σ is surface charge density on the positively charged plate and n is the normal (Fig. 7.21).

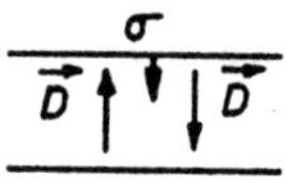

Figure 7.21.

The presence of conduction current leads to a decrease in the surface charge density σ, and hence in **D** as well. This means that conduction current will be accompanied by the displacement current whose density is

$$\mathbf{J}_D = \frac{\partial \mathbf{D}}{\partial t} = \left(\frac{\partial \sigma}{\partial t}\right)\mathbf{n} = -J\mathbf{n} = -\mathbf{J}$$

Hence it follows that, indeed

$$\mathbf{J}_t = \mathbf{J} + \mathbf{J}_D = 0$$

EXAMPLE 7.15: The space between the plates of a parallel plate capacitor in the form of circular discs is filled with a homogeneous poorly conducting medium with a conductivity σ and permittivity ε. Ignoring edge effects, find the magnitude of vector **H** between the plates at a distance r from their axes, if the electric field strength between the plates varies with time in accordance with $E = E_m \cos \omega t$.

Solution: From Maxwell's equation for circulation of vector **H**, we have

$$2\pi r H = \left(J_n + \varepsilon\varepsilon_0 \frac{\partial E_n}{\partial t}\right)\pi r^2$$

Taking into account Ohm's Law $J_n = \sigma E_n(t)$, one obtains

$$\begin{aligned} H &= \frac{r}{2}\left(E_n + \frac{\varepsilon\varepsilon_0}{\sigma}\frac{\partial E_n}{\partial t}\right) \\ &= \frac{rE_m}{2}(\sigma \cos \omega t - \varepsilon\varepsilon_0\omega \sin \omega t) \end{aligned}$$

Let us transform the expression in the parentheses to cosine. For this purpose, we multiply and divide the expression by $f = \sqrt{\sigma^2 + (\varepsilon\varepsilon_0\omega)^2}$ and then introduce angle δ through the relation $\frac{\sigma}{f} = \cos\delta$, $\varepsilon\varepsilon_0\frac{\omega}{f} = \sin\delta$. This gives $H = \frac{1}{2} r E_m \sqrt{\sigma^2 + (\varepsilon\varepsilon_0\omega)^2}\cos(\omega t + \delta)$.

EXAMPLE 7.16: Protons having the same velocity V form a beam of a circular cross section with Current I. Find the direction and magnitude of Poynting's vector **S** outside the beam at a distance r from its axis.

Solution: Fig. 7.22 shows that **S** is parallel to **V**. Let us find the magnitude of **S** : $S = EH$, where E and H depend on r. According to the Gauss theorem, we have

$$2\pi rE = \frac{\lambda}{\varepsilon_0}$$

where λ is the charge per unit length of the beam. Besides it follows from the theorem on circulation of vector **H** that

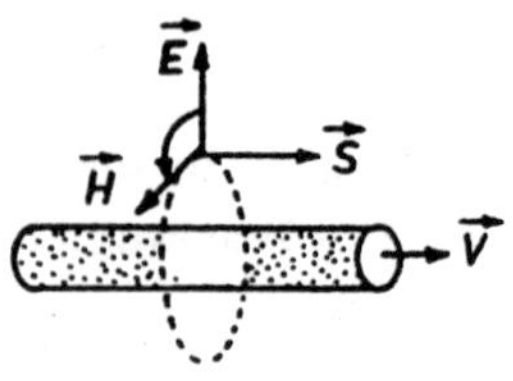

Figure 7.22.

$$2\pi rH = I$$

Now $I = \lambda V$, we obtain

$$S = EH = \frac{I^2}{4\pi^2\varepsilon_0 V r^2}$$

EXAMPLE 7.17: A parallel plate air capacitor whose plates are made in the form of discs of radius a are connected to a source of varying harmonic voltage of frequency ω. Find the ratio of the maximum values of magnetic and electric field energy inside the capacitor.

Solution: Let the voltage across the capacitor vary in accordance with the law $V = V_m \cos \omega t$ and the distance between the capacitor plates h. Then the electric energy of the capacitor is equal to

$$W_e = \frac{\varepsilon_0 E^2}{2}\pi a^2 h = \frac{\varepsilon_0 \pi a^2}{2h} V_m^2 \cos^2 \omega t \tag{1}$$

The magnetic energy can be determined through the formula

$$W_m = \int_v \frac{B^2}{2\mu_0} dv \tag{2}$$

The quantity B required for evaluating this integral can be found from the theorem on the circulation of vector **H** : $2\pi rH = \pi r^2 \frac{\partial D}{\partial t}$. Since $H = \frac{B}{\mu_0}$ and $\frac{\partial D}{\partial t} = -\varepsilon_0 \left(\frac{V_m}{h}\right) \omega \sin \omega t$, one obtains

$$B = \frac{1}{2}\varepsilon_0 \mu_0 \frac{r\omega V_m}{h} |\sin \omega t| \tag{3}$$

Substituting (3) in (2), where for dv we must take an elementary volume in the form of a ring for which $dv = 2\pi r dr h$. On performing integration one obtains

$$W_m = \frac{\pi}{16}\frac{\mu_0\varepsilon_0^2\omega^2 aV_m^2}{h}\sin^2\omega t \qquad (4)$$

The ratio of the maximum values of magnetic energy (4) and electric energy (1) is obtained as

$$\frac{(W_m)_{\max}}{(W_e)_{\max}} = \frac{1}{8}\mu_0\varepsilon_0 a^2\omega^2$$

For e.g., for $a = 6$ m , $\omega = 10000$/s, this ratio is equal to 5×10^{-15}.

QUESTIONS

1. State and Explain Faraday's laws of electromagnetic induction.
2. State and explain the Laws of Faraday and Lentz relating to electromagnetic induction. Deduce a relationship between the quantity of electricity flowing through a circuit and a magnetic flux change producing it.
3. Deduce an expression for e.m.f. induced in varying magnetic field. What is the difference between electric field so produced and that produced by charges.
4. A conducting rod is moving with a velocity a through a uniform magnetic field of induction **B**. Show that there appears inside the rod an electric field. $\mathbf{E} = -(\mathbf{V} \times \mathbf{B})$
5. Deduce Faraday's law of electromagnetism, $\nabla \times \mathbf{E} = -\frac{\partial \mathbf{B}}{\partial t}$.
6. Explain the meaning of self and mutual induction Derive an expression for self induction of a solenoid.
7. Find the self inductance of a long solenoid and calculate the energy stored per unit volume in magnetic field.
8. Show that the energy stored in a magnetic field per unit volume in free space is $\frac{B^2}{2\mu_0}$.
9. State Ampere's circuital law. Discuss why and how it was modified to include displacement current.
10. Write down Maxwell's field equations and prove Poynting's theorem relating to the flow of energy at a point in space in an electromagnetic field.

11. Write down Maxwell's field equations in differential and integral forms and explain their physical meaning.
12. Show that the total current flowing out of some volume must be equal to the rate of decrease of charge within the volume, assuming that no sources and sinks are present within the volume.
13. Obtain the Poynting theorem for the conservation of energy in an electromagnetic field and discuss the physical meaning of each term in the resulting equation.
14. Show that the law of conservation of energy for electromagnetic fields in non-conducting medium is

$$\frac{d\varepsilon_m}{dt} + \nabla \cdot \mathbf{N} = 0$$

where ε_m is the energy density and **N** is Poynting vector.

15. State Maxwell's equations for a system of charges and currents. Show that the energy density and Poynting vector of electromagnetic field are given by

(i) $\varepsilon_m = \frac{1}{2}[\mathbf{E} \cdot \mathbf{D} + \mathbf{B} \cdot \mathbf{H}]$ and (ii) $\mathbf{N} = [\mathbf{E} \times \mathbf{H}]$

PROBLEMS

1. A car having a vertical aerial of 1 m is moving with a uniform speed of 80 km/hour in the east-west direction. If the value of $H = 0.36$ Oersted show that the induced e.m.f. in the aeril is $8 \times 10^{-4} V$.

2. A rectangular loop of wire with dimensions a and b and lying in the x-y plane, is moving with a uniform velocity V in the direction of the positive y-axis in a static magnetic vector field pointing in the z-direction and varying sinusoidally in the y-direction as $B_z = B_0 \sin ky$. Show that the e.m.f. in the moving loop is $-2B_0 aV \sin \frac{kb}{2} \cos kvt$.

3. A railway line 1.2 m wide runs along the magnetic meridian. The vertical component of earth's magnetic field is 0.5 Oersted. Show that the e.m.f. that will exist between rails when a train runs on the line at a speed of 60 km/hour is 1 mV.

4. If a potential difference of 90 V is developed in a coil, when the current in it is changing at a rate of 3 amperes/s. Calculate the self inductance of the Coil. **[Ans.** 30 henry]

5. A long solenoid of cross-sectional area 4×10^{-4} m^2 is would with 1000 turns/m and the current in its windings is increased at the rate of 100 amp/s. Show that the e.m.f. induced in it is $16\pi \times 10^{-6}$ V.

6. A coil of 20 turns is wound on a solenoid of length 1 m, area of cross-section 10 sq. cms and having 1000 turns. Show that the mutual inductance of the system is 25.1 m henry.

7. A horizontal radial magnetic field $B = 1.1 \frac{Wb}{m^2}$ exists in the annular air gap between a vertical hollow cylinder and a coaxial cylindrical core. A thin circular ring of aluminium wire falls in the gap with the plane of the ring horizontal and its centre on the axis of the coaxial cylindrical system, under the influence of gravity and electromagnetic forces. If the resistivity of the aluminium is $\sigma^{-1} = 3.1 \times 10^{-8} \Omega$-m and its density is $\rho = 2.7 \times 10^3$ kg/m^3, then show that the terminal velocity of the ring is 6.8×10^{-4} m/s.

8. The magnetic flux through a circular coil, perpendicular to its plane is varying according to the following relation $\phi = (6t^2 + 7t + 1)$, where ϕ is in milli Webers and t is in seconds. Show that the e.m.f. induced in the coil when $t = 2$ sec. is 0.03 V.

9. A railway track has two parallel rails at separation 1 m. A train passes at a speed of 90 km/hour. Calculate the e.m.f. induced between the rails. Horizontal component of earth's field is 0.5 gauss and angle of dip $\phi = 45°$. **[Ans.** 1.25 mV]

10. A small magnet of magnetic moment m is rotating about its centre with an angular velocity ω. Show that $\boldsymbol{\omega} \times \mathbf{m} = \frac{d\mathbf{m}}{dt}$ and further show that the motion of the magnet gives rise to an electric field $\mathbf{E} = \frac{\mu_0}{4\pi r^3}[(\mathbf{r} \cdot \mathbf{m})\omega - (\mathbf{r} \cdot \boldsymbol{\omega})m]$.

11. Starting from the equation of continuity show for a conducting medium obeying Ohm's law, $\mathbf{J} = \sigma\mathbf{E}$ and using Gauss's law that $\frac{\partial \rho}{\partial t} + \frac{\sigma}{\varepsilon} = 0$. From this, show that if initially there is charge density ρ_0 at any point inside the conducting medium, the charge

density at a later time is given by $\rho = \rho_0 \exp\left[\left(-\frac{\sigma}{\varepsilon}t\right)\right]$, where σ is conductivity and $\frac{\varepsilon}{\sigma}$ is called the relaxation time.

12. Obtain equation of continuity $\nabla \cdot \mathbf{J} + \frac{\partial \rho}{\partial t} = 0$ from Maxwell's equations.

13. A conducting rod MN of length L rotates with angular velocity ω about an axis perpendicular to its length and passing through a point C, where $MC = L/3$. A uniform magnetic induction $\mathbf{B}$ is parallel to $\boldsymbol{\omega}$. Show that the potential difference $V_M - V_N = B\omega L^2/6$.

SHORT ANSWER QUESTIONS

1. A rectangular closed loop moves horizontally in a uniform magnetic field. Will there be any induced current in the loop if the loop is completely in the magnetic field?

   [**Ans.** There will be no induced current in the loop]

2. What is an ideal inductor?

   [**Ans.** An ideal inductor is one that has zero resistance]

3. Name the quantity that plays an identical role in an electrical circuit as is played by inertia in mechanics. [**Ans.** Inductance]

4. A quantitative measure of coupling of two coils is given by the coefficient of coupling $K = \sqrt{\frac{M}{L_1 L_2}}$, where L_1 and L_2 are the self-inductance of the two coils. Can K exceed unity?

   [**Ans.** No, K cannot be more than Unity]

5. Is the electric field defined through (i) $\nabla \times \mathbf{E} = -\frac{\partial \mathbf{B}}{\partial t}$ and (ii) $\oint_c \mathbf{E} \cdot \mathbf{dl} = -\frac{\partial}{\partial t}\int_s \mathbf{B} \cdot \mathbf{dS}$ conservative?

   [No, the electric field defined through (i) and (ii) is non conservative as can be seen from the fact that its line integral for a closed path is not zero. This must be contrasted with the electrostatic field **E** which is conservative.]

6. A straight wire carrying a uniform current density over its circular cross-section. If we consider only the magnetic flux within the wire, then what will be the self inductance per unit length?

$$\left[\textbf{Ans. } \frac{\mu_0}{8\pi}\right]$$

7. What the Poynting vector $\mathbf{N} = [\mathbf{E} \times \mathbf{H}]$ represent?

   [**Ans.** The power transmitted through unit area held normal to **N**]

8. Does Lenz's Law violate the law of conservation of energy?

   [**Ans.** No. Lenz's Law is in confirmity with the law of conservation of energy]

9. Does Lenz's Law hold for an open circuit?

   [**Ans.** Yes, it can still hold]

10. How will you define displacement current at any point?

$$\left[\textbf{Ans. } \mathbf{J} = \frac{\partial \mathbf{D}}{\partial t}\right]$$

11. The current density in a given region is $\mathbf{J} = J_0 \hat{i} \exp(-x^2)$. What will be time rate of change of the charge density at the point (0, 0, 0) [zero].

12. In problem 11, what will the change of the charge density at the point (1,0,0)? [**Ans.** $2J_0 \exp(-1)$]

OBJECTIVE QUESTIONS

1. The dimensional formula for induced e.m.f. is

   (a) $ML^2T^{-2}A^{-1}$ (b) $ML^2T^{-3}A^{-1}$
   (c) $ML^2T^{-3}A^{-2}$ (d) $ML^2T^{-2}A^{-2}$ **[b]**

2. The dimensional formula for the coefficient of self inductance is

   (a) ML^2T^{-2} (b) $ML^2T^{-2}A^{-1}$
   (c) $ML^2T^{-2}A^{-2}$ (d) $ML^2T^{-3}A^{-2}$ **[c]**

3. The dimensional formula for magnetic flux is

   (a) ML^2T^{-2} (b) $ML^2T^{-2}A^{-1}$
   (c) $ML^2T^{-2}A^{-2}$ (d) $MLT^{-2}A^{-2}$ **[b]**

4. Whenever the flux linked with a circuit changes there is an induced e.m.f. in the circuit. This e.m.f. would last in the circuit

   (a) for a very short time

(b) for ever
(c) for a very long time
(d) as long as the flux in the circuit changes **[d]**

5. The unit of magnetic permeability is

(a) Weber (b) Weber/m^2
(c) henry-metre (d) henry/m **[d]**

6. The condition for electric field **E** to be conservative is

(a) $\oint \mathbf{E}\cdot d\mathbf{l}=0$ (b) $\nabla\times\mathbf{E}=0$

(c) $\oint_S \mathbf{E}\cdot d\mathbf{S}=\frac{q}{\varepsilon_0}$ (d) $\nabla\times\mathbf{E}=\delta$ **[a]**

7. Magnetic flux density **B** and vector potential **A** are related by

(a) $\mathbf{A}=\nabla\times\mathbf{B}$ (b) $\mathbf{A}=\nabla\cdot\mathbf{B}$
(c) $\mathbf{B}=\nabla\cdot\mathbf{A}$ (d) $\mathbf{B}=\nabla\times\mathbf{A}$ **[d]**

8. Maxwell's Curl equation for static magnetic field is

(a) $\nabla\times\mathbf{B}=0$ (b) $\nabla\times\mathbf{B}=\mu_0\mathbf{J}$
(c) $\nabla\cdot\mathbf{B}=\mu_0\mathbf{J}$ (d) $\nabla\cdot\mathbf{B}=\frac{\mu_0}{\mathbf{J}}$ **[b]**

9. Displacement vector **D** and electric field strength **E** are related by

(a) $\mathbf{D}=\frac{\mathbf{E}}{\varepsilon}$ (b) $\mathbf{D}=\varepsilon\mathbf{E}$
(c) $\mathbf{D}=\varepsilon E^2$ (d) none of the above **[d]**

10. Hysterisis losses

(a) are caused by high frequency a.c. currents
(b) usually increase with direct current in a coil
(c) are zero in an iron core because it is a conductor
(d) none of the above **[b]**

11. The electromagnetic momentum density, **g** is

(a) $\mathbf{g}=[\mathbf{D}\times\mathbf{E}]$ (b) $\mathbf{g}=[\mathbf{D}\times\mathbf{B}]$

(c) $\mathbf{g}=[\mathbf{B}\times\mathbf{A}]$ (d) $[\mathbf{g}]=\left[\frac{\mathbf{B}}{H}\right]$ **[b]**

12. The dimensions of the vector $[\mathbf{E}\times\mathbf{H}]$ are

(a) $\frac{\text{energy}}{\text{area}\times\text{time}}$ (b) $\frac{\text{energy}}{\text{time}}$

(c) $\frac{\text{energy}}{\text{area}}$ (d) $\frac{\text{energy}}{\text{length}\times\text{time}}$ **[a]**

13. Gauge Transformation equations are

(a) $$A' = A + \nabla\Psi$$

and $$\phi' = \phi - \frac{\partial \Psi}{\partial t}$$

(b) $$A' = A + \Psi$$

and $$\phi' = \phi - \Psi$$

(c) $$\mathbf{A} = \mathbf{A}'$$

and $$\phi' = \phi$$

(d) none of the above

(Symbols have usual meanings.) **[a]**

Appendix 1

Moving Coil Galvanometer

1. Introduction

A galvanometer is an electrical device which measures the current. The moving coil galvanometer is based on the following principle: When a conductor carrying current, is placed in a magnetic field it experiences a force in a direction given by Flemings left hand rule and of magnitude given by formula

$$F = IlB \sin \theta \tag{1}$$

where I is the current-flowing through the conductor, l-is its length, B, the magnetic field in which the conductor is placed and θ is the angle between direction of current and magnetic field. In vector notation the force in the conductor is given by.

$$\mathbf{F} = (I\mathbf{l}) \times \mathbf{B} \tag{2}$$

when $(I\mathbf{l})$ is called current element. This force rotates the suspended or pivoted coil against a restoring couple due to either a torsion fibre in sensitive instrument of the lamp and scale pattern, or a hair spring in more robust pointer instruments.

A moving coil galvanometer consists of a rectangular coil of insulated wire of many turns wound on rectangular frame. The frame C is suspended from a phospher-bronze fibre F or an jewelled bearings in less sensitive pointer instrument of pivoted type, so that it is within the magnetic field B provided by horse shoe form of permanent magnet with pole pieces NS shaped so as to have a cylindrical gap between them. The axis of rotation of the coil coincides with the principal axis of symmetry of the magnetic field. To increase the magnetic flux in the gap, and to ensure that the sides of the coil parallel to the axis of rotation are always perpendicular to the lines of force a cylindrical soft iron core, "*I*", is arranged within the coil so that the lines of force are radial every where in the narrow angular gap. The phospher bronze suspension fibre, often a flat strip commonly acts as a current lead, whereas a second lead is provided by a very spring wire attached to the bottom of the coil Fig 1(a). The moving system is completely insulated from the metal case which is best earthed. Copper thermals are fitted to minimise thermo-electric emfs. The coil can be adjusted to zero position by turning the torsion head at the top of the instrument.

The galvanometers are classified as *dead-beat* and *ballistic* type. For an instrument required only for the measurement of current, a steady deflection proportional to the current is required. Hence the oscillation of the pointer or spot of light on scale before coming to rest is not required. As such the coil, for such use of galvanometer is wound on conducting frame (like aluminium) to make the damping high because as the coil oscillates in magnetic field, the induced emf (and hence induced current) according to Lenz's law in opposite direction will prevent the coil from oscillating. Such a galvanometer is called *dead beat-galvanometer* and the damping so produced is called electromagnetic damping.

On the other hand ballistic galvanometers measure charge (= current × time). In this case it is essential that the whole of the charge is passed through the coil before the coil be given to move, i.e. the time required by the charge to pass through coil must be short compared to the time required for the deflection. In such case the coil receives an impulse and a "*throw*" is recorded. To achieve this, a coil of high moment of inertia about the axis of suspension is necessary. In addition the coil must swing freely without significant damping. To achieve this the coil is wound on a non-conducting frame made of ivory, bamboo or ebonite. So that the damping due to eddy currents (electromagnetic damping) is reduced. Such a galvanometer is called *ballistic galvanometer*. The damping in these galvanometers is made as small as possible and for the residual damping that exists, the damping correction is employed.

However, such a ballistic galvanometer of lamp and scale variety can be damped by connecting a resistance across its terminals. The movement of the coil in magnetic field will induce a current in the now closed circuit which will be directed so as to try to oppose coil movement. Thus after the ballistic deflection has been recorded, the coil can be brought to rest by short-circuiting the galvanometer terminals.

2. Theory of moving coil galvanometer

Fig. 7.1(a) represents the front view of the galvanometer and Fig. 7.1(b) represents the top (sectional) view. *AD* is the length of the coil (l) and *AB* is the breadth (b) of the coil. $A'B'$ represents the deflection position of the coil. Let the angle of deflection be θ. Since the magnetic field used is radial, hence the direction of magnetic field is parallel to the plane of coil at very position and perpendicular to the length *AD* of the coil. The length side conductors *AD* and *BC* experience force *IlB* in the directions shown in Fig. 7.1(b), where *B* is magnetic flux density. These forces constitute a couple and the restoring couple is provided by the torsion in the suspension fibre. In equilibrium position th-ese two couples, are equal and opposite. If there are n of turns then total deflecting couple; is given by

$$\text{deflecting couple} = C\theta.$$

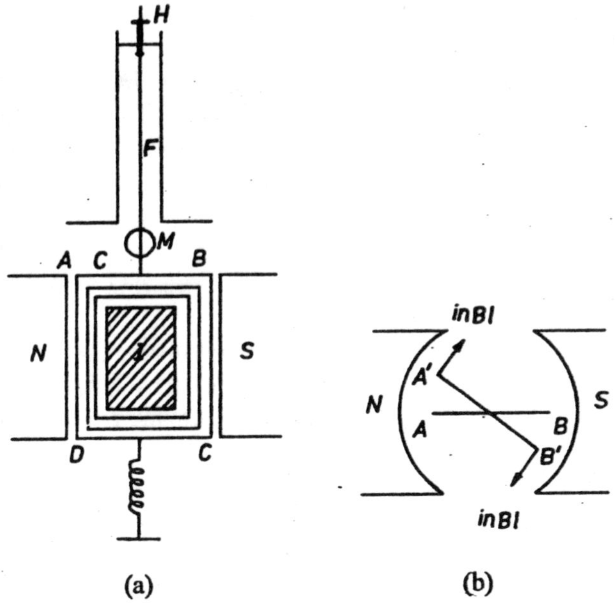

Figure 7.1a.

If C in the twisting couple per unit twist of suspension wire (Phosphar-bronze) then for a twist θ, the restoring couple is

$$\text{restoring couple} = C\theta$$

For equilibrium

$$nIlBb = C\theta \tag{3}$$

or

$$I = \frac{C\theta}{nlBb} = k\theta \tag{4}$$

when $k = \dfrac{C}{nlBb} = \dfrac{C}{nAB}$. Here $A = lb$ the area of coil.

Here k is constant and is called the *reduction factor* of the moving coil galvanometer.

It is clear from equation (1) that in moving coil galvanometer, the current is directly proportional to the deflection ($I \propto \theta$). Usually lamp and scale arrangement is used to measure the deflection in the case of suspended moving coil galvanometer.

Sensitivity: Equation (3) gives

$$\theta = \frac{InAB}{C}$$

A galvanometer is said to be sensitive if for a given small value of the current I, the deflection θ should be large. Thus the sensitivity of a moving coil galvanometer can be increased by increasing n, A and B and decreasing the value of C. C is decreased by using suspension of phospher-bronze or quartz of rectangular cross-section. The quartz fibres are silvered to make them conducting.

Advantages:

(i) A moving coil galvanometer can be made very sensitive by increasing n, A and B and decreasing C.

(ii) Flux density B between the pole pieces is made very high by using a soft iron core. Due to high value of B external fields do not affect the deflection of the coil. As such the moving coil galvanometer can be used in any position.

(iii) The deflection of the coil is directly proportional to the current and hence a linear scale can be used to measure the current.

(iv) In case of dead beat galvanometer the coil is wound on metallic frame, which provides the electromagnetic damping due to which the coil comes to rest-quickly.

The suspended type moving coil galvanometers are very sensitive and in experiments where sensitivity is not required pivoted type galvanometers are used. In these the coil is pivoted between two bearings and pointer is attached to the moving coil. The controlling couple is provided with the help of a spring.

Ballistic Galvanometers: As already discussed, the galvanometers are used to measure current. The ballistic galvanometers are used to measure charge viz. discharge of a condenser or in experiments employing the principle of electromagnetic induction. We have already discussed their construction.

3. Theory of moving coil ballistic galvanometer

In ballistic galvanometer whole of the charge to be measured is passed through the coil in a time insignificant, compared with the time of deflection. A ballistic throw, due to the electrical impulse is recorded. In case the damping is negligibly small, then the magnitude of the first throw is readily calculated in terms of charge Q and the galvanometer constant. If the coil is set in oscillations, the period of free oscillations is given by

$$T = 2\pi\sqrt{\frac{I}{C}} \tag{4}$$

where I is the moment of inertia of the coil about its axis of suspension and C is the twisting couple permit-twist of the suspension fibre. T must be very large in comparison to the time of passage of the charge.

If the instantaneous current is I through the coil of n turns (length l, bread the b), then the force experienced by the length side conductor of the coil = Binl. Where B is the magnetic flux density. The couple acting on the coil then is

$$\tau = Binlb = BinA \qquad A = l \times b(\text{the area of coil})$$

This couple acts for a short time $t(t \ll T)$ during which the whole of the charge Q is passed. The corresponding impulse therefore is

$$\text{Impulse} = \int_0^l nABidt = nABit = nABQ$$

If ω is the angular velocity of the coil then angular momentum = $I\omega$ and the impulse is equal to the change in angular momentum

$$\therefore \qquad I\omega = nABQ \tag{5}$$

The initial kinetic energy of the coil is $\frac{1}{2}I\omega^2$. This kinetic energy is used up in doing the work necessary to twist the suspension fibre through an angle θ_0, where θ_0 is the first ballistic through when the damping, is zero This work is one

$$\omega = \int_0^{\theta} C\theta\, d\theta = \frac{1}{2}C\theta_0^2$$

$$\therefore \quad I\omega^2 = \frac{1}{2}C\theta_0^2$$

or

$$\omega = \sqrt{\frac{C\theta_0^2}{I}} = \theta_0\sqrt{\frac{C}{I}} \tag{6}$$

Equating the value of ω obtained from equation (5) and (6) we have

$$\frac{nABQ}{I} = \theta_0\sqrt{\frac{C}{I}}$$

$$\therefore \quad Q = \frac{IQ_0}{nAB}\sqrt{\frac{C}{I}} = \frac{Q_0}{nAB}\sqrt{IC} \tag{7}$$

But

$$T = 2\pi\sqrt{\frac{I}{C}} \quad \therefore \quad \sqrt{I} = \frac{T}{2\pi}\sqrt{C}$$

Eliminating I, we have

$$Q = \frac{Q_0}{nAB}\sqrt{C}\cdot\frac{T}{2\pi}\sqrt{C} = \frac{CTQ_0}{2\pi nAB} \tag{8}$$

Equation (8) gives the charge in terms of the ballistic deflection θ_0 and the constants of the galvanometer c, t, n, A and B

$$\therefore \quad Q = K'\theta_0 \tag{8a}$$

Here $K' = \dfrac{CT}{2\pi nAB}$ and is called *ballistic reduction factor* where as $K = \dfrac{C}{nAB}$ is called *current reduction factor*.

In determining the charge Q by using a ballistic galvanometer it is necessary to know these constants or eliminate some of them by a determination of the deflection per unit current of the galvanometer given by equation (3). Thus if a known steady current i produces a recorded deflection θ_D, then

$$i = \frac{C}{nAB}\theta_D$$

and equation (8) becomes

$$Q = \frac{i}{\theta_D}\frac{T\theta_0}{2\pi} = \frac{iT}{2\pi}\frac{\theta_0}{Q_D} \tag{9}$$

Here $\frac{i}{\theta_D}$ is determined experimentally. T is determined easily by counting the time of same known oscillations of the freely swinging galvanometer coil. Hence charge Q corresponding to a ballistic deflection θ_0 is found.

4. General theory of moving coil galvanometer

An instantaneous current i passing through galvanometer at any instant t, if produces deflection θ; then the equation of motion for the coil can be written taking into account the following factors.

(1) The retarding couple due to the moment of inertia of the oscillating system $= I\frac{d^2\theta}{dt^2}$

(2) The retarding couple due to damping, which is assumed to be proportional to the angular velocity of the coil $= a\frac{dQ}{dt}$. Here a is constant called damping constant. This damping is due to an Viscosity, elastic hysteresis in suspension fibre and electromagnetic damping caused by opposing currents induced in neighbouring masses of metal as the coil moves and in the frame of the coil itself, if it is of conducting metal.

(3) The restoring couple due to the torsion in the suspension fibre $= C\theta$. Here C is twisting couple per unit twist.

(4) The deflecting couple when a current i passes through the coil $= nABi$

(5) The electromagnetic damping, due to current induced in the coil itself when it moves in the magnetic field $= nAB\omega$. Where ω is angular velocity $= \frac{d\theta}{dt}$. Hence this couple $= nAB\omega$.

(6) An opposing e.m.f. $e = -L\frac{di}{dt}$ where L the induction of the galvanometer and its circuit.

If now the galvanometer is in a closed circuit of total resistance R, where R includes the galvanometer resistance and the external resistance, then, when a constant e.m.f. E is applied to the circuit terminals, the current i at any time t, during the motion is given by

$$i = \frac{E - nAB\frac{dQ}{dt} - L\frac{di}{dt}}{R}$$

$$= \frac{E - G\cdot\frac{dQ}{dt}}{R}$$

where $G = nAB$ called galvanometer constant, and when current is constant $\frac{di}{dt} = 0$.

Hence the deflecting Torque $= naBi = Gi = G\left(\frac{E - G\frac{d\theta}{dt}}{R}\right)$

Now equating the total deflecting torque equal to the total restoring torque, we get the differential equation of motion for the coil

$$I\frac{d^2\theta}{dt^2} + a\frac{d\theta}{dt} + C\theta = Gi = \frac{G}{R}\left(E - G\cdot\frac{d\theta}{dt}\right)$$

and re-arranging the terms

$$I\frac{d^2\theta}{dt^2} + \left(a + \frac{G^2}{R}\right) + C\theta = \frac{GE}{R}$$

or

$$\frac{d^2\theta}{dt^2} + \left(\frac{a}{I} + \frac{G^2}{IR}\right)\frac{d\theta}{dt} + \frac{C\theta}{I} = \frac{G}{RI}E \tag{10}$$

When the galvanometer is used for the measurement of a steady current i or e.m.f. E, then Gi and $\frac{GE}{R}$ will be continuously applied couples. The resultant final deflection θ_D will be given by equation (3).

When a moving galvanometer is used as a ballistic galvanometer, the whole charge passes through the coil before the coil begins to move. Therefore

$$\frac{E}{R} = I = 0 \quad \text{or} \quad \frac{GE}{RI} = 0$$

Hence equation (10) reduces to

$$\frac{d^2\theta}{dt^2} + 2b\frac{d\theta}{dt} + \omega_0^2 C = 0 \tag{11}$$

When $\left(\frac{a}{I} + \frac{G^2}{IR}\right) = 2b$ and $\frac{C}{I} = \omega_0^2$

The general solution of equation (11) is

$$\theta = Ae^{\left(-b+\sqrt{b^2-\omega_0^2}\right)t} + Be^{\left(-b-\sqrt{b^2-\omega_0^2}\right)t} \tag{12}$$

There are three cases of interest depending upon the relative values of b and ω_0. (a) When $b^2 > \omega_0^2$. Equation (12) gives that the deflection increases from zero to the maximum and the movement of the coil is non-oscillatory or a periodic. This is called *over damped* motion of galvanometer.

(b)When $b^2 = \omega_0^2$: The motion of the coil is just non-oscillatory and is called *critically damped.* In this case.

$$\frac{1}{4}\left(\frac{a}{I}+\frac{G^2}{IR}\right)^2 = \frac{C}{I}$$

Here the first term within the parenthesis is neglected because resistance R of the circuit provides appreciable damping. In that case

$$\left(\frac{G^2}{2IR}\right)^2 = \frac{C}{I}$$

(c) When $b^2 < \omega_0^2$

The motion of the coil is oscillatory and equation (12) can be written as

$$\theta = e^{-bt}\{A\, r^{i\omega t} + B\, e^{-i\omega t}\} \tag{13}$$

where $\omega = \sqrt{\omega_0^2 - b^2}$ and $i = \sqrt{-1}$ and the frequency of oscillation is

$$\begin{aligned} f &= \frac{\omega}{2\pi} = \frac{\sqrt{\omega_0^2 - b^2}}{2\pi} \\ &= \frac{1}{2\pi}\sqrt{\frac{C}{I} - \left(\frac{a}{2I}+\frac{G^2}{2IR}\right)^2} \end{aligned} \tag{14}$$

and time period

$$T = 2\pi \frac{1}{\sqrt{\frac{C}{I} - \left(\frac{a}{2I}+\frac{G^2}{2IR}\right)^2}} \tag{14a}$$

5. Correction for damping in Ballistic galvanometer

There are two possible ways for using galvanometer ballastically for charge measurement. Either the galvanometer is used with as little damping as possible. So the ballistic throw is recorded whilst the galvanometer is on open circuit-or where the circuit resistance is very high, i.e. the circuit resistance across the galvanometer is such that the galvanometer is not quite dead beat so that the time of deflection is desirably long. In this case the effect of damping must be accurately assessed. In general first of these methods is used using long period galvanometer. The second method is generally used for measurement of magnetic field. We will only consider the first procedure.

If the galvanometer is in an open circuit during the ballistic deflection then $R = \infty$ and time period becomes

$$T = 2\pi \frac{1}{\sqrt{\frac{C}{I} - \frac{a^2}{4I^2}}}$$

If the damping factor "a" is negligible, then

$$T = 2\pi\sqrt{\frac{I}{C}}$$

i.e. vibrations are free. In this ideal case, the rate at which the amplitude as oscillation dies a way is also zero. Because

$$\text{Amplitude } \theta_0 = e^{-bt}$$

or $$\frac{d\theta_0}{dt} = -be^{-bt} = 0 \quad \because \quad b = 0$$

i.e. the amplitude of the swing remains constant and is a measure of the charge. In practice, though R may be accounted as infinite yet a is necessarily of finite magnitude, though small and the amplitude of successive swings decay in accordance with

$$e^{-bt} = e^{-\frac{at}{2I}}$$

The equation (13) of motion of the coil can be conveniently written as

$$\theta = A_0 e^{-pt} \sin\left(\frac{2\pi}{T}t + \alpha\right) \tag{15}$$

where $T = 2\pi/\omega$, and A_0 and α are constants and can be determined from initial conditions: when $t = 0$, $\theta = 0$ and when $t = T/4$, $\theta = A_0 \exp(-bT/4)\sin\left(\frac{\pi}{2} + \alpha\right) = \theta_1$, the first peak throw on one side of zero.

When $t = \frac{3T}{4}$, $\theta = A_0 \exp(-3bt/4) \sin\left(\frac{\pi}{2} + \alpha\right)$ the second peak throw on the other side of zero.

$$\therefore \quad \frac{\theta_1}{\theta_2} = \exp(bt/2) = \frac{\theta_2}{\theta_3} = \frac{\theta_3}{\theta_4} \cdots = d \text{ (say)} \tag{16}$$

where θ_3 is the third peak throw, θ_4 is fourth peak throw and so on. d is called the *decrement*. The *logarithmic decrement* $\lambda = \log d = \frac{bT}{2}$. If there were no damping the first throw θ_1 would have been θ_0, where

$$\frac{\theta_0}{\theta_1} = \frac{A_0}{A_0 e^{-\frac{bt}{4}}} = e^{\frac{bT}{4}} = e^{\frac{\lambda}{2}}$$

$$\therefore \quad \theta_0 = \theta_1 e^{\frac{\lambda}{2}} = \theta_1\left(1+\frac{\lambda}{2}+\frac{\lambda^2}{4.2!}+\cdots\right)$$

If λ is small, i.e. the damping is small then

$$\theta_0 = \theta_1\left(1+\frac{\lambda}{2}\right) \tag{17}$$

In order to determine charge Q by ballistic galvanometer the equation (8) and (9) are used where $\theta_0 = \theta_1\left(\frac{\lambda}{2}\right)$. Equation (8) now becomes

$$Q = \frac{CT}{2\pi nAB}\cdot\theta_1\left(1+\frac{\lambda}{2}\right) = \frac{CT}{2\pi G}\theta_1\left(1+\frac{\lambda}{2}\right) \tag{18}$$

where $G = nAB$ and equation (10) becomes

$$Q = \frac{iT}{2\pi\theta_D}\theta_1\left(1+\frac{\lambda}{2}\right) \tag{18(a)}$$

In order to find the logarithmic decrement the first ten (say) peak throws (deflections) on either side of zero, i.e. $\theta_1, \theta_2, \theta_3, \cdots, \theta_{10}$ are observed. Then

$$\frac{\theta_1}{\theta_{10}} = \frac{\theta_1}{\theta_2}\cdot\frac{\theta_2}{\theta_3}\cdots\frac{\theta_9}{\theta_{10}} = e^{9\lambda}$$

$$\therefore \quad \lambda = \frac{1}{9}\log\frac{\theta_1}{\theta_{10}} \tag{19}$$

Hence $\frac{\lambda}{2}$ and the correct throw θ_0 is determined.

6. Sensitivity and Factor of merit

The Current Sensitivity of galvanometer is defined as the deflection in millimeters produced on a scale 1 m away by a current of $1\mu A$.

Current Senstivity

$$\frac{i}{\theta_D} = \frac{C}{nAB}$$

The *Voltage Sensitivity* is defined as the deflection obtainable on a scale 1 m away when the voltage applied to the galvanometer is $1\mu V$.

The *quantity* or *charge sensitivity* of a ballistic galvanometer is defined as the correct throw obtained in millimeters at 1 m. When the charge passed through the galvanometer is 1 micro Coulomb.

Charge Sensitivity

$$\frac{Q}{\theta_0} = \frac{T}{2\pi}\frac{C}{nAB}$$

$$= \frac{T}{2\pi} \times \text{Current Senstivity}$$

The current sensitivity of a galvanometer is proportion at to (a) the square of the undamped period, (b) the square root of the coil resistance, the coil winding dimensons being constant. The more correct rule in case (b) is a two-fifths power rather than a square root.

From equation (3) we have

$$\frac{\theta_D}{i} = \frac{nAB}{C}$$

For given value of n, A, and B, $\frac{\theta_D}{i} \propto \frac{1}{C}$.

But time period $T = 2\pi\sqrt{\frac{I}{C}}$.

As such for given value of I-the moment of inertia we have

$$\frac{\theta_D}{i} \propto \frac{1}{C} \propto T^2$$

i.e.

$$S_i \propto T^2 \tag{20}$$

which is statement (a).

For statement (b) we have

$$S_i \propto \frac{\theta_D}{L}$$

or

$$S_i \propto \frac{nAB}{C}$$

which gives, that for given values of C, A and B, S_i is proportional to n the number of turns. If the number of turns is increased, keeping cross-sectional area of winding itself constant, then the length of the winding will increase n times, but the cross-sectional area of the wire must be reduced by n. Hence the resistance of the winding R_G will correspondingly increase by n^2. Hence

$$S_i \propto n \quad \text{and} \quad n \propto \sqrt{R_G}$$
$$\therefore \quad S_i \propto \sqrt{R_G}$$

In view of the dependence of current sensitivity on T and R_G, a comparision of galvanometer performance is made, with a galvanometer whose period had been 1 sec. and resistance of 1Ω. The factor resulting, which compares the standard of performance of galvanometer from the design point of view is called *the factor of merit.*

$$\text{The factor of merit} = \frac{100D}{T^2 R_G^{\frac{1}{2}}}$$

$$\text{or more exactly the factor merit} = \frac{100D}{T^2 R_G^{\frac{2}{5}}}$$

where T is undamped periodic time in seconds, R_G is galvanometer resistance in Ohms and D is deflection in mm/μA at a scale distance of 1 m.

It is readily seen from equation (9) that sensitivity of galvanometer per micro Coulomb $\frac{2\pi}{T} \times$ Current Senstivity:

7. Capacitance by ballistic galvanometer

The galvanometer constant $G = \frac{C}{nBA}$ is determined by sending a constant known current through it. This is called caliberation. Fig. 7.2. shows the circuit for caliberation. The current i is given by

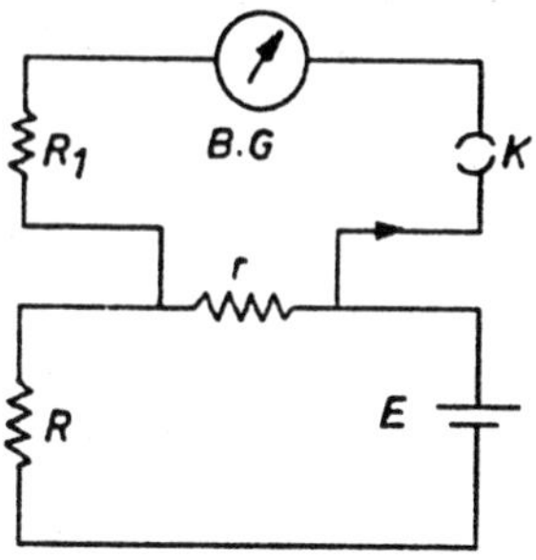

Figure 2.

$$i = \frac{rE}{R+r} \quad \frac{1}{R_1 + R_g}$$
$$= \frac{C}{nBA} \cdot \theta = K\theta$$

where R_1 is a high resistance and R_g is the resistance of the galvanometer.

Fig. 7.3. shows a circuit which can be used for comparision of the capacitances C_1 and C_2 or then absolute values may be determined. It is possible to connect either C_1 or C_2 in the circuit. One of the capacitors is charged by pressing the Key K and then releasing it so that a ballistic throw θ_1 is observed. Care should be taken that the capacitor is fully charged.

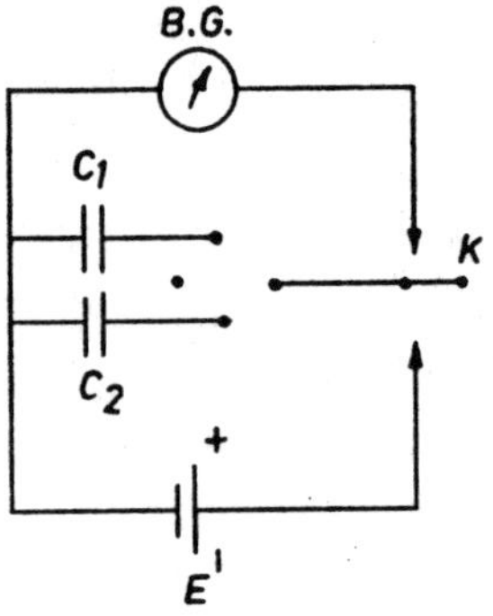

Figure 3.

$$q = E'C_1 = \frac{T}{2\pi} \cdot \frac{rE}{(R+r)(R_1+R_g)} \times \frac{\theta_1}{\theta} \left(1+\frac{\lambda}{2}\right)$$

or $$C_1 = \frac{T}{2\pi} \cdot \frac{rE}{(R+r)(R_1+R_g)} \times \frac{E}{E'} \cdot \frac{\theta_1}{\theta} \left(1+\frac{\lambda}{2}\right)$$

i.e. $$C_1 \propto \theta_1$$

Similarly for C_2 when first-throw is θ_2, we have

$$C_2 \propto \theta_2$$

or $$\frac{C_1}{C_2} = \frac{\theta_1}{\theta_2}$$

where E and E' are kept same. When resistances are taken in Ωs and T in seconds the capacity is obtained in farads.

QUESTIONS

1. What is ballistic galvanometer? Give the theory of a moving coil ballistic galvanometer. **(Delhi 1996)**
2. Give the theory of a ballistic galvanometer of the suspended coil type. How is the damping accounted for?
3. Describe the construction of a suspended coil ballistic galvanometer. Develop its working formula. How may it be converted into a dead heat galvanometer? **(Delhi 1989)**

4. Describe, with necessary theory, the construction and working of a moving coil galvanometer. What are the factors on which the sensitivity of a moving coil galvanometer depends.

   (Delhi 1998; Punjab 1999)

5. Give the theory of a suspended coil ballistic galvanometer. Describe an experimental arrangement to determine the constant of such a galvanometer. **(Delhi 1997)**

6. Describe the construction and working of a moving coil galvanometer. Explain how it may be used to measure electric charge.

7. Give the theory and construction of a moving coil galvanometer. In what respect does it differ from a dead beat galvanometer

8. Show that the ratio of the charge sensitivity and the current sensitivity of a moving coil galvanometer is $\frac{T}{2\pi}$. Why is the coil wound an a non-metallic frame in a moving coil ballistic galvanometer? **(Lucknow 1998)**

9. Give the theory of moving coil ballistic galvanometer. How can it be critically damped.

10. Describe a moving coil ballistic galvanometer and obtain an expression connecting the quantity of electricity through it and the first throw. How will you determine the charge density.

PROBLEMS

1. The current sensitivity of a ballistic galvanometer is 2.2×10^{-9} amperes for a deflection of 1 mm. On a scale kept at a distance of 1 meter. Calculate the charge sensitivity of the galvanometer if time period is 6.2 sec.

   [**Ans.** Charge Sensitivity $= \frac{T}{2\pi} \times$ Current Sensitivity $=$

   $\frac{6.2 \times 2.2 \times 10^{-9}}{2\pi} = 2.17 \times 10^{-9}$ Coul/mm]

2. A condenser charged to $2V$ is discharged through a ballistic galvanometer when the corrected throw is 9.6 cm, and the current sensitivity 2.2×10^{-8} amp.per cm. and the periodic time is 12 seconds. Calculate the capacity of the condenser **(Punjab 1995)**

[**Hint**: $q = \frac{T}{2\pi} \times$ Current Senstivity $\times$ throw $= \frac{12}{2\pi} \times 2.2 \times 10^{-8} \times$ 9.6 and $C = \frac{Q}{V} = \frac{12}{2\pi \times 2} \times 2.2 \times 10^{-8} \times 9.6 = 2.016 \times 10^{-7}$ farad].

3. The successive throws on the same side of the mean position for a free by oscillating coil are observed to be 25, 24.9, 24.8 cm on a scale kept at a distance of one meter. Calculate the lagarithmic decrement.

   [**Hint**: $\lambda = \frac{1}{n-1} \log_e \frac{\theta_1}{\theta_0} = \frac{1}{4} \log 25 \frac{25}{24.8} = 0.002$]

4. A capacitor is charged to a potential difference of one volt; then discharged through a ballistic galvanometer. The throw obtained is 31.1 cm, on a scale kept at 100 cm. Time period of free oscillations is 20 sec. and lagarithmic decrement 0.02. The figure of merit of galvanometer is 10^{-11}. What is the capacitance of the capacitor. [**Ans.** 1000 pF]

5. The successive throws to the right and left of the mean position for the coil of a ballistic galvanometer are 25, 24.9, 24.8 cm. Calculate the logarithmic decrement. [**Ans.** 0.0039]

Chapter 8

Electromagnetic Waves

8.1 INTRODUCTION

We established in Chapter 7 that a varying electric field sets up a magnetic one which, generally speaking, is also varying. This varying magnetic field sets up an electric field, and so on. Thus if we use oscillating charges to produce a varying (alternating) electromagnetic field, then in the space surrounding the charges a sequence of mutual transformations of an electric field and a magnetic field propagating from point to point will appear. This process will be periodic in both time and space and consequently, will be a wave. This may propagave in vacuum with velocity

$$C = \frac{1}{\sqrt{\varepsilon_0 \mu_0}} \approx 3 \times 10^8 \ m/s$$

which corresponds to the velocity of light in vacuum. When we were dealing with the phenomenon of electromagnetic induction in Chapter 7, we indicated the possibility of transmitting a signal from one place to another using a time dependent electromagnetic field. Near the end of 19th century, the German physicist proved beyond any doubt that the electromagnetic field does propagate in vacuum with a velocity equal to C. The property of these electromagnetic waves discovered by Hertz have been examined experimentally with great care. The large body of information that we have accumulated about the properties of electromagnetic waves, such as facts about their production, propagation, and absorption, has opened the door to marvellous world of communications that we know today. Before Hertz performed his experiment the existence of electromagnetic waves had been predicted by Maxwell as a result of a careful analysis of the equations of the electromagnetic field. Maxwell's equations lead to the following important conclusion about the existence of a new (in principle) physical phenomenon. Electromagnetic field may exist independently without electric charges and currents. A change in its state in this case necessarily has a wave nature. Such fields are called '*electromagnetic waves*'. In a vacuum these waves always propagate at a velocity equal to the velocity of light C. Now we shall show that the existence of electromagnetic waves follows from Maxwell's equations.

8.2 PLANE ELECTROMAGNETIC WAVES IN NON-CONDUCTING MEDIA

Let us assume that ε, μ and σ are constants. The curl of Maxwell's equation $\nabla \times \mathbf{E} + \dfrac{\partial \mathbf{B}}{\partial t} = 0$ reads as

$$\nabla \times \nabla \times \mathbf{E} = -\nabla \times \frac{\partial \mathbf{B}}{\partial t} = -\mu \frac{\partial}{\partial t}(\nabla \times \mathbf{H}) \tag{8.1}$$

or

$$\nabla(\nabla \cdot \mathbf{E}) - \nabla^2 \mathbf{E} = -\mu \frac{\partial \mathbf{J}}{\partial t} - \mu \varepsilon \frac{\partial^2 \mathbf{E}}{\partial t^2} \tag{8.2}$$

In obtaining the above equation we have used the well known vector identity to transform the L.H.S. and Maxwell's equation $\nabla \times \mathbf{H} = \mathbf{J} + \dfrac{\partial \mathbf{D}}{\partial t}$ to transform the R. H. S. of Eq. (8.1). Rewritting Eq. (8.2), we have

$$\nabla^2 \mathbf{E} - \mu \varepsilon \frac{\partial^2 \mathbf{E}}{\partial t^2} - \mu \sigma \frac{\partial \mathbf{E}}{\partial t} = \nabla \left(\frac{\rho}{\varepsilon_0} \right) \tag{8.3}$$

In a region in which there is no free charge, i.e., $\rho = 0$, one obtains

$$\nabla^2 \mathbf{E} - \mu \varepsilon \frac{\partial^2 \mathbf{E}}{\partial t^2} - \mu \sigma \frac{\partial \mathbf{E}}{\partial t} = 0 \tag{8.4}$$

Assuming $E(\mathbf{r}, t) = E(\mathbf{r}) \exp(-i\omega t)$, one can compare the relative magnitudes of the second and the third term in Eq. (8.4). Obviously,

$$\frac{\left| \mu \sigma \frac{\partial \mathbf{E}}{\partial t} \right|}{\left| \mu \varepsilon \frac{\partial^2 \mathbf{E}}{\partial t^2} \right|} \simeq \frac{\sigma}{\varepsilon \omega} = \frac{1}{\omega \tau} = \frac{T}{2\pi \tau} \tag{8.5}$$

we must note that in obtaining (8.5) we have replaced $\frac{\varepsilon}{\sigma}$ by τ (relaxation time) and $\omega = \frac{2\pi}{T}$, T being the period of oscillation. If $\tau \ll T$, which is the case for a conducting medium, the third term in Eq. (8.4) is dominant and one can write (8.4) as

$$\nabla^2 \mathbf{E} - \mu\sigma \frac{\partial \mathbf{E}}{\partial t} = 0 \tag{8.6}$$

a diffusion equation.

If $\tau \gg T$, one can neglect the term involving σ in Eq. (8.4). One obtains for the non-conducting medium

$$\nabla^2 \mathbf{E} - \mu\varepsilon \frac{\partial^2 \mathbf{E}}{\partial t^2} = 0 \tag{8.7}$$

One can obtain similar equations for $\vec{H}$, viz.,

$$\nabla^2 \mathbf{H} - \mu\varepsilon \frac{\partial^2 \mathbf{H}}{\partial t^2} - \mu\sigma \frac{\partial \mathbf{H}}{\partial t} = 0 \tag{8.8}$$

and

$$\nabla^2 \mathbf{H} - \mu\varepsilon \frac{\partial^2 \mathbf{H}}{\partial t^2} = 0 \tag{8.9}$$

One can see that these equations are of the type of wave equation. We have the general wave equation:

$$\nabla^2 \psi - \frac{1}{v^2} \frac{\partial^2 \psi}{\partial t^2} = 0 \tag{8.10}$$

where v is the velocity of the propagation of the wave. Comparing (8.7), (8.9) and (8.10), we note that the velocity in our case is

$$v = \frac{1}{\sqrt{\varepsilon\mu}} \tag{8.11}$$

This reveals that the fields generated by moving charges can leave the source and travel through space in the form of waves. For vacuum or free space, $\varepsilon = \varepsilon_0 = 8.8542 \times 10^{-12}\ C^2/\mathrm{N}-\mathrm{m}^2$ and $\mu = \mu_0 = 4\pi \times 10^{-7}\ \mathrm{H/m}(= N - s^2/C^2)$. Thus

$$v = \frac{1}{\sqrt{\varepsilon_0\mu_0}} = 2.99794 \times 10^8 \mathrm{m/s} \approx 3 \times 10^8\ \mathrm{m/s} \tag{8.12}$$

Thus v represents the speed of propagation of electromagnetic waves in free space. This agrees very closely with the experimental value of the velocity of light in free space ($c = 2.99794 \times 10^8\ \mathrm{m/s}$). This result is of immense importance in physics. This unites the subjects of optics and electromagnetism.

We should note that eq. (8.10) is a scalar equation, while eqs. (8.6) and (8.9) are vector equations. This reveals that latter eqs. are valid for each component of **E** and **H**.

The simplest type of wave that is a solution of (8.6) or (8.9) is a plane wave. A plane wave is one in which the wave amplitude-the field vector component-is constant over all points of a plane normal to the direction of propagation. This plane constitutes a wave-front which advances with a velocity v in a direction normal to itself. The field vector components that lie in given plane are functions of perpendicular distance of the plane from the origin and also of time. If one chooses the coordinate system in such a way that the direction of wave propagation coincides with, say, x-axis. We can see that the wave function in this case does not depend on y and z. We can now write the one dimensional wave equation

$$\frac{\partial^2 \psi}{\partial x^2} - \frac{1}{v^2}\frac{\partial^2 \psi}{\partial t^2} = 0 \tag{8.13}$$

Eq. (8.13) has a general solution

$$\begin{aligned} \psi(x,t) &= A\exp[i(kx-\omega t)] + B\exp[-t(kx+\omega t)] \\ &= f(x-vt) + g(x+vt) \end{aligned} \tag{8.14}$$

A and B appearing in (8.14) are constants (usually complex) and $k = \frac{\omega}{v}$, where v is the *phase* velocity of the wave. Eq. (8.14) represents waves travelling to the right and to the left with velocity v. Let us assume, that the plane wave fields are of the form.

$$E(x,t) = E_0 \exp[i(kx-\omega t)] \tag{8.15}$$

$$H(x,t) = H_0 \exp[i(kx-\omega t)] \tag{8.16}$$

We have assumed that **E** ad **H** are in phase. One can find the justification for the said assumption in the following argument. Let us consider that α be the phase difference between **E** and **H**, i.e.

$$\mathbf{E} = \mathbf{E}_0 \exp[i(kx-\omega t)]$$

and

$$\mathbf{H} = \mathbf{H}_0 \exp[i(kx-\omega t+\alpha)]$$

Now, the fields **E** and **H** have to satisfy the Maxwell's equation,

$\nabla \times \vec{E} = -\frac{\partial \mathbf{B}}{\partial t}$. Therefore, we have

$$\frac{\partial E_z}{\partial x} = \frac{\partial B_y}{\partial t}; \quad \text{and} \quad \frac{\partial E_y}{\partial x} = -\frac{\partial B_z}{\partial t}$$

($\frac{\partial}{\partial y}$ and $\frac{\partial}{\partial z}$ components of **E** are zero since **E** travels in x direction).

From the second equation, we have

$$ikE_{oy}\exp[i(kx - \omega t)] = i\omega\mu H_{oz}\exp[i(kx - \omega t + \alpha)]$$

considering the real parts, one obtains

$$kE_{oy}\cos(kx - \omega t) = \omega\mu H_{oz}\cos(kx - \omega t + \alpha)$$

we may easily see that the above is true for all x ant t. We can further see that the only value of α which satisfies this condition is zero. This means **E** and **H** are in phase (Fig. 8.1). One finds the ratio of amplitudes of **E** and **H** as

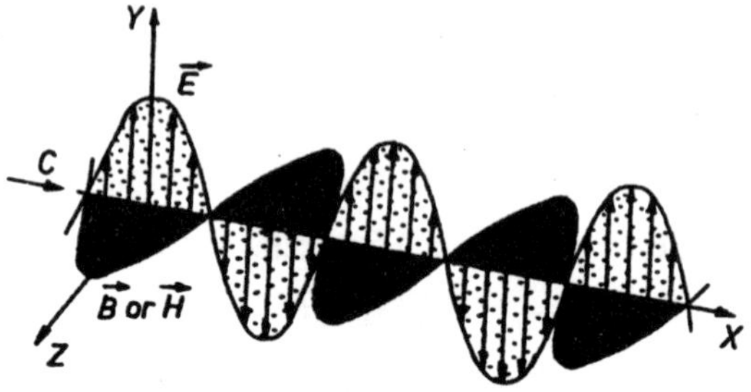

Figure 8.1.

$$\frac{E}{H} = \frac{\mu\omega}{k} = \mu v = Z_0 \tag{8.17}$$

Z_0 is called the intrinsic impedance of the medium, i.e. it has the dimensions of impedance. For free space,

$$Z_0 = \mu_0 V_0 = 376.7\Omega \quad (\because \quad \frac{E_0}{H_0} = \mu_0 V_0).$$

If **E** and **H** are in time phase, Z_0 is a pure resistance. This is the case for free space and all lossless dielectric media. To emphasize the fact that the impedance is a pure resistance, one can speak of the intrinsic resistance (instead of impedance) of free space and of lossless dielectric media. If **E** and **H** are not in time phase as in conducting media, the ratio of **E** to **H** is complex, so that the more general term intrinsic impedance must be used in connection with such media.

As the wave propagates, the two vectors **E** and **H** may both change their direction in $y-z$ plane without violating any of the conditions established so far.

For the propagation of wave in any arbitrary direction, one obtains

$$\mathbf{E}(\mathbf{r},t) = \mathbf{E}_0 \exp[i(\mathbf{k}\cdot\mathbf{r}-\omega \mathrm{t})] \tag{8.18}$$

and

$$\mathbf{H}(\mathbf{r},t) = \mathbf{H}_0 \exp[i(\mathbf{k}\cdot\mathbf{r}-\omega t)] \tag{8.19}$$

where $\mathbf{E}_0$ and $\mathbf{H}_0$ are vector constant in time and $\mathbf{k} = \hat{a}_k|\mathbf{k}|$ is the propagation vector, and $\hat{a}_k$ is the unit vector in the direction of propagation. We note that **E** and **H** are real and hence we are interested only in the real part of Eqs. (8.18) and (8.19). Moreover, **E** and **H** obtained above must also satisfy Maxwell's equations. However, we must note that this is not automatic. Maxwell's equations do not completely determine the electromagnetic field. Using Eq. (8.18), Maxwell's eq. $\nabla\cdot\mathbf{D} = \rho$ takes the form

$$\nabla\cdot\mathbf{E}_0 \exp[i(\mathbf{k}\cdot\mathbf{r}-\omega t)] = 0 \quad (\because \quad \rho = 0)$$

This means

$$\mathbf{k}\cdot\mathbf{E} = 0 \tag{8.20}$$

Similarly, substituting (8.19) in Maxwell's eq. $\nabla\cdot\mathbf{B} = 0$, one obtains

$$\mathbf{k}\cdot\mathbf{H} = 0 \tag{8.21}$$

Eqs. (8.21) and (8.22) reveals that both **E** and **H** are perpendicular to the propagation vector **k**. This means that such a wave is *transverse wave*. Obviously, electromagnetic plane waves are wholly *transverse* in character. Now, we substitute Eq. (8.18) in Maxwell's equation $\nabla\times\mathbf{E} + \dfrac{\partial\mathbf{B}}{\partial t} = 0$, one obtains

$$\begin{aligned}
\nabla\times\mathbf{E} &= \nabla\times\mathbf{E}_0 \exp[i(\mathbf{k}\cdot\mathbf{r}-\omega t)] \\
&= \nabla\exp[i(\mathbf{k}\cdot\mathbf{r}-\omega t)]\times\mathbf{E}_0 \\
&= i\mathbf{k}\times\mathbf{E} = -\frac{\partial\mathbf{B}}{\partial t} = -\mu\frac{\partial\mathbf{H}}{\partial t} = i\omega\mu\mathbf{H}
\end{aligned} \tag{8.22}$$

This means $\mathbf{k}\times\mathbf{E} = \omega\mu\mathbf{H}$.

Relation (8.22) reveals that **H** is perpendicular both to **k** and **E**. This means, **E** and **H**, which relate to the propagation of the wave, in addition to being perpendicular to the direction of propagation are also perpendicular to one another. The vector **E** × **H** points along the direction of propagation. We can see that the vectors, **E**, **H** and **k** constitute a right handed orthogonal set (Fig. 8.2). This right handed relation is an intrinsic property of an electromagnetic wave independent of the choice of coordinate system.

We have seen that the velocity of the electromagnetic magnetic waves in a non conducting natural non ferromagnetic medium is the same as the velocity of light in free space. The vectors **E** and **B** always oscillate in phase, the instantaneous values of E and B at any point being connected through the relation $E = vB$ or $\sqrt{\varepsilon\varepsilon_0}\, E = \sqrt{\mu\mu_0} H$. This means that E and H (or B) simultaneously attain their maximum values, vanish etc.

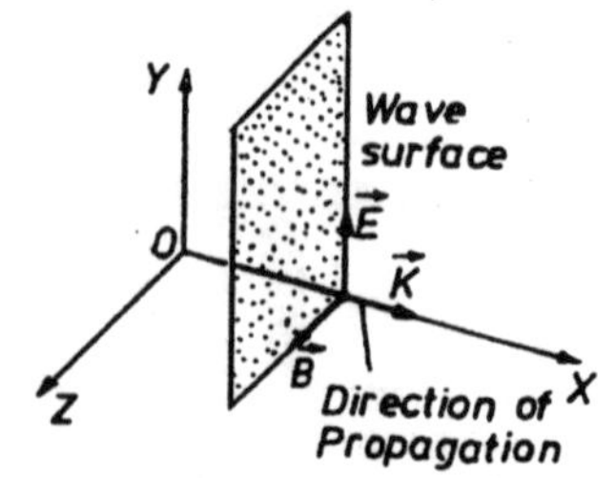

Figure 8.2. Orientation of the electric and magnetic vector fields relative to the direction of propagation of plane electromagnetic wave

Maxwell's brilliant success in the development of the electromagnetic theory of light was due to his understanding of the possibility of existence of electromagnetic waves, following from differential equations discussed in Chapter 7.

Electromagnetic waves cover a wide range of frequencies or wavelengths, and may be classified according to their main source. The classification does not have very sharp boundaries, since different sources may produce waves in overlapping ranges of frequencies. Fig. 8.3. relate the various sections of the electromagnetic spectrum in terms of energy, frequency and wavelength. Light is simply a form of electromagnetic radiation. x-rays, ultraviolet rays, infra red rays, radiowaves and microwaves are all electromagnetic radiations. They differ from each other only in the order of magnitude of their wavelength, i.e. frequency. Obviously, all these travel with the same speed, i.e. speed of light in free space.

So far we have considered only the propagation of electromagnetic waves in vacuum. Experiments reveal that the velocity of propagation of an electromagnetic wave through matter is different from its value of propagation in vacuum. The phase velocity of electromagnetic waves

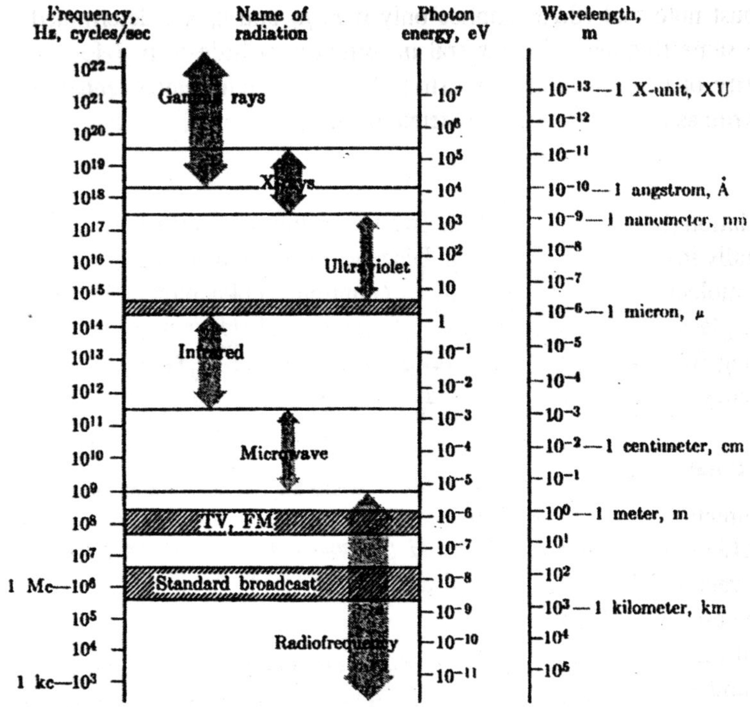

Figure 8.3. The electromagnetic spectrum.

in a material medium is given by

$$v = \frac{q}{\sqrt{\varepsilon\mu}} = \frac{1}{\sqrt{\varepsilon_0\varepsilon_r\mu_0\mu_r}} = \frac{v_0}{\sqrt{\varepsilon_r\mu_r}} \tag{8.23}$$

which is less that V_0 i.e. $v < v_0$.

The ratio between the velocity of electromagnetic waves in vacuum c, and in matter, v. is called the *absolute index of refraction* of the substance, designated by n. It is a useful concept for describing the properties of materials in relation to electromagnetic waves. Thus

$$n = \frac{c}{v} = \frac{\sqrt{\varepsilon\mu}}{\sqrt{\varepsilon_0\mu_0}} = \sqrt{\frac{\varepsilon\mu}{\varepsilon_0\mu_0}} \tag{8.24}$$

But $\frac{\varepsilon}{\varepsilon_0} = \varepsilon_r$ and $\frac{\mu}{\mu_0} = \mu_r$, where ε_r and μ_r are the relative permittivity and permeability of the medium. Then

$$n = \frac{c}{v} = \sqrt{\varepsilon_r\mu_r} \tag{8.25}$$

we must note that (8.25) applies only if n, μ_r and ε_r are determined at the same frequency. In general μ_r differs very little from 1 for the majority of substances that transmit electromagnetic waves, and one can write as a satisfactory approximation,

$$n = \sqrt{\varepsilon_r} \tag{8.26}$$

Relation (8.26) is termed as Maxwell's relation and is varified experimentally in case of air, hydrogen, benzene and carbon etc. having non-polar molecules. But for substances containing polar molecules e.g., water, glass etc., ε_r is dependent on frequency and decreases with increasing frequency. Obviously, refractive index exhibit variation with frequency—a phenomenon termed as *dispersion.*

8.3 Polarization

The direction **E** in Eq. (8.18) is constant in time. A wave represented by (8.18) is said to be *linearly polarized* wave. The plane of electric wave vector **E** is taken as the plane of polarization. We can see that in the case of (8.15), it is $y-z$ plane. One can express a linearly polarized monochromatic plane wave as a superposition of two linearly independent solutions of the wave equation, e.g.

$$\mathbf{E} = (\hat{a}_y E_{oy} + \hat{a}_z E_{oz}) \exp[i(kx - \omega t)] \tag{8.27}$$

where $\hat{a}_y$ and $\hat{a}_z$ are the unit vectors along y and z respectively. These unit vectors are called *polarization vectors.* We must note that the amplitudes E_{oy} and E_{oz} are complex amplitudes. One can always express any complex quantity as the product of a real quantity and a complex phase factor. Thus we have

$$E_{oy} = E_0^y \exp(i\alpha)$$

or

$$E_{oz} = E_0^z \exp(i\beta) \tag{8.28}$$

E_0^y and E_0^z in (8.28) are the real amplitudes. One finds the two independent solutions as

$$\begin{aligned} \mathbf{E}_y &= \hat{a}_y E_0^y \exp[i(kx - \omega t + \alpha)] \\ \mathbf{E}_z &= \hat{a}_z E_0^z \exp[i(kx - \omega t + \beta)] \end{aligned} \tag{8.29}$$

$$\therefore \quad \begin{aligned} \mathbf{E} &= \mathbf{E}_y + \mathbf{E}_z \\ &= (\hat{a}_y E_0^y \exp(i\alpha) + \hat{a}_z E_0^z \exp(i\beta)) \exp[i(kx - \omega t)] \end{aligned} \tag{8.30}$$

or $$\mathbf{E} = (\hat{a}_y E_0^y + \hat{a}_z E_0^z \exp[i(\beta - \alpha)]) \exp[i(kx - \omega t + \alpha)] \quad (8.31)$$

Particular cases of interest are:

(i) $\alpha = \beta$, i.e. $\mathbf{E}_y$ and $\mathbf{E}_z$ have the same phase or their phase differ by an integral multiple of π, i.e. $\beta = \alpha \pm m\pi$, where $m = 0, 1, 2, \ldots$, then

$$\mathbf{E} = (\hat{a}_y E_0^y \pm \hat{a}_z E_0^z) \exp[i(kx - \omega t + \alpha)] \quad (8.32)$$

Eq.(8.32) represents a linearly polarized wave. One finds that the resultant polarization vector oscillates along a line making an angle $\theta = \tan^{-1} \dfrac{E_0^z}{E_0^y}$ (Fig. 8.4).

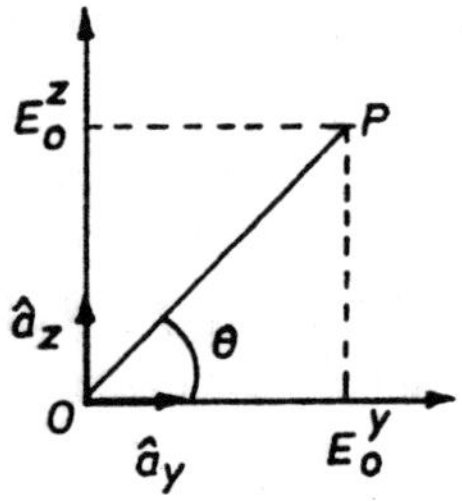

Figure 8.4.

(ii) We can see that the amplitudes of two vectors are equal, i.e. $E_0^y = E_0^z = E_0$ but their phases differ by $\dfrac{\pi}{2}$. Therefore

$$\mathbf{E} = E_0(\hat{a}_y \pm i\hat{a}_z) \exp[i(kx - \omega t + \alpha) \quad (8.33)$$

One can write the components of $\mathbf{E}$ (only the real part) as

$$\begin{aligned} E_y &= E_0 \cos(kx - \omega t + \alpha) \\ E_z &= \pm E_0 \sin(kx - \omega t + \alpha) \end{aligned} \quad (8.34)$$

Obviously, (8.33)is a wave progressing along the x-axis in which the vector $\mathbf{E}$ has a constant length $\mathbf{E}_0$ and is a function of time only changing its direction continuously. One obtains from (8.34),

$$\frac{E_y^2}{E_0^2} + \frac{E_z^2}{E_0^2} = 1 \quad (8.35)$$

This means that the vector, traces out a circle at a frequency ω. We know that such a wave is said to be *circularly polarized* Fig. 8.5. One can determine the direction of rotation by the sign of E_z in (8.34). If we assume that the observer is facing the oncoming wave and sign is plus, then the rotation is counter clockwise. Such a wave has *positive helicity* and is said to have *left circular polarization.* When the rotation is clockwise i.e. the sign is minus the wave has the *negative helicity*, i.e. the wave is *right circularly polarized.*

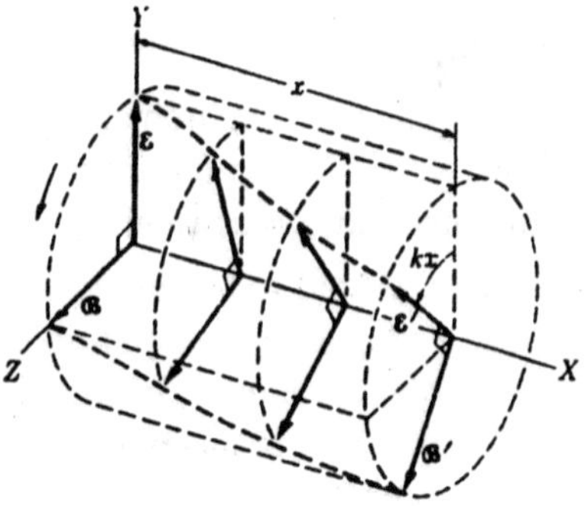

Figure 8.5. Circularly polarized electromagnetic wave. The E and H fields rotate around the direction of propagation.

(iii) When $E_0^y \neq E_0^z$ and $\beta = \alpha \pm \frac{\pi}{2}$, one obtains the components of **E** as

$$\begin{aligned} E_y &= E_0^y \cos[i(kx - \omega t + \alpha)] \\ E_z &= \mp E_0^z [i \sin(kx - \omega t + \alpha \pm \pi/2)] \end{aligned}$$

Now, considering the real parts, one obtains

$$\begin{aligned} E_y &= E_0^y \cos(kx - \omega t + \alpha) \\ E_z &= \mp E_0^z \sin(kx - \omega t + \alpha) \end{aligned}$$

$$\therefore \quad \frac{E_y^2}{(E_0^y)^2} + \frac{E_z^2}{(E_o^z)^2} = 1 \qquad (8.36)$$

We can see that the resultant vector in this case traces an ellipse and one finds that the wave is *elliptically polarized.*

An arbitrary plane wave can be represented as the combination of two plane waves linearly polarized in mutually perpendicular planes.

The amplitude a of a linearly polarized monochromatic electromagnetic wave is the maximum value of $|\mathbf{E}|$. Obviously, $a = |\mathbf{E}|_{max}$.

The intensity I of an electromagnetic wave is a quantity numerically equal to the energy transported by the wave per unit time through a surface of unit area perpendicular to the direction of propagation. The intensity I is related to the Poynting vector **N** as

$$I = |\vec{N}| = \frac{1}{T}\left|\int_0^T \mathbf{N}dt\right| \tag{8.36(a)}$$

where T is the period of the wave.

For a linearly polarized plane monochromatic wave, $I \propto a^2$, where a is the amplitude of the wave. For an arbitrary plane wave in a homogenous nonabsorbing medium, I = constant.

An electromagnetic wave is said to be *spherical* if its intensity depends only on the distance r from a certain point called the wave centre. It follows from the law of conservation of energy that for a spherical wave in a homogeneous nonabsorbing medium $I = \text{constant} / r^2$.

Real electromagnetic waves are not monochromatic, if for no other reason than that they always are of limited extent in space and of limited duration in time. Such waves can be represented as a totality of monochromatic waves which is called a *wave train* or *packet*. As it is propagated in a dispersive medium the shape of a wave train is distorted because of the different phase velocities of monochromatic components of the train. The conception of the phase velocity is therefore insufficient to characterize the propagation of a wave train and its rate of energy transport, i.e. the rate of propagation of a signal.

8.4 ENERGY FLUX IN A PLANE ELECTRO MAGNETIC WAVE

We have studied about the Poynting vector $\mathbf{N} = \mathbf{E} \times \mathbf{H}$ in Chapter 7. The Poynting vector represents the rate at which the energy is radiated across unit area. Now we will calculate the Poynting vector for a plane wave in which the field vector **E** and **H** are expressed in terms of complex amplitudes.

The Poynting vector $\mathbf{N} = \mathbf{E} \times \mathbf{H}$ gives the instantaneous rate of energy flow. We know that **E** and **H** vary harmonically with time and therefore one can obtain the average energy flow by taking the average of $\mathbf{N} = \mathbf{E} \times \mathbf{H}$ over a complete period, i.e.

$$< N > = < \text{Realpart } \mathbf{E} \times \text{Realpart } \mathbf{H} > \tag{8.37}$$

Since **E** and **H** are assumed to be complex quantities and therefore one can express them as

$$\mathbf{E} = (\mathbf{E}_1 + i\mathbf{E}_2)\exp(-i\omega t)$$

and
$$\mathbf{H} = (\mathbf{H}_1 + i\mathbf{H}_2)\exp(-i\omega t) \tag{8.38}$$

where E_1, E_2, H_1 and H_2 are all real. Now

$$\text{Real } \mathbf{E} = \mathbf{E}_1 \cos\omega t + \mathbf{E}_2 \sin\omega t$$
$$\text{Real } \mathbf{H} = \mathbf{H}_1 \cos\omega t + \mathbf{H}_2 \sin\omega t$$

$$\therefore \quad \text{Real } \mathbf{E} \times \text{Real}\mathbf{H} = (\mathbf{E}_1 \times \mathbf{H}_1)\cos^2\omega t + (\mathbf{E}_2 \times \mathbf{H}_2)\sin^2\omega t + \{(\mathbf{E}_1 \times \mathbf{H}_1) + (\mathbf{E}_2 \times \mathbf{H}_1)\}\sin\omega t \cos\omega t$$

Over a complete period of oscillation, we have

$$< \cos^2\omega t > = < \sin^2\omega t > = \frac{1}{2} \text{ and } < \sin\omega t \cos\omega t > = 0$$

$$\therefore \quad < \text{Real } \mathbf{E} \times \text{Real}\mathbf{H} > = \frac{1}{2}\{(\mathbf{E}_1 \times \mathbf{H}_1) + (\mathbf{E}_2 \times \mathbf{H}_2)\} \tag{8.39}$$

Now we compute Real $(\mathbf{E} \times \mathbf{H}^*)$. We have

$$\mathbf{E} = (\mathbf{E}_1 + i\mathbf{E}_2)\exp(-i\omega t) = (\mathbf{E}_1 + i\mathbf{E}_2)(\cos\omega t - i\sin\omega t)$$

and $$\mathbf{H}^* = (\mathbf{H}_1 - i\mathbf{H}_2)\exp(i\omega t) = (\mathbf{H}_1 - i\mathbf{H}_2)(\cos\omega t + i\sin\omega t)$$

$$\begin{aligned}
\therefore \quad \text{Real } (\mathbf{E} \times \mathbf{H}^*) &= (\mathbf{E}_1 \times \mathbf{H}_1)\cos^2\omega t + (\mathbf{E}_1 \times \mathbf{H}_1)\cos\omega t \sin\omega t \\
&+ (\mathbf{E}_2 \times \mathbf{H}_2)\cos^2\omega t - (\mathbf{E}_2 \times \mathbf{H}_1)\cos\omega t \sin\omega t \\
&- (\mathbf{E}_1 \times \mathbf{H}_2)\cos\omega t \sin\omega t + (\mathbf{E}_1 \times \mathbf{H}_1)\sin^2\omega t \\
&+ (\mathbf{E}_2 \times \mathbf{H}_1)\cos\omega t \sin\omega t + (\mathbf{E}_2 \times \mathbf{H}_2)\sin^2\omega t \\
&= (\mathbf{E}_1 \times \mathbf{H}_1) + (\mathbf{E}_2 \times \mathbf{H}_2)
\end{aligned} \tag{8.40}$$

Obviously $< \text{Real } \mathbf{E} \times \text{Real } \mathbf{H} > = \frac{1}{2}\text{Real } (\mathbf{E} \times \mathbf{H}^*)$

$$\therefore \quad < N > = \frac{1}{2} \text{ Real } (\mathbf{E} \times \mathbf{H}^*) \tag{8.41}$$

we note that the time factors are automatically cancelled in the product. Thus one can obtain the time averaged energy from the field vectors.

We have $\mathbf{k} \times \mathbf{E} = \omega\mu\mathbf{H}$ (from Eq. (8.22))

$$\therefore \quad H^* = \frac{\mathbf{k} \times \mathbf{E}^*}{\omega\mu} \tag{8.42}$$

$$\therefore \quad \mathbf{E}\times\mathbf{H}^* = \frac{1}{\omega\mu}[\mathbf{E}\times(\mathbf{k}\times E^*)] = \frac{1}{\omega\mu}[(\mathbf{E}\cdot\mathbf{E}^*)\mathbf{k}-(\mathbf{E}\cdot\mathbf{k})\mathbf{E}^*]$$

$$= \frac{1}{\omega\mu}|E_0^2|\mathbf{k}$$

$$\therefore \quad \text{Average of } \mathbf{N}, \textit{i.e.}, <\mathbf{N}> = \frac{|\mathbf{E}_0^2|}{2\omega\mu}\mathbf{k}$$

$$= \frac{1}{2}|E_0|^2\left(\frac{\varepsilon}{\mu}\right)^{1/2}\hat{a}_k \qquad (8.43)$$

The above result has been obtained by using $\mathbf{k} = k\hat{a}_k$, where $\hat{a}_k$ is the unit vector in the direction of propagation and $k = \frac{\omega}{v} = \omega\sqrt{\varepsilon\mu}$.

Let us consider that the wave is moving in the x-direction, i.e. energy flow is in the x-direction and is the same at all points for which z-coordinate is the same. A rectangular box with length c along the x-axis and a and b being its other two sides is shown in Fig. 8.6. We have the energy density in an electromagnetic field $= \frac{1}{2}(\mathbf{E}\cdot\mathbf{D}+\mathbf{B}\cdot\mathbf{H})$. One obtains the total energy as

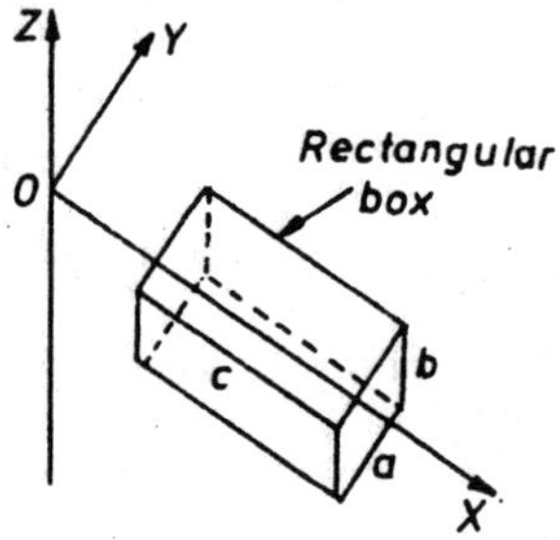

Figure 8.6.

$$U = \frac{1}{2}\int_v(\mathbf{E}\cdot\mathbf{D}+\mathbf{B}\cdot\mathbf{H})d\tau \qquad (8.44)$$

Let us assume that the fields **E** and **H** act along y and z-axis respectively. We obtain

$$\mathbf{E} = \hat{a}_y E_0 \exp[i(kx-\omega t)] = \hat{a}_y E_0\cos(kx-\omega t)$$

and

$$\mathbf{H} = \hat{a}_z H_0\cos(kx-\omega t)$$

$$\therefore \quad U = \frac{1}{2}\int_0^c\int_0^a\int_0^b[\varepsilon_0 E_0^2\cos^2(kx-\omega t)$$
$$+\mu_0 H_0^2\cos^2(kx-\omega t)]dx\,dy\,dz$$
$$= \frac{ab}{2}\int_0^c[\varepsilon_0 E_0^2\cos^2(kx-\omega t)+\mu_0 H_0^2\cos^2(kx-\omega t)]dx$$

Let T is the period of oscillation. One obtains the average energy in the box as

$$\begin{aligned} <U> &= \frac{ab}{2T}\int_0^T\int_0^c [\varepsilon_0 E_0^2\cos^2(kx-\omega t)+\mu H_0^2\cos^2(kx-\omega t)]\,dx\,dt \\ &= \frac{ab}{2}\int_0^c \frac{1}{2}[\varepsilon_0 E_0^2+\mu_0 H_0^2]dx = \frac{abc}{4}(\varepsilon_0 E^2+\mu_0 H_0^2) \qquad (8.45)\end{aligned}$$

Now, we obtain the time averaged energy density associated with a wave as

$$\begin{aligned} <U'> &= \frac{1}{4}(\mathbf{E}\cdot\mathbf{D}^*+\mathbf{B}\cdot\mathbf{H}^*) = \frac{1}{4}(\varepsilon E^2+\mu H^2) \\ &= \frac{1}{4}\left[\varepsilon|E_0|^2+\mu\frac{\mathbf{k}\times\mathbf{E}}{\omega\mu}\cdot\frac{\mathbf{k}\times\mathbf{E}^*}{\omega\mu}\right] \\ &= \frac{1}{4}[\varepsilon|E_0|^2+\varepsilon|E_0|^2] = \frac{1}{2}\,\varepsilon|E_0|^2 \qquad (8.46)\end{aligned}$$

$$\therefore\quad <N> = \frac{1}{2}\sqrt{\frac{\varepsilon}{\mu}}\hat{a}_k|E_0|^2 = <U'>\frac{\hat{a}_k}{\sqrt{\varepsilon\mu}} = v<U'>\hat{a}_k \qquad (8.47)$$

Eq. (8.47) reveals that the time averaged energy flow is in the direction of propagation of the wave and is equal to the phase velocity of the wave multiplied by the average energy density. Obviously, the energy flows with the same velocity as the wave itself. One can now summarize important results regarding electromagnetic waves in free space as

(i) In free space electromagnetic waves travel with the speed of light.

(ii) The electromagnetic field vectors **E** and **H** are mutually perpendicular and they are also perpendicular to the direction of propagation of electromagnetic waves. This shows that electromagnetic waves are transverse in nature.

(iii) **E** and **H** field vectors are in the same phase.

(iv) The direction of the flow of electromagnetic energy is along the direction of wave propagation and the energy flow per unit area per second is given by (8.47).

(v) The electrostatic energy density is equal to the magnetic energy density and the energy density associated with the electromagnetic wave in free space propagates with the speed of light.

$$\text{Energy density} = \frac{1}{2}\varepsilon E^2+\frac{1}{2}\mu H^2 = \varepsilon E^2 = \mu H^2\ \mathrm{J/m^3}.$$

8.5 PRESSURE AND MOMENTUM OF AN ELECTROMAGNETIC WAVE

Let us consider a plane polarized wave travelling in the z-direction. We must note that the only non-zero quantities involved are E_x, H_y and J_x and only variation in z-coordinate is permitted. In this case, Maxwell's equations

$$\nabla \times \mathbf{E} = -\frac{\partial \mathbf{B}}{\partial t} \quad \text{and} \quad \nabla \times \mathbf{H} = \mathbf{J} + \frac{\partial \mathbf{D}}{\partial t}$$

reduces to

$$\frac{\partial E_x}{\partial z} = -\frac{\partial B_y}{\partial t} \quad \text{and} \quad -\frac{\partial H_y}{\partial z} = J_x + \frac{\partial D_x}{\partial t} \tag{8.48}$$

We note that any force on a body is due to the action of the wave on currents and charges induced within the body. One obtains the force per unit area as

$$\mathbf{F} = \rho \mathbf{E} + \mathbf{J} \times \mathbf{B} \tag{8.49}$$

we have $\nabla \cdot \mathbf{E} = 0$, $\rho = 0$. Thus the only component of the force is

$$\begin{aligned} F_z = J_x B_y &= -B_y \left(\frac{\partial H_y}{\partial z} + \frac{\partial D_x}{\partial t} \right) \\ &= -B_y \frac{\partial H_y}{\partial z} + D_x \frac{\partial B_y}{\partial t} - \frac{\partial}{\partial t}(D_x B_y) \\ &= -B_y \frac{\partial H_y}{\partial z} - D_x \frac{\partial E_x}{\partial z} - \frac{\partial}{\partial t}(D_x B_y) \\ &= -\frac{1}{2}\frac{\partial}{\partial z}(\varepsilon_0 E_x^2 + \mu_0 H_y^2) - \mu_0 \varepsilon_0 \frac{\partial}{\partial t}(E_x H_y) \\ &= -\frac{1}{2}\frac{\partial}{\partial z}(\varepsilon_0 E_x^2 + \mu_0 H_y^2) - \mu_0 \varepsilon_0 \frac{\partial N}{\partial t} \end{aligned}$$

Let us consider that the currents are confined to a region $Z_1 < Z < Z_2$, then the total force per unit area, i.e. mean pressure on the region is obtained as

$$\mathcal{F}_z = \left[\frac{1}{2}(\varepsilon_0 E_x^2 + \mu_0 H_y^2) \right]_{z_1} - \left[\frac{1}{2}(\varepsilon_0 E_x^2 + \mu_0 H_y^2) \right] - \mu_0 \varepsilon_0 \int_{z_1}^{z_2} \frac{\partial N}{\partial t} dz \tag{8.50}$$

we can see that the fields E_x, H_y in the first two terms of (8.50) refer to the values just outside the region considered. Obviously, we can consider these two terms as an action from the outside of the region and clearly these represents an external force F_e. We also note the term

$\frac{1}{2}(\varepsilon_0 E_x^2 + \mu_0 H_y^2)$ represent the pressure due to radiation. One can easily show that

$$\frac{1}{2}(\varepsilon_0 E_x^2 + \mu_0 H_y^2) = \frac{N}{c}$$

Now, we can express (8.50) as

$$F_z = F_e - \frac{dP}{dt}$$

where,

$$p = \mu_0 \varepsilon_0 \int_{z_1}^{z_2} N dz \tag{8.51}$$

In accordance with the Newton's second law of motion, one can interpret p as a momentum. Thus one can attribute to the electromagnetic wave a momentum equal to

$$\mu_0 \varepsilon_0 \mathbf{N} = \frac{\mathbf{N}}{c^2} \tag{8.52}$$

we can see that $\frac{\mathbf{N}}{c^2}$ has the dimensions of $\mathrm{m}^{-2}\ \mathrm{kg\ s}^{-1}$, which corresponds to momentum per unit volume. Since electromagnetic radiation propagates with velocity c, we may use the relation between energy (E), momentum (p); $p = \frac{vE}{c^2}$, (with $v = c$). Since momentum must have the direction of propagation, we may write

$$\mathbf{p} = \frac{E}{c}\hat{a}_n = \frac{\mathbf{N}}{c^2} \tag{8.53}$$

where $\hat{a}_n$ is the unit vector in the direction of propagation.

If an electromagnetic wave has momentum, it also has angular momentum. The angular momentum per unit volume is

$$\mathbf{L} = \mathbf{r} \times \mathbf{p} = \mathbf{r} \times \frac{\mathbf{N}}{c^2} \tag{8.54}$$

This is called the "orbital" angular momentum of radiation. In addition, electromagnetic radiation possesses an intrinsic angular momentum or spin, similar to the spin of fundamental particles. For circularly polarized plane waves it can be shown that the spin has a component along the direction of propagation equal to $\mp E/\omega$ ($E \rightarrow$ energy), depending on whether the polarization is clockwise or counterclockwise. For a linearly polarized wave, the average value of the component of the spin along the direction of propagation is zero.

Therefore, when a charged particle absorbs or emits electromagnetic radiation, not only do its energy and momentum change but also its angular momentum changes accordingly, a result has been verified experimentally both directly and indirectly. In summary we conclude that *an electromagnetic wave carries momentum and angular momentum as well as energy.*

The result is not surprising. An electromagnetic interaction between two electric charges means an exchange of energy and momentum between the charges. This is accomplished by means of the electromagnetic field, which is the carrier of the energy and momentum exchanges.

8.6 PLANE ELECTROMAGNETIC WAVES IN A CONDUCTING MEDIUM

We have Maxwell's equation

$$\left.\begin{aligned} \nabla\cdot\mathbf{D} &= \rho \\ \nabla\cdot\mathbf{B} &= 0 \\ \nabla\times\mathbf{E} &= -\frac{\partial\mathbf{B}}{\partial t} \\ \text{and}\quad \nabla\times\mathbf{H} &= \mathbf{J}+\frac{\partial\mathbf{D}}{\partial t} \end{aligned}\right\} \tag{8.55}$$

Let us assume that medium is linear and isotropic and is characterised by permittivity ε, permeability μ and conductivity σ, but not any charge or any current other than that determined by Ohm's law. Thus we have

$$\mathbf{D}=\varepsilon\mathbf{E},\quad \mathbf{B}=\mu\mathbf{M},\quad \mathbf{J}=\sigma\mathbf{E}\quad\text{and}\quad \rho=0$$

Using the above, Maxwell's equation (8.55) takes the form

$$\left.\begin{aligned} \nabla\cdot\mathbf{E} &= 0 && \text{(i)} \\ \nabla\cdot\mathbf{H} &= 0 && \text{(ii)} \\ \nabla\times\mathbf{E} &= -\mu\frac{\partial\mathbf{H}}{\partial t} && \text{(iii)} \\ \text{and}\quad \nabla\times\mathbf{H} &= \sigma\mathbf{E}+\varepsilon\frac{\partial\mathbf{E}}{\partial t} && \text{(iv)} \end{aligned}\right\} \tag{8.56}$$

Taking curl of 8.56 (iii), one obtains

$$\nabla\times\nabla\times\mathbf{E}=-\mu\frac{\partial}{\partial t}(\nabla\times\mathbf{H})$$

Substituting $\nabla \times \mathbf{H}$ from 8.56(iv), one obtains

$$\nabla \times \nabla \times \mathbf{E} = -\mu \frac{\partial}{\partial t}\left(\sigma \mathbf{E} + \varepsilon \frac{\partial \mathbf{E}}{\partial t}\right)$$

or

$$\nabla \times \nabla \times \mathbf{E} = -\sigma\mu \frac{\partial \mathbf{E}}{\partial t} - \varepsilon\mu \frac{\partial^2 \mathbf{E}}{\partial t^2} \tag{8.57}$$

Similarly, if we take the curl of 8.56(d) and substitute $\nabla \times \mathbf{E}$ from 8.56(c), one obtains

$$\nabla \times \nabla \times \mathbf{H} = -\sigma\mu \frac{\partial \mathbf{H}}{\partial t} - \varepsilon\mu \frac{\partial^2 \mathbf{H}}{\partial t^2} \tag{8.58}$$

Now using vector identity

$$\nabla \times \nabla \times \mathbf{A} = \nabla(\nabla \cdot \mathbf{A}) - \nabla^2 \mathbf{A}$$

and keeping in mind 8.56 (i) and 8.56 (ii) (i.e. $\nabla \cdot \mathbf{E} = 0$ and $\nabla \cdot \mathbf{H} = 0$), Eqs. (8.57) and (8.58) takes the form

$$\nabla^2 \mathbf{E} - \sigma\mu \frac{\partial \mathbf{E}}{\partial t} - \varepsilon\mu \frac{\partial^2 \mathbf{E}}{\partial t^2} = 0 \tag{8.59}$$

$$\nabla^2 \mathbf{H} - \sigma\mu \frac{\partial \mathbf{H}}{\partial t} - \varepsilon\mu \frac{\partial^2 \mathbf{H}}{\partial t^2} = 0 \tag{8.60}$$

Eqs. (8.59) and (8.60) represent wave equations governing electromagnetic fields **E** and **H** in a homogeneous isotropic conducting medium of conductivity σ. We can see that these equations are vector equations of identical form. This means that each of the six components of **E** and **H** separately satisfies the same scalar wave equation of the form

$$\nabla^2 \psi - \sigma\mu \frac{\partial \psi}{\partial t} - \varepsilon\mu \frac{\partial^2 \psi}{\partial t^2} = 0 \tag{8.61}$$

where ψ is a scalar, and holds good for any one of components of **E** and **H**. Eq. (8.61) contain damping terms proportional to the conductivity of the medium. Writing (8.61) for **E** and **H**, we have

$$\nabla^2 \mathbf{E} - \varepsilon\mu \frac{\partial^2 \mathbf{E}}{\partial t^2} - \sigma\mu \frac{\partial \mathbf{E}}{\partial t} = 0 \tag{8.62}$$

and

$$\nabla^2 \mathbf{H} - \varepsilon\mu \frac{\partial^2 \mathbf{H}}{\partial t^2} - \sigma\mu \frac{\partial \mathbf{H}}{\partial t} = 0 \tag{8.63}$$

Let us find the plane wave solution of Maxwell's equations for a conducting medium. Let us assume that field vectors **E** and **H** vary harmonically with time, i.e.

$$\mathbf{E}(\mathbf{r}, t) = E_0 \exp[i(\mathbf{k} \cdot \mathbf{r} - \omega t)] \tag{8.64}$$

and
$$\mathbf{H}(\mathbf{r},t) = H_0 \exp[i(\mathbf{k}\cdot\mathbf{r} - \omega t)] \tag{8.65}$$

Substituting (8.64) in (8.62), one obtains

$$-k^2\mathbf{E}(r,t) + \varepsilon\mu\omega^2\mathbf{E}(r,t) + i\sigma\mu\omega\mathbf{E}(r,t) = 0$$

or

$$[k^2 - \varepsilon\mu\omega^2 - i\sigma\mu\omega]\mathbf{E}(r,t) = 0 \tag{8.66}$$

$$\therefore \quad k^2 = \varepsilon\mu\omega^2\left[1 + \frac{i\sigma}{\varepsilon\omega}\right] \tag{8.67}$$

we can see that the first term corresponds to the displacement current and the second to the conduction current.

We know that for any wave there is a fundamental relationship between the wave number k and wave frequency ω. This relationship is known as *dispersion relation*. The dispersion relation for electromagnetic waves in vacuum is

$$k = \frac{\omega}{v} \tag{8.68}$$

For a conducting medium k is given by (8.67). We can see that k is complex. For a free space, $\sigma = 0$ and we have $k^2 = \varepsilon\mu\omega^2 = \frac{\omega^2}{v^2}$. This relation will provide us with information regarding the nature of the propagation of electromagnetic waves inside a medium.

As stated above k is complex in a conducting medium. We may write assuming that σ is real

$$k = \alpha + i\beta \tag{8.69}$$

so that
$$k^2 = \alpha^2 - \beta^2 + 2i\alpha\beta \tag{8.70}$$

Comparing (8.70) with (8.67), one obtains

$$\left.\begin{aligned} \alpha^2 - \beta^2 &= \varepsilon\mu\omega^2 \\ 2\alpha\beta &= \sigma\mu\omega \end{aligned}\right\} \tag{8.71}$$

Solving these equations, one obtains

$$\alpha = \omega\sqrt{\frac{\varepsilon\mu}{2}}\left[1 + \left\{1 + \left(\frac{\sigma}{\varepsilon\omega}\right)^2\right\}^{1/2}\right]^{1/2} \tag{8.72}$$

$$\beta = \omega\sqrt{\frac{\varepsilon\mu}{2}}\left[-1 + \left\{1 + \left(\frac{\sigma}{\varepsilon\omega}\right)^2\right\}^{1/2}\right]^{1/2} \tag{8.73}$$

In order to make the solutions yield the proper form for k in free space, we have to consider positive square root only.

Since $k = \alpha + i\beta$, one can write Eqs. (8.64) and (8.65) as

$$\mathbf{E} = \mathbf{E}_0 \exp(-\beta r) \exp[i(\alpha r - \omega t)] \tag{8.74}$$

and

$$\mathbf{H} = \mathbf{H}_0 \exp(-\beta r) \exp[i(\alpha r - \omega t)] \tag{8.75}$$

From Eqs. (8.74) and (8.75) it is obvious that a plane wave cannot propagate in a conducting medium without attenuation. The field amplitudes are spatially attenuated due to the presence of term $(-\beta r)$. The quantity β is a measure of attenuation and is known as *absorption coefficient*. Also in last exponential terms in (8.74) and (8.75) we have replaced k by α, therefore we conclude that the field vectors are propagated in the conducting medium with speed $\left(v = \frac{\omega}{k}\right)$ given by

$$v = \frac{\omega}{\alpha} = \frac{1}{\sqrt{\mu\varepsilon}} \left[\frac{\sqrt{\{1 + (\frac{\sigma}{\varepsilon\omega})^2\}} + 1}{2} \right]^{-1/2} \tag{8.76}$$

Now, we can consider the form of propagation vector $k = \alpha + i\beta$ in the two particular cases of interest:

(i) $\frac{\sigma}{\omega} \ll 1$ (For a poor conductor): We obtain $\alpha = \sqrt{\mu\varepsilon\omega}$ and $\beta = \frac{\sigma}{2}\sqrt{\frac{\mu}{\varepsilon}}$

$$\therefore \quad k + \alpha + i\beta = \sqrt{\mu\varepsilon\omega} + i\frac{\sigma}{\varepsilon}\sqrt{\frac{M}{\varepsilon}} \tag{8.77}$$

This is correct to first order in $\frac{\sigma}{\omega\varepsilon}$. In this limit $\alpha \gg \beta$ and the attenuation of wave determine by β is independent of frequency, aside from the possible variation of conductivity.

(ii) $\frac{\sigma}{\omega\varepsilon} \gg 1$ (For a good conductor): Obviously α and β are approximately equal, i.e.

$$\alpha \approx \beta = \omega\sqrt{\mu\varepsilon}\sqrt{\frac{\sigma/\omega\varepsilon}{2}} = \sqrt{\frac{\mu\sigma\omega}{2}} \tag{8.78}$$

$$\therefore \quad k \approx \alpha + i\beta = (1 + i)\sqrt{\frac{\mu\sigma\omega}{2}} \tag{8.79}$$

we have kept only lowest order term in $\frac{\omega\varepsilon}{\sigma}$.

One can immediately obtain a qualitative picture of the behaviour of the medium at high and low frequencies from (8.73). As $\omega \to \infty$, $\beta \to 0$, i.e. at very high frequencies the electromagnetic waves will be able to travel through the medium. At lower frequencies β will be finite and obviously there will be appreciable damping of the electromagnetic wave.

8.7 SKIN DEPTH OR PENETRATION DEPTH

We can see that the term $i\sigma\mu\omega$ in Eq. (8.67) arises from the term involving $\frac{\partial \mathbf{E}}{\partial t}$ in Eq. (8.62), i.e. from the conduction current; while the term $\varepsilon\mu\omega^2$ arises from the term involving $\frac{\partial^2 \mathbf{E}}{\partial t^2}$ in the same equation, i.e. from the displacement current. In almost all conducting media the conduction current dominates the displacement current and obviously, one can neglect the middle term in Eq. (8.62). Thus, for a good conducting medium

$$\nabla^2 E = \sigma\mu \frac{\partial E}{\partial t} \tag{8.79}$$

One finds the attenuated solution of (8.79) as

$$\mathbf{E} = \mathbf{E}_0 \exp(-\beta r) \exp[i(\alpha r - \omega t)] \tag{8.80}$$

For a good conductor, if the frequency is not too high, we have $\frac{\sigma}{\varepsilon\omega} \gg 1$. From Eqs. (8.72) and (8.73), we have

$$\alpha = \beta = \sqrt{\frac{\omega\sigma\mu}{2}} = \frac{1}{\delta}, \quad \text{where } \delta = \sqrt{\frac{2}{\mu\sigma\omega}} \tag{8.81}$$

The term $\delta = \frac{1}{\beta} = \sqrt{\frac{2}{\mu\sigma\omega}}$ measures the depth at which electromagnetic wave entering a conductor is damped to $\frac{1}{e} = 0.369$ of its initial amplitude at the surface. This depth is known as the *skin depth* or the *penetration depth*. Using (8.81), one can write (8.80) as

$$\mathbf{E} = \mathbf{E}_0 \exp\left(-\frac{r}{\delta}\right) \exp\left[i\left(\frac{r}{\delta} - \omega t\right)\right] \tag{8.82}$$

It is obvious from (8.81) that the skin depth decreases with increasing frequency and goes to zero as the conductivity approaches infinity. δ is small for good conductors at high frequency currents. For copper at 60 cycles δ is 8.5×10^{-3}, but at 1 MHz, it has dropped to 6.6×10^{-5} m. At 30 GHz frequency, δ is found to be 3.8×10^{-7} m. This shows that why in high frequency circuits current flows only on the surface of the conductors. The importance of the skin depth is that it measures the depth to which an electromagnetic wave can penetrate a conducting medium. Obviously, the conducting sheets which are used as electromagnetic shields must be thicker than the skin depth. The magnitudes of skin depth in different types of materials is listed below

Material	Frequency	δ (Penetration depth)
Sea water	30 kHz	0.1m
Aluminium	50 Hz	1.25×10^{-2} m
Silver	100 MHz	10^{-7} m

We must note that the penetration depth is inversely proportional to the square root of the frequency, therefore a thin sheet of conducting material can act as a *low-pass filter* for electromagnetic waves. One can even make a good conductor from a poor conductor by thin coating of silver or copper. The relatively higher value of δ in the case of sea water explains why radio communication with submerged submarine becomes difficult at depth of several metres.

8.8 REFLECTION AND REFRACTION OF PLANE ELECTROMAGNETIC WAVES AT A PLANE SURFACE

Let us consider two non-conducting ($\sigma = 0$) dielectric media marked as "1" and "2". These are characterized by μ, ε_1 and μ_2, ε_2 and separated by a plane boundary - plane x as shown in Fig. 8.7.

Let us consider that a plane electromagnetic wave is incident obliquely on the plane boundary. One can see that there will be in general both reflected wave and transmitted wave. Let us determine the fraction of the energy of the incident wave reflected and transmitted.

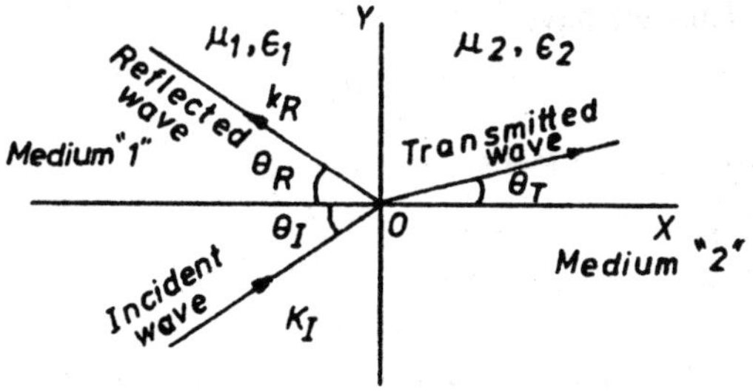

Figure 8.7. Reflection and refraction of plane electromagnetic wave at a plane interface.

One can express the fields for incident, reflected and transmitted waves as

$$\mathbf{E}_I = \mathbf{E}_{oI} \exp[i(\mathbf{k}_I \cdot \mathbf{r} - \omega t)], \mathbf{H}_I = \frac{\mathbf{K}_I \times \mathbf{E}_I}{\omega_I \mu_1} \tag{8.83}$$

$$\mathbf{E}_R = \mathbf{E}_{oR} \exp[i(\mathbf{k}_R \cdot \mathbf{r} - \omega t)], \mathbf{H}_R = \frac{\mathbf{K}_R \times \mathbf{E}_R}{\omega_R \mu_1} \tag{8.84}$$

$$\mathbf{E}_T = \mathbf{E}_{oT} \exp[i(\mathbf{k}_T \cdot \mathbf{r} - \omega t)], \mathbf{H}_T = \frac{\mathbf{K}_T \times \mathbf{E}_T}{\omega_T \mu_2} \tag{8.85}$$

where the subscripts I, R, and T in Eqs. (8.83) to (8.85) represent incident, reflected and transmitted waves. We must note that E_{oI}, E_{oR} and E_{oT} are time independent scalar amplitudes and these may be complex. One can find the relationship between these quantities by making the total fields obey the boundary conditions in the plane $x = 0$. We can easily see that the tangential components of **E** and **H** can be continuous across the boundary at all points and at all times only if the exponentials are the same at the boundary for all three fields. Obviously,

$$\omega_I = \omega_R = \omega_T \tag{8.86}$$

This means that the frequency remain unchanged in the reflected and transmitted waves, and

$$\mathbf{k}_I \cdot \mathbf{r} = \mathbf{k}_R \cdot \mathbf{r} = \mathbf{k}_T \cdot \mathbf{r} \tag{8.87}$$

Relation (8.87) reveals that all the propagation vectors are coplanar. Let us choose **r** to lie in the boundary plane, i.e. $\hat{a}_n \cdot \mathbf{r} = 0$, where $\hat{a}_n$ is a unit vector normal to the plane, and in the plane of the propagation vector, one finds

$$k_I \sin\theta_I = K_R \sin\theta_R = k_T \sin\theta_T \tag{8.88}$$

we can see that $\mathbf{K}_I$ and $\mathbf{K}_R$ vectors are in the same medium. This means

$|\mathbf{K}_I| = |\mathbf{K}_R|$. Thus we have

$$\theta_I = \theta_R \tag{8.89}$$

This expresses *law of reflection.*

Now we have

$$\begin{aligned} k_I \sin\theta_I &= k_T \sin\theta_T \\ \text{or} \qquad \frac{\sin\theta_I}{\sin\theta_T} &= \frac{k_I}{k_T} = \sqrt{\frac{\varepsilon_2\mu_2}{\varepsilon_1\mu_1}} \quad (\because \quad k = \omega\sqrt{\varepsilon\mu}) \end{aligned} \tag{8.90}$$

One can assume $\mu_1 = \mu_2$ for non-magnetic materials. We have

$$\frac{\sin\theta_I}{\sin\theta_T} = \sqrt{\frac{\varepsilon_2}{\varepsilon_1}} = \frac{n_2}{n_1} = n_{12} \tag{8.91}$$

This is *Snell's law* and expresses the law of refraction. Here n_1 and n_2 are the refractive indices of media "1" and "2" respectively. n_{12} is known as the *relative refractive index* of the second substance with respect to the first one. Thus the relative refractive index of two substances equals the ratio of their absolute refractive indices. Obviously, (8.89) and (8.91) are the simpler laws of geometrical optics.

Now we obtain the relationship between the various field vectors.

The divergence equations $\nabla\cdot\mathbf{D} = \rho$ and $\nabla\cdot\mathbf{B} = 0$ can be obtained by applying divergence operator to the remaining Maxwell's equations involving **E** and **H**. One can see that the boundary conditions on D_n and B_n are automatically satisfied provided the conditions on E_I and H_T are met. One can express the conditions as

$$(\mathbf{E}_I + \mathbf{E}_R) \times \hat{a}_n = \mathbf{E}_T \times \hat{a}_n \tag{8.92}$$

and

$$(\mathbf{H}_I + \mathbf{H}_R) \times \hat{a}_n = \mathbf{H}_T \times \hat{a}_n \tag{8.93}$$

Eq. (8.93) can be expressed as

$$(\mathbf{k}_I \times \mathbf{E}_I + \mathbf{k}_I \times \mathbf{E}_R) \times \hat{a}_n = (\mathbf{k}_T \times \mathbf{E}_T) \times \hat{a}_n \quad (\because \quad \mu_1 = \mu_2) \tag{8.94}$$

We now consider two separate situations: (a) **E** is polarized perpendicular to the plane of incidence, i.e. the plane defined by **k** and $\hat{a}_n$ and (b) **E** is polarized parallel to the plane of incidence.

One can obtain the general case of arbitrary polarization by appropriate linear combination of the above two situations.

(a) **E** *polarized perpendicular to the plane of incidence:* Fig. 8.8 exhibit the field vectors corresponding to reflection and refraction with **E** polarized perpendicular to the plane of incidence. We can see that the electric field vectors are directed away from the viewer. The conditions (8.92) and (8.94) yields

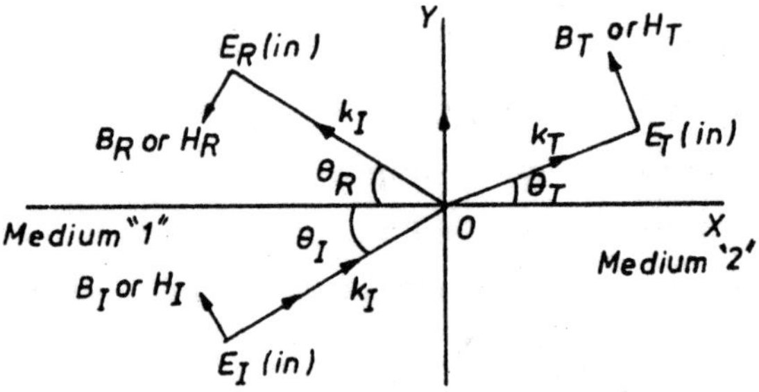

Figure 8.8. Reflection and refraction when **E** polarized $\perp^r$ to the plane of incidence.

$$E_{oI} + E_{oR} = E_{oT} \tag{8.95}$$

and

$$k_1 E_{oI} \cos\theta_I - k_1 E_{oR} \cos\theta_R = k_T E_{oT} \cos\theta_T$$

i.e.

$$(E_{oI} - E_{oR}) \cos\theta_I = \frac{k_T}{k_I} E_{oT} \cos\theta_T \tag{8.96}$$

where E_{oI}, E_{oR} and E_{oT} represents the scalar amplitudes of the incident, reflected and transmitted electromagnetic waves respectively. Solving (8.95) and (8.96), one obtains

$$\frac{E_{oR}}{E_{oI}} = \frac{\cos\theta_I - \frac{k_T}{k_I}\cos\theta_T}{\cos\theta_I + \frac{k_T}{k_I}\cos\theta_T} = \frac{\cos\theta_I - \frac{n_2}{n_1}\cos\theta_T}{\cos\theta_I - \frac{n_2}{n_1}\cos\theta_T}$$

Using Snell's law, one obtains

$$\frac{E_{oR}}{E_{oI}} = \frac{\cos\theta_I - \frac{\sin\theta_I}{\sin\theta_T}\cos\theta_T}{\cos\theta_I + \frac{\sin\theta_I}{\sin\theta_T}\cos\theta_T}$$
$$= \frac{\sin(\theta_T - \theta_I)}{\sin(\theta_T - \theta_I)} \tag{8.97}$$

and

$$\frac{E_{oT}}{E_{oI}} = \frac{2\cos\theta_I}{\cos\theta_I + \frac{k_T}{k_I}\cos\theta_T} = \frac{2\cos\theta_I \sin\theta_T}{\sin(\theta_I + \theta_T)} \tag{8.98}$$

we obtain the ratio of the amplitudes of the reflected and incident waves from Eq. (8.98). Eqs. (8.97) and (8.98) represents *Fresnel's*

equations for non-conducting media when electric vector **E** is perpendicular to the plane of incidence. We have following situations:

(i) $n_2 > n_1$ i.e. $\frac{n_1}{n_2} < 1$, i.e. the ratio in (8.98) is negative. This indicates that the reflection of the electromagnetic wave results in a phase change of π, i.e. the electric vector of the reflected wave oscillates 180° out of phase with that in the incident wave. The ratio $\frac{E_{oT}}{E_{oI}}$ is always positive.

One can define the reflection coefficient R as the energy flux reflected from the interface divided by the flux incident on it. Therefore

$$R_{\perp} = \frac{\hat{a}_n \cdot \langle N_R \rangle}{\hat{a}_n \cdot \langle N_I \rangle} = \frac{|\mathbf{E}_R \times \mathbf{H}_R|}{|\mathbf{E}_I \times \mathbf{H}_I|} = \frac{E_{oR}^2}{E_{oI}^2} \tag{8.99}$$

where $\mathbf{N}_R$ and $\mathbf{N}_I$ are the Poynting vectors and subscripts $\perp$ indicates that E is polarized perpendicular to the plane of incidence. Thus, we have

$$R_{\perp} = \frac{\sin^2(\theta_T - \theta_I)}{\sin^2(\theta_T + \theta_I)} \tag{8.100}$$

Similarly, one can obtain the transmission coefficient as

$$\begin{aligned} T_{\perp} = \frac{\hat{a}_n \cdot \langle N_T \rangle}{\hat{a}_n \cdot \langle N_I \rangle} &= \frac{|E_{oT}|^2}{|E_{oI}|^2} \frac{n_2 \cos\theta_T}{n_1 \cos\theta_I} \\ &= \frac{4\cos^2\theta_I \sin^2\theta_T}{\sin^2(\theta_I + \theta_T)} \frac{\sin\theta_I \cos\theta_T}{\sin\theta_T \cos\theta_I} \\ &= \frac{4\cos\theta_I \cos\theta_T \sin\theta_I \sin\theta_T}{\sin^2(\theta_I + \theta_T)} \\ &= \frac{\sin 2\theta_I \sin 2\theta_T}{\sin^2(\theta_I + \theta_T)} \end{aligned} \tag{8.101}$$

From (8.100) and (8.101), we have

$$R_{\perp} + T_{\perp} = 1 \tag{8.102}$$

From Eqs. (8.97) and (8.98), we have for normal incidence

$$R_{\perp} = \left(\frac{n_1 - n_2}{n_1 + n_2}\right)^2 \quad \text{and} \quad T_{\perp} = \frac{n_2}{n_1}\left(\frac{2n_1}{n_1 + n_2}\right)^2 \tag{8.103}$$

(ii) $\frac{n_1}{n_2} > 1$: when an electromagnetic wave is incident on the interface of two dielectrics from a denser medium, then it can be shown that refracted ray is deviated away from the normal and reflected and incident wave are in the same phase.

(b) E *Vector is Polarized Parallel to the Plane of Incidence:* In this case electric field vectors of all in the three waves are parallel to the plane of incidence (Fig. 8.9). The boundary conditions give

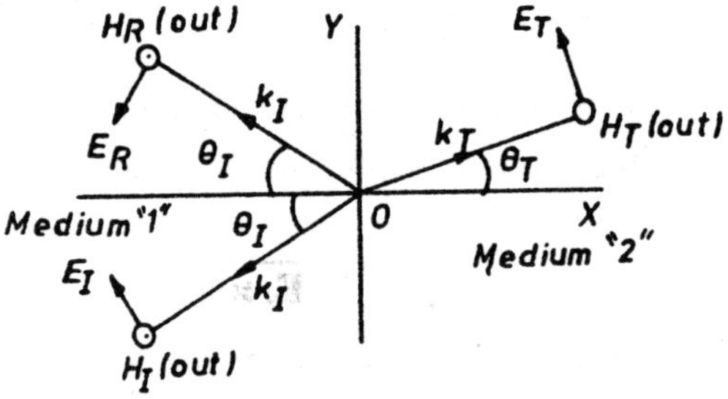

Figure 8.9. Reflection and refraction of electromagnetic wave with polarization parallel to the plane of incidence.

$$E_{oI}\cos\theta_I - E_{oR}\cos\theta_R = E_{oT}\cos\theta_T$$

or

$$(E_{oI} - E_{oR})\cos\theta_I = E_{oT}\cos\theta_T \quad (\because \quad \theta_I = \theta_R) \quad (8.104)$$

and

$$k_I E_{oI} + k_I E_{oR} = k_T E_{oT}$$

or

$$E_{oI} + E_{oR} = \frac{n_2}{n_1} E_{oT} \quad (8.105)$$

Solving (8.104) and (8.105), one obtains

$$\frac{E_{oR}}{E_{oI}} = \frac{\cos\theta_I - \left(\frac{n_1}{n_2}\right)\cos\theta_T}{\cos\theta_I + \left(\frac{n_1}{n_2}\right)\cos\theta_T} = \frac{\tan(\theta_I - \theta_T)}{\tan(\theta_I + \theta_T)} \quad (8.106)$$

and

$$\frac{E_{oT}}{E_{oI}} = \frac{2\sin\theta_T\cos\theta_I}{\sin(\theta_I + \theta_T)\cos(\theta_I - \theta_T)} \quad (8.107)$$

Relations (8.106) and (8.107) represent *Fresnel's equations* for the case when **E** vector is polarized parallel to the plane of incidence.

Comparison of (8.97) and (8.106) indicates that there is an important distinction between the two states of polarization of E. Considering only simple case, i.e. when $n_2 = n_1$, i.e. $\theta_T = \theta_I$, we note from (8.97) that $\frac{E_{oR}}{E_{oI}} \neq 0$ for any angle of incidence θ_I lying between 0 and $\frac{\pi}{2}$. However, when E is polarized in the plane of incidence, Eq. (8.107) gives

$$\frac{E_{oR}}{E_{oI}} = 0 \tag{8.108}$$

Obviously, $R_{\parallel} = 0$ when $\theta_I + \theta_T = \frac{\pi}{2}$.

Thus, when the wave is incident at an angle $\theta_I = \frac{\pi}{2} - \theta_T$, it crosses the interface without suffering reflection. This angle is called as *Brewster's angle* (θ_B) or polarizing angle. One can find the magnitude of θ_B by applying Snell's law, i.e.

$$\frac{\sin\theta_B}{\sin\theta_T} = \frac{\sin\theta_B}{\sin\left(\frac{\pi}{2} - \theta_B\right)} = \tan\theta_B = \frac{n_2}{n_1}$$

$$\therefore \quad \theta_B = \tan^{-1}\left(\frac{n_2}{n_1}\right) \tag{8.109}$$

Thus if an unpolarised electromagnetic wave consisting of a mixture of both types of radiation falls on the interface at Brewster's angle, only the radiation with its electric vector at right angles to the plane of incidence will be reflected and the reflected radiation will be polarized.

An important application of Brewster's angle can be found in gas lasers. The windows in a gas laser are arranged at Brewster's angle. In a gas laser, mirror outside the glass windows are as shown in Fig. 8.10. It is reported that at normal incidence about 92% of the incident intensity is transmitted through a glass window and about 8% is lost in each transport. We know that in a laser there are a large number of traverses and obviously, very little of light will be left after a few traverses. To overcome this difficulty the windows in gas lasers are arranged at Brewster's angle. This helps in perfect transmission of the electric field component polarized parallel to the plane of incidence, i.e. it suffers negligible loss even after many traverses. The emerging laser beam is 100% purely linearly polarized as the component polarized perpendicular to the plane of incidence is partly reflected and partly transmitted each time it strikes the surface and after a number of traverses it is completed eliminated.

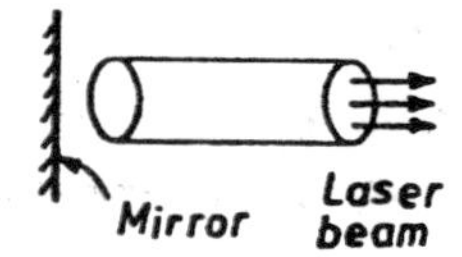

Figure 8.10.

8.9 TOTAL INTERNAL REFLECTION

Let us consider that a plane monochromatic electromagnetic wave is incident on the interface from a denser medium (say glass) to the rarer medium, i.e. $n_1 > n_2$. We have from Snell's law

$$\sin\theta_T = \frac{n_1}{n_2}\sin\theta_I = \sqrt{\frac{\varepsilon_1}{\varepsilon_2}}\sin\theta_I \tag{8.110}$$

obviously $\theta_T > \theta_I$. We have assumed the media to be non-magnetic, i.e. $n_1 = \sqrt{\frac{\varepsilon_1}{\varepsilon_0}}$ and $n_2 = \sqrt{\frac{\varepsilon_2}{\varepsilon_0}}$. Let θ_I is gradually increased starting from zero. This means θ_T will also increase until its attains the value $\frac{\pi}{2}$. Let the value of θ_I corresponding to this situation be θ_c. From (8.110), we have

$$\sin\theta_c = \frac{n_2}{n_1} \tag{8.111}$$

when $\theta_I = \theta_c$, there will only be reflected wave and no transmitted wave. The angle θ_c is called the *critical angle*. Obviously, the wave incident at $\theta_I = \theta_c$, the refracted wave is propagated parallel to the surface.

Now let us see what happens if θ_c is increased further, i.e. when $\theta_I > \theta_c$. Let us express θ_T in terms of θ_I and θ_c we have

$$\cos\theta_T = \sqrt{1-\sin^2\theta_T}$$

But $$\sin\theta_T = \frac{n_1}{n_2}\sin\theta_I = \frac{\sin\theta_I}{\sin\theta_c} \quad \text{(using (8.111))}$$

$$\therefore \quad \cos\theta_T = \sqrt{1-\frac{\sin^2\theta_I}{\sin^2\theta_c}} \tag{8.112}$$

One can see that the value of $\cos\theta_T$ decreases as θ_I is increased: it becomes zero at $\theta_I = \theta_c$ and for values of θ_I greater than θ_c, $\cos\theta_T$ becomes an imaginary number, since sine of any real angle can never be greater than unity.

Now, we determine the amplitude of the reflected electric vector when $\theta_I > \theta_c$. For $\theta_I > \theta_c$, one can write

$$\cos\theta_T = \sqrt{1 - \frac{\sin^2\theta_I}{\sin^2\theta_c}} = i\sqrt{\frac{\sin^2\theta_I}{\sin^2\theta_c} - 1} = iQ \tag{8.113}$$

when **E** is polarized perpendicular to the plane of incidence, we have

$$\frac{E_{oR}}{E_{oI}} = \frac{\cos\theta_I - \frac{n_2}{n_1}\cos\theta_T}{\cos\theta_I + \frac{n_2}{n_1}\cos\theta_T} = \frac{\cos\theta_I - \frac{n_2}{n_1}iQ}{\cos\theta_I + \frac{n_2}{n_1}iQ}$$

$$\therefore \quad \left|\frac{E_{oR}}{E_{oI}}\right|^2 = 1 \quad \text{This means} |E_{oR}| = |E_{oI}|$$

Similarly, one finds when **E** is polarized parallel to the plane of incidence,

$$\left|\frac{E_{oR}}{E_{oI}}\right|^2 = 1, i.e. |E_{oR}| = |E_{oI}|$$

Obviously, the wave is totally reflected. This phenomenon is called *total internal reflection.*

We must note that there is a change of phase on reflection, i.e. if the incident wave is polarized in a plane intermediate between the plane of incidence and the plane normal to it, the two components will not be in phase after reflection and the electromagnetic wave will be elliptically polarized.

One can conclude that when $\theta_I > \theta_c$, there is no refracted wave and all the energy is reflected. To verify it, we compute the average rate of energy flow across the boundary.

We have the rate of energy flow $= \langle \mathbf{N} \rangle \cdot \hat{a}_n$

$$= \frac{1}{2}R_e(\mathbf{E}_T \times \mathbf{H}_T^*) \cdot \hat{a}_n = \frac{1}{2}R_e\left(\mathbf{E}_T \times \frac{\mathbf{k}_T \times \mathbf{E}_T^*}{\mu\omega}\right) \cdot \hat{a}_n$$

$$= \frac{1}{2\mu\omega}R_e[(\mathbf{E}_T \cdot \mathbf{E}_T^*)\mathbf{k}_T - (\mathbf{E}_T \cdot \mathbf{k}_T)\mathbf{E}_T^*] \cdot \hat{a}_n$$

$$= \frac{1}{2\mu\omega}R_e(\mathbf{E}_T \cdot \mathbf{E}_T^*)\mathbf{k}_T \cdot \hat{a}_n \quad (\text{Since } \mathbf{E}_T \text{is perpendicular to} \mathbf{k}_T)$$

$$= \frac{1}{2\omega\mu}R_e|E_{oT}|^2\mathbf{k}_T\cos\theta_T = \frac{1}{2\mu\omega}R_e iQk_T|E_{oT}|^2 \tag{8.114}$$

one can see that (8.114) is purely imaginary. This means $\langle \mathbf{N} \rangle \cdot \hat{a}_n = 0$. Obviously, this justifies our assumption.

Now, the question arises, if $n_1 > n_2$ and the wave is incident at an angle greater than θ_c, will there be field on the other side of the boundary? Although there is no energy flow across the surface, but the field does exist on the other side of the surface. We can easily verify it.

Let us consider the equation for the refract wave. We have

$$\mathbf{E}_T = \mathbf{E}_{oT} \exp[i(\mathbf{k}_T \cdot \mathbf{r} - \omega t) \tag{8.115}$$

Eq. (115) takes the following for the coordinate axes shown in Fig. 8.7,

$$\begin{aligned} \mathbf{E}_T &= \mathbf{E}_{oT} \exp[i(k_T x \cos\theta_T + k_T y \sin\theta_T - \omega t)] \\ &= \mathbf{E}_{oT} \exp[-k_T x Q \exp\{i(k_T y \sin\theta_T - \omega t)\}] \end{aligned} \tag{8.116}$$

Eq. (116) clearly reveals that the field does exist on the far side of the surface in the medium 2, but it rapidly attenuated in the (Fig. 8.9).

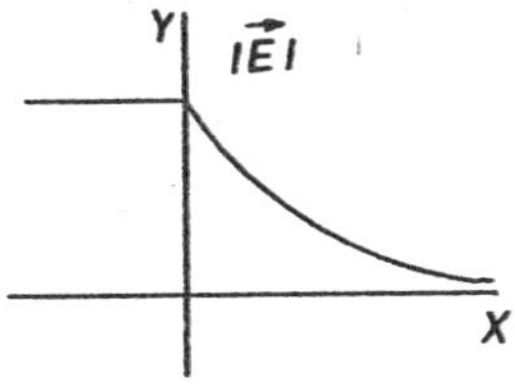

Figure 8.11.

The skin depth or penetration depth, i.e. how far the wave penetrate the medium is given by

$$x = \frac{1}{k_T Q} = \frac{1}{k_T}\left(\frac{\sin^2\theta_I}{\sin^2\theta_c} - 1\right)^{-1/2} \tag{8.117}$$

Let us consider that wave is going from glass to air. We have μ for glass = 1.5. Thus the critical angle for the glass is $\theta_c = \sin^{-1} 2/3 \approx 42^\circ$. Obviously if light is incident internally from the glass at an angle greater than 42°, say 45°, then there will be total internal reflection and we have

$$x = \frac{1}{k_T}\left(\frac{1}{2} \times \frac{9}{4} - 1\right)^{-1/2} = \frac{2\sqrt{2}}{k_T} = \frac{\lambda}{2\pi} 2\sqrt{2} = 0.45\,\lambda$$

This means that the field becomes negligible beyond distances of the order of a few wavelengths. Attenuation results in an increase in the moleculer internal energy, mainly rotational molecular motion, or in the translational kinetic energy. This means that the molecular charges in the medium oscillate due to interaction with the incident wave and give rise to a radiation field. The forward wave in this field interferes destructively with the original wave and gives rise to a small transmission.

The question arises, how one can account for the decaying field on the far side of the boundary, when there is no transport of energy across the boundary. We must note that the energy does flow in the second medium since the component of the field in the medium in finite, but during the later part of the cycle the flow is in the opposite direction and the energy is returned to the first medium.

8.10 REFLECTION FROM A CONDUCTING PLANE, i.e. METALLIC REFLECTION

Let us consider the reflection and refraction at the interface separating a conducting medium "1" be characterised by permittivity ε_1, permeability μ_1 and refractive index n_1, while the conducting medium '2' by permeability μ_2, permittivity ε_2, refractive index n_2 and conductivity σ_2. Let a plane monochromatic electromagnetic wave be incident on the interface from medium '1' in medium '2'. Since the case of oblique incidence is more involved, we shall confine to the simple case of normal incidence only.

One can write for incident, reflected and transmitted waves as

$$\left.\begin{aligned} E_I &= E_{oI}\exp\{\mathbf{k}_I\cdot\mathbf{r}-\omega t)\} \\ H_I &= \frac{\mathbf{k}_I\times\mathbf{E}_I}{\omega\mu_1} \end{aligned}\right] \tag{8.117}$$

$$\mathbf{E}_R = \mathbf{E}_{oR}\exp[i(\mathbf{k}_I\cdot\mathbf{r}-\omega t)];\ \mathbf{H}_R = \frac{\mathbf{k}_I\times\mathbf{E}_R}{\omega\mu_1} \tag{8.118}$$

and

$$\mathbf{E}_T = \mathbf{E}_{oT}\exp[i(\mathbf{k}_T\cdot\mathbf{r}-\omega t)];\ \mathbf{H}_T = \frac{\mathbf{k}_T\times\mathbf{E}_T}{\omega\mu_2} \tag{8.119}$$

In conducting medium '2' the propagation vector $\mathbf{k}_T$ is given by

$$k_T^2 = \varepsilon_2\mu_2\omega^2\left[1+\frac{i\sigma}{\varepsilon_2\omega}\right] \qquad 8.119(a)$$

The boundary conditions gives

$$E_{oI}-E_{oR} = E_{oT} \tag{8.120}$$

$$k_I(E_{oI}+E_{oR}) = k_T E_{oT} \tag{8.121}$$

we know that k_T is complex and hence E_{oR} and E_{oT} cannot both be real. This means, one should expect phase shifts other than 0 or π in the reflected and transmitted waves. Solving (8.120) an (8.121), one obtains

$$E_{oR} = \frac{k_T - k_I}{k_T + k_I} E_{oI} \tag{8.122}$$

and
$$E_{oT} = \frac{2k_I}{k_T + k_I} E_{oI} \tag{8.123}$$

Substituting $k_I = \omega\sqrt{\varepsilon_1\mu_1}$ and k_T from 8.119 (a), one obtains

$$E_{oR} = \frac{\omega\sqrt{\varepsilon_2\mu_2}\left(1+\frac{i\sigma}{\omega\varepsilon_2}\right)^{1/2} - \omega\sqrt{\varepsilon_1\mu_1}}{\omega\sqrt{\varepsilon_2\mu_2}\left(1+\frac{i\sigma}{\varepsilon_2\omega}\right)^{1/2} + \omega\sqrt{\varepsilon_1\mu_1}} E_{oI} \tag{8.124}$$

and
$$E_{oT} = \frac{2\omega\sqrt{\varepsilon_1\mu_1}}{\omega\sqrt{\varepsilon_2\mu_2}\left(1+\frac{i\sigma}{\omega\varepsilon_2}\right)^{1/2} + \omega\sqrt{\varepsilon_1\mu_1}} \tag{8.125}$$

Special Cases

(i) *Perfect Conductor:* In the case of perfect conductor we have $\sigma = \infty$. This means

$$E_{oI} = E_{oR} \quad \text{and} \quad E_{oT} = 0 \tag{8.126}$$

Obviously, the reflection is complete.

(ii) *Very good conductor:* When the conductor is not a perfect conductor but a very good conductor, we have $\dfrac{\sigma}{\varepsilon_2\omega} \gg 1$. The approximation adopted earlier, yields,

$$k_T = \alpha + i\beta = (1+i)\sqrt{\frac{\omega\sigma\mu_2}{2}} = \frac{1+i}{\delta} \tag{8.127}$$

where $\delta = \sqrt{\frac{2}{\omega\sigma\mu_2}}$ is the skin depth. Using (8.127), (8.122) gives

$$E_{oR} = \frac{\frac{1+i}{\delta} - \omega\sqrt{\varepsilon_1\mu_1}}{\frac{1+i}{\delta} + \omega\sqrt{\varepsilon_1\mu_1}} E_{oI} = \frac{\left[\frac{1}{\delta} - \omega\sqrt{\varepsilon_1\mu_1}\right] + \frac{i}{\delta}}{\left[\frac{1}{\delta} + \omega\sqrt{\varepsilon_2\mu_2}\right] + \frac{i}{\delta}} E_{oI} \tag{8.128}$$

one obtains the reflection coefficient R as

$$R = \frac{|E_{oR}|^2}{|E_{oI}|^2} = \frac{[1 - \omega\sqrt{\varepsilon_1\mu_1}\delta]^2 + 1}{[1 + \omega\sqrt{\varepsilon_1\mu_1}\delta]^2 + 1}$$

Since $\dfrac{\sigma}{\varepsilon_2\omega} \gg 1$, $\omega\sqrt{\varepsilon_1\mu_1}\delta \ll 1$ and therefore, we have

$$R \simeq 1 - 2\omega\sqrt{\varepsilon_1\mu_1} \tag{8.129}$$

$$= 1-2\sqrt{\frac{2\omega\varepsilon_1\mu_1}{\sigma\mu_2}} \tag{8.130}$$

If $\mu_1 = \mu_2$, i.e. the magnetic permeabilities are equal then, one obtains

$$R = 1-2\sqrt{\frac{2\omega\varepsilon_1}{\sigma}} \tag{8.131}$$

$$= 1-\frac{2}{n}, \quad \text{where} \quad n = \sqrt{\frac{\sigma}{2\omega\varepsilon_1}} \tag{8.132}$$

This equation shows that at low frequencies for substances of high conductivity the reflection coefficient (R) is close to unity and obviously, almost all the energy will be reflected. This shows that why metals are opaque to light. The little energy that flows into the metal is rapidly dissipated by the heat loss associated with the induced currents. One can obtain the measure of the energy transmitted into the conducting medium by calculating the transmission coefficient (T). T gives the ratio of the energy transmitted to the energy incident. We have

$$T = 1-R = 2\sqrt{\frac{\omega\varepsilon_1}{\sigma}} = \frac{2}{n} \tag{8.133}$$

For example taking $\sigma = 6\times 10^7 (\Omega - m)^{-1}$ for copper and $\nu = 10^{10}/$sec, we obtain $T = 3\times 10^{-4}$. This is extremely small for direct measurements to be made. We must note that the deviation of reflection coefficient R from unity becomes significant for extremely good conductors like copper, silver and aluminium only for radiations of very short wave length. These materials absorbs waves strongly, i.e. the skin depth is very small. Concluding we can say that all waves which are strongly absorbed are also strongly reflected by metals, i.e. all *good conductors are good absorbers and good reflectors.*

The total energy in the medium is

$$\mathbf{E} = \mathbf{E}_I + \mathbf{E}_R$$

where $\mathbf{E}_I = \hat{a}_n E_{oI} \exp[i(k_I x - \omega t)]$ and $\mathbf{E}_R = -\hat{a}_n E_{oI} \exp[i(k_I x - \omega t)]$ we must note that the amplitudes in both the waves are taken equal because the reflection coefficient for metallic surfaces is almost unity. Therefore, we have

$$\mathbf{E} = \hat{a}_n \mathbf{E}_{oI}\, e^{-i\omega T}\{e^{ik_I x} - e^{-ik_I x}\} \tag{8.134}$$

Considering the real parts only, we have

$$\mathbf{E} = 2\hat{a}_n E_{ol} \sin \omega t \sin k_l x \tag{8.135}$$

Obviously, the field in the medium is represented by a standing wave.

EXAMPLE 8.1: The earth receives 2 cal/min/sq. cm Solar energy. What are the amplitudes of electric and magnetic field of radiation?

Solution: We have

$$|\mathbf{N}| = |\mathbf{E} \times \mathbf{H}| = EH \sin 90° = EH. \tag{1}$$

The energy flux per unit area per second at the surface of the earth is

$$2\text{cal/min/sq cm} = \frac{2 \times 4.2 \times 10^4}{60} \text{ J/m}^2/\text{s}^2 \tag{2}$$

comparing (1) and (2) yields

$$EH = \frac{2 \times 4.2 \times 10^4}{60} = 1400 \tag{3}$$

But

$$\frac{E}{H} = \sqrt{\frac{\mu_0}{\varepsilon_0}} = \sqrt{\frac{4\pi \times 10^{-7}}{8.85 \times 10^{-12}}} = 376.6 \tag{4}$$

Multiplying (3) and (4), one obtains

$$E^2 = \frac{1400 \times 376.6}{\sqrt{1400 \times 376.6}} \text{ or } E = 726.1 \text{ V/m} \tag{5}$$

Using (5), eq (3) yields

$$H = \frac{1400}{E} = \frac{1400}{726.1} = 1.928 \text{ amp} - \text{turn/m}.$$

$$\therefore E_o = E\sqrt{2} = 726.1 \times \sqrt{2} = 1027 \text{ V/m}$$

and

$$H_o = H\sqrt{2} = 1.928 \times \sqrt{2} = 2.73 \text{ amp} - \text{turn/m}$$

EXAMPLE 8.2: Assuming that the radiation due to sunlight is falling normally on to a non-reflecting totally absorbing slab and that it is of the order of 1 kW/m^2. Show that the force exerted on the slab is $\sim 10^{-5}$ N/m^2.

Solution: The slab is totally absorbing. Obviously the second term in Eq. (8.50) is zero. Moreover when time variations are sinusoidal the average of $\frac{dN}{dt}\left(\text{i.e. } \frac{d}{dt}(E_x H_y)\right)$ is zero. Now we have the mean pressure on the slab as

$$\bar{T}_z = \left[\overline{\frac{1}{2}(\varepsilon_0 E_x^2 + \mu_0 H_y^2)}\right]_{z_1} = \varepsilon_0 E_x^2 \quad \text{(using (8.17)}$$

Let us assume $E_x = E_0 \cos(\omega t - kz)$

$$\therefore \quad \bar{T}_z = \varepsilon_0 E_0^2 \overline{\cos^2(\omega t - kz)} = \frac{1}{2}\varepsilon_0 E_0^2 \qquad (1)$$

The momentum of the incident wave

$$p = \mu_0 \varepsilon_0 E_x H_y = \mu_0 \varepsilon_0 \sqrt{\frac{\varepsilon_0}{\mu_0}} = \frac{\varepsilon_0}{C} E_0^2 \cos^2(\omega t - kz)$$

$\therefore$ The mean momentum per unit volume is given by

$$\bar{p} = \frac{1}{2}\frac{\varepsilon_0}{c} E_0^2. \qquad (2)$$

The momentum given by (2) is totally absorbed. One can say that the momentum per unit area carried by the wave and lost in the slab in one second is that contained in a cylinder of unit cross sectional area and length equal to C, i.e. $\overline{cp} = \frac{1}{2}\varepsilon_0 E_0^2$. We know that this is equal to the force per unit area.

The power carried by the wave in $1\frac{kw}{m^2}$ is.

$$\frac{1}{2}\sqrt{\frac{\varepsilon_0}{\mu_0}} E_0^2 = 10^3$$

$$\therefore \quad \frac{1}{2}\varepsilon_0 E_0^2 = 10^3\sqrt{\mu_0 \varepsilon_0} = 0.33 \times 10^{-5}$$

obviously the mean force per unit area $\sim 10^{-5}\ \text{N/m}^2$.

EXAMPLE 8.3: A plane electromagnetic wave travelling in positive z-direction in an unbounded lossless dielectric medium with relative permeability $\mu_r = 1$ and relative permittivity $\varepsilon_r = 3$ has a peak electric field intensity $E_0 = 6$ V/m. Determine (i) the speed of the wave (ii) the intrinsic impedance of the medium (iii) the peak magnetic intensity (H_o) and (iv) the peak Poynting vector $\mathbf{N}(z,t)$.

Solution: We have $E_0 = \sqrt{(E_{ox}^2 + E_{oy}^2)} = 6$ V/m, $\varepsilon_r = 3$ and $\mu_r = 1$.

(i) The speed of electromagnetic wave

$$v = \frac{1}{\sqrt{\mu\varepsilon}} = \frac{1}{\sqrt{\mu_r \mu_o \varepsilon_r \varepsilon_o}} = \frac{1}{\sqrt{\mu_r \varepsilon_r}} \frac{1}{\sqrt{\mu_0 \varepsilon_0}} = \frac{c}{\sqrt{\mu_r \varepsilon_r}}$$

$$= \frac{3 \times 10^8}{\sqrt{1 \times 3}} = 1.73 \times 10^8 \text{ m/p}$$

(ii) Impedance of medium

$$Z = \sqrt{\frac{\mu}{\varepsilon}} = \sqrt{\frac{\mu_r \mu_0}{\varepsilon_r \varepsilon_o}} = \sqrt{\left(\frac{\mu_0}{\varepsilon_0}\right)} \sqrt{\frac{\mu_r}{\varepsilon_r}}$$

$$= \sqrt{\left(\frac{4\pi \times 10^{-7}}{8.86 \times 10^{-12}}\right)} \sqrt{\frac{1}{3}} = \frac{376.6}{\sqrt{3}} = 217.6\ \Omega.$$

(iii) Peak value of magnetic field

$$H_0 = \frac{E_o}{Z} = \frac{6}{217.6} = 2.76 \times 10^{-2} \text{ A/m}$$

(iv) Poynting vector $\mathbf{N} = [\mathbf{E} \times \mathbf{H}]$

$$\therefore \quad \text{Peak poynting vector} = E_o H_o = \frac{E_o^2}{Z} = \frac{(6)^2}{217.6} = 0.165 \text{ W/m}^2$$

EXAMPLE 8.4: A surface wave whose magnetic field strength is perpendicular to its line of propagation travels along the interface of two dielectrics with permittivities ε_1 and $-|\varepsilon_2|$. Show that the dispersion relation is

$$k = \left[\frac{\omega^2 \varepsilon_1 |\varepsilon_2|}{c^2(|\varepsilon_2| - \varepsilon_1)}\right]^{1/2}$$

Solution: Let us consider that the interface of two dielectrics lies in the $X - Y$ plane at $Z = 0$. Let the wave is propagating along the x-axis and the magnetic field acts along the y-axis. Let the medium with permittivity ε_1 lies in the region $z > 0$ and that with permittivity $-|\varepsilon_2|$ in the region $z < 0$. One can describe the wave by the solution of the equation

$$\nabla^2 H_y + \frac{\omega^2}{c^2} H_y = 0$$

This has the form

$$H_{1y} = H_{01}\, e^{ik_1 x - f_1 z}, \quad \text{where} \quad f_1 = \sqrt{k^2 - \frac{\omega^2 \varepsilon_1}{c^2}} \quad \text{for} \quad z > 0 \quad \text{(i)}$$

and

$$H_{2y} = H_{02}\, e^{ik_1 x + f_2 z}, \quad \text{where} \quad f_2 = \sqrt{k^2 - \frac{\omega^2 |\varepsilon_2|}{c^2}} \quad \text{for} \quad z < 0 \quad \text{(ii)}$$

From boundary conditions at $z = 0$, one obtains $H_{1y} = H_{2y}$ and $E_{1x} = E_{2x}$. Applying these boundary conditions, we obtain

$$H_{01} = H_{02} \quad \text{and} \quad \frac{f_1}{\varepsilon_1} = \frac{f_2}{|\varepsilon_2|} \quad \text{(iii)}$$

Using (i) and (ii), (iii) yields

$$\frac{k^2}{\varepsilon_1^2} - \frac{\omega^2}{c^2 \varepsilon_1} = \frac{k^2}{|\varepsilon_2|^2} - \frac{\omega^2}{c^2 |\varepsilon_2|}$$

$$\therefore \quad k = \sqrt{\frac{\omega^2 \varepsilon_1 |\varepsilon_2|}{c^2 (|\varepsilon_2| - \varepsilon_1)}} \quad \text{(iv)}$$

EXAMPLE 8.5: The constitution parameter for aluminium are: $\mu_r = 1$, $\varepsilon_r = 1$ and $\sigma = 3.54 \times 10^7$ mho/m. Show that the skin depth in aluminium is 0.01 mm for the frequency 71.5 MHz.

Solution:

$$\delta = \sqrt{\frac{2}{\mu \sigma \omega}} = \sqrt{\frac{2}{\mu \sigma 2\pi \nu}} = \sqrt{\frac{1}{\mu \sigma \pi \nu}}$$

$$\therefore \quad \nu \text{ (frequency)} = \frac{1}{\delta^2 \pi \mu \sigma}$$

$$= \frac{1}{(0.01 \times 10^{-3})^2 \times \pi \times 4\pi \times 10^{-7} \times 3.54 \times 10^7}$$

$$= 71.5 \times 10^6 \text{ Hz} = 71.5 \text{ MHz}$$

EXAMPLE 8.6: Calculate the energy flux dW passing during time dt through a unit area perpendicular to the direction of propagation of the electromagnetic wave.

Solution: If we know the values of **E** and **B** in the region of location of the unit area, then

$$dW = wc\,dt \quad (1)$$

where w is the energy density $= \varepsilon_0 \frac{E^2}{2} + \mu_0 \frac{H^2}{2}$. For an electromagnetic wave, we have

$$\varepsilon_0 E^2 = \mu_0 H^2$$

This means that the electric energy density in the electromagnetic wave at any instant is equal to the magnetic energy density at the same point, so one can write for the energy density

$$\omega = \varepsilon_0 E^2$$

Then

$$dW = \varepsilon_0 E^2 c\,dt = \sqrt{\frac{\varepsilon_0}{\mu_0}} E^2\, dt$$

One can also obtain dW with the help of Poynting vector. We have

$$dW = N dt = EH dt = \sqrt{\frac{\varepsilon_0}{\mu_0}} E^2 dt$$

Thus both expressions lead to the same result.

EXAMPLE 8.7: Given $\mu = \mu_0$, $\varepsilon_r = 2.3$ and $\sigma = 2.56 \times 10^{-4}$ mho/m, find the phase velocity and the magnitude of the attenuation constant of plane waves in polyethelene at 10 GHz frequency.

Solution: (i) Phase velocity,

$$V = \frac{1}{\sqrt{\mu\varepsilon}} = \frac{\sqrt{\varepsilon_0/\varepsilon}}{\sqrt{\mu_0 \varepsilon_0}} = c\sqrt{\frac{\varepsilon_0}{\varepsilon}}$$

$$= 3 \times 10^8 \times \sqrt{\frac{1}{2.3}} = 1.97 \times 10^8 \text{ m/s}$$

we have

$$\frac{\sigma}{\omega\varepsilon} = \frac{2.56 \times 10^{-4}}{2\pi \times 10^{10} \times 2.3 \times 8.85 \times 10^{-12}} = 2 \times 10^{-4}$$

Obviously this is much smaller that 1. We have

$$\beta = \omega\sqrt{\frac{\varepsilon\mu}{2}}\left[-1 + \left\{1 + \left(\frac{\sigma}{\varepsilon\omega}\right)^2\right\}^{1/2}\right]^{1/2}$$

$$\approx \omega\sqrt{\frac{\varepsilon\mu}{2}}\left(\frac{\sigma^2}{2\varepsilon^2\omega^2}\right)^{1/2} = \frac{1}{2}\sigma\sqrt{\frac{\mu}{\varepsilon}} = \frac{1}{2}\frac{\sigma}{\omega\varepsilon}\omega\sqrt{\mu\varepsilon} = \frac{1}{2}\frac{\sigma}{\omega\varepsilon}\frac{\omega}{v}$$

$$= \frac{1}{2} \times 2 \times 10^{-4} \times 2\pi \times 10^{10} \times \frac{1}{1.97 \times 10^8} = 3.1 \times 10^{-2}.$$

EXAMPLE 8.8: Calculate the degree of polarisation for ordinary light (index 1.5) at an angle of 45°.

Solution: We have from Snell's law

$$\frac{\sin\theta_I}{\sin\theta_R} = \frac{n_2}{n_1} = {}_1n_2 = n \text{ (say)}$$

$$\therefore \sin\theta_R = \frac{\sin\theta_I}{n} = \frac{\sin 45^\circ}{1.5} = \frac{1}{\sqrt{2}\times 1.5} \quad \left| \begin{aligned} \theta_1 &= 45^\circ \\ n &= 1.5 \end{aligned} \right.$$

$$\therefore \quad \theta_R = \sin^{-1}\left(\frac{2}{3\sqrt{2}}\right) = \sin^{-1}\left(\frac{2\sqrt{2}}{6}\right) = \sin^{-1}(0.4713) = 28.1^\circ$$

$$\therefore \quad R_\perp = \frac{\sin^2(\theta_I - \theta_R)}{\sin^2(\theta_I + \theta_R)} = \frac{\sin^2(45^\circ - 28.1^\circ)}{\sin^2(45^\circ + 28.1^\circ)} = \frac{\sin^2 16.9^\circ}{\sin^2 73.1^\circ}$$

$$= \frac{(0.2294)^2}{(0.9568)^2} = 0.09337$$

$$R_{11} = \frac{\tan^2(\theta_I - \theta_R)}{\tan^2(\theta_I + \theta_R)} = \frac{\tan^2 16.9^\circ}{\tan^2 73.1^\circ} = \frac{(0.3038)^2}{(3.2914)^2} = 0.00852$$

$$\therefore \quad \text{Degree of polarisation} = \frac{R_\perp - R_\parallel}{R_\perp + R_\parallel} = \frac{0.09337 - 0.00852}{0.09337 + 0.00852}$$

$$= \frac{0.08485}{0.10189} = 0.833 = 83.5\%$$

EXAMPLE 8.9: Fig. 8.12 shows that if the media are interchanged the angles of incidence and refraction are reversed. If $R'_\parallel$ and $I'_\parallel$ denote the amplitude reflection and transmitted coefficient corresponding to

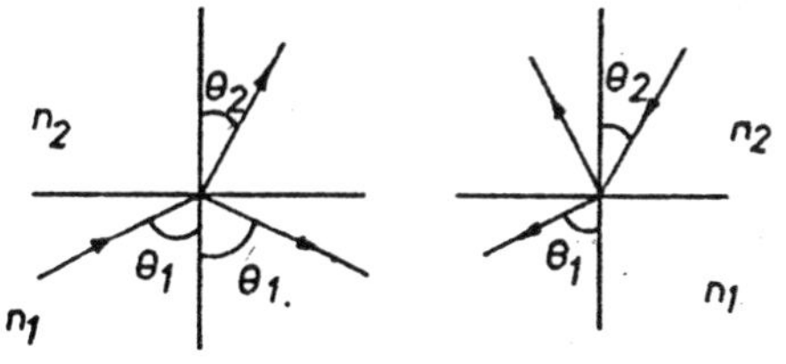

Figure 8.12. When the media are interchanged, the angle of incidence and refraction are reversed.

Fig. 8.12(b), then show that

$$1 + R_\parallel R'_\parallel = T_\parallel T'_\parallel$$

Solution:

One can show that

$$R_{\|} = \frac{\varepsilon_2 \sin\theta_2 \cos\theta_1 - \varepsilon_1 \sin\theta_1 \cos\theta_2}{\varepsilon_2 \sin\theta_2 \cos\theta_1 + \varepsilon_1 \sin\theta_1 \cos\theta_2} \tag{1}$$

one can obtain $R'_{\|}$ by replacing ε_1 by ε_2, θ_1 by θ_2 and θ_2 by θ_1 in Eq. (8.1) (corresponding to fig. 8.12), one obtains

$$R'_{\|} = R_{\|}$$

Thus

$$\begin{aligned} 1 + R_{\|} R'_{\|} &= 1 - R_{\|}^2 \\ &= 1 - \frac{(\varepsilon_2 \sin\theta_2 \cos\theta_1 - \varepsilon_1 \sin\theta_1 \cos\theta_2)^2}{(\varepsilon_2 \sin\theta_2 \cos\theta_1 + \varepsilon_1 \sin\theta_1 \cos\theta_2)^2} \\ &= \frac{4\varepsilon_1 \varepsilon_2 \sin\theta_1 \cos\theta_1 \sin\theta_2 \cos\theta_2}{(\varepsilon_2 \sin\theta_2 \cos\theta_1 + \varepsilon_1 \sin\theta_1 \cos\theta_2)^2} \end{aligned} \tag{2}$$

one obtains

$$T_{\|} = \frac{2\varepsilon_1 \sin\theta_1 \cos\theta_1}{\varepsilon_2 \sin\theta_2 \cos\theta_1 + \varepsilon_1 \sin\theta_1 \cos\theta_2} \tag{3}$$

Making the above mentioned replacements in eq (3), one obtains

$$T'_{\|} = \frac{2\varepsilon_2 \sin\theta_2 \cos\theta_2}{\varepsilon_1 \sin\theta_1 \cos\theta_2 + \varepsilon_2 \sin\theta_2 \cos\theta_1} \tag{4}$$

Multiplying (3) and (4), one obtains (2). Thus

$$1 + R_{\|} R'_{\|} = T_{\|} T'_{\|}$$

EXAMPLE 8.10: A 1 MHz (300 m wavelength) plane electromagnetic wave travelling in a normally dispersive, lossless medium has a phase velocity at this frequency of 3×10^8 m/s. The phase velocity as a function wavelength is given by $v = k\sqrt{\lambda}$ where k is constant. Show that the group velocity is 1.5×10^8 m/s.

Solution: The group velocity v_g is given by

$$\begin{aligned} v_g &= v - \lambda \frac{dv}{d\lambda} \\ &= v - \frac{k}{2}\sqrt{\lambda} = k\sqrt{\lambda} - \frac{k}{2}\sqrt{\lambda} \\ &= v\left(1 - \frac{1}{2}\right) = \frac{v}{2} = \frac{3 \times 10^8}{2} = 1.5 \times 10^8 \text{ m/s.} \end{aligned}$$

EXAMPLE 8.11: For sea water with $\sigma =$ mho s/m and $\varepsilon_r = 80$ find the distance upto which a radio signal can be transmitted at 25 kHz and 25 MHz, if the range is taken to be the distance at which 90% of the wave amplitude is attenuated.

Solution: Let us consider that a electromagnetic wave is propagating in the x-direction with a propagation constant λ. Then for a plane wave linearly polarized in y-direction, we have

$$E_y = E_0 \exp(-\gamma x) \quad \text{where} \quad \gamma = \alpha + j\beta$$

$$\therefore \quad E_y = E_0 e^{-\alpha x} e^{-j\beta x}$$

Attenuation, therefore, is

$$e^{-\alpha x} = 0.1$$

This yields $x = \dfrac{2.30}{\alpha}$

(i) $f = 25$ kHz

$$\alpha = \omega\sqrt{\frac{\varepsilon\mu}{2}}\left[1+\left\{1+\left(\frac{\sigma}{\varepsilon\omega}\right)^2\right\}^{1/2}\right]^{1/2}$$

$$= 0.715 \quad \therefore \quad x = \frac{2.33}{0.715} = 3.22 \text{ m}$$

(ii) $f = 25$MHz,

$$\text{one obtains } \alpha = 22.6 \quad \therefore \quad x = \frac{2.33}{22.6} = 10.4 \text{ cm.}$$

QUESTIONS

1. Derive the electromagnetic wave equation from Maxwell's field equations. Consider plane wave solutions of this equation and prove that the energy density associated with such a wave in a stationary homogeneous non-conducting medium propagates with the same speed with which the fields do.
2. Explain the propagation of electromagnetic waves in a conducting medium and hence defining skin depth explain why radiocommunication with submerged submarine becomes increasing difficult at several skin depths.

3. Discuss propagation of electromagnetic waves in an isotropic dielectric medium show that the waves are of transverse nature. Also show that the wave energy is equally shared between electric and magnetic fields.
4. Discuss the propagation of plane electromagnet waves in an isotropic dielectric medium. Show that electric and magnetic field vectors (**E** and **H**) are mutually perpendicular.
5. Explain the propagation of electromagnetic waves in a conducting medium. Explain why in high frequency circuits current flows only on the surface of conductors.
6. Explain (i) why the speed of electromagnetic waves in isotropic dielectric is more than that in free space? (ii) why electromagnetic wave can easily penetrate in a dielectric medium but cannot do so in a conducting medium.
7. Using electromagnetic equations, obtain an expression for field energy and momentum densities.
8. A plane wave travelling in a homogeneous isotropic dielectric is incident obliquely at the plane surface of another dielectric. Determine the reflection and transmission coefficients if the incident wave is polarised with its electric vector (i) perpendicular to the plane of incidence and (ii) in the plane of incidence.
9. Obtain the boundary conditions satisfied by the electromagnetic field vectors **E, D, B** and **H** on the plane interface between two media.
10. Derive Fresnel's equations for reflection and refraction of electromagnetic waves at a plane boundary separating two dielectric media. Discuss important consequences of these relations.
11. Discuss metallic reflection and refraction. Obtain an expression for the reflection power of a metallic surface.
12. Derive Fresnel's equations for reflection and refraction of electromagnetic wave at a plane boundary separating two media when the incident wave is polarised with **E** vector parallel to the plane of incidence. Hence show that there is an angle of incidence for which there is no reflected wave. What is the name of this angle?
13. Show on the basis of electromagnetic theory that all good conductors are good absorbers and good reflectors.

14. Derive Fresnel's equations for reflection and refraction of electromagnetic waves at a plane boundary separating two dielectric media. Discuss the phenomenon of total internal reflection.
15. What are the pecularities of metallic reflection Give its theory in brief and describe method for its verification.

PROBLEMS

1. A plane electromagnetic wave travelling in z-direction in an unbounded loss less dielectric medium $k_m = 1$ and $k_e = 1$ has a peak electric field strength $E = 6$ V/m where k_m is relative permeability of the medium and k_e is relative permittivity (or dielectric constant) of the medium. Determine (i) the velocity of wave (ii) intrinsic impedance of the medium (iii) peak value of magnetic field strength and (iv) peak value of Poynting vector.

   [**Ans.** (i) $v = 1.7 \times 10^8$ m/s (ii) 217.65 Ω (iii) 2.76×10^{-2} A/m (iv) 0.65 W/m^2]
2. Given the equation of a wave $E = E_0 \cos(500x - 6000t)$, Show that the wave length is $\dfrac{\pi}{250}$ while the speed of propagation is 12.
3. A laser beam of monochromatic light has a uniform intensity over its cross-sectional area of 0.2 mm^2. If the total power of the beam is 500 mV, calculate the amplitude of the magnetic field vector of the plane wave that can represent the beam.

   [**Ans.** $B_0 = 1.447 \times 10^{-4}\ T$]
4. Electromagnetic radiation from the sun falls on the earth's surface at the rate of 1.4×10^3 W/m^2. Assuming that this radiation can be considered as plane waves, estimate the magnitude of the electric and magnetic field amplitudes in the wave.
5. Show that the average value of the Poynting vector of a plane harmonic wave is $\dfrac{1}{2} c\varepsilon_0 E_0^2$ or $\dfrac{\varepsilon_0 B_0}{2\mu_0}$.
6. Radio waves received by a radio set have an electric field of maximum amplitude equal to 0.1 V/m. Assuming that the wave can be considered as plane, calculate (i) the amplitude of the magnetic field (ii) the average intensity of the wave (iii) the average energy

density (iv) assuming that the radio set is 1 km from the broadcasting station and that the station radiates energy isotropically determine the power of the station.

[**Ans.** (i) $\frac{1}{3} \times 10^{-9}$ T (ii) 1.33×10^{-6} W/m^2 (iii) 4.42×10^{-15} J/m^3 (iv) 167 W]

6. A wave is moving in a conducting medium show that the average energy in the magnetic field is greater than that in the electric field.
7. A monochromatic linearly polarized plane wave is moving in an isotropic non-conducting medium show that the time average of its energy density is distributed equally between the magnetic and the electric fields.
8. The wave length of low frequency radio waves in sea water is 3×10^3 m, the electric conductivity of which is $\sim 4/\Omega - m$. Find the skin depth. [**Ans.** 0.796 m]
9. A good conductor is roughly defined as one for which $\frac{\sigma}{\omega\varepsilon}$ is to or greater than 10. If σ and ε for sea water are 4 mho/m and $\left(\frac{9 \times 10^{-9}}{4\pi}\right)$ F/m, determine the frequency range for which the sea water is good conductor. [**Ans.** $\nu \leq 88$ mHz]
10. A good dielectric is roughly defined as one for which $\frac{\sigma}{\omega\varepsilon}$ is equal to or less than 0.1. If conductivity σ and permeability ε of fresh water are respectively 10^{-3} mho/m and $\frac{9 \times 10^{-9}}{4\pi}$ F/m; determine the frequency range for which the fresh water is a good dielectric. [**Ans.** $\nu \geq 2.2$ mHz.]
11. A laser beam has a diameter of 10^{-3} m and its power is 1 watt. Show that the amplitude of the magnetic field in the beam in free space is 10^{-4}.
12. Electromagnetic waves are incident normally on a metal surface of refractive index n and extinction coefficient defined by the relation $\frac{n_2}{n_1} = {}_1n_2 = n(1+ik)$. Show that the ratio of the reflected intensity to incident intensity is $= \frac{n^2+k^2-2n+1}{n^2+k^2+2n+1}$.
13. Compute the coefficients of reflection and transmission for electromagnetic waves in the visible region for crown glass at an angle

of incidence equal to 30°. Given $n_{\text{glass}} = 1.52$ and $n_{\text{air}} \approx 1$.
[Hint: $\sin\theta_I = 1.52\sin\theta_R$, For $\theta_I = 30°$ $\theta_R = 19°12'$
$\therefore$ $R_{\parallel} = 0.165$, $R_{\perp} = -0.248$, $T_{\parallel} = 0.442$ and $T_{\perp} = 0.752$]

14. Linearly polarized light wave is incident on a glass plate ($n = 1.5$) with an angle of incidence of 45°. Determine the coefficients of reflection and refraction if the internal field of the incident wave is (a) in the plane of the incidence, (b) normal to the plane of incidence. **[Ans.** (a) −0.0791 (b) 0.7194]

15. A plane travelling wave in free space has an average Poynting vector of $1\,\text{W/m}^2$. Show that the average energy density is $3.33 \times 10^{-9}\ \text{J/m}^3$.

16. In a perfect dielectric medium the electric field progressing in the z-direction is given by the equation $E_x = E_{xo}\cos(\omega t - \beta z)$ and the associated magnetic field by the equation $H_y = \frac{E_{xo}}{\eta}\cos(\omega t - \beta z)$, where E_{xo} is the peak value of E_x at $t = 0$ and $z = 0$ and η is the intrinsic impedance of the dielectric. Show that the average power through any area S normal to the z-axis is given by, $P_{z,\omega} = \frac{1}{2}\frac{E_{xo}^2}{\eta}S$.

17. A plane electromagnetic wave falls perpendicular on a plane surface separating a medium of index n_1 from a medium of index n_2. Show that the coefficient of reflection and refraction are $R = \frac{(n_1 - n_2)}{(n_1 + n_2)}$ and $T = \frac{2n_1}{(n_1 + n_2)}$.

SHORT ANSWER QUESTIONS

1. What is absolute index of refraction?
[Ans. $n = \frac{c}{v} = \frac{1}{\sqrt{\varepsilon_0\mu_0}}\sqrt{\varepsilon\mu} = \sqrt{\frac{\varepsilon\mu}{\varepsilon_0\mu_0}} = \sqrt{\varepsilon_r\mu_r}$]
2. Is gamma rays are electromagnetic radiations?
[Ans. Yes, these waves are of nuclear origin]
3. Is there are electromagnetic waves in cosmic radiations?
[Ans. Yes, there are electromagnetic waves of shorter wavelengths or longer frequencies]

4. Why does not one see the portion other than visible one of the electromagnetic spectrum?

   [**Ans.** The retina of the eye is sensitive only to colours in the visible region, i.e. wavelength lying between 3900 Å. and 7800 Å. This region corresponds to visible part of the spectrum]

5. On what factors the velocity of electromagnetic wave depends?

   [**Ans.** (i) Permittivity and (ii) Permeability of medium, $c = \frac{1}{\sqrt{\varepsilon\mu}}$]

6. Do electromagnetic waves have same velocity in all transparent media?

   [**Ans.** No, as the refractive index is different for different media $n = \frac{c}{v}$, v is different for different media]

7. An electromagnetic wave carries momentum what it signifies?

   [**Ans.** An electromagnetic interaction between two electric charges means an exchange of energy and momentum between the charges.]

8. What is the average value of the Poynting vector $\mathbf{N} = [\mathbf{E} \times \mathbf{H}]$ for the wave represented by $\mathbf{E} = \hat{i}E_o \sin(kz - \omega t)$?

   [**Ans.** $< N >= \hat{k}\frac{c}{2}\varepsilon_0 E^2$]

9. What will be the total electromagnetic energy in a cube of side 5 km in the path of the wave? [**Ans.** 8837×10^{-13} J.]

10. What do you understand by the dispersion of a electromagnetic wave when propagating in matter?

    [**Ans.** When a electromagnetic wave propagates through matter suffer dispersion, i.e a pulse containing several frequencies will be distorted because each component will travel with a different velocity]

OBJECTIVE QUESTIONS

1. The ratio $\sqrt{\frac{\mu}{\varepsilon}}$ for a medium is called

   (a) Characteristic impedance, (b) Reactance,
   (c) Refractive Index, (d) Frequency. **[a]**

2. For a electromagnetic wave propagated in a good dielectric having $\left(\frac{\sigma}{\omega t}\right) \gg 1$, the attenuation factor α and phase shift factor β are given by

   (a) $\alpha = \frac{\sigma}{2}\sqrt{\frac{\mu}{\varepsilon}}, \ \beta = \sqrt{\omega\mu\varepsilon}$

   (b) $\alpha = \sqrt{\frac{\mu}{\varepsilon}}, \ \beta = \frac{1}{2}\sqrt{\omega\mu\varepsilon}$

   (c) $\alpha = \sqrt{\frac{\mu}{\varepsilon}}, \beta = \sqrt{\frac{\mu\omega}{\varepsilon}}$

   (d) $\alpha = \frac{\sigma}{2}, \ \beta = \sqrt{\mu\varepsilon}.$ **[a]**

3. The value of $\sqrt{\frac{\mu}{\varepsilon}}$ for a free space is about

   (a) 380 Ω, (b) 38 Ω, (c) 3.8 Ω, (d) 3800 Ω. **[a]**

4. The intrinsic impedance of free space is

   (a) $z_0 = \sqrt{\frac{\mu_0}{\varepsilon_0}}$, (b) $z_0 = \frac{\mu_0}{\varepsilon_0}$, (c) $z_0 = \mu_0\varepsilon_0$,

   (d) $z = \sqrt{\mu_0\varepsilon_0}$. **[a]**

5. In a uniform plane wave **E** and **H** are related by

   (a) $\frac{\mathbf{E}}{\mathbf{H}} = \sqrt{\frac{\mu}{\varepsilon}}$, (b) $\frac{\mathbf{E}}{\mathbf{H}} = \sqrt{\frac{\varepsilon}{\mu}}$, (c) $\frac{\mathbf{E}}{\mathbf{H}} = \frac{\mu}{\varepsilon}$, (d) $\frac{\mathbf{E}}{\mathbf{H}} = \frac{\varepsilon}{\mu}$. **[a]**

6. In a perfect dielectric, wave propagation occurs:

   (a) with small attenuation.
   (b) with large attenuation.
   (c) with zero attenuation.
   (d) none of the above. **[c]**

7. An electromagnetic wave is incident on a perfect conductor is

   (a) fully transmitted.
   (b) entirely reflected.
   (c) partially reflected and partially transmitted.
   (d) none of the above. **[b]**

8. For a good conductor, the depth of the penetration of electromagnetic waves is given by

   (a) $\delta = \sqrt{\frac{2}{\omega\sigma\mu}}$, (b) $\delta = \sqrt{\frac{1}{\omega\sigma\mu}}$, (c) $\delta = \sqrt{\frac{2}{\omega\sigma^2\mu}}$,

   (d) $\delta = \frac{1}{\sqrt{\omega\sigma}}$. **[a]**

9. For a plane good conductor skin depth varies
   (a) directly as the conductivity.
   (b) inversely as the conductivity.
   (c) inversely as the square root of conductivity.
   (d) directly as the square root of conductivity. **[c]**

10. For a good conductor, the skin depth varies
   (a) inversely as frequency, ω (b) inversely as $\sqrt{\omega}$,
   (c) directly as ω, (d) directly as $\sqrt{\omega}$. **[b]**

11. The dependence of the phase velocity of an electromagnetic wave in a medium on the frequency of the wave is called
   (a) Reflection, (b) Refraction, (c) Polarization, (d) Dispersion. **[d]**

12. The relation between the amplitude of a linearly polarized monochromatic electromagnetic wave and the vector **E** is
   (a) $q = |\mathbf{E}|_{max}$, (b) $a = |E^2|$
   (c) $a = \left|\frac{\mathbf{E}}{2}\right|_{max}$ (d) $a = |\mathbf{E}|_{max}^{1/2}$ **[a]**

Chapter 9

Scattering and Dispersion of Electromagnetic Waves

9.1 INTRODUCTION

We have read that the electromagnetic waves undergo geometrical reflection, refraction and propagate through waveguides. In addition to specular reflections, the electromagnetic waves incident on targets are scattered in various directions. The scattering of electromagnetic radiation by a free electron has certain pecularities in comparison to the scattering by bound electrons or molecules. Scattering is a double process by which an electron absorbs energy from an electromagnetic wave and reradiates it as scattering radiation. We must keep in mind that an electromagnetic wave carries energy and momentum, and if some energy E is removed from the wave, a corresponding amount of momentum $P = \frac{E}{c}$ must also be removed from the wave. We must note that a free electron cannot absorb an amount of energy E and at the same time acquire a momentum $P = \frac{E}{c}$ because the relation between kinetic energy and momentum for an electron in $E_K = c\sqrt{m_0c^2 - p^2c^2} - m_0c^2$ at high energies and $E_K = \frac{p^2}{2m_e}$ lower energies. Either of these is incompatible with the relation $P = \frac{E}{c}$ as required.by conservation of energy. However in the case of bound electron, the energy and momentum absorbed are shared by both the electron and the ion forming the remaining part of the atom, and it is always possible to split both energy and momentum in correct proportions. However, the ion, having a much larger mass, carries (along with some momentum) only a small fraction of the energy available, and usually one does not consider it all. In the case of a free electron there is no other particle with which the electron shares the energy and the momentum and no absorption or scattering should be possible.

Experiments, however tells a different story. When we analyse the electromagnetic radiation that has passed through a region where free electrons are present, we observe that, in addition to the incident radiation there is another sort of radiation present of *different* frequency. This new radiation is interpreted as the radiation scattered by the free electrons The frequency of the scattered radiation is *smaller* than the incident frequency and accordingly the wavelength of the scattered radiation is *longer* than the incident wavelength The wavelength of the scattered radiation is also different for each direction of scattering. This interesting phenomenon is called the *compton effect* after A. H. compton, who first observed and analyzed it in the early 1920's. Compton effect is an example of inelastic scattering. Other examples of *inelastic scattering* are Raman effect and Tyndall effect. Rayleigh scattering is an example of *elastic scattering*. In this process the photons of the radiation are reflected, i.e. they bounce off the atoms and molecules with unchanged energy and momentum. Theoretical analysis shows that one can "explain" the scattering of electromagnetic radiation by a free electron by identifying the process with collision of a free electron and a particle of zero rest mass, which is a photon of energy $h\nu$. However, this is quantum explanation of scattering of electromagnetic waves by a free charge.

So far we have considered only the propagation of electromagnetic waves in vacuum. Experiments reveal that the velocity of propagation of an electromagnetic wave through matter is different from its velocity of propagation in vacuum. When an electromagnetic wave propagates through matter, even if there are no free charges and currents, it induces certain charges and currents in the substance as a result of polarization and magnetization of matter as discussed in previous chapters. If the substance is homogeneous and isotropic we have seen that the net effect of the polarization and magnetization of the medium by the electromagnetic wave is to replace the constants ε_0 and μ_0 in the Maxwell's equations by the electric permittivity ε and the magnetic permeability μ characteristic of the material. Obviously the velocity of the wave now becomes $v = \frac{1}{\sqrt{\varepsilon\mu}}$. The ratio $\frac{c}{v}$ is called the absolute index of refraction of the substance designated by n. We have

$$n = \frac{c}{v} = \frac{1}{\sqrt{\varepsilon_0\mu_0}}\sqrt{\varepsilon\mu} = \sqrt{\frac{\varepsilon\mu}{\varepsilon_0\mu_0}} \tag{9.1}$$

But $\frac{\varepsilon}{\varepsilon_0} = \varepsilon_r$ (relative permittivity) and $\frac{\mu}{\mu_0} = \mu_r$ (relative permeability). Then

$$n = \frac{c}{\mathbf{v}} = \sqrt{\varepsilon_r\mu_r} \tag{9.2}$$

In general μ_r differs very little from 1 for the majority of substances that transmit electromagnetic waves and we can write as a satisfactory approximation

$$n = \sqrt{\varepsilon_r} \tag{9.3}$$

This relation affords a simple experimental method for determining the relative permittivity of the substance if the index of refraction is obtained independently. One obtains the following relation between n, ε_r, number of electrons per unit volume (N) and wave frequency (ω),

$$n^2 = \varepsilon_r = 1 + \frac{Ne^2}{m\varepsilon_0}\left(\sum_\alpha \frac{f_\alpha}{\omega_\alpha^2 - \omega^2}\right) \tag{9.4}$$

The quantities designated by f_i are called the oscillator strengths of the substance. They are all positive and smaller than one, and represent the relative proportion in which each of the frequencies of the spectrum contribute to the polarizability of the atom. They satisfy the relation $\Sigma_\alpha f_\alpha = 1$. Obviously, the index of refraction depends on the wave frequency, and hence also on the wavelength (Fig. 9.1). $\omega_1, \omega_2, \ldots$ are the

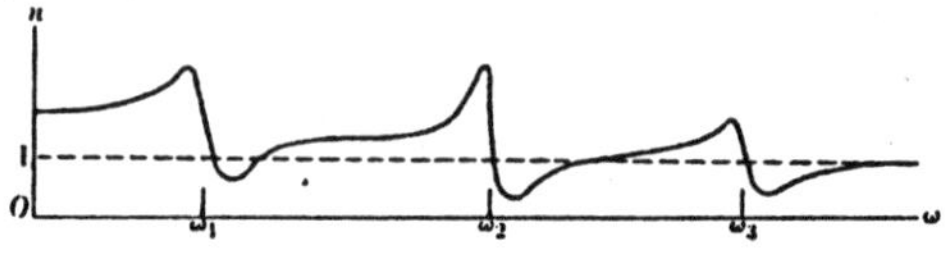

Figure 9.1. Variation of the index of refraction with frequency and wavelength.

characteristic frequencies of the emission spectrum of the substance consequently the phase velocity $\mathbf{v} = \frac{c}{n}$ of the electromagnetic wave in matter also depends on the frequency of the radiation. Therefore, electromagnetic wave suffer dispersion when propagating in a medium That is a pulse containing several frequencies will be distorted because each component will travel with a different velocity. The group velocity v_g is given by

$$v_g = v + k\frac{dv}{dk}$$

we have
$$\frac{dv}{dk} = \frac{dv}{d\omega}\cdot\frac{d\omega}{dk} = v_g\frac{dv}{d\omega}$$

Since, $v_g = \dfrac{d\omega}{dk}$ and $v = \dfrac{c}{n}$, we have

$$\frac{dv}{d\omega} = -\frac{c}{n^2}\frac{dn}{d\omega}$$

∴
$$v_g = v - \frac{v_g c_k}{n^2}\cdot\frac{dn}{d\omega}$$

Solving, one obtains

$$v_g = \frac{v}{1+\left(\frac{ck}{n^2}\right)\left(\frac{dn}{d\omega}\right)} = \frac{c}{n+\omega\left(\frac{dn}{d\omega}\right)} \quad \left(\text{where } k = \frac{\omega}{v} = \frac{\omega n}{c}\right) \tag{9.5}$$

When $\dfrac{dn}{d\omega}$ is positive, the group velocity is smaller than the phase velocity. Such a situation is called *normal dispersion* But if $\dfrac{dn}{d\omega}$ is negative then the group velocity is larger than the phase velocity and *anomalous dispersion* results. The possibility exists in this case that the group velocity is larger than c, and that an electromagnetic pulse can be transmitted at a velocity larger than c. This is apparently in contradiction to the results obtains from Lorentz transformation and the principle of relativity.

A careful analysis of the transmission of an electromagnetic signal made by Brillouin, Sommerfeld and others, revealed that it is impossible to transmit a signal with a velocity greater than c. Fig. 9.2 shows the variation of the phase velocity v, the group velocity v_g and the signal velocity v_σ near and characteristic frequency ω. The signal velocity practically coincides with the group velocity except near the characteristic frequency and is never larger than c even in the region of anomalous dispersion.

We have said that electromagnetic waves appear to propagate in matter with a phase velocity different from their propagation velocity in vacuum. But that difference seems to stem from the fact that the

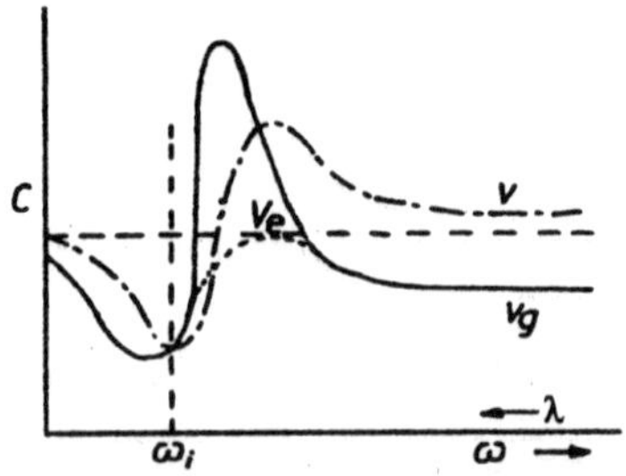

Figure 9.2. Phase, group and signal velocities of an electromagnetic pulse in a dispersion medium.

permittivity and permeability of matter are different from those of vacuum. This difference (in the permittivity and permeability) is, in turn, a consequence of the electric and magnetic permeability of matter under the action of the incoming electromagnetic wave. Thus when an electromagnetic wave falls on a piece of matter it induces oscillations in the charged particles of the atoms or molecules which then emit *secondary* or "*scattered*" waves. These scattered waves are superposed on the original wave, giving a resultant wave. The phases of the secondary waves are in general different from that of the original wave, since a force oscillator is not always in phase with the driving force. A detailed analysis indicates that this phase difference affects the resultant wave in such a way that this wave appears to have a phase velocity different from that which a wave in vacuum has. This result is particularly satisfying since, from the atomic point of view, all charges, both free and sound, are equivalent, and the electromagnetic waves they emit must all propagate with the velocity c. It is the wave resulting from the superposition of their individual waves, which have different phases, that as a consequence of phase differences, appears to have a different velocity of propagation.

9.2 SCATTERING CROSS SECTION AND SCATTERING AMPLITUDE

The scattering process is described in terms of a quantity known as scattering cross-section. Let us consider that a beam of electromagnetic radiation moving along the z-direction with a flux of $I_0/\text{cm}^2/\text{sec}$. When the radiation is incident on a target, it is scattered in a solid angle $d\Omega$. The conservation of intensity of electromagnetic wave flux in the scattering process demands that

$$I\,d\Omega = I_0\,d\sigma \tag{9.6}$$

where $d\sigma$ is the effective cross-sectional area which scattering target presents, perpendicular to the incident flux of electromagnetic radiation. The effective area (perpendicular to the incident beam of radiation) governs the solid angle in which incident radiations are scattered. The ratio $\dfrac{d\sigma}{d\Omega}$ is known as differential scattering and presents the basic feature of the scattering process. The total scattering cross-section is defined as

$$\sigma = \int_0^{2\pi} \frac{d\sigma}{d\Omega}\, d\Omega \tag{9.7}$$

The scattering process holds good equally for waves and particles. The scattering process can best be explained and defined from quantum-mechanical considerations.

9.3 SCATTERING FROM A FREE CHARGE PARTICLE (THOMSON SCATTERING)

The incidence and scattering of electromagnetic radiation on a free charge particle of charge q and mass m constitutes a fundamental process of scattering phenomena. Let us suppose that plane monochromatic polarized electromagnetic wave is incident on a free charged particle of mass m and carrying a charge q. The wave is of the form

$$\mathbf{E} = \mathbf{E}_0 \exp\left[i(\mathbf{k}\cdot\mathbf{r} - \omega t)\right] \tag{9.8}$$

The force experienced by the charge particle is given by

$$\mathbf{F} = q\mathbf{E} \tag{9.9}$$

or

$$m\frac{dv}{dt} = q\mathbf{E}_0 \exp\left[i(\mathbf{k}\cdot\mathbf{r} - \omega t)\right] \tag{9.10}$$

Obviously, (9.10) is the equation of motion of the charged particle. The displacement of the charged particle under the influence of the incident field is small compared to the wavelength. This means that the velocity acquired by the particle as a result of this interaction is much smaller that $c(v \ll c)$. Therefore we have restricted to the non-relativistic limit and ignored the second term in the expression for the Lorentz force given by

$$\mathbf{F} = q(\mathbf{E} + \mathbf{v} \times \mathbf{B}) \tag{9.11}$$

The oscillating charge in the presence of electro magnetic field confirms to an electric dipole whose dipole moment is written as

$$\mathbf{p}(t) = q\mathbf{r}(t),$$

Therefore, we have

$$\ddot{\mathbf{p}} = q\ddot{\mathbf{r}}(t) = \frac{q^2}{m}\mathbf{E}(t) \tag{9.12}$$

The time average of the power radiated per unit solid angle is obtained as (see ch. 11)

$$\left\langle \frac{dW}{d\Omega} \right\rangle = \frac{q^2(\dot{u})^2}{16\pi^2\varepsilon_0 c^3}\sin^2\theta \tag{9.13}$$

$$= \frac{\langle \ddot{p}^2 \rangle \sin^2\theta}{16\pi^2\varepsilon_0 c^3} = \frac{q^4 E_0^2 \sin^2\theta}{16\pi^2 m^4 \varepsilon_0^3} \tag{9.14}$$

We note that the rate of energy radiation is independent of frequency of the incident electromagnetic field. One can obtain the Poynting vector of the incident electromagnetic radiation as

$$N_i = \frac{1}{2}E_0 H_0 \hat{k} = \frac{E_0^2}{2Z_0}\hat{k} = \frac{1}{2}\varepsilon_0 c E_0^2 \hat{k} \quad \left(\text{Here } z_0 = \frac{\mu_0}{\varepsilon_0}\right). \tag{9.15}$$

A target or an obstacle appearing in the direction of incident electromagnetic energy flow intercepts a part of the electromagnetic energy which is redirected in different directions. This phenomena is termed as scattering of electromagnetic waves. This is characterized by certain cross-section. We have differential scattering cross-section

$$\frac{d\sigma}{d\Omega} = \frac{\text{Scattered energy per unit solid angle per unit time}}{\text{Incident energy per unit area per unit time}}$$

$$= \frac{\left\langle \frac{dW}{d\Omega} \right\rangle}{\text{Incident flux}} = \frac{\left\langle \frac{dW}{d\Omega} \right\rangle}{\sqrt{\frac{\varepsilon_0}{\mu_0}}E_0^2} = \left(\frac{q^2}{4\pi\varepsilon_0 mc^2}\right)^2 \sin^2\theta \tag{9.16}$$

where θ is the angle between the induced dipole moment $\mathbf{p}$ and the direction of the out going radiation, i.e. θ is the angle between $\mathbf{E}$ and the unit vector $\hat{a}_r$ as shown in Fig. 9.3.

Now, we consider a more general case. Supposing that incident light is unpolarized let us find the angular distribution of scattered light. For this, we will have to average over all possible azimuthal orientations of field E, i.e. we have to average over the angle Ψ (Fig. 9.3). From

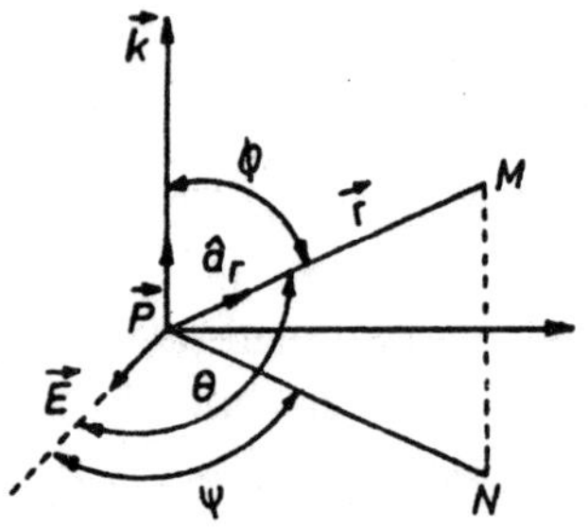

Figure 9.3.

Fig. 9.3, we have

$$\cos\theta = \cos\Psi\sin\phi$$

or

$$\sin^2\theta = 1-\cos^2\Psi\sin^2\Psi$$

Let $\overline{\sin^2\theta}$ represent the average of $\sin^2\theta$ over all angles Ψ then we have

$$\overline{\sin^2\theta} = 1-\overline{\cos^2\theta}\sin^2\theta = 1-\frac{1}{2}\sin^2\phi = \frac{1}{2}(1+\cos^2\phi)$$

∴ For unpolarized electromagnetic radiation, we have

$$\left\langle\frac{d\sigma}{d\Omega}\right\rangle_{\text{unpolarized}} = \left(\frac{q^2}{4\pi\varepsilon_0 mc^2}\right)^2\overline{\sin^2\theta}$$

$$= \left(\frac{q^2}{4\pi\varepsilon_0 mc^2}\right)^2\left(\frac{1+\cos^2\phi}{2}\right) \quad (9.17)$$

ϕ is called the scattering angle, as it is the angle between the directions of the incident radiation and the scattered radiation (Fig. 9.3).

One can obtain the total scattering cross-section by integrating over the entire solid angle. One obtains

$$\langle\sigma\rangle_{\text{unpol.}} = \int\frac{d\sigma}{d\Omega}d\Omega = \left(\frac{q^2}{4\pi\varepsilon_0 mc^2}\right)^2\int\frac{1+\cos^2\phi}{2}\sin\phi\, d\phi\, d\Psi$$

$$= \frac{8\pi}{3}\left(\frac{q^2}{4\pi\varepsilon_0 mc^2}\right)^2 \quad (9.18)$$

For electron $q = e$ and

$$\langle\sigma\rangle = \frac{8\pi}{3}\left(\frac{e^2}{4\pi\varepsilon_0 mc^2}\right)^2 = \frac{8\pi}{3}r_0^2 \quad (9.19)$$

where $r_0 = \left(\frac{e^2}{4\pi\varepsilon_0 mc^2}\right)^2 = \left(\frac{\mu_0 e^2}{4\pi m}\right) = 2.8178 \times 10^{-15}$ m and is known as *classical radius of electron* and the corresponding cross-section is known as *Thomson scattering cross-section.* We must note that the scattering of unpolarized plane electromagnet wave from free charges, in general and electrons in particular play an important role in the distribution of scattered waves.

9.4 SCATTERING OF RADIATION BY A BOUND CHARGE (RAYLEIGH SCATTERING)

Let us consider the case when an electron system is subjected to a monochromatic electromagnetic radiation. Let us assume that the bound electron is allowed to move with damped oscillations. We can write the equation of motion of such a electron as

$$F = eE = m(\ddot{x} + l\dot{x} + \omega_0^2 x) \tag{9.20}$$

where $E = E_0$ represents the applied field, l measures the damping force and ω_0 is the frequency of bound electron. We can rewrite (9.20) as

$$\frac{d^2x}{dt^2} + l\frac{dx}{dt} + \omega_0^2 x = \frac{e}{m}E_0 \exp(-i\omega t) \tag{9.21}$$

one can easily see that the steady state solution of (9.21) is

$$x(t) = \frac{eE_0}{m} \frac{\exp(-i\omega t)}{[(\omega_0^2 - \omega^2) - il\omega]} \tag{9.22}$$

For free electron ω_0 and l both are zero and therefore, we have

$$x(t) = -\frac{eE_0}{m} \frac{\exp(-i\omega t)}{\omega^2} \tag{9.23}$$

The negative sign in (9.23) is not of importance here as one is not concerned with the relative phases. We note that (9.22) and (9.23) are similar. This suggests that one can convert the results for the free charge into the corresponding results for the bound charge by simply multiplying each term in E_0 by $\frac{\omega^2}{[(\omega_0^2 - \omega^2) - il\omega]}$.

Now we obtain the time average of the power per unit solid angle radiated by a bound charge

$$\left\langle \frac{dW}{d\Omega} \right\rangle_{\text{bound}} = \frac{e^4 E_0^2}{16\pi^2 m^2 \varepsilon_0 c^2} \frac{\omega^4}{[(\omega_0^2 - \omega^2)^2 + \omega^2 l^2]} \sin^2\theta \tag{9.24}$$

Differential Scattering cross-section for a bound charge is obtained as

$$\left\langle \frac{d\sigma}{d\Omega} \right\rangle_{\text{bound}} = \frac{\left\langle \frac{dW}{d\Omega} \right\rangle}{\sqrt{\frac{\mu_0}{\varepsilon_0}} E_0^2} = \frac{e^4}{16\pi^2 m^2 \varepsilon_0^2 c^4} \frac{\omega^4 \sin^2\theta}{(\omega_0^2 - \omega^2)^2 + \omega^2 l^2} \tag{9.25}$$

The total scattering cross-section is given by

$$\begin{aligned} \sigma_{\text{bound}} &= \frac{e^4}{6\pi\varepsilon_0 m^2 c^4} \frac{\omega^4}{[(\omega_0^2 - \omega^2)^2 + \omega^2 l^2]} \\ &= \frac{\frac{8\pi}{3} r_0^2 \omega^4}{(\omega_0^2 - \omega^2)^2 + \omega^2 l^2} \qquad \left(\text{where } r_0 = \frac{e^2}{4\pi\varepsilon_0 mc}\right) \\ &= \sigma \frac{\omega^4}{(\omega_0^2 - \omega^2)^2 + \omega^2 l^2} \end{aligned} \tag{9.26}$$

where σ is the scattering cross section for a free electron. When $\omega = \omega_0$ one obtains the maximum value of scattering cross-section as

$$\sigma_{\text{bound}} = \sigma \frac{\omega^2}{l^2} \tag{9.27}$$

when $\omega \ll \omega_0$ and $\omega l \ll \omega_0^2$ one obtains

$$\sigma_{\text{bound}} = \sigma \frac{\omega^4}{\omega_0^4} \tag{9.28}$$

This scattering is known as *Rayleigh scattering*. In this case (9.28) shows that the amount of scattering light is proportional to $\left(\frac{1}{\lambda}\right)^4$ where λ is the wavelength of incident radiation. We must note that for the result to be applicable, the size of the particles causing scattering must be small in comparison to the wavelength of the scattered radiation

$$\sigma_{\text{bound}} = \frac{8\pi r_0^2}{3} \left(\frac{\omega_0}{l}\right)^2 \tag{9.29}$$

This scattering is known as *resonant scattering* In this case the total scattering cross section is maximum. In this case the natural frequency of the electron as an oscillator becomes equal to the frequency of incident radiation. The resonance scattering occurs when the sodium vapour is illuminated with the characteristic yellow sodium radiation.

If $\omega \gg \omega_0$ and $l \to 0$ we have total scattering cross section

$$\sigma_{\text{bound}} = \frac{8\pi}{3} r_0^2$$

i.e. *Thomson scattering* occurs and the charol behaves as if it was free.

Fig. 9.4 shows a plot of σ_{bound} bound as a function of ω,

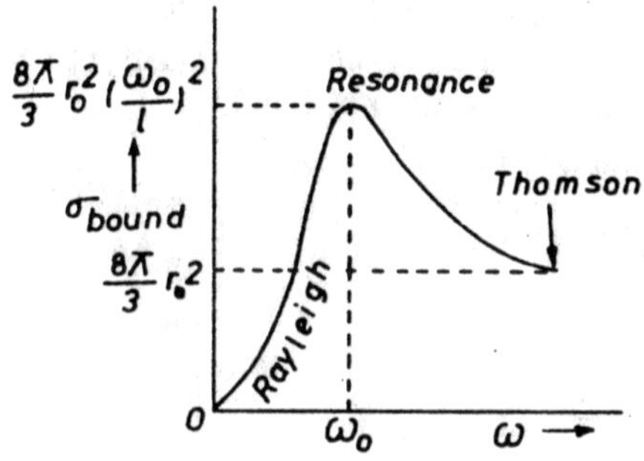

Figure 9.4.

Rayleigh scattering helps to explain the following

(i) *Sky appears blue*: According to Rayleigh scattering

$$\sigma_{bound} \propto \frac{1}{\lambda^4}$$

i.e., shorter wave length is scattered more strongly. The light coming from sun is scattered by air molecule and as we know that shorter visible wavelength corresponds to blue light, hence it is scattered strongly. One observes this scattered light which is blue. This is why the sky appears blue.

(ii) *Red colour at the tie of sunset ad sunrise*: At sunset or sunrise, the sun and its neighbouring portion appears red because sunlight travels a greater length of earth's atmosphere and strikes directly to the observer's eye. During this process the blue light is almost scattered from the direct light reaching to observer's eye, i.e. leaving only the red light to be seen.

(iii) *Use of Red light for Danger signals*: Human eye is most sensitive to yellow green colour even then red light is used for danger signals. The red light has longest wavelength in the visible region and scatters least. Obviously, red light can travel longer distances in atmosphere in comparison to the other colours and can be seen from larger distances. This is way red light is used for danger signals.

9.5 RADIATION DAMPING OR RADIATION REACTION

The motion of charged particles in external force fields necessarily involves the emission of radiation whenever the charged are accelerated. The emitted radiation carries off energy, momentum, and angular momentum and so must influence the subsequent motion of the charged particles. Consequently the motion of the sources of radiation is determined in part, by the manner of emission of the radiation. A correct treatment must include the reaction of the radiation on the motion of the sources. The reaction of the radiation on the motion of the particle is called *radiation damping* or *radiation reaction.*

We now examine that how one can include the reactive effects of radiation in the equation of motion of a charged particle. We have the classical equation of motion for a particle of mass m and charge e acted on by a conservative external force F as

$$F_e = m\frac{d\mathbf{v}}{dt} \tag{9.30}$$

While writing (9.30) we have neglected the emission of radiation. As the particle is accelerated, it will emit radiation. One will have to take account of the reaction of this radiation field on the particle. Obviously, we will have to modify (9.30) by adding a reaction force F_r i.e.

$$m\frac{d\mathbf{v}}{dt} = F_e + F_r \tag{9.31}$$

The energy radiated per unit time by a charge particle of charge e moving with a speed $\mathbf{v}(\ll c)$, which undergoes acceleration $a\left(=\frac{d\mathbf{v}}{dt}\right)$ is given by

$$W = \frac{e^2(\dot{\mathbf{v}})^2}{6\pi\varepsilon_0 c^2} \tag{9.32}$$

The energy must be conserved. Obviously the work done by the force F_r on the particle in a time interval, say t_1 to t_2, be equal to the negative of the energy retarded in this time interval. We have

$$\int_{t_1}^{t_2} \mathbf{F}_r \cdot \mathbf{v} dt = -\frac{e^2}{6\pi\varepsilon_0 c^3}\int_{t_1}^{t_2} \dot{\mathbf{v}} \cdot \dot{\mathbf{v}} dt \tag{9.33}$$

Integrating the right hand side by parts one obtains

$$\int_{t_1}^{t_2} \mathbf{F}_r \cdot \mathbf{v} dt = -\frac{e^2}{6\pi\varepsilon_0 c^3}[\mathbf{v}\cdot\dot{\mathbf{v}}]_{t_1}^{t_2} + \frac{e^2}{6\pi\varepsilon_0 c^3}\int_{t_1}^{t^2} \mathbf{v}\cdot\ddot{\mathbf{v}} dt$$

When the motion of the charge particle is periodic or if the time interval is short, then we can consider that the state of the system will be approximately the same at t_2 as at t_1. This means, one can neglect the integrated term in (9.34). Eq. (9.34) reduces to

$$\int_{t_1}^{t_2}\left(\mathbf{F}_r - \frac{e^2}{6\pi\varepsilon_0 c^3}\ddot{\mathbf{v}}\right)\cdot \mathbf{v}\,dt \simeq 0 \tag{9.34}$$

This leads to

$$F_r = \frac{e^2}{6\pi\varepsilon_0 c^3}\ddot{v} = m\ddot{v}\tau \tag{9.35}$$

where

$$\tau = \frac{e^2}{6\pi\varepsilon_0 mc^3} \tag{9.36}$$

Thus the modified equation (9.31) becomes

$$m\frac{dv}{dt} = F_e + m\tau\frac{d^2v}{dt^2}$$

or

$$m(\dot{v} - \tau\ddot{v}) = F_e \tag{9.37}$$

Eq. (9.37) is sometimes called the *Abraham-Lorentz equation of motion.* It can be considered as an equation that includes some approximate and time-averaged way the reactive effect of the emission of radiation.

If the external force F_e is a linear restoring force of the type $\mathbf{F}_e = -m\gamma\mathbf{r}$ (here this γ is different than earlier used), one can express (9.37) as

$$m\ddot{\mathbf{r}} - m\tau\dddot{\mathbf{r}} + m\gamma\mathbf{r} = 0 \tag{9.38}$$

Considering the radiation damping terms small, the modified equation of motion then reads

$$m\ddot{\mathbf{r}} + m\gamma\mathbf{r} = 0 \tag{9.39}$$

The solution of (9.39) is

$$r = r_0\exp[-i\gamma^{\frac{1}{2}}t] = r_0\exp(-i\omega_0 t) \tag{9.40}$$

where $\omega_0^2 = \gamma$ has been used. we have

$$\dot{\mathbf{r}} = -i\omega_0\mathbf{r}(t)$$

and

$$\dddot{\mathbf{r}}(\mathbf{t}) = i\omega_0^3\mathbf{r}(t) = -\omega_0^2\dot{\mathbf{r}}(t) \tag{9.41}$$

Using (9.41), (9.38) becomes

$$m\ddot{\mathbf{r}} + m\tau\omega_0^2\dot{\mathbf{r}} + m\gamma\mathbf{r} = 0$$

or
$$m[\ddot{\mathbf{r}}+l\dot{\mathbf{r}}+\gamma\mathbf{r}] = 0 \tag{9.42}$$
where
$$l=\omega_0^2\tau=\frac{e^2\gamma}{6\pi\varepsilon_0 mc^3} \tag{9.43}$$

This shows that the characteristic time is of the order of $\frac{1}{(\omega_0\tau)\omega_0}$. This time is very long compared to the orbital period $\frac{2\pi}{\omega_0}$, provided $\omega_0\tau \ll 1$.

The equations for radiative effects can be used to discuss practical problems such as the moderation time of the mu(μ) or pi(π) meson is cascading from an orbit of very large quantum number around a nucleus down to the low-lying orbits. Over most of the time interval the quantum numbers are sufficiently-large that the classical description of continuous motion is an adequate approximation.

9.6 DISPERSION

We have read that all media show some dispersion. When the refractive index of a medium varies with frequency the medium is said to be dispersive and the variation of the refractive index n of the medium with wavelength λ (i.e., frequency) constitute the phenomenon of dispersion. In the absence of dispersion an arbitrary wave train travels without distortion. Dispersion is represented by $\frac{dn}{d\lambda}$. It is generally observed that n decreases with λ known as *normal dispersion*. But in the small wavelength ranges, there is often increase of refractive index due to an increased absorption of the passing radiation through the medium called as *anomalous dispersion*.

Dispersion in Gases (Lorentz Theory)

Let us consider the propagation of electromagnetic waves in a dilute gas in which the mutual interactions between the constituent particles are negligible. As the electromagnetic wave passes through the gas the electrons in the molecules will be displaced from their equilibrium position and molecules will be polarized, i.e. this induces dipole moment in the gas molecules. As the density of gas is low, one can neglect the difference between the applied electric field and the local field. We assume that the electrons are bound by a linear restoring force and that the damping is proportional to the velocity: over an tom or a molecule an electric field **E** is constant in space, i.e.

$$\mathbf{E} = \mathbf{E}_0 \exp[-(\omega t - \mathbf{k}\cdot\mathbf{r})]$$

or

$$\mathbf{E} \approx \mathbf{E}_0 \exp(-i\omega t) \tag{9.44}$$

The equation of motion for an ith electron of charge-e and mass m bound by a harmonic force and acted on by an electric field **E** is

$$m[\ddot{\mathbf{r}}_\alpha + l_\alpha \mathbf{r}_\alpha + \gamma_\alpha \mathbf{r}_\alpha] = e\mathbf{E} \tag{9.45}$$

where l_α measures the damping force. We must remember that the γ used in preceding chapters is different from this γ. We can rewrite (9.45) as

$$\ddot{\mathbf{r}}_\alpha + l_\alpha \dot{\mathbf{r}}_\alpha + \gamma_\alpha \mathbf{r}_\alpha = \frac{e}{m}\mathbf{E}_0 \exp(-i\omega t) \tag{9.46}$$

Magnetic force effects are neglected in (9.46). We make an additional approximation that the amplitude of oscillation is small enough to permit evaluation of the electric field at the average position of the electron. The steady state solution of (9.46) is

$$r_\alpha(t) = \frac{\left(\frac{e}{m}\right) E_0}{(\omega_\alpha^2 - \omega^2) - i l'_\alpha \omega} \exp(-i\omega t) \tag{9.47}$$

where

$$\omega_\alpha^2 = r_{\alpha/m} \quad \text{and} \quad l'_\alpha = l_{\alpha/m} \tag{9.48}$$

If the field varies harmonically in time with frequency ω as $e^{-i\omega t}$, the dipole moment contributed by one electron is

$$\mathbf{p}_\alpha = e\mathbf{r}_\alpha(t) = \frac{\left(\frac{e^2}{m}\right)\mathbf{E}}{(\omega_\alpha^2 - \omega^2) - i l'_\alpha \omega} \tag{9.49}$$

If we suppose that there are N molecules per unit volume with z electrons per molecule in the gas and a fraction f_α of these have the characteristic resonance frequency ω_α. One obtains the total dipole moment per unit volume as

$$\mathbf{P} = \sum_\alpha N f_\alpha \mathbf{p}_\alpha = \mathbf{E}\sum_\alpha \frac{\left(\frac{e^2}{m}\right) N f_\alpha}{(\omega_\alpha^2 - \omega^2) - i l'_\alpha \omega} \tag{9.50}$$

where the *oscillator strengths* f_α satisfy the sum rule

$$\sum_\alpha f_\alpha = 1$$

We have $\mathbf{P} = \varepsilon_0 \chi \mathbf{E}$ and therefore the electric susceptibility χ is given by

$$\chi = \sum_{\alpha} \frac{\left(\frac{e^2}{m\varepsilon_0}\right) N f_\alpha}{(\omega_\alpha^2 - \omega^2) - i l'_\alpha \omega} \tag{9.51}$$

The dielectric constant ε is given by

$$\varepsilon = 1 + \chi = 1 + \sum_{\alpha} \frac{\left(\frac{e^2}{m\varepsilon_0}\right) N f_\alpha}{(\omega_\alpha^2 - \omega^2) - i l'_\alpha \omega} \tag{9.52}$$

The electrical polarizability is given by

$$\frac{\mathbf{P}}{\mathbf{E}} = \frac{Ne^2}{m} \sum_{\alpha} \frac{f_\alpha}{(\omega_\alpha^2 - \omega^2) - i l'_\alpha \omega} \tag{9.53}$$

The refractive index n is equal to $\sqrt{\varepsilon}$ and therefore

$$n^2 = \varepsilon = 1 + \sum_{\alpha} \frac{\left(\frac{e^2}{m\varepsilon_0}\right) N f_\alpha}{(\omega_\alpha^2 - \omega^2) - i l'_\alpha \omega} \tag{9.54}$$

The relation (9.54) is known as *dispersion relation* for dilute gases. Relation (9.52) shows that the dielectric constant is complex. This means refractive index is also complex. The significance of complex refractive index is that there is an absorption of energy in the medium. The damping constant l'_α is generally small and therefore one can take (or ε) approximately real for most frequencies for values of $\omega < \omega_\alpha$, one can see that all terms in the sum (9.54) are positive and ε is greater than 1. For value of $\omega > \omega_\alpha$, one finds more and more negative terms occur in the sum and finally is less that 1. However the behaviour of ω_α is peculiar in the neighbourhood. We find that the real part vanishes and the term is imaginary and large. Fig. 9.5 Shows the variation of the real and imaginary parts of n^2. These situations are discussed below:

(i) Normal Dispersion

In the region remote from the natural frequencies of oscillators, i.e. the absorption frequencies of the medium, the term $l'_\alpha \omega$ in the denominator of (9.54) is so small that one can easily neglect it in comparison to $\omega_\alpha^2 - \omega^2$. Eq (9.54) reduces to

$$n^2 = 1 + \frac{1}{4\pi\varepsilon_0} \left(\frac{4\pi N e^2}{m}\right) \sum_{\alpha} \frac{f_\alpha}{\omega_\alpha^2 - \omega^2} \tag{9.55}$$

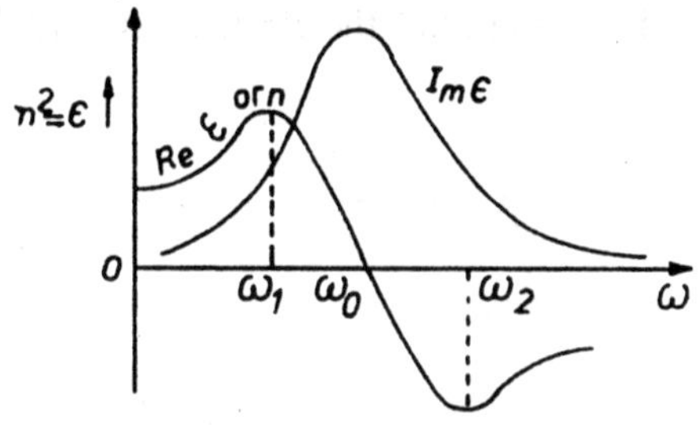

Figure 9.5.

This shows that the refractive index is real and increases with frequency of the incident wave. This is called normal dispersion. We have $\omega = 2\pi\nu$. Using this value in (9.55), one obtains

$$\begin{aligned} n^2 &= 1+\frac{Ne^2}{m\varepsilon_0}\sum_\alpha \frac{f_\alpha}{\left(\frac{4\pi^2C^2}{\lambda_\alpha^2}-\frac{4\pi^2C^2}{\lambda^2}\right)} \\ &= 1+\frac{Ne^2}{\varepsilon_0 m(4\pi^2c^2)}\sum_\alpha \frac{f_l\lambda^2\lambda_\alpha^2}{(\lambda^2-\lambda_\alpha^2)} \\ &= n^2 = 1+\sum_\alpha \frac{A_\alpha\lambda^2}{(\lambda^2-\lambda_\alpha^2)} \end{aligned} \tag{9.56}$$

where $$A_\alpha = \frac{Ne^2\lambda_\alpha^2 f_\alpha}{\varepsilon_0 m 4\pi^2 c^2} \tag{9.57}$$

Relation (9.57) is known as *Sellmeier's equation.* If we suppose that $\lambda \gg \lambda_\alpha$ then (9.56) can be expressed as

$$\begin{aligned} n^2 &= 1+\sum_\alpha A_\alpha\left(1-\frac{\lambda_\alpha^2}{\lambda^2}\right)^{-1} \\ &= 1+\sum_\alpha A_\alpha+\sum_\alpha A_\alpha\frac{\lambda_\alpha^2}{\lambda^2}+\sum_\alpha A_\alpha\frac{\lambda_\alpha^4}{\lambda^4}+\cdots \\ &= A+\frac{B}{\lambda^2}+\frac{C}{\lambda^4}+\cdots \end{aligned} \tag{9.58}$$

where $A = 1+\Sigma_\alpha A_\alpha$; $B = \Sigma_\alpha A_\alpha\lambda_\alpha^2$; $C = \Sigma_\alpha A_\alpha\lambda_\alpha^4$.

Relation (9.58) is called *Cauchy's dispersion relation.* The coefficient A is the coefficient of refraction while the coefficient B is called the coefficient of dispersion. From (9.58) we have

$$2n\frac{dn}{d\lambda} = -\frac{2B}{\lambda^3}-\frac{4C}{\lambda^5}$$

or
$$\frac{dn}{d\lambda} = -\frac{1}{\lambda^3 n}\left[B+\frac{2C}{\lambda^2}\right]$$
$$= -\frac{1}{\lambda^3}\left(B+\frac{2C}{\lambda^2}+\cdots\right)\left(A+\frac{B}{\lambda^2}+\cdots\right)^{-\frac{1}{2}}$$
$$= -\frac{BA}{\lambda^3} = -\frac{k}{\lambda^3} \tag{9.59}$$

Relation (9.59) gives the *dispersive power*. Obviously, the dispersive power varies inversely as the cube of wavelength The minus sign indicates that the slope of the dispersive power is negative.

(ii) Anomalous Dispersion

When the frequency of the electromagnetic wave is nearly equal to the natural frequency, one obtains the phenomenon of anomalous dispersion. For simplicity, let us consider that there is only one natural frequency, i.e. $\omega_0 = \omega_\alpha$, then Eq. (9.54) retains complex refractive say n^* index and becomes

$$n^{*^2} = 1+\frac{1}{4\pi\varepsilon_0}\frac{4\pi Ne^2}{m}\frac{1}{(\omega_0^2-\omega^2)-il'\omega}$$

$\therefore$
$$n^* = \left[1+\frac{1}{4\pi\varepsilon_0}\frac{4\pi Ne^2}{m}\frac{1}{\omega_0^2-\omega^2-il\omega}\right]^{\frac{1}{2}}$$
$$= 1+\frac{1}{4\pi\varepsilon_0}\frac{2\pi Ne^2}{m}\frac{1}{\omega_0^2-\omega^2-il\omega} \tag{9.60}$$

Multiplying numerator and denominator of right hand side by $(\omega_0^2-\omega^2+il\omega)$, one obtains

$$n^* = 1+\frac{1}{4\pi\varepsilon_0}\frac{2\pi Ne^2}{m}\left[\frac{\omega_0^2-\omega^2+il\omega}{(\omega_0^2-\omega^2)^2+l^2\omega^2}\right]$$

or
$$n^* = 1+\frac{2\pi Ne^2}{4\pi m\varepsilon_0}\left[\frac{\omega_0^2-\omega^2}{(\omega_0^2-\omega^2)^2+l^2\omega^2}\right]$$
$$+i\frac{2\pi Ne^2}{4\pi m\varepsilon_0}\left[\frac{l\omega}{(\omega_0^2-\omega^2)^2+l^2\omega^2}\right] \tag{9.61}$$

Replacing n^* by $n(1+i\kappa)$ and comparing real and imaginary parts in (9.61), one obtains

$$n = 1+\frac{N}{2\varepsilon^2}\left(\frac{e^2}{m}\right)\frac{\omega_0^2-\omega^2}{(\omega_0^2-\omega^2)^2+l^2\omega^2}$$

and
$$n\kappa = \frac{N}{2\varepsilon^2}\left(\frac{e^2}{m}\right)\frac{l\omega}{(\omega_0^2-\omega^2)^2+l^2\omega^2} \tag{9.62}$$

Fig. 9.6 represents variation of real and imaginary parts of n^* (i.e. n and $n\kappa$) as a function of ω. From figure it is clear that at low frequencies, n (real) is slightly greater than 1 and increases as ω increases and reaches a maximum and in the neighbourhood of resonance frequency ω_0, n decreases rapidly and at $\omega = \omega_0$ it approaches unity and continues to decrease until it becomes minimum at certain frequency greater than but in the vicinity of resonance frequency. When frequency further increases, n again increases and approaches a value slightly less than 1, for every large value of frequency. Obviously, in the neighbourhood of absorption region the refractive index decreases with increasing frequency and this is the anomalous behaviour giving rise to the term *anomalous dispersion.*

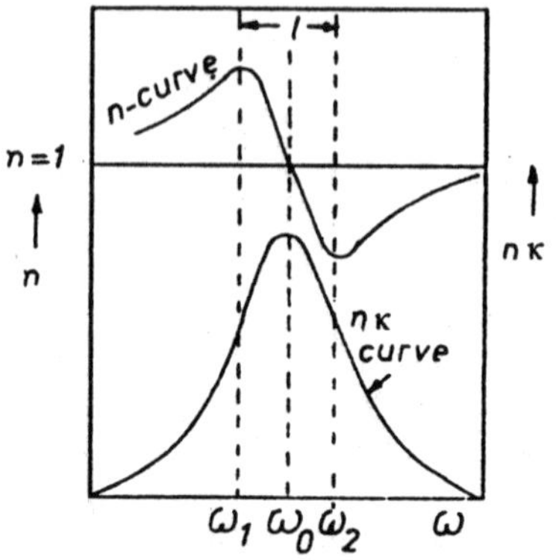

Figure 9.6. Plot of $n\kappa$ and n as a function of ω.

The imaginary part ($n\kappa$) corresponds to absorption of electromagnetic waves propagating through the gas. From Fig. 9.6, it is evident that κ is maximum when $\omega = \omega_0$, where n is 1. It has a width at half maximum approximately equal to l. This means that in the region where n changes rapidly the gas is relatively highly absorbing.

We must note that in the case of a real gas, there exist many resonant frequencies and corresponding damping coefficients, the complex refractive index is given by

$$n^* = 1 + \frac{1}{4\pi\varepsilon_0}\frac{2\pi Ne^2}{m}\sum_{\kappa}\frac{f_\kappa(\omega_\kappa^2-\omega^2-il\omega)}{(\omega_\kappa^2-\omega^2)^2+l^2\omega^2} \tag{9.63}$$

Similarly, one can consider the behaviour of real and imaginary parts. R.W. Wood in 1904 provided the experimental demonstration of anamolous dispersion of sodium vapour in the neighbourhood of yellow D-line.

9.7 DISPERSION IN LIQUIDS AND SOLIDS

When the medium is not dilute, i.e. one is dealing with liquids or solids, one finds that the local field is not equal to the external field. Obviously one can not neglect the interaction among the molecules within the liquid or gas. One will have to replace the electric field vector **E** in (9.50) by

$$\mathbf{E}_{\text{eff}} = \mathbf{E} + \frac{\mathbf{P}}{3\varepsilon_0} \tag{9.64}$$

Thus, one obtains the total dipole moment per unit volume as

$$\mathbf{P} = \left(\mathbf{E} + \frac{\mathbf{P}}{3\varepsilon_0}\right) \sum_\alpha \frac{\left(\frac{e}{m}\right)^2 N f_\alpha}{(\omega_0^2 - \omega^2) - i l_\alpha \omega} \tag{9.65}$$

Since $\mathbf{D} = \varepsilon_0 \mathbf{E} + \mathbf{P} = \varepsilon \mathbf{E}$, we have

$$\frac{\mathbf{P}}{\mathbf{E} + \frac{P}{3\varepsilon_0}} = \frac{\varepsilon - \varepsilon_0}{\varepsilon + 2\varepsilon_0} 3\varepsilon_0$$

$$\therefore \quad \frac{\varepsilon - \varepsilon_0}{\varepsilon + 2\varepsilon_0} = \sum_\alpha \frac{\left(\frac{e^2}{m}\right) N f_\alpha}{3\varepsilon_0\{(\omega_\alpha^2 - \omega^2 - \imath l_\alpha\}} \tag{9.66}$$

or

$$\frac{\frac{\varepsilon}{\varepsilon_0} - 1}{\frac{\varepsilon}{\varepsilon_0} + 2} = \frac{n^2 - 1}{n^2 + 2} = \sum_\alpha \frac{\left(\frac{e^2}{m}\right) N f_\alpha}{3\varepsilon_0\{(\omega_\alpha^2 - \omega^2) - i l \omega\}} \tag{9.67}$$

Equation (9.67) shows that ε_r(i.e. n) is again complex and depends on the frequency. Particular cases of interest are:

(i) Clausius Mossotti Relation

If M is the molecular weight of the substance, ρ its density and N_0 the Avogadro's number, then we have

$$\text{Number of molecules per unit volume } N = \frac{N_0 \rho}{M}. \tag{9.68}$$

Using (9.68), Eq. (9.67) becomes

$$\frac{\varepsilon_{r-1}}{\varepsilon_{r+2}} = \frac{n^2-1}{n^2+2} = \frac{N_0 e^2}{3m\varepsilon_0}\sum_{\alpha}\frac{f_\alpha}{(\omega_\alpha^2 - \omega^2 - il_\alpha\omega)} \tag{9.69}$$

For a given frequency

$$\left(\frac{\varepsilon_{r-1}}{\varepsilon_{r+2}}\right)\frac{M}{\rho} = \left(\frac{n^2-1}{n^2+2}\right)\frac{M}{\rho} = \text{Constant} \tag{9.70}$$

This is known as *Clausius-Mossotti relation.*

(ii) Lorentz-Lorentz Relation

For transparent materials one can neglect the damping force, i.e. $l_\alpha \to 0$. This mean that the refractive index is real. Eq. (9.69) reduces to

$$\left(\frac{n^2-1}{n^2+2}\right)\frac{M}{\rho} = \text{Constant} \tag{9.71}$$

The relation (9.71) shows that now the refractive index depends on the density of the medium. This is known as *Lorentz-Lorentz relation.* For gases $n \approx 1$ and therefore one may approximate

$$n^2 - 1 = (n-1)(n+1) \approx 3$$

and $(n^2+2) \approx 3$. Therefore, for gases Lorentz-Lorentz relation takes the approximate from

$$\frac{2}{3}(n-1)\frac{M}{\rho} = \text{Constant} \tag{9.72}$$

9.8 MEDIUM CONTAINING FREE ELECTRONS

When a medium contain free electrons, there is no current flow due to random motion. When the electric field is applied the electrons acquire an additional velocity component and a current results. The electrons in the medium collide with the atoms of the medium and are scattered. Obviously damping results which depends upon the velocity of the electrons. One can write the equation of motion of an electron in the form

$$m\dot{\mathbf{v}}_\alpha + ml_\alpha\mathbf{v}_\alpha = eE_0\exp(-i\omega t) \tag{9.73}$$

We must note that here we have assumped that the field varies harmonically with time. One can easily verify that the solution of (9.73) is

$$v_\alpha = \frac{E_0 e\exp(-i\omega t)}{m(l_\alpha - i\omega)} \tag{9.74}$$

Now, the current density **J** is obtained as

$$\mathbf{J} = \sum_{\alpha} e\mathbf{v}_{\alpha} = Ne\mathbf{v} = \frac{Ne^2\mathbf{E}_0 \exp(-i\omega t)}{m(l - i\omega)} \tag{9.75}$$

While obtaining (9.75), we have assumed that there are N electrons per unit volume moving with the common velocity **v** we have

$$\mathbf{J} = \sigma\mathbf{E}$$

$$\therefore \quad \sigma = \frac{\mathbf{J}}{\mathbf{E}} = \frac{Ne^2}{m(l - i\omega)} \tag{9.76}$$

EXAMPLE 9.1: Calculate the value and dimension of a Thomson scattering cross section.

Solution: We have

$$\sigma_T = \frac{8\pi}{3}\left(\frac{1}{4\pi\varepsilon_0}\frac{e^2}{mc^2}\right)^2$$

$$\therefore \quad \sigma_T = \frac{8}{3} \times 3.14 \times \left[9 \times 10^9 \times \frac{1.6 \times 10^{-19}}{9.1 \times 10^{-31} \times (3 \times 10^8)^2}\right]^2$$

$$= 6.65 \times 10^{-20}\ \text{m}^2$$

where $e = 1.6 \times 10^{-19}$ c ; $m = 9.1 \times 10^{-31}$ kg ; $c = 3 \times 10^8$ m/s.

We have $r_0^2 = \left(\frac{1}{4\pi\varepsilon_0}\frac{e^2}{mc^2}\right)$.

This means the dimensions of σ_T are the same as that of r_0^2. We have

$$r_0 = \frac{1}{4\pi\varepsilon_0}\frac{e^2}{mc^2} \quad \text{and} \quad F = \frac{1}{4\pi\varepsilon_0}\frac{e^2}{r^2} \quad \text{(Coulomb's Law)}$$

$$\therefore \quad [r_0] = \left[\frac{Fr^2}{mc^2}\right] = \frac{[MLT^{-2}][T^2]}{M[(LT^{-1})^2]}$$

$$= \frac{[ML^3T^{-2}]}{[ML^2T^{-2}]} = [M^\circ L' T^\circ]$$

Thus the dimensions of σ_T are $[M^\circ L^2 T^\circ]$, i.e. that of area.

EXAMPLE 9.2: Prove that the two frequencies at which the refractive index is maximum and minimum in gases are separated by a distance l.

Solution: We have

$$n^* = 1+\frac{2\pi Ne^2}{4\pi m\varepsilon_0}\left[\frac{\omega_0^2-\omega^2}{(\omega_0^2-\omega^2)^2+l^2\omega^2}\right]$$

$$+\frac{i2\pi Ne^2}{4\pi m\varepsilon_0}\left[\frac{l\omega}{(\omega_0^2-\omega^2)^2+l^2\omega^2}\right]$$

$$= n+i\kappa$$

where n is refractive index and κ the coefficient of absorption. The refractive index will be maximum and minimum according to the condition

$$\frac{\partial}{\partial\omega}(n)=0$$

or $$\frac{\partial}{\partial\omega}=\left[1+\frac{2\pi Ne^2}{4\pi\varepsilon_0 m}\left\{\frac{\omega_0^2-\omega^2}{(\omega_0^2-\omega^2)^2+l^2\omega^2}\right\}\right]$$

or $$\frac{\partial}{\partial\omega}\left[\frac{\omega_0^2-\omega^2}{(\omega_0^2-\omega^2)^2+l^2\omega^2}\right]=0$$

We have, near absorption band $\omega\to\omega_0$, i.e.

$$(\omega_0^2-\omega^2)=(\omega_0+\omega)(\omega_0-\omega)\simeq 2\omega_0(\omega_0-\omega)$$

$$\therefore\quad \frac{\partial}{\partial\omega}\left[\frac{2\omega_0(\omega_0-\omega)}{4\omega_0^2(\omega_0-\omega)^2+l^2\omega^2}\right]=0$$

Solving, one obtains

$$\omega=\omega_0\pm\frac{l}{2}$$

$$\therefore\quad \omega_1=\omega_0-\frac{l}{2}\quad\text{and}\quad\omega_2=\omega_0+\frac{l}{2}$$

$$\therefore\quad \omega_2-\omega_1=l$$

Obviously, the frequencies at which refractive index is maximum and minimum are separated by a distance l on the frequency axis.

EXAMPLE 9.3: Show that in the case of gases the maximum and minimum values of refractive index n occurs at the positions where the absorption coefficient reaches half its maximum value.

Solution: We have to prove that $\omega_0 - \omega_1 = \frac{l}{2}$. The absorption coefficient κ is given by

$$\kappa = \frac{2\pi Ne^2}{4\pi m\varepsilon_0}\left[\frac{l\omega}{(\omega_0^2-\omega^2)^2+l^2\omega^2}\right]$$

Now, the values of absorption coefficient at ω_1 and ω_2 are expressed as

$$(\kappa)\omega_1 = \frac{2\pi Ne^2}{4\pi m\varepsilon_0}\left[\frac{l\omega_1}{(\omega_0^2-\omega_1^2)^2+l^2\omega_2^2}\right]$$

$$(\kappa)\omega_2 = \frac{2\pi Ne^2}{4\pi m\varepsilon_0}\left[\frac{l\omega_2}{(\omega_0^2-\omega_2^2)^2+l^2\omega_2^2}\right]$$

Since ω_1 lies near ω_0, i.e. $\frac{l}{\omega_0} \ll 1$, one obtains

$$(\kappa)\omega_1 = \frac{2\pi Ne^2}{4\pi\varepsilon_0 m}\left[\frac{l\omega_0}{(\omega_0-\omega_1)^2 4\omega_0^2+l^2\omega_0^2}\right]$$

Now, using $\omega_0 - \omega_1 = \frac{l}{2}$, one obtains

$$(\kappa)\omega_1 = \frac{2\pi Ne^2}{4\pi\varepsilon_0 m}\left[\frac{l\omega_0}{\frac{l^2}{4}4\omega_0^2+l^2\omega_0^2}\right]$$

$$= \frac{2\pi Ne^2}{4\pi\varepsilon_0 m}\left(\frac{1}{2l\omega_0}\right) = \frac{A}{2\omega_0 l} \qquad \left(\text{here } A = \frac{2\pi Ne^2}{4\pi m\varepsilon_0}\right)$$

Similarly, we obtain

$$(\kappa)\omega_2 = \frac{A}{2\gamma\omega_0}$$

$$\therefore \quad (\kappa)\omega_1 = \omega_2 = \frac{A}{2\gamma\omega_0}$$

Absorption coefficient (κ) is maximum, when $\omega = \omega_0$

$$\therefore \quad (\kappa)_{\max} = \frac{2\pi Ne^2}{4\pi\varepsilon_0 m}\left[\frac{l\omega_0}{l^2\omega_0^2}\right] = A\left[\frac{1}{\omega_0 l}\right]$$

$$\therefore \quad (\kappa)\omega_1 = \omega_2 = \frac{1}{2}(\kappa)_{\max}$$

EXAMPLE 9.4: Determine the damping constant l using the relation $\sigma = \dfrac{Ne^2}{m(l - i\omega)}$. Given $N \simeq 8 \times 10^{28}$ atoms/m^2 at normal temperature and low frequency conductivity $\sigma \simeq 6 \times 10^7$ mho/m

Solution: We have

$$\sigma = \frac{Ne^2}{ml}$$

$$\therefore \quad l = \frac{Ne^2}{m\sigma} = \frac{8 \times 10^{28} \times (1.6)^2 \times 10^{-38}}{9 \times 10^{-31} \times 6 \times 10^7}$$

This shows that for the frequencies of the order of 10^{11} or less the conductivity of the metals are essentially real and do not depend upon the frequency. We must note that for infrared frequency and beyond the conductivity of metals is complex and varies with frequency in accordance with Eq. (9.76)

EXAMPLE 9.5: Compute the group velocity for very high frequency electromagnetic radiation such as x-rays

Solution: If ω is much larger than the characteristic frequency ω_α we may neglect ω_α's and have

$$n^2 = 1 - \frac{Ne^2}{m\varepsilon_0\omega^2}\sum_\alpha f_\alpha = 1 - \frac{Ne^2}{m\varepsilon_0\omega^2}$$

Since $\Sigma_\alpha f_\alpha = 1$. Using the approximation $(1 - x)^{\frac{1}{2}} = 1 - \frac{1}{2}x + \cdots$ for $x \ll 1$, we have

$$n = 1 - \frac{Ne^2}{2\varepsilon_0 m\omega^2}$$

which is smaller than 1, giving $\mathbf{v} > c$. Now $\dfrac{dn}{d\omega} = \dfrac{Ne^2}{\varepsilon_0 m\omega^2}$. Substituting in

$$n^2 = \varepsilon_r = 1 + \frac{Ne^2}{m\varepsilon_0}\left(\sum_\alpha \frac{f_\alpha}{\omega_\alpha^2 - \omega^2}\right)$$

one obtains

$$V_g = \frac{c}{n + \omega\left(\frac{dn}{d\omega}\right)} = \frac{c}{1 - \left(\frac{Ne^2}{2\varepsilon_0 m\omega^2}\right) + \omega\left(\frac{Ne^2}{\varepsilon_0 m\omega^3}\right)}$$

$$= \frac{c}{1 + \left(\frac{Ne^2}{2\varepsilon_0 m\omega^2}\right)}$$

Therefore, although the phase velocity **v** is larger than c because n is smaller than 1, group velocity V_g is smaller than c. For the phase velocity we have

$$v = \frac{c}{n} = \frac{c}{1 - \left(\frac{Ne^2}{2\varepsilon_0 m\omega^2}\right)} \approx c\left[1 + \left(\frac{Ne^2}{2\varepsilon_0 m\omega^2}\right)\right]$$

Obviously,

$$\mathbf{v}_g \mathbf{v} = c^2$$

This relation although not of general validity, is satisfied to a very good approximation over a wide range of frequencies.

QUESTIONS

1. Define scattering cross section. Discuss different types of scattering by considering the incidence of electromagnetic wave on bound electrons. Why does the sky appears blue.
2. What is the difference between Thomson, Rayleigh and resonance scattering of radiations? Derive the expression for the differential cross-sections for Thomson scattering of radiation by electrons.
3. Outline Thomson's theory of scattering of x-rays by free electron and discuss its success and limitations.
4. What is the effect of radiation damping on the monochromatic nature of radiations emitted by a harmonic oscillator?
5. Show that the scattering cross-section of electromagnetic radiation by a bound electron depends upon the inverse fourth power of the incident wave length.
6. The intensity of x-rays scattered by a electron at a distance r from it and making an angle ϕ with the original direction is given by

$$I_c = \left[\frac{e^2}{4\pi\varepsilon_0 mc^2}\right] \frac{I}{r^2} \frac{1}{2} \left(1 + \cos^2\phi\right).$$

Using the above result

(i) Obtain an expression for the coefficient of scattering and its dimensions

(ii) Calculate the intensity of scattered radiation I_s in terms I_e for n-electrons (a)when they scatter independently (b) when all of them act as a single scattering centre.

(iii) Plot schematic curves to illustrate the variations in I_s with ϕ in the two cases in (ii) above for $n = 3$

7. Obtain the total scattering cross-section for scattering of a plane monochromatic waves by a charge carrying out small vibrations under an elastic force (oscillator type)
8. Explain the following on the basis of Rayleigh scattering:

   (i) Red light is used for danger signals while the eye is most sensitive to yellow green

   (ii) The clear sky appear blue

   (iii) The rising or setting sun appears red.
9. Find the differential scattering cross-section for the scattering of an elliptically polarized wave by a free charged particle.
10. What do you understand by dispersion? Discuss it in the case of gaseous medium having both real and complex refractive index.
11. Explain normal and anomalous dispersion. Derive Sellmeier's equation for the refractive index of a dielectric medium.
12. What do you understand by anomalous dispersion. Give Lorentz's theory of dispersion and explain the connection between absorption and anomalous dispersion.
13. Give Lorentz's theory of dispersion and show that in the case of non-polar dielectrics

$$\frac{n^2-1}{n^2+1}\frac{M}{\rho} = \text{constant}$$

where M is the molecular weight of the dielectric, ρ its density and n the refractive index.

14. Explain dispersion. From the theory of dispersion show that in the case of transparent gaseous substance in the region remote from the natural frequencies the index of refraction is governed approximately by the relation

$$n^2 = A + \frac{B}{\lambda^2} + \frac{C}{\lambda^4}$$

where A, B, and C are constants, λ is the wavelength of incident light. Obtain dispersive power relation for the medium.

15. Write short notes on the following
    (i) Rayleigh scattering
    (ii) Thomson scattering
    (iii) Scattering of electromagnetic waves
    (iv) Normal and anomalous scattering
16. Show that for gases, we may write

$$n \approx 1 + \frac{Ne^2}{2m\varepsilon_0} \sum_\alpha \left(\frac{f_\alpha}{\omega_\alpha^2 - \omega^2} \right)$$

For only one resonant frequency, the expression becomes

$$n \approx 1 + \frac{Ne^2}{2m\varepsilon_0(\omega_\alpha^2 - \omega^2)}$$

PROBLEMS

1. The index of refraction of hydrogen gas at STP is $n = 1 + 1.400 \times 10^{-4}$ at $\lambda = 5.46 \times 10^{-7}$ m and $n = 1 + 1.547 \times 10^{-4}$ at $\lambda = 2.54 \times 10^{-7}$. Assuming a single resonant frequency, compute this frequency and the number of electronic oscillators per unit volume.
   [**Ans.** (i) 2.11×10^{-16}/s, (ii) 3.80×10^{25}]
2. Consider a gas whose molecules behave as dipole oscillators, with a restoring constant of $\kappa = 3 \times 10^2$ kg/s^2. The oscillating particles are electrons. Compute their characteristic frequency. Write the index of refraction of the gas as a function of the frequency, assuming that the gas is at STP. Obtain the values of the index for $\lambda = 5 \times 10^{-7}$ m and $\lambda = 10^{-2}$ m.
   [**Ans.** 1.82×10^{-16}/s, $n = \dfrac{1 + 6.05 \times 10^{28}}{(3.29 \times 10^{-32} - \omega^2)}$ and 1.00019, 1.00018]
3. For the D-lines of sodium the following values of constants in the dispersion formula are found

$$\omega_0 = 3 \times 10^{16}, \ \alpha = 2 \times 10^{10} \quad \text{and} \quad \frac{Ne^2}{m\varepsilon_0} = 10^{23}$$

Plot the refractive index (n) and absorption coefficient as a function of frequency of electromagnetic waves. Find the maximum and minimum values of n. Also find the maximum values of absorption coefficient and the half width of the absorption band.

4. Referring to the problem 1 compute the index of refraction of hydrogen for $\lambda = 4 \times 10^{-7}$ m, a pressure of 10 atm and a temperature of 300 K.

SHORT ANSWER QUESTIONS

1. What is absolute index of refraction?

   [**Ans.** The ratio between the velocity of electromagnetic waves in vacuum c, and in matter, v, is called the absolute index of refraction of the substance usually designated by n]

2. What is the relation between the absolute index of refraction (n) and relative permittivity (ε_r) of the medium.

   [**Ans.** $n = \sqrt{\varepsilon_r}$]

3. What is classical electron radius?

   [**Ans.** $r_0 = \dfrac{e^2}{4\pi\varepsilon_0 mc^2}$, where symbols have usual meanings]

4. Why sun appears red at sunset?

   [**Ans.** As the light passes through the atmosphere it looses heavily from the blue end of the spectrum due to scattering]

5. In the absence of atmospheric scattering how the sky will appear?

   [**Ans.** Dark]

6. What is radiation reaction?

   [*Ans.* The reaction of the radiation on the motion of the particle is called radiation reaction or radiation damping]

7. When a dispersion is said to be normal? [**Ans.** $\dfrac{dn}{d\omega} > 1$]

8. What is the condition for anomalous dispersion?

   [**Ans.** $\dfrac{dn}{d\omega} < 1$. The ordering of prismatic colours is reversed]

9. What the condition $\frac{dn}{d\omega} > 1$ indicates?

[**Ans.** Dispersion is normal]

10. How will you interpret the regions where $I_m \varepsilon > 0$?

[**Ans.** The regions where $I_m\varepsilon$ is large are called regions of *resonant absorption*]

11. What is the significance of the condition $I_m\varepsilon < 0$?

[**Ans.** This shows that energy is given to the wave by medium resulting into the amplification of waves. Examples are masers and lasers]

OBJECTIVE QUESTIONS

1. Thomson scattering coefficient of an unpolarized plane electromagnetic wave scattered by a free charge particle having charge q and mass m is

   (a) $\langle\sigma\rangle = \frac{8\pi}{3}\left(\frac{q^2}{4\pi\varepsilon_0 mc^2}\right)$ (b) $\langle\sigma\rangle = \frac{8\pi}{3}\left(\frac{q^2}{4\pi\varepsilon_0 mc^2}\right)^2$

   (c) $\langle\sigma\rangle = \frac{8\pi}{3}\left(\frac{q^2}{4\pi\varepsilon_0 mc^2}\right)^{\frac{3}{2}}$ (d) $\langle\sigma\rangle = \frac{8\pi}{3}\left(\frac{q^2}{4\pi\varepsilon_0 mc^2}\right)^{\frac{1}{2}}$ **[b]**

2. The maximum value of scattering coefficient for the scattering of radiation by a bound charge of frequency ω_0 occurs at ($\omega \to$ frequency of a free electron)

   (a) $\omega > \omega_0$ (b) $\omega < \omega_0$

   (c) $\omega = \omega_0$ (d) $\omega = \frac{\omega_0}{2}$ **[c]**

3. If ω is the frequency of a free electron, ω_0 the frequency of bound charge and σ is the scattering cross-section, then Rayleigh scattering cross section is given by

   (a) $\sigma_{\text{bound}} = \sigma\frac{\omega_0^4}{\omega^4}$ (b) $\sigma_{\text{bound}} = \sigma\frac{\omega^4}{\omega_0^4}$

   (c) $\sigma_{\text{bound}} = \sigma\frac{\omega_0^2}{\omega^2}$ (d) $\sigma_{\text{bound}} = \sigma\frac{\omega^2}{\omega_0^2}$ **[b]**

4. The dimension of Thomson Scattering coefficient are

   (a) $[M^\circ L^2 T^\circ]$ (b) $[M' L' T^\circ]$

   (c) $[M^\circ L^3 T^\circ]$ (d) $[M^\circ L^{\frac{3}{2}} T^{\frac{1}{2}}]$ **[a]**

5. If r_0 is classical electron radius, ω free electron frequency and ω_0 being the frequency of bound charge, then the resonant scattering cross-section is

   (a) $\sigma = \frac{8}{3}\pi r_0^2 \left(\frac{\omega_0}{\omega}\right)^4$ (b) $\sigma = \frac{8}{3}\pi r_0^3 \left(\frac{\omega_0}{\omega}\right)^2$

   (c) $\sigma = \frac{8}{3}\pi r_0^2 \left(\frac{\omega_0}{\omega}\right)^2$ (d) $\sigma = \frac{8}{3}\pi r_0^2 \left(\frac{\omega_0}{\omega}\right)$ **[c]**

6. To a close approximation, dispersion varies inversely as:

   (a) wavelength of radiation

   (b) square of the wavelength of radiation

   (c) cube of the wavelength of the radiation

   (d) none of the above **[c]**

7. The Lorentz-Lorentz formula for the dispersion in liquids and solids is (symbols have usual meanings)

   (a) $\frac{n^2-1}{n^2+2}\frac{1}{\rho} = A^2$ (b) $\frac{n^2-1}{n^2+2}\frac{1}{\rho^2} = A$

   (c) $\frac{n^2-1}{n^2+2}\frac{1}{\sqrt{\rho}} = A$ (d) $\frac{n^2-1}{n^2+2}\frac{1}{\rho} = A$ **[d]**

8. The dispersion relation for a dilute gas is (symbols have usual meanings)

   (a) $n^2 = 1 + \sum_\alpha \frac{\left(\frac{e^2}{m\varepsilon_0}\right) N f_\alpha}{(\omega_\alpha^2 - \omega^2) - i l_\alpha \omega}$

   (b) $n^2 = \sum_\alpha \frac{\left(\frac{e^2}{m\varepsilon_0}\right) N f_\alpha}{(\omega_\alpha^2 - \omega^2)^2 - i l_\alpha \omega^2}$

   (c) $n^2 = \sum_\alpha \frac{\left(\frac{e^2}{m\varepsilon_0}\right)^2 N f_\alpha}{(\omega_\alpha^2 - \omega^2)^2 - i l_\alpha \omega^2}$

   (d) $n^2 = \sum_\alpha \frac{\left(\frac{e^2}{m\varepsilon_0}\right) N f_\alpha}{\omega_\alpha^2 - \omega^2}$. **[a]**

9. Molecular refractivity of a substance with molecular weight W is given by (symbols have usual meanings)

   (a) $A_m = \frac{W}{\rho}\left(\frac{n^2-1}{n^2+2}\right)^2$ (b) $A_m = \frac{W}{\rho}\left(\frac{n^2-1}{n^2+2}\right)$

   (c) $A_m = \frac{W}{\rho}\left(\frac{n^2-1}{n^2+2}\right)^{\frac{1}{2}}$ (d) $A_m = \sqrt{\frac{W}{\rho}}\left(\frac{n^2-1}{n^2+2}\right)$ **[b]**

Chapter 10

Wave Guides

10.1 INTRODUCTION

We have read that a conductive medium, such as metal for example intensively absorbs electromagnetic waves. When an electrical disturbance passes along a conductor, it acts up varying electrical and magnetic fields in the neighbourhood of the conductor. Since the boundary conditions must be satisfied at every instant at the interface of the conductor and the surrounding dielectric, the disturbance is propagated at the same speed in the conductor and in the dielectric. In this chapter we shall discuss the behaviour of such waves in the vicinity of the boundaries so configured as to guide that wave along a certain path. Such systems or devices are called *wave guides* or hollow metal tubes with highly conductive walls and these can transmit energy only in higher-order modes, i.e. only short waves can propagate in a wave guide, equal in length to or shorter than the maximum linear dimension of the tube cross-section. Wave guides are extensively used for transmission of the very high frequencies where the attenuation caused by the wave guide is smaller than that for any other form of transmission line. At much higher (infrared) frequencies, dielectric optical fibers are exploited in the telecommunications industry, optical fibers are made of grass or quartz, although other materials, such as nylon, are being tested. When the fibers are arranged in bundles, an image can be transmitted from one point to another. Acoustical waveguides are also very common. The air ducts in the heating system of house, for example, act as acoustical wave guides which are capable of transmitting the noises from the furnace or sounds from one room to another. The inner ear is essentially an acoustical wave guide.

10.2 ELECTROMAGNETIC WAVE PROPAGATION BETWEEN TWO PARALLEL CONDUCTING PLATES

Let us consider the propagation of electromagnetic wave between two parallel conducting planes of infinite extent separated by a distance of the order of a wavelength. The free space is assumed between the two conducting parallel plates. The propagation of the electromagnetic wave through the system of two parallel plates obey the Maxwell's field equations for free-space and also satisfy appropriate boundary conditions at all points on the conducting planes.

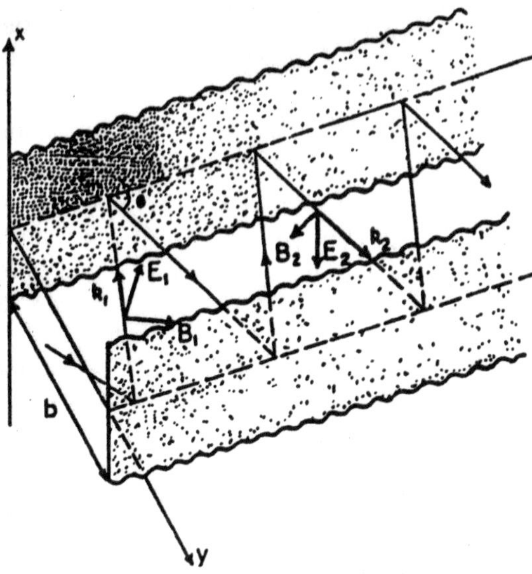

Figure 10.1. Propagation of electromagnetic wave between two parallel infinite and conduction plates.

Fig. 10.1 shows two parallel, infinite and conducting plates. One plate is in the plane $y = 0$ and the other in the plane $y = b$. Since the space between the plates is a vacuum, we have $\varepsilon = \varepsilon_0, \mu = \mu_0, \sigma = 0$ and $\rho = 0$. Let a plane polarized electromagnetic wave is introduced in the region between the plates, the walls will reflect the wave to and fro and as a result the wave will propagate in the direction parallel to the plates. As stated above, the fields must obey Maxwell's equations in free space between the plates and must satisfy certain boundary conditions at all points on the wall. The boundary conditions to be satisfied are:

(i) The tangential component of **E** must be zero at all points on the wall, i.e.

$$\mathbf{n} \times \mathbf{E} = 0 \tag{10.1}$$

(ii) The normal component of the magnetic field must be zero at all points on the wall, i.e.

$$\mathbf{n} \cdot \mathbf{E} = 0 \tag{10.2}$$

where **n** is the surface normal.

One can obtain derived boundary conditions from the two boundary conditions given by (10.1) and (10.2). Substituting for **E** from Maxwell's equation in (10.1), one obtains

$$\mathbf{n} \times (\nabla \times \mathbf{B}) = 0 \tag{10.3}$$

and likewise substituting for **B** from Maxwell's equation on (10.2), one obtains

$$\mathbf{n} \cdot (\nabla \times \mathbf{E}) = 0 \tag{10.4}$$

These conditions imply that the normal component of the electric field vector, **E** is not necessarily zero, since there can be some charges on the conducting surfaces and likewise the tangential component of the magnetic field vector. **B** is not necessarily zero, since there can be some surface currents in the perfectly conducting plates. Since the electromagnetic waves are transverse, with their electric (**E**) and magnetic field (**H**) vectors being transverse to the direction of propagation (**k**), they are called TEM waves.

Let us consider that a linearly plane wave is propagating between the plates in a direction given by the propagation vector **k** in the plane perpendicular to the x-axis and making an angle θ with Y-axis as shown in Fig. 10.1. If **E** is in the x-direction, **B** must be in the plane OYZ (since **B** is perpendicular to **E**). Obviously, $\mathbf{B} = \hat{a}_y B_y + \hat{a}_z B_z$, has a component in the direction of wave, i.e. z-direction; whereas **E** does not. We call such a wave as *transverse electric* (TE) wave. If one chooses **B** to be along the x-axis, then **E** will have a longitudional component in the z-direction. Such a wave is called as *transverse magnetic* (TM) wave. We must note that the waves propagating within closed conducting regions are either TE or TM modes. The expression for the plane wave electric field is written as

$$E_1 = \hat{a}_x E_{ol} \exp[i(\mathbf{k}_i\mathbf{r} - \omega t)] \tag{10.5}$$

where ω and $\mathbf{k}$ are the angular frequency and wave vector of the wave in free-space respectively. The vector $\mathbf{k}_1$ as affected by propagation through the parallel plates is expressed as

$$\mathbf{k}_1 = -\hat{a}_x k \sin\theta + \hat{a}_z \cos\theta \tag{10.6}$$

we find that

$$\mathbf{k}_1 \cdot \mathbf{r} = -ky \sin\theta + kz \cos\theta \tag{10.7}$$

Using (10.7), one can rewrite (10.5) as

$$\mathbf{E}_1 = \hat{a}_x E_{0l} \exp[-i(\omega t - kz \cos\theta + ky \sin\theta)] \tag{10.8}$$

when this wave strikes the conductor plane at $y = 0$, it is totally reflected with a phase change of 180°. For the reflected wave, we have

$$\mathbf{k}_2 = \hat{a}_y k \sin\theta + \hat{a}_z k \cos\theta \tag{10.9}$$

Taking account of the appropriate phase change of totally reflected wave from the perfectly conducting surface, one can write

$$\mathbf{E}_2 = -\hat{a}_x E_0 \exp[-i(\omega t - kz \cos\theta - ky \sin\theta)] \tag{10.10}$$

one can see that the specularly reflected wave goes on propagating until it is reflected from the bottom plane $y = b$. The resultant electric field of the two field vectors $\mathbf{E}_1$, and $\mathbf{E}_2$ is obtained as

$$\mathbf{E} = \mathbf{E}_1 + \mathbf{E}_2 = \hat{a}_x 2i \sin(ky \sin\theta) E_0 \exp[-i(\omega t - kz \cos\theta)] \tag{10.11}$$

one finds that at each point between the two plates there is an interference between the incident and reflected wave so that the resultant field and the associated energy always flows parallel to the z-axis.

At the boundary (i.e. at $y = 0$) $E = 0$. One can see that this condition will be satisfied if $E_1 + E_2 = 0$, i.e. $E_1 = -E_2$ and $\mathbf{E}$ is given by (10.11).

The waves propagating at an arbitrary angle θ do not satisfy simultaneously the boundary conditions at both the planes. For example, the wave reflecting from the upper surface satisfies the boundary condition given by (10.1). However, at the lower surface, i.e. at $y = b$, the boundary condition is satisfied by the resultant of the incident and reflected waves. For $\mathbf{E} = 0$ at $y = b$, one obtains

$$d_1 + d_2 = n\lambda \tag{10.12}$$

which leads to

$$2d \sin\theta = n\lambda \tag{10.13}$$

where n takes only integer values. It cannot take value $n = 0$ for which all the components of the wave vanish. There will, thus, be several acceptable fields corresponding to the different values of n. The different waves that correspond to different values of n are called *modes*. For a given wave frequency the maximum number of wave modes satisfying the boundary condition (and since $\cos\theta \leq 1$) are governed by the following condition,

$$\frac{n\pi}{kb} = \frac{n\lambda}{2b} \leq 1 \tag{10.14}$$

Obviously, (10.14) restricts the maximum value of n for a given frequency of radiation. If $\frac{n\pi}{kb} \geq 1$, i.e. $\frac{n\pi}{2b} \geq 1$ the boundary conditions are not satisfied and the transverse propagation between the parallel plates is not possible. Obviously, the system of parallel plates behaves like a filter and is characterized by a cut-off frequency below which transverse electric waves do not propagate. Each mode has its own cut-off wavelength λ_c. One can obtain the cut-off frequencies for n order modes by assigning different values to n and taking the equality sign in (10.14). One can write the general equation for cut-off frequency as

$$f_c = \frac{nc}{2b} \tag{10.15}$$

or

$$\lambda_c = \frac{2b}{n} \tag{10.16}$$

we note that the cut-off frequency of each mode increases as the order of mode increases. f_c also depends upon the separation of plates.

From (10.13), we get

$$\sin\theta = \frac{n\lambda}{2b} \tag{10.17}$$

$$\cos\theta = \sqrt{1-\sin^2\theta} = \sqrt{1-\frac{n^2\lambda^2}{4b^2}} \tag{10.18}$$

Using (10.17) and (10.18), we obtain from (10.11) the form of electric field of the n-mode waves propagating along the z-axis.

$$\mathbf{E} = \hat{a}_x 2iE_0 \sin\left(\frac{n\pi y}{b}\right) \exp[-i(\omega t - k_g z)] \tag{10.19}$$

where

$$k_g = \frac{2\pi}{\lambda g} = \left(k^2 - \frac{n^2\pi^2}{b^2}\right)^{1/2} \tag{10.20}$$

The wave number k_g is known as the *guide wave number* and the corresponding λ_g is known as *guide wave length.* λ_g is given by

$$\lambda_g = \frac{2\pi}{k_g} = \frac{2\pi}{\left(k^2 - \frac{n^2\pi^2}{b^2}\right)^{1/2}} = \frac{\lambda}{\left(1 - \frac{f_c}{f}\right)^2} \tag{10.21}$$

The electric field **E** given by (10.19) does not have a component in the z-direction and is thus said to be formed by transverse electric waves and is generally denoted by TE. The waves of mode n are usually designated by TE_n, where n is the mode number-one can obtain the magnetic field in TE_n waves using the Maxwell's relation,

$$\nabla \times \mathbf{E} = -\frac{\partial \mathbf{B}}{\partial t} + i\omega\mathbf{B} \tag{10.22}$$

The phase velocity of the wave is given by

$$V_p = \frac{\omega}{k_g} = c\frac{k}{k_g} = c\left(1 - \frac{n^2\pi^2}{k^2b^2}\right)^{-1/2} \tag{10.23}$$

one can see from (10.23) that the phase velocity of the wave exceeds the velocity of light ($V_p > c$) in free space. Question arises, whether this contradict Einstein's principle of special relativity? We can see that this does not. We must note that the phase velocity V_p is the velocity of nodes of the wave-the velocity of the maxima in the sinusoidal wave, or more generally, V_p is the velocity of the plane of constant phase-and not of energy. The later can be obtained from the group velocity, which is note the same as phase velocity. The group velocity associated with the phase velocity is given by

$$V_g = \frac{d\omega}{dk_g} \tag{10.24}$$

where $V_0 = \dfrac{d\omega}{dk}$, is the group velocity of the wave in the medium, filling the waveguide. The energy of the group of waves is moving with this velocity. From Eq. (10.20), we have

$$k_g^2 = k^2 - \frac{n^2\pi^2}{b^2} \tag{10.25}$$

Differentiating with respect to k_g, one obtains

$$k_g = k\frac{dk}{dk_g} = \frac{k}{c}\frac{d\omega}{dk_g}$$

$$V_g = c\frac{k_g}{k} \tag{10.26}$$

which is less than c since $k_g \le k$. Multiplying Eqs. (10.23) and (10.26), we get

$$V_g V_g = c^2 \tag{10.27}$$

Obviously, the phase velocity may be greater than c, the energy velocity can never exceed the velocity of light. We see than that, even if it is empty, an electromagnetic wave guide acts as a dispersive medium with an index of refraction less than one. First three modes of propagation of a wave between two parallel reflecting planes is shown in Fig. 10.2.

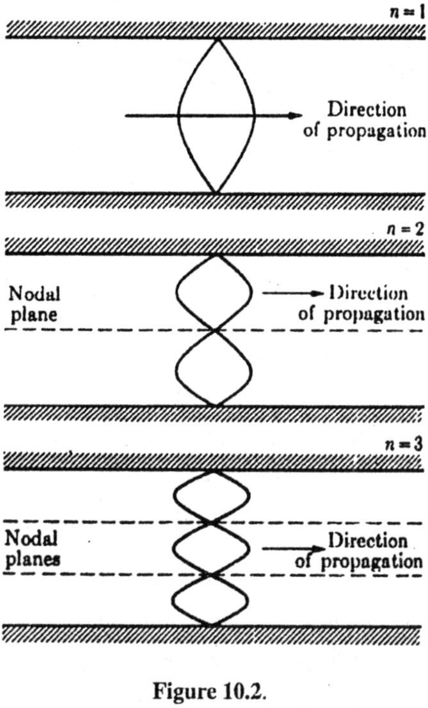

Figure 10.2.

One can obtain another set of solutions in the similar way in which the magnetic field acts along the x-axis and it does not have component in the direction of propagation of wave. These are designated as *transverse magnetic waves* or TM_n waves. We must note that these different TM_n modes have cut off wavelengths at the same value as the TE_n modes given by (10.16). In practice wave guides have either a rectangular or a circuilar cross-section. The two shapes yield similar results with respect to phase velocity along the axis of the guide and cut off frequency.

A simple example of parallel plane wave guides in the optical region is two parallel mirrors, such as those found in some barber shops.

10.3 PROPAGATION OF ELECTROMAGNETIC WAVES IN GUIDES OF ARBITRARY CROSS-SECTION

We now consider the propagation of electromagnetic waves inside a hollow conductor of any uniform cross-section, the walls of which are assumed to be perfect conductors. Let us obtain the general solution of the wave equation for the electric and magnetic fields within the wave guide, subject to the requirement that they satisfy Maxwell's equations and appropriate boundary conditions.

We have the wave equation for the electric field

$$\left(\nabla^2 - \frac{1}{C^2}\frac{\partial^2}{\partial t^2}\right)\mathbf{E} = 0 \tag{10.28}$$

The plane wave solution of (10.28) is of the form

$$\mathbf{E} = \mathbf{E}_0(x,y)\exp[i(k_g r - \omega t)] \tag{10.29}$$

one can easily write similar equations for the magnetic field. Substituting (10.29) in (10.28), one obtains

$$\left(\frac{\partial^2}{\partial x^2} + \frac{\partial^2}{\partial y^2} - k_g^2 + \frac{\omega^2}{c^2}\right) E_0(x,y) = 0 \tag{10.30}$$

Let us define a transverse Laplacian operator as

$$\nabla_T^2 = \frac{\partial^2}{\partial x^2} + \frac{\partial^2}{\partial y^2} = \nabla^2 - \frac{\partial^2}{\partial z^2} \tag{10.31}$$

Using (10.31), one can write (10.30) as

$$\left(\nabla_T^2 + k_c^2\right) E_0(x,y) = 0 \tag{10.32}$$

where

$$k_c^2 = -k_g^2 + \frac{\omega^2}{C^2} \tag{10.33}$$

One finds that it is more convenient, if we express fields $\mathbf{E}$ and $\mathbf{B}$ in terms of the components parallel (E_z, B_z) and transverse (E_T, B_T) to the axis of the conductor, i.e. z-axis. We have

$$\mathbf{E} = \mathbf{E}_z + \mathbf{E}_T \quad \text{and} \quad \mathbf{B} = \mathbf{B}_z + \mathbf{B}_T$$

where

$$\mathbf{E}_z = \hat{a}_z E_{0z}(x,y)\exp[i(k_g z - \omega t)] \tag{10.34}$$

and

$$\begin{aligned}\mathbf{E}_T &= \mathbf{E}_{0T}\exp[i(k_g z - \omega t)] \\ &= \{\hat{a}_x E_{0x}(x,y) + \hat{a}_y E_{0y}(x,y)\exp[i(k_g z - \omega t)\}\end{aligned} \tag{10.35}$$

one can write similar equations for B_z and B_T. The electric and magnetic fields must satisfy Maxwell's equations

$$\nabla\cdot\mathbf{E} = 0,\ \text{i.e.}\ \frac{\partial E_{0x}}{\partial x} + \frac{\partial E_{0y}}{\partial y} + ik_g E_{0z} = 0 \tag{10.36}$$

$$\nabla\cdot\mathbf{B} = 0,\ \text{i.e.}\ \frac{\partial B_{0x}}{\partial x} + \frac{\partial B_{0y}}{\partial y} + ik_g B_{0z} = 0 \tag{10.37}$$

$$\nabla\times\mathbf{E} = -\frac{\partial\mathbf{B}}{\partial t} = i\omega\mathbf{B}$$

i.e.,

$$\frac{\partial E_{0z}}{\partial y} - ik_g E_{0y} = i\omega B_{0x} \tag{10.38}$$

$$ik_g E_{0x} - \frac{\partial E_{0z}}{\partial x} = i\omega B_{0y} \tag{10.39}$$

$$\frac{\partial E_{0y}}{\partial x} - \frac{\partial E_{0x}}{\partial y} = i\omega B_{0z} \tag{10.40}$$

and

$$\nabla\times\mathbf{B} = \mu\frac{\partial\mathbf{D}}{\partial t} = -\frac{i\omega}{c^2}\mathbf{E}$$

$$\frac{\partial B_{0z}}{\partial y} - ik_g B_{0y} = -\frac{i\omega}{c^2}E_{0x} \tag{10.41}$$

$$ik_g B_{0x} - \frac{\partial B_{0z}}{\partial x} = -\frac{i\omega}{c^2}E_{0y} \tag{10.42}$$

$$\frac{\partial B_{0y}}{\partial x} - \frac{\partial B_{0x}}{\partial y} = -\frac{i\omega}{c^2}E_{0z} \tag{10.43}$$

Solving (10.39) and (10.41) for E_{0x}, one obtains

$$E_{0x} = \frac{i}{k_c^2}\left(k_g\frac{\partial E_{0z}}{\partial x} + \omega\frac{\partial B_{0z}}{\partial y}\right) \tag{10.44}$$

Similarly, one obtains

$$E_{0y} = \frac{i}{k_c^2}\left(k_g\frac{\partial E_{0z}}{\partial y} - \omega\frac{\partial B_{0z}}{\partial x}\right) \tag{10.45}$$

$$\therefore\quad E_{0T} = \hat{a}_x E_{0x} + \hat{a}_y E_{0y}$$

$$= \frac{i}{k_c^2} k_g \left(\hat{a}_x \frac{\partial E_{0z}}{\partial x} + \hat{a}_y \frac{\partial E_{0z}}{\partial y} \right) + \frac{i\omega}{k_c^2} \left(\hat{a}_x \frac{\partial B_{0z}}{\partial y} - \hat{a}_y \frac{\partial B_{0z}}{\partial x} \right)$$

$$= \frac{i}{k_c^2} [k_g \nabla_T E_{0z} - \omega \hat{a}_z \times \nabla_T B_{0z}] \tag{10.46}$$

Here

$$\nabla_T = \hat{a}_x \frac{\partial}{\partial x} + \hat{a}_y \frac{\partial}{\partial y}$$

One can obtain similar structural equations for B_{0x} and B_{0y}. We note that these equations reveal that all the transverse components may be completely specified in terms of longitudional components.

Now, we examine whether TEM waves can propagate inside a hollow conductor.

We have for TEM waves, $E_{0z} = B_{0z} = 0$. Therefore, from Eqs. (10.36) and (10.37), one obtains

$$\frac{\partial E_{0x}}{\partial x} + \frac{\partial E_{0y}}{\partial y} = 0; \quad \frac{\partial B_{0x}}{\partial x} + \frac{\partial B_{0y}}{\partial y} = 0 \tag{10.47}$$

and from Eqs. (10.39) and (10.42), one obtains

$$\frac{\partial E_{0y}}{\partial x} - \frac{\partial E_{0x}}{\partial y} = 0; \quad \frac{\partial B_{0y}}{\partial x} - \frac{\partial B_{0x}}{\partial y} = 0 \tag{10.48}$$

Taking the partial derivatives of (10.47) and (10.48) with respect to x and y respectively and combining them suitably, one obtains

$$\nabla_T^2 E_{0x} = 0; \quad \nabla_T^2 B_{0x} = 0 \tag{10.49}$$

The components of **E** satisfy Laplace's equation. This means that the surface of the guide is an equipotential surface and, hence, **E** inside the conductor is zero. Obviously, there cannot exist, therefore, a TEM mode within a hollow wave guide with perfectly conducting walls. The result obtained is valid for a singly connected surface. Thus for a TEM mode to exist, it is necessary to have two or more unconnected surfaces. One can see that a TEM mode is a dominant mode in a coaxial cable.

Eqs. (10.32) and (10.49) reveal that for TEM waves $k_c = 0$, i.e. $k_g = k$. Obviously, the wave number is real for all frequencies. This means that there is no cut off frequency for TEM waves.

The boundary conditions to be satisfied by the fields **E** and **B** are

$$E_{\text{tangential}}\, \hat{a}_n \times \mathbf{E} = 0 \quad \text{and} \quad \mathbf{B}_{\text{normal}} = \hat{a}_n \cdot \mathbf{B} = 0 \tag{10.50}$$

One can see that boundary conditions given by (10.50) together with the wave equations for the field vectors **E** and **B** lead to the eigen value problems. Obviously, for any given frequency ω, only certain values of k will be constant with the boundary conditions and the wave equation.

10.4 RECTANGULAR WAVE GUIDE

The rectangular wave guide is the most commonly used wave guide and can be obtained by imposing two conducting metal plates with a separation a along the x-axis as shown in Fig. 10.3. The configuration thus formed is a hollow metal pipe of rectangular cross-section which allows the wave propagation along the z-axis. This guide is an example of the theory developed in section 10.3.

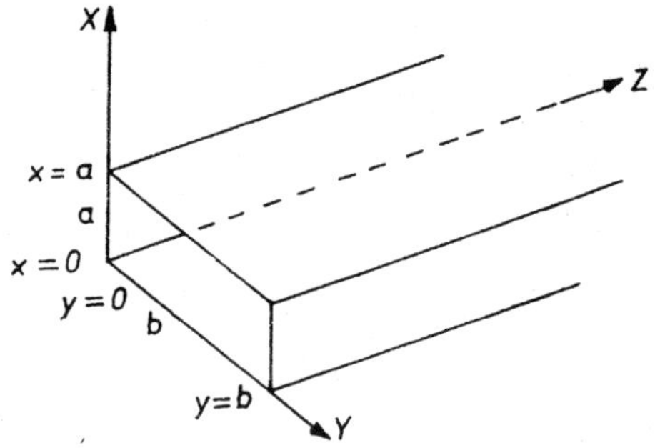

Figure 10.3. A rectangular wave guide.

TE Mode Waves

For such waves $E_{0z} = 0$. Therefore, we will have to determine the field by the solution of the equation

$$(\nabla_T^2 + k_c^2)B_{0z} = 0 \tag{10.51}$$

From (10.46), one finds that B_{0z} has to satisfy besides Eq.(10.51) the following condition

$$\left.\frac{\partial B_{0z}}{\partial x}\right|_{x=o,a} = 0 \quad \text{and} \quad \left.\frac{\partial B_{0z}}{\partial y}\right|_{y=o,b} = 0 \tag{10.52}$$

This will make $E_{\text{tangential}} = 0$.

One can easily verify that the solution

$$B_{0z} = B_o \cos\left(\frac{m\pi x}{a}\right) \cos\left(\frac{n\pi y}{b}\right) \tag{10.53}$$

where m, n are integers, satisfies both (10.50) and (10.51) provided

$$k_c^2 = \pi^2\left(\frac{m^2}{a^2}+\frac{n^2}{b^2}\right) \tag{10.54}$$

i.e.

$$\omega_{mn} = \pi c\left(\frac{m^2}{a^2}+\frac{n^2}{b^2}\right)^{1/2} \tag{10.55}$$

where ω_{mn} is the cut-off frequency corresponding to the mode numbers m, n. The cut-off wavelength is given by

$$(\lambda_c)_{mn} = \frac{2\pi}{k_c} = \frac{2}{\left[\left(\frac{m}{a}\right)^2+\left(\frac{n}{b}\right)^2\right]^{1/2}} \tag{10.56}$$

we find that the different values of m and n correspond to different modes of wave propagation in the rectangular wave guide. The corresponding mode is designated as R mode. If $m = 0$ and $n = 0$, the case corresponds to TEM mode. Obviously, TE_{00} mode does not exist. In case of TE_{10} mode, $n = 1$ and $m = 0$. If $a < b$, one can obtain the lowest cut off frequency by putting $n = 1, m = 0$, i.e.

$$\omega_{01} = \frac{\pi C}{b} \tag{10.57}$$

This is called the *principal* or *dominant mode*. If free space wavelength $\lambda = 2b$, then the guide wavelength becomes infinite and the corresponding $k_c = 0$. The phase velocity of the wave becomes infinite. This has close similarity with wave propagation through ionized medium at $\omega = \omega_p$.

In terms of cut-off wave number, the separation equation

$$k_x^2 + k_y^2 + k_z^2 = k^2 \tag{10.58}$$

can be rewritten as

$$\left(\frac{m\pi}{a}\right)^2 + \left(\frac{n\pi}{b}\right)^2 = k_c^2 - k_g^2 \tag{10.59}$$

This is known as *waveguide equation* and determines the values of waveguide wave number k_g for propagating waves in terms of the integers m and n, the dimensions a and b of the wave guide and the wave number k of the wave in free-space. From Eq. (10.59), one finds that k_g is real if free-space wave number

$$k^2 \geq \left(\frac{m\pi}{a}\right)^2 + \left(\frac{n\pi}{b}\right)^2 = k_c^2 \tag{10.60}$$

and the wave propagates in the wave guide. One the other hand, k_g is imaginary for free-space wave number

$$k^2 < \left(\frac{m\pi}{a}\right)^2 + \left(\frac{n\pi}{b}\right)^2 \tag{10.61}$$

Under this condition the waves are heavily attenuated and become evanescent. One can restate these conditions in terms of k values as

$$R_e(k_z) = \sqrt{(k_c)_{mn} - k^2} \quad \text{for} \quad k < k_c \tag{10.62}$$

and

$$I_m(k_z) = \sqrt{(k_c)_{mn} - k^2} \quad \text{for} \quad k > k_c \tag{10.63}$$

In terms of wavelengths, one obtains

$$R_e(k_z) = k_c\sqrt{1 - \left(\frac{f}{f_c}\right)^2} \quad \text{and} \quad f < f_c \tag{10.64}$$

and

$$I_m(k_z) = k_c\sqrt{1 - \left(\frac{f_c}{f}\right)^2} \quad \text{and} \quad f > f_c \tag{10.65}$$

we must note that for waves propagating through the waveguide, the usual concept of wave length and phase velocity holds good. The wave length λ_g is defined as the distance in which the phase of **E** or **H** fields increase by 2π. In the case of TE mode waves, one can write the guide wavelength as

$$\lambda_g = \frac{\lambda}{\sqrt{1 - \left(\frac{f_c}{f}\right)^2}} \tag{10.66}$$

The values of f and f_c govern the range of permissible λ_g in practical guides. The f_c is function of waveguide dimensions.

Obviously, the physical size of the waveguide as compared to wavelength determines which wave modes can propagate in a given wave guide. Whenever two or more wave modes have the same cut-off frequency, they are said to be *degenerate modes*. In order to avoid *degeneracy* of wave modes in a rectangular waveguide and allow propagation of only one frequency in wave guide $\frac{b}{a} > 1$. The ideal ratio used in practice is $\frac{b}{a} = 2$. For higher values of $\frac{b}{a}$, the high power, operation of the wave guide produces arching. At frequency close to f_c, one finds that the conductor losses become rather large.

One can write the field components of TE_{01}, as

$$\begin{aligned} E_{0x} &= i\omega\frac{\partial B_{0z}}{\partial y} = -\frac{i\omega}{k_c^2}B_0\sin\left(\frac{\pi y}{b}\right)\frac{\pi}{b} \\ &= -\frac{i\omega b}{\pi}B_0\sin\left(\frac{\pi y}{b}\right) \\ E_{0y} &= 0,\ E_{0z} = 0 \\ B_{0y} &= 0,\ B_{0y} = \frac{ik_g}{k_c^2}\frac{\partial B_{0z}}{\partial y} = -\frac{ik_g b}{\pi}B_0\sin\left(\frac{\pi y}{b}\right) \\ B_{0z} &= B_0\cos\left(\frac{\pi y}{b}\right) \end{aligned} \tag{10.67}$$

Fields given by (10.67) correspond to the wave propagated in the z-direction. We must note that this is also the direction of the Poynting vector. One obtains the energy flow in the wave guide as

$$\begin{aligned} <\mathbf{N}>_{01} &= \frac{1}{2}R_e(\mathbf{E}\times\mathbf{H}^*) \\ &= \frac{1}{2\mu}R_e(\mathbf{E}\times\mathbf{B}^*) = \frac{1}{2\mu}R_e\left(-\hat{a}_y E_{0x}B_{0z}^* + \hat{a}_z E_{0x}B_{0y}^*\right) \end{aligned}$$

one finds from (10.67) that $E_{0x}B_{0z}^*$ is purely imaginary.

$$\therefore \quad <\mathbf{N}>_{01} = \frac{\hat{a}_z}{2\mu}\frac{\omega b^2}{\pi^2}B_0^2 k_g\sin^2\left(\frac{\pi y}{b}\right) \tag{10.68}$$

One can obtain the total power by integrating the above expression from $y = 0$ to $y = b$.

TM Mode Waves

One can obtain the TM_{mn} mode waves in an analogous manner by considering the characteristic *mode functions* of the form

$$\Psi_{mn}^{TE} = \sin\left(\frac{m\pi x}{a}\right)\sin\left(\frac{m\pi y}{b}\right)e^{-ik_z z}$$

where m, $n = 1, 2, 3, \ldots$. These wave functions satisfy the separation parameter Eq. (10.58). One obtains the TM_{mn} wave mode fields in free space from derivatives of wave mode functions.

$$E_x = -\frac{1}{i\omega\varepsilon_0}\frac{\partial^2\Psi}{\partial x\partial z},\ H_x = \frac{\partial\Psi}{\partial y}\ \text{at } y = 0 \text{ and } y = b \tag{10.69}$$

$$E_y = -\frac{1}{i\omega c}\frac{\partial^2\Psi}{\partial y\partial z},\ H_y = -\frac{\partial\Psi}{\partial x}\ \text{at } x = 0 \text{ and } x = a \tag{10.70}$$

and
$$E_z = -\frac{1}{i\omega\varepsilon_0}\left(\frac{\partial^2}{\partial z^2}+k^2\right)\Psi, H_z = 0 \tag{10.71}$$

Obviously, the TM_{mn} mode waves are characterized by zero H_z component and are known as *transverse magnetic* to z-axis.

TE Versus TM: Wave mode characteristics

We have seen that in the case of TE mode waves the E_z components is zero. The TE mode field components are designated by integers m and n which characterize their behaviour in x and y directions. For TE mode waves the field components do not vanish for either m or n equal to zero. However, one finds that both m and n cannot be equal to zero at a time. We have seen that the lowest order TE mode wave in the case of rectangular wave guide is the TE_{10} mode. This mode has the lowest cut-off frequency and is said to be the *dominant mode* wave in the rectangular wave guide. The subscripts m and n represent the number of half period variations of the field along the x and y-coordinates respectively. We have considered the x-coordinate to coincide with the larger transverse dimension, so the TE_{10} wave has a lower cut-off frequency than TE_{01}. Fig. 10.4 shows the electric and magnetic field configurations for TE_{10} mode in rectangular wave guide.

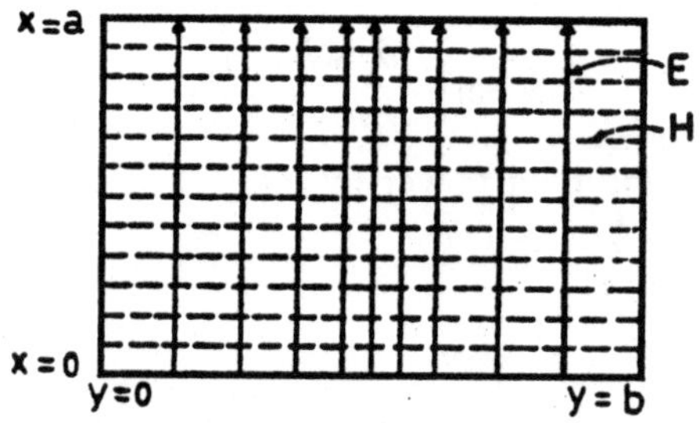

Figure 10.4.

We have seen that B_z component is zero in the case of TM mode waves. TM mode waves are also designated by integers m and n which characterize their behaviour in x-and y-directions. Further, one finds that if either m or n is zero, the field will all be identically zero which is in contrast to the behaviour of TE mode waves in rectangular waveguides. The TM mode waves obey the wave guide equation (10.59) and have identical cut-off frequencies given by (10.55). There are no magnetic charges and the B field has to vary with the x-coordinate in order to make the lines of the field continuous. As a result of these requirements, one finds that in the case of TM mode waves, the lowest possible value for either m or n is unity. The lowest cut-off frequency for the TM mode corresponds to $m = n = 1$. The wave mode corresponding to these restrictions forms the dominant mode and is designated as TM_{11}. The lowest order magnetic mode in rectangular wave guide is TM_{11} which is significantly different from the dominant electric mode TE_{10} in the rectangular wave guide. Fig. 10.5 shows **E** and **H** configuration of dominant magnetic TM_{11} in rectangular mode.

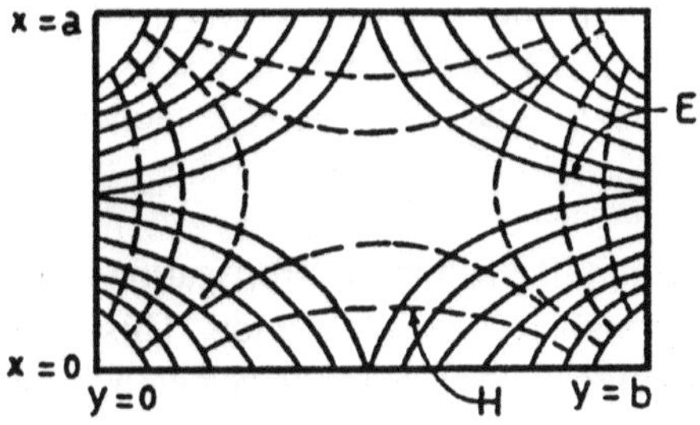

Figure 10.5.

10.5 COAXIAL WAVE GUIDE

It consists of an inner and an outer conductor. Let us find the field distribution for such a wave guide. The solution for such a waveguide should be circularly symmetric with a radial electric field and an azimuthal magnetic field (Fig. 10.6). One can write the components of the field in cylindrical coordinates as $(E_R, 0, 0)$ and $(0, H_\phi, 0)$. We can see that these satisfy the boundary conditions. Maxwell's equations are

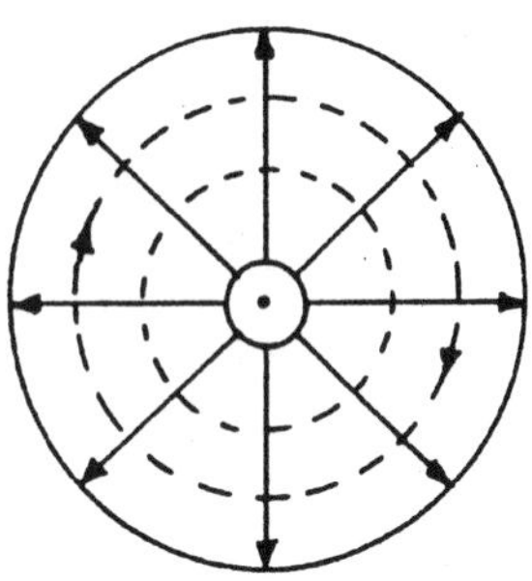

Figure 10.6.

$$\nabla \times \mathbf{H} = \mathbf{J} + \frac{\partial \mathbf{D}}{\partial t} \text{ and } \nabla \times \mathbf{E} = -\frac{\partial \mathbf{B}}{\partial t}$$

These lead to the equations:

$$-\frac{\partial H_\phi}{\partial z} = \varepsilon\frac{\partial E_R}{\partial t} = -i\omega\varepsilon E_R \tag{10.72}$$

and

$$\frac{\partial E_R}{\partial z} = -\mu\frac{\partial H_\phi}{\partial t} = i\mu\omega H_\phi \tag{10.73}$$

Differentiating (10.63), one obtains

$$\frac{\partial^2 E_R}{\partial z^2} = i\mu\omega\frac{\partial H_\phi}{\partial z} = i^2\varepsilon\omega^2 E_R$$

or

$$\frac{\partial^2 E_R}{\partial z^2} + \varepsilon\mu\omega^2 E_R = 0$$

and

$$\frac{\partial^2 E_R}{\partial z^2} + k^2 E_R = 0 \tag{10.74}$$

One can see from (10.64) that the propagation constant does not depend on the dimensions of the wave guide. This shows that the guide wavelength is equal to the free space wavelength. Obviously, the propagation is similar to that of a plane wave. We must note that there is no cut-off wavelength in a coaxial waveguide.

10.6 CYLINDRICAL WAVEGUIDE

Hollow metallic cylinders are also used as waveguides. Although rectangular waveguides are widely used in practice, cylindrical wave guides have some advantages over the rectangular ones. The propagation of electromagnetic wave in this waveguide also confirms to the scalar Helmholtz equation in cylindrical coordinates given by

$$\frac{1}{\rho}\frac{\partial}{\partial\rho}\left(\rho\frac{\partial\psi}{\partial\rho}\right) + \frac{1}{\rho^2}\frac{\partial^2\psi}{\partial\phi^2} + \frac{\partial^2\psi}{\partial z^2} + k^2\psi = 0 \tag{10.75}$$

one can show that the solution of this equation confirm with a cylindrical wave function given by

$$\psi = R(\rho)\Phi(\phi)Z(z) \tag{10.76}$$

Fig. 10.7 Shows a cylindrical waveguide and the coordinate system.

One can express the ϕ variation of cylindrical wave function as

$$\psi = \left\{\begin{matrix}\sin n\phi \\ \cos n\phi\end{matrix}\right\} J_n(k_\rho\rho)\exp(-ik_z z) \tag{10.77}$$

where $J_n(k_\rho\rho)$ is a Bessel function.

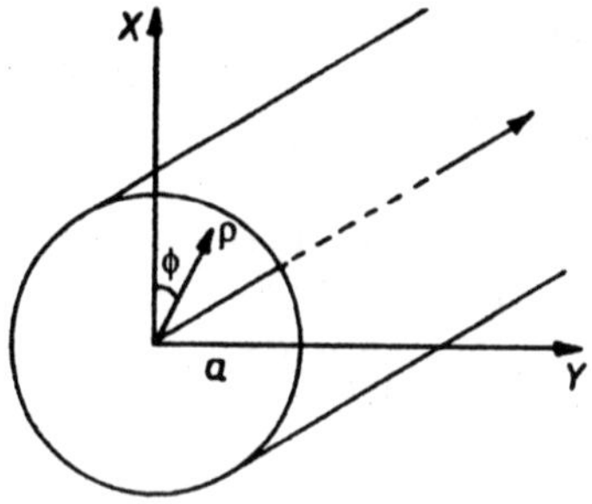

Figure 10.7. Cylindrical waveguide.

One can choose the wave function either with $\sin n\phi$ or $\cos n\phi$ in order to achieve mode degeneracy except for the case when $n = 0$. In terms of this wave function the components of wave field are given by their derivatives.

The components of TM modes waves in free space can be written as

$$E_\rho = \frac{1}{i\omega\varepsilon_0}\frac{\partial^2\psi}{\partial\rho\partial z} \quad \text{and} \quad H_\rho = \frac{1}{\rho}\frac{\partial\psi}{\partial\phi} \tag{10.78}$$

$$E_\phi = \frac{1}{i\omega\varepsilon_0}\frac{\partial^2\psi}{\partial\phi\partial z} \quad \text{and} \quad H_\phi = -\frac{\partial\psi}{\partial\rho} \tag{10.79}$$

$$E_z = \frac{1}{i\omega\varepsilon_0}\left(\frac{\partial^2}{\partial z^2} + k^2\right)\psi \quad \text{and} \quad H_z = 0 \tag{10.80}$$

The components of TE mode waves can be written in a similar way as

$$E_\rho = -\frac{1}{\rho}\frac{\partial\psi}{\partial\phi} \quad \text{and} \quad H_\rho = \frac{1}{i\omega\mu_0}\frac{\partial^2\psi}{\partial\rho\partial z} \tag{10.81}$$

$$E_\phi = \frac{\partial\psi}{\partial\phi} \quad \text{and} \quad H_\phi = \frac{1}{i\omega\mu_0}\frac{\partial^2\psi}{\partial\phi\partial z} \tag{10.82}$$

$$E_z = 0 \quad \text{and} \quad H_z = \frac{1}{i\omega\mu_0}\left(\frac{\partial^2}{\partial z^2} + k^2\right)\psi \tag{10.83}$$

The TE mode waves in cylindrical waveguide is characterized by $E_\phi = 0$ at $\rho = a$ which confirms with

$$J_n'(k_\rho a) = 0$$

or

$$\frac{dJ_n(k_\rho a)}{da} = k_\rho\left[\frac{n}{ka}J_n(k_\rho a) - J_{n+1}(k_\rho a)\right] = 0 \tag{10.84}$$

one finds that this Bessel function is oscillatory and has an infinite number of zeros. One can write the corresponding arguments as

$$k_\rho a = x'_{np} \tag{10.85}$$

where n stands for the order of the Bessel function, p is for order of zero and prime is used to indicate the confirmity with the first derivative of Bessel function. Using this with (10.78), one obtains for TE mode wave function

$$\Psi^{\text{TM}}_{np} = J_n\left(\frac{x'_{np}\rho}{a}\right)\{\sin n\phi\}\exp(-ik_z z) \tag{10.86}$$

where $n = 0, 1, 2, \ldots$ and $p = 1, 2, 3, \ldots$. The wave propagation confirming with the wave function is governed by the separation equation

$$\left(\frac{x'_{np}}{a}\right)^2 + k_z^2 = k^2 \tag{10.87}$$

one can define the propagation through wave guide in terms of *guide wave number.*

$$k_g^{\text{TE}} = \sqrt{\left(\frac{x'_{np}}{a}\right)^2 - k^2} \tag{10.88}$$

one finds that the wave propagation through cylindrical waveguide stops for $k_g = 0$. This defines the *critical wave number*, i.e.

$$k_c^{\text{TE}} = \frac{x'_{np}}{a} \tag{10.89}$$

obviously, the cut off frequencies for TE mode wave are proportional to x'_{np}. The zero of the Bessel function for x'_{11} determines the dominant wave propagation and is designated as TE_{11} mode. This TE_{11} mode is actually a pair of degenerate modes corresponding to $\sin\phi$ and $\cos\phi$ variations. This is why, there is no frequency range for single mode propagation.

We must note that the treatment of TM mode waves is analogus to TE mode. One finds the z-component of electric field in TE mode as

$$E_z = \frac{i}{\omega\varepsilon}(k^2 - k_z^2)\psi \tag{10.90}$$

This must vanish at the conducting walls $\rho = a$. This requires that

$$J_n(k\rho a) = 0 \tag{10.91}$$

One can write the TM mode wave function analogous to (10.86) as

$$\Psi_{np}^{\text{TM}} = J_n\left(\frac{x_{np}\rho}{a}\right)\left\{\begin{array}{c}\sin n\phi \\ \cos n\phi\end{array}\right\}\exp(-ik_z z) \tag{10.92}$$

Proceeding exactly in the same manner as in the case of the TE mode waves, one obtains the guide wave number as

$$k_g^{\text{TM}} = \sqrt{\left(\frac{x_{np}}{a}\right)^2 - k^2} \tag{10.93}$$

and the critical wave number as

$$(k_c)_{np}^{\text{TM}} = \frac{x_{np}}{a} \tag{10.94}$$

The dominant mode of TM waves is TM_{11}. On comparing the results for TE and TM mode waves, one finds that there exists *degeneracy* between TE_{op} and TE_{1p}. Other TE and TM modes are *nondegenerate* and have different cut-off frequencies and hence different propagation constants. One can define the z-directed characteristic wave impedances for the TE and TM modes as

$$(Z_0)^{\text{TE}} = \left(\frac{E_\rho}{H_\phi}\right) = -\frac{E_\phi}{H_\rho} = \frac{\omega\mu_0}{k_z} \tag{10.95}$$

and

$$(Z_0)^{\text{TM}} = \left(\frac{E_\rho}{H_\phi}\right) = -\frac{E_\phi}{H_\rho} = \frac{k_z}{\omega\varepsilon} \tag{10.96}$$

where Z_0 refers to the cylindrical waveguide with loss free media.

10.7 CAVITY RESONATORS

The purpose of transmission lines and wave guides is to transmit electromagnetic energy efficiently from one point to another. A resonator is an energy storage device and such a device will follow several physical analogies which will make the wave regions to functions as 'circuits' particularly at ultra high frequencies. Although an electromagnetic cavity resonator can be of any shape whatsoever, an important class of cavities is produced by placing end faces on a length of cylindrical waveguide. The closed chambers of conducting walls have close similarities with the LC circuit and are known as *Cavities* or *resonators*. The standing waves are set in the cavities.

Let us consider a plane electromagnetic wave with components E_x and H_y, travelling in the z-direction. Let us suppose that a perfect conductor is placed into the half infinite space $z > z_0$ as shown in Fig. 10.8.

One can see that the wave incident on the conductor turns back and one find the total electric field as

$$E_x = A_i \exp[i(kz-\omega t)] + A_r \exp[-i(kz+\omega t)]$$

This must satisfy the boundary condition, $E_x = 0$ at $z = z_0$. We have

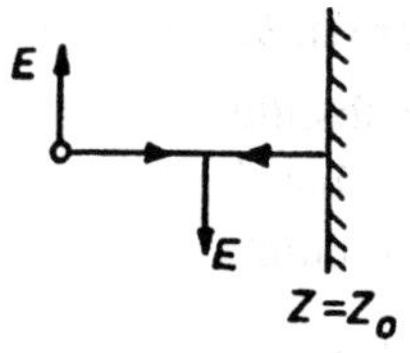

Figure 10.8.

$$A_i \exp[i(kx_0 - \omega t)] + A_r \exp[-i(kx_0 + \omega t)] = 0$$

$$\therefore \quad A_r = -A_i \exp(2ikz_0)$$

and

$$\begin{aligned} E_x &= A_i\{\exp i(kz-\omega t) - \exp(2ikz_0)\exp(-i(kz+\omega t))\} \\ &= A_i \exp(-i\omega t)\{\exp(ikz) - \exp(2ikz_0)\exp(-ikz)\} \\ &= A_i \exp(-i\omega t)\exp(ikz_0)\{\exp ik(z-z_0) - \exp(-ik(z-z_0))\} \\ &= 2iA_i \exp(-i\omega t)\exp(ikz_0)\sin k(z-z_0) \qquad (10.97) \end{aligned}$$

obviously, (10.97) represents a standing wave. We must note that the standing wave pattern does not undergo any change if an infinitely conducting plate is placed parallel to the first plate at a position of a node where the electric field is zero. Obviously, the two conducting plates, thus, constitute a resonator in which the electromagnetic energy bounces between the two plates.

Let us consider a closed box formed by placing end faces on a rectangular wave guide. Let us assume that the end surfaces are plane and perpendicular. Because of the reflection at the end faces, we can see that the waves in the cavity are standing waves, i.e. not progressive. Obviously, this cavity resonator will resonate at a frequency at which the length of the cavity is an integral multiple of half wavelength measured in wave guide.

One can easily see that the electric field components are of the form

$$\begin{aligned} E_x &= E_1 \cos(k_1x)\sin(k_2y)\sin(k_3z)\exp(-i\omega t) \\ E_y &= E_2 \sin(k_1x)\cos(k_2y)\sin(k_3z)\exp(-i\omega t) \end{aligned}$$

and

$$E_z = E_3 \sin(k_1x)\sin(k_2y)\cos(k_3z)\exp(-i\omega t) \qquad (10.98)$$

For the boundary conditions to be satisfied, one finds that k_1, k_2, k_3

must have the values given by

$$k_1 = \frac{l\pi}{a}, k_2 = \frac{m\pi}{b} \quad \text{and} \quad k_3 = \frac{n\pi}{c} \tag{10.99}$$

where a, b, c are the dimensions of the box and l, m, n are integers.

Substitution of any component in the appropriate wave equation exhibit that the fields given by (10.98) to be acceptable, the free space wave number has to satisfy the following condition

$$k^2 = \frac{\omega^2}{c^2} = \pi^2\left(\frac{l^2}{a^2} + \frac{m^2}{b^2} + \frac{n^2}{c^2}\right) \tag{10.100}$$

Obviously, there is an infinite number of resonant frequencies and, hence, the infinite number of modes of the cavity corresponding to the different values of l, m, n.

One can obtain the magnetic field components from Maxwell's relation $\nabla \times \mathbf{E} = -\frac{\partial \mathbf{B}}{\partial t}$. One can see from (10.98) and (10.99) that if $l = 0$, $m = 1$ and $n = 1$, the electric field is transverse to the direction of propagation. This mode is designated as TE_{011} mode. One can see that besides TE_{lmn} modes there are also other modes possible in the cavity. There are TM_{lmn} modes in which the magnetic field is transverse to the direction of propagation. We must note that for the same cavity, TE_{lmn} waves and TM_{lmn} waves occur at the same frequency. Thus at a particular resonant frequency the standing wave in the cavity is the sum of two resonating waves, i.e. the TE mode and TM mode.

We have mentioned above that for a definite field configuration the cavities have some discrete resonant frequencies. Obviously, if one tries to excite a particular mode of oscillation in a cavity, the right sort of fields will not be built up unless the exciting frequency is equal to the resonant frequency. In practice. one finds that appreciable excitation occurs over a narrow band of frequencies around the resonant frequency. This means that the sharp frequency of oscillation occurs portly, because of the dissipation of energy in the cavity walls. One can express the measure of these losses in terms of the "Q" of the cavity defined as

$$Q = \omega_0 \frac{\text{Stored energy in the cavity}}{\text{Power loss per cycle to the walls of the cavity}} \tag{10.101}$$

Here ω_0 is the resonance frequency, assuming no losses. One can readily estimate the power loss in a cavity by computing the time average of the Poynting vector into the wall at the surface

$$< N >= \frac{1}{2}\text{Real}\,(E_{11} \times H_{11}) \tag{10.102}$$

where E_{11} and H_{11} represents the tangential components of the electric and magnetic fields respectively.

An important practical resonant cavity is the *right circular cylinder*, perhaps with a piston to allow tuning by varying the height. The cylinder is shown in Fig. 10.9. with inner radius R and length d. For a TM mode the transverse wave equation for $\Psi = E_z$, subject to the boundary condition $E_z = 0$ at $\rho = R$, has the solution:

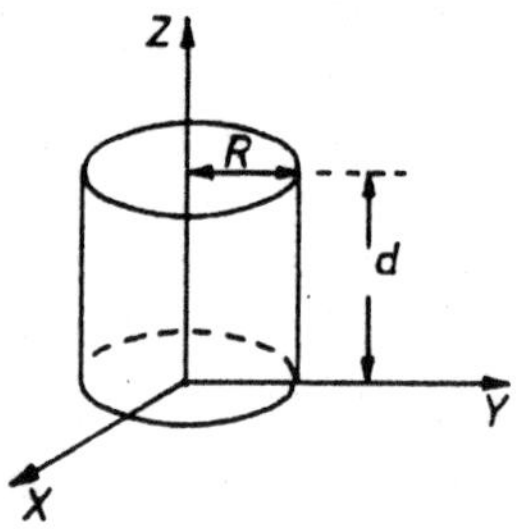

Figure 10.9. Cylindrical Cavity.

$$\psi(\rho, \phi) = E_0 J_m(\gamma_{mn}\rho)\exp(\pm im\phi) \tag{10.103}$$

where

$$\gamma_{mn} = \frac{x_{mn}}{R} \tag{10.104}$$

x_{mn} is the nth root of the equation, $J_m(x) = 0$. The integers m and n take on the values $m = 0, 1, 2, \ldots$ and $n = 1, 2, 3, \ldots$. The resonant frequencies are given by

$$\omega_{mnp} = \frac{1}{\sqrt{\mu\varepsilon}}\sqrt{\frac{x_{mn^2}}{R^2} + \frac{p^2\pi^2}{d^2}} \tag{10.105}$$

The lowest TM mode has $m = 0$, $n = 1$, $p = 0$ and so it is designated as $\text{TM}_{0,1,0}$. Its resonance frequency is

$$\omega_{010} = \frac{2.405}{\sqrt{\mu\varepsilon}R} \tag{10.106}$$

One can write the explicit expressions for the field as

$$\left.\begin{aligned} E_z &= E_0 J_0\left(\frac{2.405\rho}{R}\right)\exp(-i\omega t) \\ H_\phi &= -i\sqrt{\frac{\varepsilon}{\mu}}E_0 J_1\left(\frac{2.405\rho}{R}\right)\exp(-i\omega t) \end{aligned}\right] \tag{10.107}$$

obviously, the resonant frequency for this mode is independent of d and hence simple tuning is impossible.

For TE modes, the basic solution (10.104) still applies, but the boundary condition on H_z $\left[\left(\frac{\partial\psi}{\partial\rho}\right)|_R = 0\right]$ makes

$$\gamma_{mn} = \frac{x'_{mn}}{R}$$

where x'_{mn} is the nth root of $J'_m(x) = 0$. The resonant frequencies are given by

$$\omega_{mnp} = \frac{1}{\sqrt{\mu\varepsilon}}\left(\frac{x'^2_{mn}}{R^2} + \frac{p^2\pi^2}{d^2}\right) \qquad (10.108)$$

where $m = 0, 1, 2, \ldots$ but $n, p = 1, 2, 3, \ldots$. The lowest TE mode has $m = n = p = 1$ and is designated as $TE_{1,1,1}$. Its resonance frequency is

$$\omega_{111} = \frac{1.841}{\sqrt{\mu\varepsilon}R}\left(1 + 2.912\frac{R^2}{d^2}\right)^{1/2} \qquad (10.109)$$

while the fields can be derived from

$$\Psi = H_z = H_0 J_1\left(\frac{1.841\rho}{R}\right)\cos\phi\sin\left(\frac{\pi z}{d}\right)\exp(-i\omega t) \qquad (10.110)$$

For d large enough ($d > 2.03R$), the response frequency ω_{111} is smaller than that for the lowest-TM mode. Then the $TE_{1,1,1}$ mode is the fundamental oscillation of the cavity. Since the frequency depends on the ratio $\frac{d}{R}$, it is possible to provide easy tuning by making the separation of the end faces adjustable.

These cavities are widely used at microwave frequencies to study the nature of filling dielectrics. The resonant frequencies and strength of resonance are diagnostic features of dielectrics filling the cavities.

The quality factor can be obtained from the following expression

$$Q = \frac{1.202\eta}{\text{Real}(\eta)\left(1 + \frac{R}{d}\right)} \qquad (10.111)$$

where η is the intrinsic impedance of cavity metal. If $R = d$, the Q of the cavity depends mainly on the property of the conductor used in making the cavity.

The cavity resonators are characterized by their resonant frequencies and quality factors. These parameters depend mainly on: (i) the size of the cavity and (ii) the nature of the dielectric filling the cavity.

Any change in these parameters pertub the normal functing of the cavity and causes corresponding changes in the performance of the cavity. In practice, the perturbation in the size of a cavity is impractical and is generally not used for diagnostic purposes. However, the perturbation produced by the changes in the dielectric property of the cavity has proved to be of great diagnostic value.

Cavities are widely used as frequency meters. Cavities are also used in experiments where high microwave fields are required, e.g. in electric spin resonance experiments.

There are also open resonators which employ open structures. Such resonators consist of two reflectors facing each other. The reflection may be flat or spherical. These resonators have been found to be very useful at optical wavelengths and are widely used in lasers.

10.8 EARTH AND IONOSPHERE AS A RESONANT CAVITY

A some what unusual example of a resonant cavity is provided by the earth itself as one boundary surface and the ionosphere as the other. The region between the earth's surface and the ionosphere, which is approximately 80 km above the earth, forms a wave guide that allows the propagation of radio waves around the curve of the earth, as shown in Fig. 10.10.

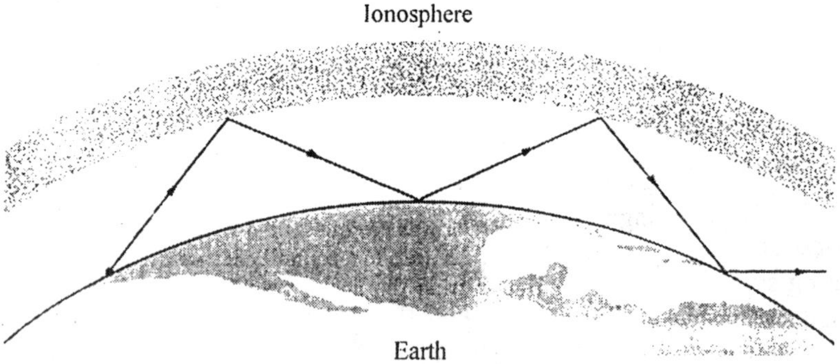

Figure 10.10. The ionosphere and the earth act as a waveguide for radio waves.

The lowest resonant modes of such a system are evidently of very low frequency, since the characteristic wavelength must of the order of magnitude of the earth's surface. In such circumstances the ionosphere and the earth both appear as conductors with real conductivities. Sea water has a conductivity of $\sigma \sim 0.1\Omega^{-1}m^{-1}$, while the ionosphere has $\sigma \sim 10^{-7} - 10^{-4}\Omega^{-1}m^{-1}$. The walls of the cavity are thus far from perfectly conducting, especially the outer one.

10.9 OPTICAL FIBER WAVE GUIDE

It consists of transparent fibres with a diameter of a few microns, called optical fibers. These fibers are made of glass or quartz, although other materials, such as nylon, are being tested. A ray entering at one end follows the axis of the fiber as a result of several reflections, emerging at the other end (Fig. 10.11). When the fibers are arranged in bundles, an image can be transmitted from one point to another.

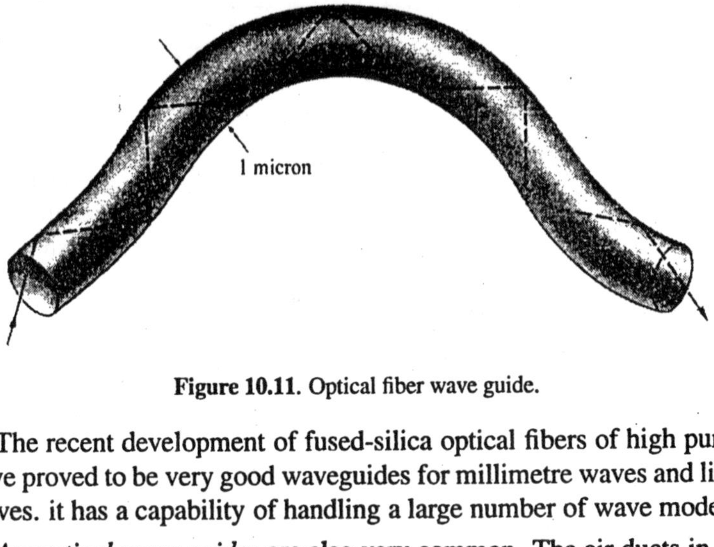

Figure 10.11. Optical fiber wave guide.

The recent development of fused-silica optical fibers of high purity have proved to be very good waveguides for millimetre waves and light waves. it has a capability of handling a large number of wave modes.

Acoustical wave guides are also very common. The air ducts in the heating system of a house, for example, act as acoustical wave guides which are capable of transmitting the noises from the furance or sounds from one room to another. The inner ear is essentially an acoustical wave guide.

10.10 DIELECTRIC WAVE GUIDES

We have seen that in hollow conducting wave guides the electromagnetic energy propagates by the process of total internal reflections from the conducting surface and their superposition from the hollow space.

One also finds a close analogy in a dielectric rod or slab. The electromagnetic energy launched in the dielectric slab or rod undergoes the process of total internal reflections at the interface between the dielectric and air (Fig. 10.12). The crical angle for given inner and outer dielectrics is given by

$$\theta_c = \sin^{-1}\sqrt{\frac{\mu_2\varepsilon_2}{\mu_1\varepsilon_1}} \tag{10.112}$$

where μ_2, ε_2 and μ_1, ε_1 are relative permeabilities and dielectric constants of outer and inner media.

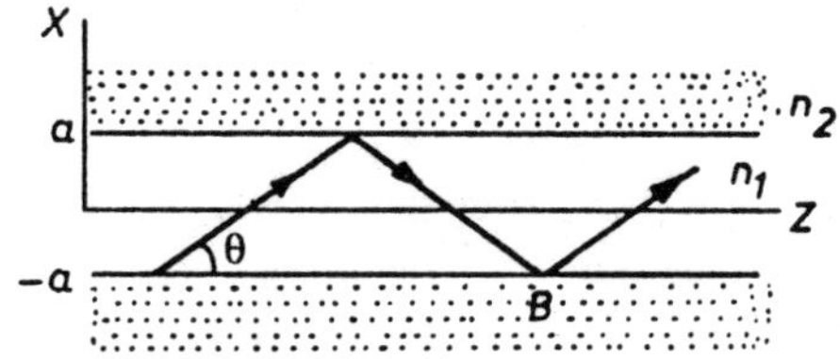

Figure 10.12. Slab dielectric wave guide.. Ray or normal to wave front at angle θ shown.

Let us now consider how a plane slab of a dielectric can serve as a wave guide.

We consider a plane dielectric sheet (A) sandwitched between two layers (B) and (C) as shown in Fig. 10.13. The outer layers B and C forming the cladding are called as substrate (B) and superstrate (C). Between these two layers there is the infinite sheet-a thin film of thickness of the order of the optical wavelength d. Usually these are made from gallium arsenide, Zinc oxide etc. Let us assume that the layers do not absorb the electromagnetic radiation incident on them, or absorb it very weakely, and their refractive indices satisfy the following condition,

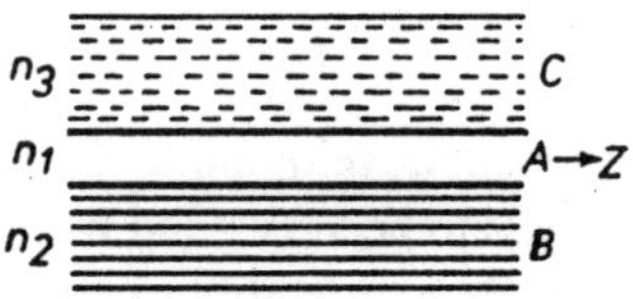

Figure 10.13.

$$n_1 > n_2 \geq n_3 \tag{10.113}$$

The condition (10.113) fulfills the condition for internal reflection.

TE-mode waves

Let us consider a TE mode propagating in the z-direction and field E to be a function of transverse coordinates. Let the components of the field be $\{0, E_y \exp i(\beta z - \omega t), 0\}$, where β is the propagation constant in the z-direction. β is given by

$$\beta = k' \cos\theta = n_1 k \cos\theta \tag{10.114}$$

$\left(\because \quad k' = \dfrac{2\pi}{\lambda_d} = \dfrac{2\pi}{\lambda} n_1\right.$, where λ_d is the wavelength in the dielectric).

One finds that the wave equation

$$\left(\nabla^2 - \frac{1}{c^2}\frac{\partial^2}{\partial t^2}E\right) = 0$$

reduces to

$$\frac{\partial^2 E_y}{\partial x^2} + (n^2k^2 - \beta^2)E_y = 0 \tag{10.115}$$

The fields must vanish at $x = \pm\infty$. The general solution of (10.101) for two media in this problem can be written in the form of sines and cosines or in terms of complex exponentials. We have

$$\begin{aligned} E_y &= A e^{-\delta x} e^{i(\beta z - \omega t)} \quad \text{for } x \geq 0 \\ &= (B\cos kx + c\sin kx)\exp i(\beta z - \omega t) \quad \text{for } 0 \geq x \geq -d \\ &= D e^{\gamma x} e^{i(\beta z - \omega t)} \quad \text{for } x \leq -d \end{aligned} \tag{10.116}$$

where $\delta^2 = \beta^2 - n_3^2k^2; k^2 = n_1^2k^2 - \beta^2; \gamma^2 = \beta^2 - n_2^2k^2$; each one of these being a positive quantity. From first of these relations, one obtains

$$\beta^2 > n_3^2k^2, \quad i.e. \quad n_1^2k^2\cos^2\theta > n_3^2k^2$$

or

$$n_1^2\cos^2\theta > n_3^2$$

and from the third relation, one obtains, $n_1^2\cos^2\theta > n_2^2$. Obviously, these satisfy conditions for internal reflection at the two boundary conditions.

One obtains the tangential components of the magnetic field H with the help of Maxwell's equations as,

$$H_z = -\frac{i\delta}{\omega\mu_0}[Ae^{-\delta x}e^{i(\beta z-\omega t)}] \qquad \text{for} \quad x \geq 0$$
$$= \frac{ik}{\omega\mu_0}[-A\sin kx + C\cos kx]\exp i(\beta z-\omega t) \qquad \text{for} \quad 0 \geq x \geq -d$$
$$= \frac{i\gamma}{\omega\mu_0}[A\cos kd - C\sin kd]\exp\{\gamma(x+d)\}$$
$$\exp\{i(\beta z-\omega t)\} \qquad \text{for} \quad x \leq -d$$

The tangential component of H has to satisfy boundary conditions, i.e. H_t must be continuous across the surface at $x = 0$ and $x = -d$. One obtains

$$-\delta A = kc \tag{10.117}$$

or

$$k[A\sin kd + C\cos kd] = \gamma[A\cos kd - C\sin kd] \tag{10.118}$$

i.e.

$$\delta A + kc = 0$$

and

$$(kA+\gamma C)\sin kd + (kc-\gamma A)\cos kd = 0$$

one can easily see that for these to have a non-trivial solution for A and C, the determinant of the system must vanish. This gives

$$\tan kd = \frac{k(\gamma+\delta)}{k^2-\gamma\delta} \tag{10.119}$$

When one try to solve (10.119) graphically or numerically, one obtains the value of β^2 which, in turn, provide the value of the angle θ at which the electromagnetic energy must be coupled into the guide.

We must note that if $\gamma = 0$, there is a loss of total internal reflection at the lower boundary and the mode will be no longer be guided.

EXAMPLE 10.1: Discuss electromagnetic waves in a plane parallel wave guide.

Solution: We have studied that electromagnetic waves in guides have certain pecularities of their own, which are due to their transverse character and to the boundary conditions at the surface of the conductor. These boundary conditions are: (i) the electric field is normal, and (ii) the magnetic field is tangential to the surface of the conductor. One possible solution of Maxwell's equations satisfying these conditions for a plane wave is $E_y = E_0 \sin(\omega t - kx)$ and $B_z = B_0 \sin(\omega t - kx)$ with $B_0 = \frac{E_0}{C}$. The lines of force of the electric field are indicated in Fig. 10.14(a) and those of the magnetic field by dots and crosses. We must note that in this case wave guide does not change the phase velocity of the wave, which propagates with the same phase velocity $c = \frac{\omega}{k}$, corresponding to propagation in free space; the wave guide limits only the wave front.

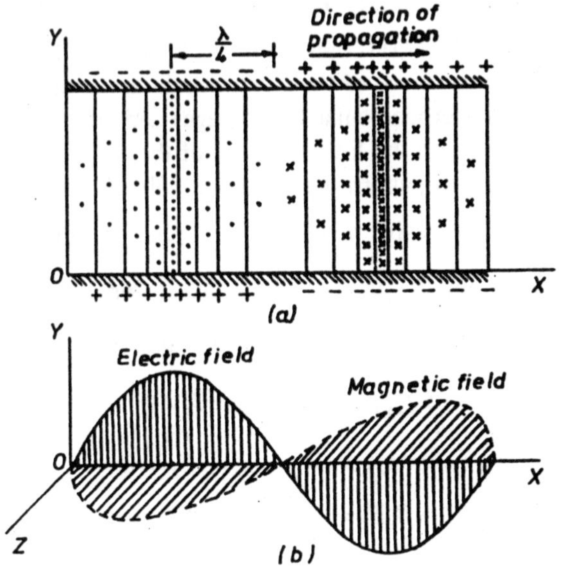

Figure 10.14. (a) Electric lines of force (vertical lines) and magnetic line of force (dots and crosses in the x-y plane for an electromagnetic wave propagating parallel to two reflecting planes parallel to XZ plane. (b) Electric and magnetic field corresponding to fields depicted in part (a).

However, Maxwell's equations also admit other solutions, which also satisfy our boundary conditions. One possible solution can be expressed as

$$\begin{aligned} E_x &= E_y = 0 \\ E_z &= E_0 \sin k_2 y \cos(\omega t - k_1 x), \\ B_x &= -\frac{k_2}{\omega} E_0 \cos k_2 y \sin(\omega t - k_1 x), \end{aligned}$$

$$B_y = -\frac{k_1}{\omega} E_0 \sin k_2 y \cos(\omega t - k_1 x), B_z = 0$$

One can easily verify this by direct substitution in Maxwell's equations. This solution is called the TE solution, because the electric field is transverse, but the magnetic field has a component along the effective direction of propagation, or x-axis. The electric and magnetic fields are, however perpendicular to each other. To satisfy the boundary conditions at both conducting planes, we must set $E_z = 0$ and $B_y = 0$ for $y = 0$ and $y = a$. The first is automatically satisfied, while the second requires that $\sin k_2 a = 0$ or $k_2 a = n\pi$, so that the condition $k_2 = \frac{n\pi}{a}$ is obtained. Fig. 10.15(a) shows the lines of force for the lowest mode, $n = 1$. The lines of force of the electric field are straight lines parallel to the planes (perpendicular to the page) and are thus indicated by dots or crosses, while the lines of force of the magnetic field are the closed curves. Each pattern occupies one-half of the effective wavelength $\frac{2\pi}{k}$, and successive patterns have a phase difference of π. The patterns travel along the guide with the phase velocity $v_p = \frac{\omega}{k_1}$.

Another possible solution of Maxwell's equations is

$$E_x = -\frac{k_2}{k_1} E_0 \sin k_2 y \sin(\omega t - k_1 x)$$

$$E_y = E_0 \cos k_2 y \cos(\omega t - k_1 x), E_z = 0$$

$$B_x = B_y = 0, B_z = \frac{\omega}{k_1 c^2} E_0 \cos k_2 y \cos(\omega t - k_1 x)$$

One can again verify this by direct substitution in Maxwell's equations. This second solution, denoted TM (for transverse magnetic), is so called because the magnetic field is transverse.

The electric field, however, has a component along the effective direction of propagation. Both fields remain perpendicular to each other. To satisfy the boundary conditions at the conducting planes, we must make $E_x = 0$ at $y = 0$ and $y = a$. The first is automatically satisfied, and the second once more require that $\sin k_2 a = 0$ or $k_2 a = n\pi$, so that again $k_2 = \frac{n\pi}{a}$ is obtained. Obviously, both modes have the same cut-off frequency. Fig. 10.15(b) again shows the lines of force for the lowest mode, $n = 1$. However, the lines of force of the *magnetic* field are now straight lines parallel to the planes (perpendicular to the page) and are

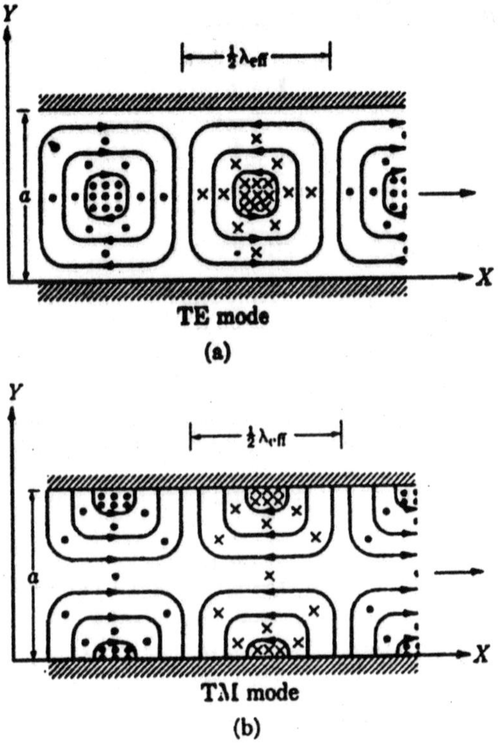

Figure 10.15. Wave guide for electromagnetic waves. (a) Electric field perpendicular to the page, or TE mode (b) Magnetic field perpendicular to the page, or TM mode.

indicated by dots or crosses, while the lines of force of the electric field correspond to the pattern shown. As in the TE case, each pattern occupies one half of the effective wavelength $\frac{2\pi}{k_1}$, and the pattern travel along the guide with the phase velocity $v_p = \frac{\omega}{k_1}$.

We must note that the general solution of Maxwell's equation satisfying the boundary conditions of this problem is a linear combination of the TE and TM modes.

EXAMPLE 10.2: Consider a parallel plane waveguide with plate separation 20 cm with the TE_{10} mode excited at 1 GHz. Determine the propagation constant (γ), the cut off frequency (f_c) and the wavelength in guide (λ), assuming $(i) \varepsilon = \varepsilon_0$ (ii) $\varepsilon_r = 4$ for the medium of propagation in the guide.

Solution: We have for TE_{mo} mode,

$$\gamma_{mo} = \sqrt{\left(\frac{m\pi}{a}\right)^2 - \omega^2\mu\varepsilon}$$

and

$$\gamma_1 = \pi\sqrt{\frac{1}{a^2} - \left(\frac{2f}{v_0}\right)^2}$$

For $f = 1$ GHz, width $a = 20$ cm, we have

$$\gamma_{10} = \gamma\sqrt{\left(\frac{1}{0.3}\right)^2 - \left(\frac{2\times 10^9}{3\times 10^8}\right)^2} = 14 \text{ radians/m}$$

$$f_c = \frac{m}{2a\sqrt{\mu\varepsilon}} = \frac{m}{2a}v_0 \quad \therefore \quad v_0 = \frac{1}{\sqrt{\mu\varepsilon}} = \frac{3\times 10^8}{\sqrt{\varepsilon_r}} \text{ m/s}$$

(i) For TE_{10} mode $(m = 1, n = 0)$ and $\varepsilon_r = 1$, we have

$$f_c = \frac{3\times 10^8}{2a} = \frac{3\times 10^8}{0.4} = 0.75 \text{ GHz}$$

$$\lambda_{10} = \frac{\lambda_0}{\sqrt{1-\left(\frac{\lambda_0}{\lambda_c}\right)^2}}, \lambda_0 = \frac{v_0}{f} = \frac{3\times 10^8}{10^9} = 0.3 \text{ m}$$

$$\therefore \quad \lambda_c = \frac{v_0}{f_c} = \frac{2a}{m} = 0.4 \text{ m}$$

$$\therefore \quad \lambda_{10} = \frac{0.3}{\sqrt{1-\left(\frac{0.3}{0.4}\right)^2}} = 45.5 \text{ cm} \quad \text{and } \gamma = 14 \text{ radians/m}$$

(ii) For TE_{10} mode and $\varepsilon_r = 4$, we have

$$\gamma = \pi\sqrt{\left(\frac{1}{0.2}\right)^2 - \left(\frac{2\times 10^9}{3\times 10^8}\right)} \times 4 = 39.3 \text{ radians/m}$$

$$f_c = \frac{v_0}{2a} = \frac{3\times 10^8}{2\times 2\times 0.2} = 0.375 \text{ GHz}$$

$$\therefore \quad \lambda_{10} = \frac{\lambda_0}{\sqrt{1-\left(\frac{\lambda_0}{\lambda_c}\right)^2}}, \quad \lambda_0 = \frac{v_0}{f} = \frac{3\times 10^8}{2\times 10^9} = 0.15 \text{ m}$$

$$\therefore \quad \lambda_c = \frac{v_0}{f_c} = 2a = 2 \times 0.2 = 0.4 \text{ m}$$

$$\therefore \quad \lambda = \frac{0.15}{\sqrt{\left(1 - \frac{0.15}{0.4}\right)^2}} = 0.16 \text{ m}$$

EXAMPLE 10.3: Find the width of a rectangular guide such that the energy of the electromagnetic waves whose free space wavelength is 3 cm. travels down the guide at 95% of the speed of the light.

Solution: The group velocity is given by

$$v_g = c\sqrt{1 + \left(\frac{\lambda_0}{\lambda_c}\right)^2}$$

where

$$\lambda_c = \frac{2}{\sqrt{\left(\frac{m}{a}\right)^2 + \left(\frac{n}{b}\right)^2}}$$

where the indices m and n specify the mode or

$$(\lambda_c)_{01} = \frac{2}{\sqrt{\frac{1}{a^2} + 0}} = 2a$$

we have $v = 0.95c$ and $\lambda_0 = 3$ cm.

$$\therefore \quad 0.95c = c\sqrt{1 - \left(\frac{3}{2a}\right)^2}$$

This gives $a = 4.8$ cm.

EXAMPLE 10.4: What guide wavelength does 10 cm radiation (free space wavelength) exhibit in a rectangular guide whose width is 6 cm. Assuming the dominant mode, determine the cut off wavelength for guide?

Solution: We have

$$\lambda_c = \frac{2}{\left[\left(\frac{m}{a}\right)^2 + \left(\frac{n}{b}\right)^2\right]^{1/2}}$$

$$= \frac{2}{\left[\left(\frac{0}{a}\right)^2 + \left(\frac{1}{6}\right)^2\right]^{1/2}} = 12 \text{ cm}$$

we have

$$\frac{1}{\lambda_{g^2}} = \frac{1}{\lambda_f^2} - \frac{1}{\lambda_{c^2}} = \frac{1}{(10)^2} - \frac{1}{(12)^2}$$

$$\therefore \quad = \frac{120}{2\sqrt{11}} \approx 18 \text{ cm}$$

EXAMPLE 10.5: Radar waves of wavelength 3 cm is propagating in a waveguide of rectangular cross-section with $a = 1$ cm, $b = 2$ cm. Find the modes and group velocity of the waves.

Solution: We have the cut off frequency

$$(\omega_c)_{mn} = \pi c \left(\frac{m^2}{a^2} + \frac{n^2}{b^2}\right)^{1/2}$$

and

$$(\omega_c)_{mn} = 2\pi \nu_{mn}$$

We have the frequency of the given radiations = 10^{10} Hz. One obtains the cut-off frequencies for different modes as

$$\nu_{01} = \frac{c}{2} \times \frac{1}{2 \times 10^{-2}} = 7.5 \times 10^9 \text{ Hz}$$

$$\nu_{10} = \frac{c}{2} \times \frac{1}{10^{-2}} = 1.5 \times 10^{10} \, Hz$$

and

$$\nu_{11} = 1.68 \times 10^{10} \, Hz$$

Obviously, the lowest cut-off frequency for TM mode, i.e. TM_{11} mode is

$$\nu_{11} = 1.68 \times 10^{10} \, Hz$$

Since only TE_{01} mode can have a frequency lower than the frequency of radiation, only TE_{01} mode can be propagated.

Now, the phase velocity of the wave

$$v_p = \frac{\omega}{k_g}$$

$$k_g = \sqrt{k_{c^2} - \left(\frac{\omega}{c}\right)^2} = 1.4/cm$$

$$\therefore \quad v_p = 4.5 \times 10^8 \text{ m/s}$$

Now, we have

$$v_p v_g = c^2$$

$$\therefore \quad v_g = 2 \times 10^8 \text{ m/s}$$

QUESTIONS

1. What is a wave guide? For transverse electric waves perfectly propagating in a rectangular waveguide with perfectly conducting walls, find: (i) the cut off wavelength (ii) the magnetic field induction (iii) the velocity with which energy is transmitted along the guide.
2. Describe the propagation of electromagnetic waves along a hollow wave guide of uniform cross-section.
3. Explain the principle of operation of wave guides. Explain TE, TM and TEM mode waves. Derive an expression for the guide wavelength of TE mode propagation in hollow rectangular waveguide. Explain how the cut-off mode arise. Show that the phase velocity of waves becomes infinite exactly at cut-off.
4. Discuss the propagation of electromagnetic waves in a rectangular waveguide in TE mode. Show that there is a certain minimum frequency below which no transmission is possible and obtain an expression for that frequency.
5. What is a wave guide? Discuss the theory of waveguides with reference to TM mode of propagation of electromagnetic waves. Show that a wave guide acts as a high pass filters.
6. What are cavity resonators? How do they resonate? How they resemble with electronic resonant circuits? Discuss in detail the theory of a rectangular cavity resonator, showing what lowest modes are possible in TE mode?
7. What is an electromagnetic cavity resonator? Describe the excitation of TEM waves in the class of cavity resonators produced by placing plane end faces on a length of a cylindrical wave guide.

PROBLEMS

1. A wave guide consists of a long tube of rectangular cross-section with sides a and b. Show that the resultant wave is described by $\xi = 4\xi_0 \sin k_2 y \sin k_3 z \cos(\omega t - k_1 x)$ and that the only frequencies transmitted along the wave guide are those satisfying $v \geq \frac{1}{2} v \sqrt{\frac{n_1^2}{a^2} + \frac{n_2^2}{b^2}}$, where n_1 and n_2 are integers. Discuss the nodal planes in the wave guide for $n_1 = 2$ and $n_2 = 3$.

2. If waves are propagating along the z-axis in a rectangular wave guide, show that the field component for the TE mode is

$$E_x = E_0 \sin\left(\frac{\pi y}{b}\right) \exp[i(\omega t - k_g z)].$$

The width of the waveguide along the y-axis is b. Also show that the E_y component of the wave is zero.

3. Show that in a rectangular waveguide TE_{00} mode cannot exist.

4. A rectangular wave guide has dimensions 2.5 cm and 5 cm. Determine guide wavelength λ_g, phase velocity and phase constant at a wavelength of 4.5 cm for dominant mode.

[**Ans.** $\lambda_g = 5$ cm, kg = 1.256/cm and $v_g = 3.3 \times 10^8$ m/s]

5. A hollow rectangular waveguide has $a = 6$ cm and $b = 4$ cm. The impressed signal has the frequency = 3 GHz. Determine TE_{10} mode and (i) cutoff wavelength (ii) guide wavelength (iii) phase constant (iv) intrinsic wave impedence (v) phase velocity and (vi) group velocity.

[**Ans.** $(\lambda_c)_{10} = 12$ cm, (ii) $\lambda_g = 18.2$ cm, (iii) kg = 0.345/cm (iv) intrinsic wave impedance $Z_E = 685\ \Omega$ (v) $v_g = 5.5 \times 10^8$ m/s and (vi) $v_z = 1.65 \times 10^8$ m/s]

6. In a rectangular cavity with square cross-section a and height b, a thin slab of dielectric with thickness d has been placed. Show that $\frac{\omega - \omega_0}{\omega_0} = -\frac{1}{2}(\varepsilon_r - 1)\frac{b}{a}$.

SHORT ANSWER QUESTIONS

1. Whether the assumption that a TEM waves can be propagated within the interior of a hollow metal pipe is in accordance with the electromagnetic theory?

   [**Ans.** No, this assumption leads to contradiction with electromagnetic theory. This shows that there can be no TEM waves within a waveguide.]

2. What is a hohlraum resonator?

   [**Ans.** If the wave guide is limited by closing the ends of a finite length of the tube in conducting surfaces then it is known as cavity or hohlraum resonator.]

3. What will be the shortest length of a simplest cavity resonator to be made from a rectangular waveguide with $a = 10.16$ mm and $b = 22.86$ mm which resonate at 10 GHz.

   [**Ans.** $\lambda_g = \dfrac{\lambda}{\left[1-\left(\frac{\lambda}{2b}\right)^2\right]^{1/2}}, \ \lambda = \dfrac{3\times 10^8}{10^{10}} = 0.03$ m

   $\therefore \ \lambda_g = 0.0396$ m. Obviously the length of the shortest cavity is 0.0198 m.]

4. In what respects cavity resonators are superior to conventional L-C circuits?

   [**Ans.** (i) The fraction of the stored energy dissipated per cycle in guide is about 1/20 the fraction dissipated per cycle in L-C circuit. (ii) Guide have resonant frequencies which range upward from a few hundred megacycles which is almost impossible in case of L-C circuit.]

5. What is a TE mode?

   [**Ans.** In transverse electric (TE) modes the electric vector of the field has no component in the direction of propagation.]

6. What is TM mode?

   [**Ans.** In transverse magnetic (TM) modes, the magnetic vector has no such component.]

7. What happens when electromagnetic waves propagates through a conductive medium e.g. metal?

   [**Ans.** The conducting medium intensively absorbs electromagnetic waves. The most prominent role in this process is played by the conduction electrons.]

8. When the wavelength of the emitter is too large, what will be the relation between the fields of the wave guide walls and the field of the incoming wave?

   [**Ans.** Both will be in antiphase. The resultant field is damped at the distance of wavelength, and the wave does not penetrate to any appreciable depth into the waveguide.]

9. What are the typical properties of electronic resonant circuits which cavity resonators show?

   [**Ans.** (i) they can store energy in oscillating electric and magnetic fields and (ii) dissipate a fraction of the stored energy in each cycle of oscillation.]

10. What constitute a complete set of fields to describe an arbitrary electromagetic disturbance in a waveguide or cavity?

    [**Ans.** The various TM and TE waves, plus a TEM wave if it can exist, consistute a complete set of fields to describe an arbitrary electromagnetic disturbance in a waveguide or cavity.]

11. The wave number k_λ is always less than the free-space value $\sqrt{\mu\varepsilon}\,\omega$, what does it indicate?

    [**Ans.** The wave length in the guide is always greater than the free-space wavelength.]

12. What will be the phase velocity at cut off. [**Ans.** Infinite.]

13. What do understand by obstacles in waveguides?

    [**Ans.** Discontinuities in the form of obstacles, dielectric slabs, diaphragms, and apertures in walls occur in the practical use of waveguides as carriers of electromagnetic energy and phase information in microwave systems.]

OBJECTIVE QUESTIONS

1. The cut-off frequency corresponding to the mode numbers m, n for rectangular cross-section of dimensions a, b wave guide is given by

   (a) $\omega_{mn} = \pi c\sqrt{\frac{m^2}{a^2}+\frac{n^2}{b^2}}$, (b) $\omega_{mn} = \left(\frac{m^2}{a^2}+\frac{n^2}{b^2}\right)$,

   (c) $\omega_{mn} = \pi\sqrt{\frac{m}{a}+\frac{n}{b}}$, (d) $\omega_{mn} = \pi c\left(\frac{m^2}{a^2}+\frac{n^2}{b^2}\right)$. [a]

2. The principal or dominant mode cut-off frequency for rectangular cross section having dimensions a, b wave guide is given by

   (a) $\omega_{mn} = \pi c\sqrt{\frac{m^2}{a^2} + \frac{n^2}{b^2}}$, (b) $\omega_{01} = \frac{\pi c}{b}$,

   (c) $\omega_{01} = \frac{\pi c}{a}$, (d) $\omega_{01} = \pi c$. **[b]**

3. Which one of the following statement is true for a coaxial wave guide

   (a) Guide wavelength is equal to the free space wavelength.
   (b) Propagation constant depend on the dimension of the wave guide.
   (c) There is a cut-off wavelength.
   (d) The propagation is not similar to that of a plane wave. **[a]**

4. Electromagnetic cavity resonators are energy storage devices a: d are used in:

   (a) Klystron, (b) Wave meters,
   (c) Band pass filter, (d) All of the above. **[d]**

5. The cavity resonators used in lasers are called

   (a) Open resonators, (b) Closed resonators,
   (c) Raman resonators, (d) None of the above. **[a]**

6. The quality factor (Q) of a cubic cavities having $a = b = c$ is

   (a) $Q = \frac{1.07 \times 10^9}{\sqrt{f}}$, (b) $Q = \frac{1}{\sqrt{f}}$,

   (c) $Q = \frac{1.07 \times 10^9}{f}$, (d) $Q = 0$. **[a]**

7. Cavity resonators are characterized by their

   (a) resonant frequencies only.
   (b) quality factors only.
   (c) resonant frequencies and quality factors.
   (d) none of the above. **[c]**

8. The critical angle for given inner and outer dielectrics in a dielectric wave guide with μ_2, ε_2 and μ_1, ε_1 are relative permeabilities and dielectric constants of outer and inner media respectively is given by

(a) $\theta_c = \cos^{-1}\sqrt{\dfrac{\mu_2\varepsilon_2}{\mu_1\varepsilon_1}}$, (b) $\theta_c = \sin^{-1}\sqrt{\dfrac{\mu_2\varepsilon_2}{\mu_1\varepsilon_1}}$,

(c) $\theta_c = \tan^{-1}\sqrt{\dfrac{\mu_2\varepsilon_2}{\mu_1\varepsilon_1}}$, (d) $\theta_c = \sin^{-1}\sqrt{\dfrac{\mu_1\varepsilon_1}{\mu_2\varepsilon_2}}$ **[b]**

Chapter 11

Electromagnetic Radiation and Radiating Systems

11.1 INTRODUCTION

We have discussed various features, e.g. properties of electromagnetic waves and their propagation in both bounded and unbounded geometries in previous chapters, but very little was said about generation of such waves. In the present chapter, we shall discuss "radiation", i.e. the unguided transport of waves and energy through empty space and also the emission of radiation by localized systems of oscillating charge and current densities. We know that visible light is also a member of electromagnetic radiations. The question arises, how visible light is produced. It is now clear that the visible light results from the rapid adjustment of the charge and current distribution within the electron clouds of atoms. We must note that the emission of radiation by atoms can only be described by quantum mechanics. We know that the breakdown of classical concepts first became evident only in the theory of radiation. However, the classical theory of radiation discussed in this chapter is of immense help in understanding the theory of radiation.

11.2 OSCILLATING CHARGE: A DIPOLE

One can visualize a charge constrained to move back and forth to form an oscillating electric *doublet* or an *electric dipole*. One can easily see that an oscillating charged particle at a given place is equivalent to an oscillating electric dipole. The harmonically oscillating charge gives rise to a harmonic current. The charged particle velocity thus undergoes a harmonic change in their acceleration which radiates electromagnetic energy in space. Obviously the number of charged particles constituting the magnitude of current thus governs the near and the far electric and magnetic field vectors of the radiated electromagnetic wave. One can easily see that in a complete cycle of oscillating electric dipole, the electric and magnetic field lines are released from the dipole and advance in free-space with the velocity of light. The cumulative forces of repulsion between the field lines radiate the electromagnetic energy at the velocity of light in free-space.

11.3 RETARDED POTENTIALS

One can easily find the relation of radiation fields to their sources if we express the fields in terms of electromagnetic potentials **A** and ϕ, i.e.

$$\mathbf{E} = -\nabla\phi - \frac{\partial \mathbf{A}}{\partial t} \quad \text{and} \quad \mathbf{B} = \nabla \times \mathbf{A} \tag{11.1}$$

We have seen that the introduction of Lorentz condition in Maxwell's equations yields the following two inhomogeneous equations for **A** and ϕ,

$$\nabla^2\phi - \frac{1}{c^2}\frac{\partial^2\phi}{\partial t^2} = -\frac{\rho}{\varepsilon_0} \tag{11.2}$$

and

$$\nabla^2\mathbf{A} - \frac{1}{c^2}\frac{\partial^2\mathbf{A}}{\partial t^2} = -\mu_0\mathbf{J} \tag{11.3}$$

We can see that the operator on the left-hand side is identical with that of the homogeneous wave equation, but the source terms **J** and ρ are appearing now on the right hand side. Now, the problem is to solve Eqs. (11.2) and (11.3) to obtain the expression for **A** and ϕ in terms of charge and current distribution.

One can obtain the solutions of (11.2) and (11.3) by analogy with solutions obtained for steady state problems in previous chapters. We have the solutions of the equations

$$\nabla^2\phi = -\frac{\rho}{\varepsilon_0} \quad \text{and} \quad \nabla^2\mathbf{A} = -\mu_0\mathbf{J}$$

as

$$\phi(\mathbf{r}) = \frac{1}{4\pi\varepsilon_0}\int_{\mathrm{v}} \frac{\rho(\mathbf{r})}{|\mathbf{r}-\mathbf{r}'|}\,d\tau$$

and

$$A(\mathbf{r}) = \frac{\mu_0}{4\pi}\int_{\mathrm{v}} \frac{\mathbf{J}(\mathbf{r})}{|\mathbf{r}-\mathbf{r}'|}\,d\tau \tag{11.4}$$

where $d\tau$ is the small volume element (Fig. 11.1). One has to compute the potentials at a point P whose position vector is $\mathbf{r}$. One can do this by integrating ρ and $\mathbf{J}$ throughout $\mathbf{v}$ by considering them as functions of $\mathbf{r}'$, the position vector of the volume element $d\tau$. We know that both the expressions of (11.4) are valid if the charges are at rest and the currents

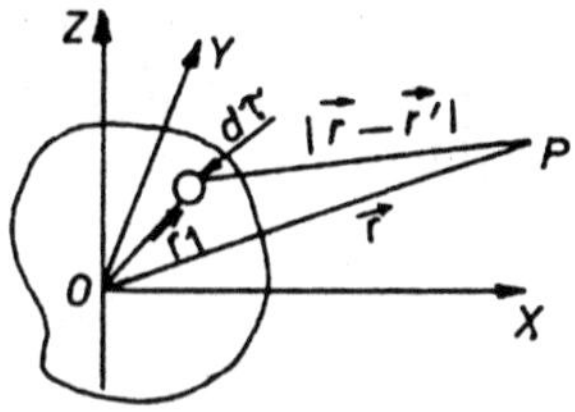

Figure 11.1.

are steady. Let us now consider a situation in which ρ changes with time t. Now, the question arises, what is the potential at P at time t? Let us consider that the charge distribution within the volume element $d\tau$ changes with time. This means the field observed at P at time t must have been "launched" by the element $d\tau$ at a time t' earlier than t. The electric fields associated with these charges propagate with the finite velocity c. Obviously, the time taken to travel the distance from $d\tau$ to P is $= \dfrac{|\mathbf{r}-\mathbf{r}'|}{c}$. This means that the contribution to the potential at the point P at a time t, due to the charge in $d\tau$ does not depend on the charge at time t but depends upon the charge at the time $t - \dfrac{|\mathbf{r}-\mathbf{r}'|}{t}$. We can see that this is the time at which the electric fields must have been propagated from the charges in $d\tau$ at $\mathbf{r}'$ in order to arrive at the time t. Thus we have

$$\phi(\mathbf{r},t) = \frac{1}{4\pi\varepsilon_0}\int_{\mathbf{v}} \frac{\rho(\mathbf{r}',t-\frac{|\mathbf{r}-\mathbf{r}'|}{c})}{|\mathbf{r}-\mathbf{r}'|}\,d\tau \qquad (11.5)$$

and

$$A(\mathbf{r},t) = \frac{\mu_0}{4\pi}\int_{\mathbf{v}} \frac{\mathbf{J}(\mathbf{r}',t-\frac{|\mathbf{r}-\mathbf{r}'|}{c})}{|\mathbf{r}-\mathbf{r}'|}\,d\tau \qquad (11.6)$$

We have obtained (11.5) and (11.6) on physical grounds. Now, the question arises, whether (11.5) and (11.6) represents the solutions of the inhomogeneous equations (11.4). To check this, let us substitute (11.5) and (11.6) in (11.4). Now, one can write

$$\phi(\mathbf{r},t) = \frac{1}{4\pi\varepsilon_0}\int_{\mathbf{v}} \frac{\rho(\mathbf{r}',t-\frac{d}{c})}{d}\,d\tau \qquad (11.7)$$

where $d = |\mathbf{r}-\mathbf{r}'|$. We can see that it is difficult to evaluate the potential at $d = 0$. To evaluate the integration in (11.7), one can divide the volume of integration into two regions, viz. (i) a small sphere of radius r_0 around the point at which ϕ is to be determined and (ii) rest of space region. One can express the potential as

$$\phi = \phi_1 + \phi_2 \tag{11.8}$$

In the case of the region of the small sphere, $\frac{d}{c} = \frac{|\mathbf{r}-\mathbf{r}'|}{c}$ is negligible and hence one finds

$$\nabla^2\phi_1 = \frac{1}{4\pi\varepsilon_0}\nabla^2 \int_v \frac{\rho(\mathbf{r},t)}{d}\,d\tau = \frac{\rho}{4\pi\varepsilon_0}\int_v \nabla^2\left(\frac{1}{d}\right) d\tau$$

we have already obtained previously that

$$\nabla^2\phi_1 = \frac{\rho}{4\pi\varepsilon_0}\int_v \nabla^2\left(\frac{1}{d}\right) d\tau = -\frac{\rho}{\varepsilon_0} \tag{11.9}$$

We find that (11.9) is independent of r_0 and, therefore, we can have the sphere as small as possible shrinking into a point.

We now determine $\nabla^2\phi_2$. One can express $\nabla^2\phi_2$ in spherical polar coordinates as

$$\nabla^2\phi_2 = \frac{1}{d}\frac{\partial^2}{\partial d^2}(d\phi_2) \tag{11.10}$$

$$= \frac{1}{4\pi\varepsilon_0}\int_v \frac{1}{d}\frac{\partial^2}{\partial d^2}\rho\left(\mathbf{r}',t-\frac{d}{c}\right) d\tau \tag{11.11}$$

Let us consider any function $f\left(\frac{d}{c}\right)$, we have

$$\frac{\partial^2 f}{\partial d^2} = \frac{1}{c^2}\frac{\partial^2 f}{\partial t^2} \tag{11.12}$$

$$\therefore \quad \nabla^2\phi_2 = \frac{1}{4\pi\varepsilon_0}\int_v \frac{1}{dc^2}\frac{\partial^2}{\partial t^2}\rho\left(\mathbf{r}',t-\frac{d}{c}\right) d\tau$$

$$= \frac{1}{c^2}\frac{\partial^2}{\partial t^2}\left[\frac{1}{4\pi\varepsilon_0}\int_v \frac{\rho\left(\mathbf{r}',t-\frac{d}{c}\right)}{d}\,d\tau\right]$$

$$= \frac{1}{c^2}\frac{\partial^2\phi}{\partial t^2} \tag{11.13}$$

$$\therefore \quad \nabla^2\phi = \nabla^2\phi_1 + \nabla^2\phi_2 = -\frac{\rho}{\varepsilon_0} + \frac{1}{c}\frac{\partial^2\phi}{\partial t^2}$$

or

$$\nabla^2\phi - \frac{1}{c^2}\frac{\partial^2\phi}{\partial t^2} = -\frac{\rho}{\varepsilon_0}$$

This shows that (11.5) is a solution of (11.2). Potentials represented by (11.5) and (11.6) are called *retarded potentials*. We must note that one will have to evaluate the integrands $\rho(\mathbf{r}')$ and $\mathbf{J}(\mathbf{r}')$ at the retarded time $t - \frac{|\mathbf{r}-\mathbf{r}'|}{c}$, where $\mathbf{r}'$ is the radius vector at the retarded time. Once the potentials are known, one can find fields **E** and **B**. However the computation of **A** and ϕ is not as simple as it appears. Difficulties arise because retarded times are different for different parts of the sources. One can determine the fields in some certain simple situations with the help of some sensible approximations. The retarted potentials have been shown to be the solution of inhomogeneous wave equation and are very useful in calculating radiated field vectors.

11.4 RADIATION FROM AN ELECTRIC DIPOLE

An electric dipole consists of two equal and opposite charges separated by a finite distance. Now, we calculate the radiation field produced by an electric oscillating dipole. One finds many important practical applications. One may consider many practical radiating systems to be made up by putting together a large number of such dipoles. A short electric dipole fed with a sinusoidal current source radiates electromagnetic energy in space.

Let us consider two small spheres at the two ends of a short wire. Supposing that the charge is transferred periodically from one sphere to the other, and the time variation being harmonic, i.e.

$$q = q_0 \exp(i\omega t) \tag{11.14}$$

where ω is the angular frequency of the oscillation and q_0 is the amplitude of the oscillating charge. Assuming that the wavelength of the radiation produced is large compared with the length of the wire l, we have

$$\lambda = \frac{2\pi c}{\omega} \gg l \quad \text{or} \quad \frac{2\pi}{\omega} = T \gg \frac{l}{c} \tag{11.15}$$

Obviously, $\frac{l}{c}$ is the time taken for a signal to propagate along the wire, from one end to the other. We can easily see that this is very much less than that over which the source current changes appreciably. This means, one can easily consider the current I to be the same at all points along its length. Thus, we have

$$I = \frac{dq}{dt} = i\omega q_0 \exp(i\omega t) \tag{11.16}$$

We know that the charge oscillating between two spheres is equivalent to an oscillating electric dipole moment **p**. We have

$$|\mathbf{p}| = ql = q_0 l \exp(i\omega t) = p_0 \exp(i\omega t) = \frac{Il}{i\omega}$$

Here $$p_0 = q_0 l \tag{11.17}$$

We are interested in studying the distribution of the radiation field of this dipole throughout space and total power radiated by it.

Let us consider the dipole shown in Fig. 11.2. We can see from the figure that the wire lies along the z-axis. The origin of the coordinate system coincides with the centre of the wire. Let us find the field values at the point *P* specified by the position vector **r**. To determine the field, we need to know the retarded potentials. We have the vector potential,

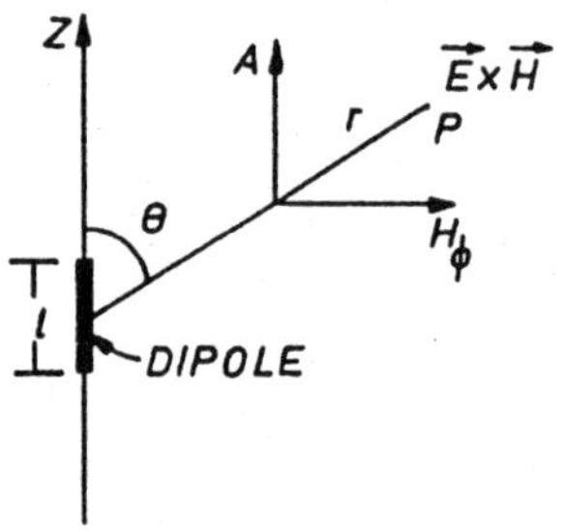

Figure 11.2. An oscillating dipole

$$\mathbf{A}(\mathbf{r},t') = \frac{\mu_0}{4\pi}\int_V \frac{J\left(\mathbf{r}',t - \frac{\mathbf{r}-\mathbf{r}'}{C}\right)}{|\mathbf{r}-\mathbf{r}'|}\, d\tau$$

The integration in the above expression is over the volume occupied by the current. We have

$$\mathbf{J}\left(\mathbf{r}',t - \frac{\mathbf{r}-\mathbf{r}'}{c}\right) d\tau = I\left(\mathbf{r}',t - \frac{\mathbf{r}-\mathbf{r}'}{c}\right) \hat{a}\, dz.$$

The current is always along the *z*-axis, and hence one may replace $\mathbf{r}'$ by *z*. Thus

$$A(\mathbf{r},t) = \frac{\mu_0}{4\pi}\int_{-\frac{l}{2}}^{\frac{l}{2}} \frac{I\left(z,t - \frac{\mathbf{r}-\hat{a}_z z}{C}\right)}{|\mathbf{r}-\hat{a}_z z|}\, a_z\, dz \tag{11.18}$$

Since $r \gg l$, one may neglect $\hat{a}_z z$ with respect to r. Now, we replace the denominator in (11.18) as r and $t - \dfrac{|\mathbf{r}-\hat{a}_z z|}{c}$ at which the current density in the wire is to be measured by the time $t - \frac{r}{c}$. One obtains,

$$A(\mathbf{r},t) = \frac{\mu_0}{4\pi}\int_{-\frac{l}{2}}^{\frac{l}{2}} \frac{I\left(z,t-\frac{r}{c}\right)}{r}\hat{a}_z dz$$

$$= \frac{\mu_0}{4\pi}\hat{a}_z l\frac{I\left(t-\frac{r}{c}\right)}{r} \tag{11.19}$$

as the current is same at all points along the wire. Eq. (11.19) reveals that vector potential **A** is everywhere parallel to l as shown in Fig. 11.2. One obtains the components of the vector potential

$$A_x(\mathbf{r},t) = 0, \qquad A_y(\mathbf{r},t) = 0$$

and

$$A_z(\mathbf{r},t) = \frac{\mu_0 l}{4\pi r} i\omega q_0 \exp\left\{i\omega\left(t-\frac{r}{c}\right)\right\}$$

$$= \frac{\mu_0}{4\pi r} i\omega p_0 \exp\left\{(i\omega t)\exp\left(\frac{-i\omega r}{c}\right)\right\}$$

$$= \frac{\mu_0}{4\pi r} i\omega p(t)\exp(-ikr) = \frac{\mu_0}{4\pi}p(t)\frac{e^{-ikr}}{r} \tag{11.20}$$

Scalar Potential

The Lorentz condition for free space is

$$\nabla\cdot\mathbf{A} + \frac{1}{c^2}\frac{\partial\phi}{\partial t} = 0 \tag{11.21}$$

Since only the z-component of A is non-zero, we have

$$\frac{\partial A_z}{\partial z} = -\frac{1}{c^2}\frac{\partial\phi}{\partial t}$$

$$\therefore \quad -\frac{1}{c^2}\frac{\partial\phi}{\partial t} = \frac{\mu_0}{4\pi}p(t)\frac{\partial}{\partial r}\left(\frac{e^{-ikr}}{r}\cos\theta\right) \quad \left(\because \frac{\partial r}{\partial z} = \cos\theta\right)$$

$$\frac{\partial\phi}{\partial t} = -\frac{1}{4\pi\varepsilon_0}p(t)\cos\theta\frac{\partial}{\partial r}\left(\frac{e^{-ikr}}{r}\right)$$

This gives

$$\phi = -\frac{1}{4\pi\varepsilon_0}p(t)\cos\theta\left(\frac{e^{-ikr}}{r}\right) \tag{11.22}$$

One can also express (11.22) as

$$\phi = \frac{l}{4\pi\varepsilon_0}\left[\frac{z}{r^3}q\left(t-\frac{r}{c}\right) + \frac{z}{cr^2}I\left(t-\frac{r}{c}\right)\right] \tag{11.23}$$

Obviously, we have determined **A** and ϕ and we can now determine **E** and **B**.

Electric Field Strength E

One can express the electric field strength **E** as

$$E = -\nabla\phi - \frac{\partial \mathbf{A}}{\partial t} \tag{11.24}$$

Now, we have the components of **A** in spherical polar coordinates as

$$\begin{aligned} A_r &= \frac{\mu_0}{4\pi r} i\omega p_0 \cos\theta \exp\left[i\omega\left(t - \frac{r}{c}\right)\right] \\ &= \frac{\mu_0}{4\pi r} p \cos\theta e^{-ikr} \end{aligned} \tag{11.25}$$

$$\begin{aligned} A_\theta &= \frac{\mu_0}{4\pi r} i\omega p_0 \sin\theta \exp\left[i\omega\left(t - \frac{r}{c}\right)\right] \\ &= \frac{\mu_0}{4\pi r} p \sin\theta e^{-ikr} \end{aligned} \tag{11.26}$$

and

$$A_\theta = 0 \tag{11.27}$$

Using (11.25) to (11.27), one obtains the components of **E** as

$$E_r = -\frac{\partial\phi}{\partial r} - \frac{\partial A_r}{\partial t} = \frac{ipk}{2\pi\varepsilon_0 r} \cos\theta \left[1 - \frac{i}{kr}\right] \frac{e^{-ikr}}{r} \tag{11.28}$$

$$\begin{aligned} E_\theta &= -\frac{1}{r}\frac{\partial\phi}{\partial\theta} - \frac{\partial A_\theta}{\partial t} \\ &= \frac{pk^2}{4\pi\varepsilon_0} \sin\theta \left[1 - \frac{i}{kr}\left(1 - \frac{i}{kr}\right)\right] \frac{e^{-ikr}}{r} \end{aligned} \tag{11.29}$$

and $$E_\phi = 0 \tag{11.30}$$

Magnetic Field Strength H

Using $\mathbf{B} = \nabla \times \mathbf{A}$, and (11.25) to (11.27), one obtains

$$H_r = 0,\ H_\theta = 0 \quad \text{and} \quad H_\phi = -\frac{cpk^2}{4\pi} \sin\theta \left[1 - \frac{i}{kr}\right] \frac{e^{-ikr}}{r} \tag{11.31}$$

The relations (11.28) to (11.31) are known as the *Hertz's relations* for the oscillating dipole. From (11.31) it is obvious that:

(i) The magnetic induction (B) varies as $\sin\theta$. This indicates that the magnetic induction **B** is minimum along the axis of the dipole and maximum along the equatorial plane.

(ii) The magnetic induction B_ϕ contains two terms, one varying as r^{-2} and other varying as r^{-1}. The term varying as r^{-1} dominates for large values of r and obviously, this term is responsible for radiation field. The term varying as r^{-2} dominates for small values of r and hence this term is responsible for induction field.

One can imagine that the space is divided into two regions:

(i) $|r| \ll \lambda$ This region is called as the *near zone* region. One finds that the field in this case is exactly the same as that for an electrostatic field.

(ii) $|r| \gg \lambda$. This region is called the *radiation zone*. One can see that fields in this region varies as r^{-1}.

We must note that the region were $kr \simeq 1$ is quite complicated.

We now examine the relations (11.28) to (11.31) in the near zone and the radiation zone.

(i) *Near Zone region*: $|r| \ll \lambda$, i.e. $kr \ll 1$, We have

$$E_r \simeq \frac{p\cos\theta}{2\pi\varepsilon_0 r^3},\ E_\theta \simeq \frac{p\sin\theta}{4\pi\varepsilon_0 r^3} \quad \text{and} \quad E_\phi = 0$$

$$H_r = 0,\ H_\theta = 0 \quad \text{and} \quad H_\phi = \frac{i\omega p\sin\theta}{4\pi r^2} \tag{11.32}$$

One finds that expressions (11.32) are equivalent to the field of an electrostatic dipole. We obtain

$$\frac{\omega\mu_0}{k}\frac{|H|}{|E|} \simeq \frac{\omega\mu_0}{k}\varepsilon_0\omega r = kr \ll 1$$

Obviously, the electric field dominates in this zone.

(ii) *Radiation Zone*: $kr \gg 1$, we have

$$E_r = 0,\ E_\theta = -\frac{pk^2}{4\pi\varepsilon_0}\sin\theta\frac{e^{-ikr}}{r},\ E_\phi = 0$$

$$H_r = 0,\ H_\theta = 0,\ H_\phi = -\frac{cpk^2}{4\pi}\sin\theta\frac{e^{-ikr}}{r} \tag{11.33}$$

While obtaining these results, we have neglected the terms involving $\frac{1}{r^2}$ and its higher orders. One obtains from (11.33), the two components E_θ and H_ϕ are in time phase in the far field approximation.

$$E_\theta = \sqrt{\frac{\mu_0}{\varepsilon_0}} H_\phi \tag{11.34}$$

The intrinsic impedance of free space can thus be defined as

$$Z = \frac{E_\theta}{H_\phi} = \frac{1}{\varepsilon_0 c} = 376.7\Omega \tag{11.35}$$

Fig. 11.3 shows the directions of the fields. We can see that the fields are mutually perpendicular. For the radiation zone $E_r = 0$ and we have the field distribution in spherical with the electric field as circles of longitude and the magnetic field as the circles of latitude. One obtains that the field is maximum at the equator and zero at the poles. Since the magnetic field is transverse everywhere, the radiation from an oscillating electric dipole is generally TM. We can see that at large distances $E_r \to 0$ and the electric field also is transverse. Obviously, the waves are TEM.

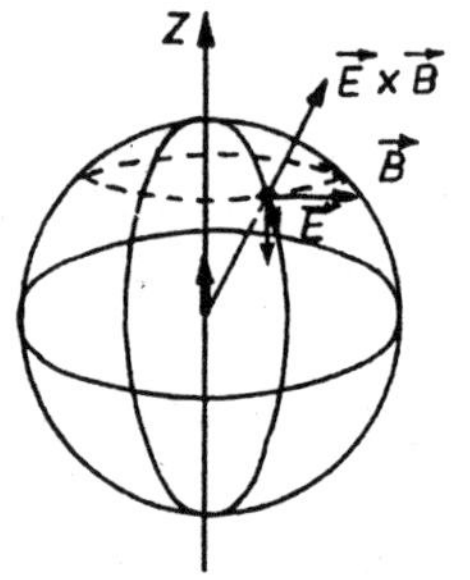

Figure 11.3. Directions of the fields.

Quasi-Stationary Field

The low frequency radiation pattern has shown special features and is generally referred to as quasi-stationary field.

For very low frequency radiation $\omega \to 0$, the field components are non-zero in the vicinity of the dipole and decrease rather fast with increasing distance.

Poynting vector $\langle N \rangle$ and Radiated Power (W)

The energy flow takes place due to radiation fields. We have the time averaged Poynting vector $\langle N \rangle$ as

$$\langle N \rangle = \frac{1}{2} R_e(\mathbf{E} \times \mathbf{H}^*)$$

Using the values obtained for radiation zone in (11.33), one obtains

$$\langle N \rangle = \frac{cp_0^2 k^4}{32\pi^2 \varepsilon_0} \frac{\sin^2\theta}{r^2} \hat{a}_n \tag{11.35}$$

Expression (11.35) represents the average energy radiated per second per unit area in a normal direction by an oscillating dipole.

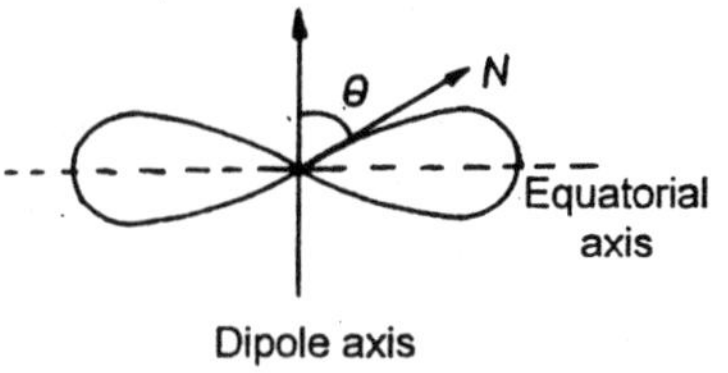

Figure 11.4.

From (11.35) it is obvious that the time averaged Poynting vector varies as $\sin^2\theta$. This means that the flow of energy is zero along the dipole axis and maximum along the equatorial axis and shown in Fig. 11.4

The total radiated power is given by

$$\begin{aligned} W &= \int_0^{2\pi} \int_0^{\pi} \frac{cp_0 k^4}{32\pi^2 \varepsilon_0} \frac{\sin^2\theta}{r^2} r^2 \sin\theta \, d\theta \, d\phi \\ &= \frac{cp_0^2 k^4}{32\pi^2 \varepsilon_0} 2\pi \frac{4}{3} = \frac{cp_0^2 k^4}{12\pi\varepsilon_0} = \frac{cp_0^2}{12\pi\varepsilon_0} \left(\frac{2\pi}{\lambda}\right)^4 \end{aligned} \tag{11.36}$$

From (11.36) it is clear that the total power radiated by an oscillating dipole is inversely proportional to the fourth power of the wavelength λ. Examples of electric dipole radiators are the emission of radiation by an excited atom and electromagnetic waves transmitted by an antenna.

Radiation Resistance

An ideal antenna should radiate all the power delivered to it. When a current $I = I_0 \exp(i\omega t)$ passes through a circuit containing a resistance R, the average rate of dissipation of energy is written as

$$\text{Rate of energy dissipated} = \frac{1}{2} I_0^2 R \tag{11.37}$$

where I_0 is the amplitude of the current. We can express the expression for power dissipated in the case of dipole, i.e. Eq. (11.36) as

$$\begin{aligned} \frac{cp_0^2}{12\pi\varepsilon_0} \left(\frac{2\pi}{\pi}\right)^4 &= \frac{p_0^2 \omega^4}{12\pi\varepsilon_0 c^3} = \frac{1}{12\pi\varepsilon_0} \frac{I_0^2 l^2 \omega^2}{c^3} \\ &= \frac{1}{12\pi\varepsilon_0} \frac{4\pi l^2}{\lambda^2} \cdot \frac{1}{c} I_0^2 \end{aligned}$$

$$= \frac{1}{2}\left(\frac{2\pi}{3}\right)\sqrt{\frac{\mu_0}{\varepsilon_0}}\left(\frac{l}{\lambda}\right)^2 I_0^2 \tag{11.38}$$

Comparing (11.38) with (11.37), one obtains the radiation resistance as

$$R_r = \frac{2\pi}{3}\sqrt{\frac{\mu_0}{\varepsilon_0}}\left(\frac{l}{\lambda}\right)^2 \tag{11.39}$$

It is obvious from (11.39) that the radiation resistance of an electric dipole depends on its length as compared to the wavelength of radiation. We must note that (11.39) is valid provided $l \ll \lambda$.

The radiation resistance governs the radiation efficiency of antennas. If a part of the power delivered to the antenna is lost, the corresponding loss resistance is denoted by R_{loss}. One can define the antenna efficiency as

$$\eta = \frac{\text{Power radiated}}{\text{Power input}} = \frac{R_r}{R_r + R_{\text{loss}}} \tag{11.40}$$

Minimizing the radiation losses enhances the radiation efficiency.

11.5 ELECTRIC QUADRUPOLE RADIATION

A quadrupole is a combination of two electric dipoles. Let us imagine two electric dipoles, each of amplitude p, aligned along a given straight line and oscillating exactly out of phase so that there is zero net dipole moment. This system possess quadrupole moment which varies with time and hence it will emit radiation, called quadrupole radiation. One can derive the quadrupole field by considering it to be the superposition of two dipole fields. Fig. 11.5 Shows the quadrupole.

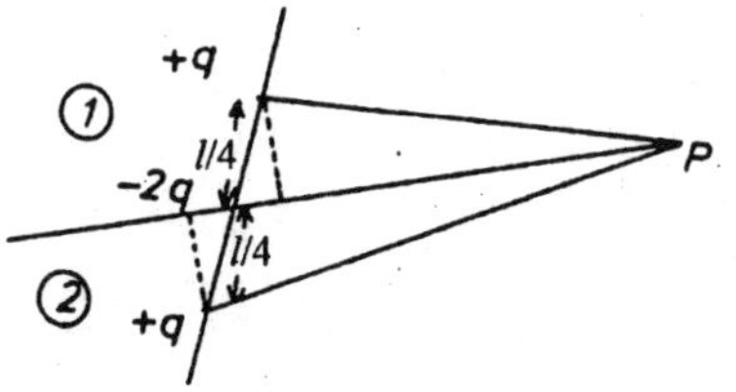

Figure 11.5.

The electric field due to a single dipole located at the origin, at a point P in the radiation zone is

$$E_\theta = -\frac{pk^2}{4\pi\varepsilon_0}\sin\theta\frac{e^{-ikr}}{r} = -\frac{k^2 p_0}{4\pi\varepsilon_0 r}e^{i\omega t'}\sin\theta \tag{11.41}$$

The dipoles are at distances $\frac{l}{4}$ and $-\frac{l}{4}$ from the origin (Fig. 11.5). From figure it is clear that the dipole (1) is close to the point P than is the origin by approximately a distance $\left(\frac{l}{4}\right)\cos\theta$. Obviously, dipole (2) is farther by the same amount. One finds the phases of two dipoles relative to origin and therefore $\frac{kl}{4}\cos\theta$ and $-\frac{kl}{4}\cos\theta-\pi$, where $k=\frac{\omega}{c}$ and $-\pi$ is introduced because two dipoles are oscillating out of phase. Therefore the field component of the quadrupole is

$$
\begin{aligned}
E_\theta &= -\frac{k^2 p_0}{4\pi\varepsilon_0 r}\sin\theta e^{i\left(\omega t'-\frac{kl}{4}\cos\theta\right)}-\frac{k^2 p_0}{4\pi\varepsilon_0 r}\sin\theta e^{i\left\{\omega t'-\left(-\frac{kl}{4}\cos\theta-\pi\right)\right\}} \\
&= -\frac{k^2 p_0\sin\theta}{4\pi\varepsilon_0 r}\left[\exp\left(-i\frac{kl}{4}\cos\theta\right)-\exp\left(i\frac{kl}{4}\cos\theta\right)\right]\exp(i\omega t) \\
&= \frac{k^2 p_0\sin\theta}{4\pi\varepsilon_0 r}2i\sin\left(\frac{kl}{4}\cos\theta\right)\exp(i\omega t) \qquad (11.42)
\end{aligned}
$$

Using H_ϕ, one can obtain Poynting vector

$$
\begin{aligned}
\langle N\rangle &= \frac{1}{2}R_e(\mathbf{E}\times\mathbf{H}^*)=\frac{1}{2}\sqrt{\frac{\varepsilon_0}{\mu_0}}|E_0|^2\hat{a}_n \\
&= \frac{1}{2}\sqrt{\frac{\varepsilon_0}{\mu_0}}\frac{k^4 p_0^2\sin^2\theta}{16\pi^2\varepsilon_0^2 r^2}\sin^2\left(\frac{kl}{4}\cos\theta\right)\cdot 4 \qquad (11.43)
\end{aligned}
$$

Average Power Radiated Per Unit Solid Angle

$$
\begin{aligned}
\left\langle\frac{dW}{d\Omega}\right\rangle &= \langle N\rangle\cdot\hat{a}_n r^2=\frac{\sqrt{\varepsilon_0}k^4 p_0^2\sin^2\theta}{8\sqrt{\mu_0}\pi^2\varepsilon_0^2}\sin^2\varepsilon_0^2\sin^2\left(\frac{kl}{4}\cos\theta\right) \\
&= \frac{\omega^4 p_0^2}{8\pi^2\varepsilon_0 c^3}\sin^2\theta\sin^2\left(\frac{kl}{4}\cos\theta\right) \qquad (11.44)
\end{aligned}
$$

Special Cases

(i) If $l=\lambda$, i.e. $kl=2\pi$, then

$$\left\langle\frac{dW}{d\Omega}\right\rangle=\omega^4 p_0^2\sin^2\theta\sin^2\left(\frac{\pi}{2}\cos\theta\right) \qquad (11.45)$$

The quadrupole moment, $Q=2p_0 l$. Therefore

$$\left\langle\frac{dW}{d\Omega}\right\rangle = \frac{\omega^4 Q^2}{8\pi^2\varepsilon_0 c^3 4l^2}\sin^2\theta\sin^2\left(\frac{\pi}{2}\cos\theta\right)$$

$$= \frac{\omega^6 Q^2}{128\pi^4\varepsilon_0 c^5}\sin^2\theta\sin^2\left(\frac{\pi}{2}\cos\theta\right) \quad (11.46)$$

(ii) *If $l \to 0$ to be very small*: Obviously, as p_0 increases, Q remains constant. One finds that the dipole pair now reduces to a point quadrupole located at the origin. Since kl is very small, one may write (11.42) as

$$E_\theta = 2i\frac{k^2 p_0}{4\pi\varepsilon_0 r}\sin\theta\frac{kl}{4}\cos\theta e^{i\omega t'} \quad (11.47)$$

Hence

$$\left\langle\frac{dW}{d\Omega}\right\rangle = \frac{k^6 p_0^2 l^2 c}{128\pi^2\varepsilon_0}\sin^2\theta\cos^2\theta$$

$$= \frac{\omega^6 p_0^2 l^2}{128\pi^2\varepsilon_0 c^5}\sin^2\theta\cos^2\theta \quad (11.48)$$

One finds that the maximum power is radiated at $\theta = 45°$.

11.6 RADIATION FROM LINEAR ANTENNA

A practical antenna is a centre-fed thin linear antenna having a length comparable with free space wavelength. The variation of electric field or magnetic vector in space gives an idea of antenna radiation pattern. The antenna used in radio or television transmission, are not short compared to the wavelength of the radiation they transmit. This means that the current in the antenna is not constant and the variation of its amplitude have to be taken into account. The current amplitude of such an antenna decreases uniformly from a maximum at the centre to minimum at the ends. A simple centre driven linear antenna is shown in Fig. 11.6.

Let the antenna (usually a very thin wire) is to be oriented along the z-axis and has a length l. The current is fed into the antenna via a coaxial cable transmission line. We assume the antenna to have a small gap at its centre so that it can be suitably excited. The current along the antenna vanishes at the end points and is an even function of z. The current density **J** is a radial current and is assumed to vary harmonically in time and space along the antenna. To a good approximation, one can write the current density as

$$\mathbf{J}(r,t) = \hat{a}_z I_0 \exp(i\omega t)\sin\left(\frac{kl}{2} - k|z|\right)\delta x\delta y \quad (11.49)$$

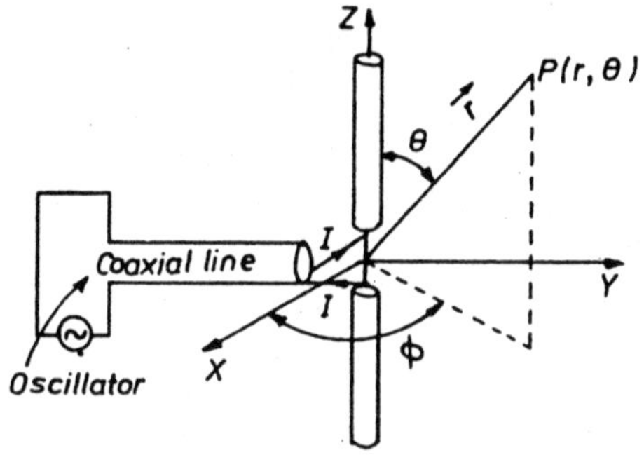

Figure 11.6. Linear, centre-fed antenna.

The delta functions in (11.49) ensures that the current flows in z-direction only. We can easily verify that (11.49) satisfies the required conditions. The exact form of the current at the gap, i.e. the input signal is

$$I(t) = I_0 \exp(i\omega t) \sin\frac{kl}{2} \tag{11.50}$$

where I_0 is the current at the current maximum. The vector potential at a point P is given by

$$\begin{aligned} \mathbf{A}(\mathbf{r},t) &= \frac{\mu_0}{4\pi}\int \frac{\mathbf{J}(\mathbf{r}',t')}{|\mathbf{r}-\mathbf{r}'|}dr' \quad \text{where} \quad t' = t - \frac{|\mathbf{r}-\mathbf{r}'|}{c} \\ &= \frac{\mu_0}{4\pi}\hat{a}_z I_0 \int_{-\frac{l}{2}}^{\frac{l}{2}} \frac{\left(t - \frac{|\mathbf{r}-\mathbf{r}'|}{C}\right)\sin\left(\frac{kl}{2} - k|z|\right)}{|\mathbf{r}-\mathbf{r}'|}dr' \\ &= \frac{\mu_0}{4\pi}\hat{a}_z I_0 \int_{-\frac{l}{2}}^{\frac{l}{2}} \frac{\exp\left\{i\omega\left(t - \frac{|\mathbf{r}-\mathbf{z}|}{C}\right)\right\}\sin\left(\frac{kl}{2} - k|z|\right)}{|\mathbf{r}-\mathbf{z}|} \end{aligned} \tag{11.51}$$

When the point P is far away, one can always replace the denominator by r and we may substitute $|\mathbf{r}-\mathbf{z}|$ in the numerator. Here θ is the angle the vector $\mathbf{r}$ make with z-axis. Thus, we have

$$\begin{aligned} A(\mathbf{r},t) &= \frac{\mu_0}{4\pi}\hat{a}_z \frac{I_0}{r}\exp(i\omega t)\exp(-ikr) \\ & \int_{-\frac{l}{2}}^{\frac{l}{2}} \sin\left(\frac{kl}{2} - k|z|\right)\exp(-ikz\cos\theta)dz \end{aligned}$$

On integerating, one obtains

$$A(\mathbf{r},t) = \frac{\mu_0}{2\pi}\hat{a}_z I_0 \exp(i\omega t)\frac{\exp(-ikr)}{kr}$$

$$\left[\frac{\cos\left(\frac{kl}{2}\cos\theta\right)-\cos\left(\frac{kl}{2}\right)}{\sin^2\theta}\right] \tag{11.52}$$

We can easily calculate from (11.52) *electric field strength* and *magnetic field strength.*

We have the *average rate of power* flow as

$$\begin{aligned}\langle N\rangle &= \frac{1}{2}R_e(\mathbf{E}\times\mathbf{H}^*) = \frac{\hat{a}_n}{2c\mu_0}|E_0|^2\\ &= \frac{1}{2}\sqrt{\frac{\mu_0}{\varepsilon_0}}\hat{a}_n\frac{I_0^2}{4\pi^2}\frac{1}{r^2}\left[\frac{\cos\left(\frac{kl}{2}\cos\theta\right)-\cos\left(\frac{kl}{2}\right)}{\sin\theta}\right]^2 \end{aligned}\tag{11.53}$$

Now, the *average power radiated* into *unit solid angle* is obtained as

$$\begin{aligned}\left\langle\frac{dW}{d\Omega}\right\rangle &= \frac{4\pi r^2\langle N\rangle}{4\pi}\hat{a}_n = r^2\langle N\rangle\hat{a}_n\\ &= \sqrt{\frac{\mu_0}{\varepsilon_0}}\frac{I_0^2}{4\pi}\left[\cos\frac{\left(\frac{kl}{2}\cos\theta\right)-\cos\frac{kl}{2}}{\sin\theta}\right]^{-2}\end{aligned}\tag{11.54}$$

Obviously, the angular distribution of radiated power depends upon the value of $\frac{kl}{2}$. Let us consider the case of half wave antenna $\left(l=\frac{\lambda}{2}\right)$, i.e. $kl=\pi$. (Half wave antennas are one of the most widely used form of practical antennas. It is a straight conductor whose length is equal to a half of the free space wavelength). When a periodic current of the form $I_0\cos\omega t$ is fed at the centre of an antenna, a standing wave pattern is formed along the length of the conductor. One can express the current pattern set in the antenna as $\left(I=I_0\cos\left(\frac{2\pi l}{\lambda}\right)\cos\omega t\right)$.

We have

$$\left\langle\frac{dW}{d\Omega}\right\rangle = \sqrt{\frac{\mu_0}{\varepsilon_0}}\frac{I_0^2}{4\pi^2}\left[\frac{\cos\left(\frac{\pi}{2}\cos\theta\right)}{\sin\theta}\right]^2 \tag{11.55}$$

One can obtain the average total power $\langle W\rangle$ radiated by a half wave antenna from (11.53) as

$$\begin{aligned}\langle W\rangle &= \frac{1}{2}\sqrt{\frac{\mu_0}{\varepsilon_0}}\frac{I_0^2}{4\pi^2}\frac{1}{r^2}\int_0^{2\pi}\int_0^{\pi}\left[\frac{\cos\left(\frac{\pi}{2}\cos\theta\right)}{\sin\theta}\right]^2 r^2\sin\theta d\theta d\phi\\ &= \sqrt{\frac{\mu_0}{\varepsilon_0}}\frac{I_0^2}{4\pi}\int_0^{\pi}\frac{\cos^2\left(\frac{\pi}{2}\cos\theta\right)}{\sin\theta}d\theta\end{aligned}$$

This integral cannot be solved analytically and from the tables we find its value. We have

$$\langle W \rangle \simeq 73.1 \frac{I_0^2}{2} \tag{11.56}$$

Thus we find that the radiation resistance of a half wave antenna is 73.1Ω, i.e.

$$R_r = 73.1\Omega$$

Comparing this value with the corresponding value for an electric dipole ($R_r = 0.08\Omega$), we find that the half wave antenna is much more efficient radiator.

When l is equal to the integral number of half wavelengths of the driving oscillations

$$\left\langle \frac{dW}{d\Omega} \right\rangle = \sqrt{\frac{\mu_0}{\varepsilon_0}} \frac{I_0^2}{8\pi^2} \left[\frac{\cos^2\left(\frac{m\pi}{2}\cos\theta\right)}{\sin^2\theta} \right] \tag{11.57}$$

for odd values of m

$$= \sqrt{\frac{\mu_0}{\varepsilon_0}} \frac{I_0^2}{8\pi^2} \left[\frac{\sin^2\left(\frac{m\pi}{2}\cos\theta\right)}{\sin^2\theta} \right] \tag{11.58}$$

for even values of m

11.7 ANTENNA ARRAY

It is a set of suitable spaced, similarly oriented (or aerials) to radiate energy predominently in some given direction. Antenna arrays make use of interference phenomenon that occur between the radiations from the different elements of the array. Accordingly the resultant field developed by an antenna array at a distant point will be the superposition of field contributions from each constitutents antenna of the array. Two types of antenna arrays are commonly used. These are:

(i) *Broad-side-array —Arrays of n sources*: It exhibits the marked directional properties in the horizontal plane and in a direction normal to the line of array, i.e. the beam is radiated at angle $\theta = 90°$ to the axis of the array.

(ii) *End-fire-array—Arrays of n sources*: It exhibits the marked directional properties in the horizontal plane and in a direction along the line of array, i.e. the beam is radiated at angle $\theta = 0°$ to the axis of the array.

Let us consider a linear (an antenna is said to be linear when the elements of the array are spaced equally along a straight line) antenna array consisting of n isotropic sources separated by a distance d apart (Fig. 11.7). With respect to the source 1 at the left end, the phase of

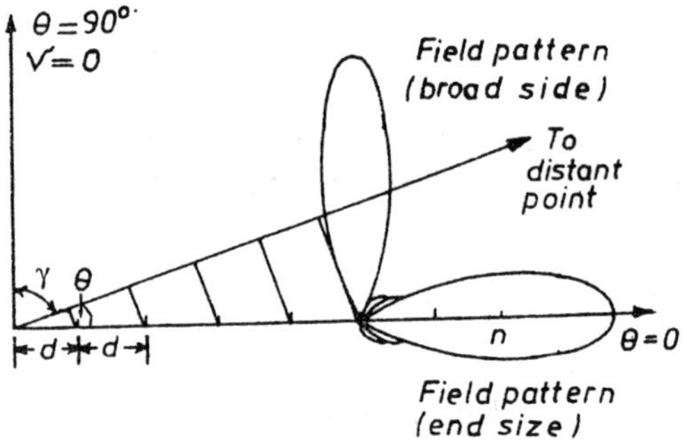

Figure 11.7. Antenna array of n isotropic point radiation with equal spacing, same amplitude and a constant phase difference.

each of the source vary by an angle $\phi = \frac{2\pi}{\lambda}(d\cos\theta + \delta)$, where δ is the progressive phase difference between two consecutive sources. If $\mathbf{E}_1$, $\mathbf{E}_2$, $\mathbf{E}_3$, ldots are the fields due to antennas 1, 2, 3, ldots respectively, then the resultant far-field at an angle θ is given by

$$
\begin{aligned}
E &= E_0[1 + \exp(i\phi) + \exp(2i\phi) + \exp(3i\phi) + \cdots + \exp\{i(n-1)\phi\}] \\
&= E_0 \sum_{n=1}^{N} \exp i(n-1)\phi \qquad (11.59)
\end{aligned}
$$

Multiplying (11.59) by $\exp(i\phi)$ and subtracting the resulting equation from (11.59), one obtains

$$
E = E_0 \frac{[1 - \exp(i\phi)]}{[1 - \exp(i\phi)]} = E_0 \frac{\sin(n\phi/2)}{\sin(\phi/2)} \Big/ (n-1)\phi/2 \qquad (11.60)
$$

We see that the resultant signal acquires a phase difference of $(n-1)\frac{\phi}{2}$. If the phase is referred to the central point of the array, then $\phi \to 0$. The electric field at a distant point is characterized by the individual elements of arrays and $\frac{\sin(n\phi/2)}{\sin(\phi/2)}$ is called the array factor. Eq. (11.60) takes the form

$$E = E_0 \frac{\sin\left(\frac{n\phi}{2}\right)}{\sin\left(\frac{\phi}{2}\right)} \tag{11.61}$$

For $\phi \to 0$, one finds that the resultant field is maximum and expressed as

$$E = nE_0 \tag{11.62}$$

One can express the normalized field as

$$E_n = \frac{E}{nE_0} = \frac{1}{n} \frac{\sin\left(\frac{n\phi}{2}\right)}{\sin\left(\frac{\phi}{2}\right)} \tag{11.63}$$

From Fig. 11.7, we can see that the path difference between radiations from the consecutive antennas at distant point is $d\cos\theta$. Thus the phase difference between fields at distant point due to this path difference is

$$\delta = \frac{2\pi}{\lambda}(d\cos\theta) \tag{11.64}$$

If α is the progressive phase shift between elements, i.e. α is the net angle by which the current in any element leads the current in the preceding element, then the resultant phase difference Ψ between the fields from two consecutive antennas is given by

$$\Psi = \frac{2\pi}{\lambda}(d\cos\theta) + \alpha \tag{11.65}$$

Broadside array: For broad side array the maximum power radiated by the array is concentrated perpendicular to the plane of the array (Fig. 11.7). In the direction of the pattern maximum radiation, the normalized electric field as given by (11.63) is maximum. If $\sin\frac{n\phi}{2} \neq 0$, we must have $\sin\frac{\phi}{2} \to 0$ or $\phi = (kd\cos\theta + \delta) = 0$

$$\text{or} \qquad kd\cos\theta = -\delta \tag{11.66}$$

For broad side array, we have $\theta_{max} = 90^\circ$, and therefore $\delta = 0$. This means the electric fields in all elements of the array are in phase. One obtains the position of null for broad side array is given by $\sin\frac{n\phi}{2} \to 0$

and $\sin\frac{\phi}{2} \neq 0$. This corresponds to

$$\phi = \frac{2m\pi}{n} = k(d\cos\theta_0 + \delta)$$

Expressing in terms of a new angle measured from the vertical, one obtains

$$\xi_0 = (90 - \theta_0) = \sin^{-1}\left(\frac{m\lambda}{nd}\right) \tag{11.67}$$

For large array, $nd \gg m\lambda$ and we have

$$\xi_0 = \frac{m\lambda}{nd} = \frac{m}{(L/\lambda)} \qquad m = 1, 2, 3, \ldots \tag{11.68}$$

where L is the length of the array. The first null corresponds to $m = 1$ and the beam width between first nulls (BWFN) on either sides is obtained as

$$2\xi_0 = \frac{2}{L/\lambda} \tag{11.69}$$

The half power beam width (HPBW) is obtained as

$$\frac{1}{L/\lambda} = \frac{\lambda}{L} \text{ radian} \tag{11.70}$$

Obviously HPBW is half of the BWFN.

We note that the width of the beam depends inversely on the product and which is virtually the length of the broad-side antenna array.

End-fire-Array: For an end fire array the maximum radiation is along the line of the array at $\theta = 0$. Preceeding exactly the same way, one obtains,

$$\phi = (kd\cos\theta + \delta) = 0$$

or

$$kd = -\delta \tag{11.71}$$

For maximum power concentration, we have

$$n\phi = \pm 2m\lambda \dashrightarrow 0$$

and

$$\phi = \pm\frac{m2\pi}{n} = (kd\cos\theta_0 + \delta) \to 0 \tag{11.72}$$

Using (11.71) in (11.72), one obtains

$$(\cos\theta_0 - 1) = \pm\frac{2m\pi}{n(kd)} = \pm\frac{m}{nd/\lambda}$$

or
$$\frac{\theta_0}{2} = \sin^{-1}\left(\pm\frac{m}{2nd/\lambda}\right) \tag{11.73}$$

We have $nd \gg m\lambda$ for a long array. Thus we have

$$\theta_0 = \pm\sqrt{\frac{2m}{nd/\lambda}} \simeq \sqrt{\frac{2m}{L/\lambda}} \tag{11.74}$$

Obviously, the beam width between the first two nulls on either side of the maximum corresponds to $m \to 1$. This is given by

$$\text{BWFN} = 2\theta_0 = \pm 2\sqrt{\frac{2}{L/\lambda}} \tag{11.75}$$

Obviously, the width of the radiated beam depends inversely on the product and which is virtually the length of the antenna array. From (11.75) it is evident that the beam width is narrower for the high frequency of an antenna array of a given length.

11.8 LIENARD-WIECHART POTENTIALS

We now consider the application of retarted potentials (11.5) and (11.6) to compute the radiation from a single charged particle, say, an electron, in arbitrary motion. We have seen that the calculation of the potentials depends upon the position and velocity of the charge at the retarded time $t - \frac{|\mathbf{r}-\mathbf{r}'|}{c}$ and therefore it is essential to know the details of the motion f the charge. Fig. 11.8 shows a trajectory of the electron described by the radius vector $r_e t'$. We can see that the calculation of the potentials given by (11.5) and (11.6) involves a retarded time integration over the entire volume enclosing the charge. We must note that how the charge is distributed geometrically within the electron is not known to us. We know that the particle has a certain total charge. Let us assume that the electron has a finite radius (the assumption that the electron has zero physical extent can lead into difficulties). However, we shall consider only those properties which are independent of the magnitude of the-radius. One can express the retarded potential for an electron in terms of delta-function as

$$\phi(r,t) = \frac{e}{4\pi\varepsilon_0}\int_{-\infty}^{\infty} \frac{\delta\{t' - \left(t - \frac{\mathbf{r}-\mathbf{r}_e t}{c}\right)\}}{|\mathbf{r}-\mathbf{r}_e|} dt' \tag{11.76}$$

To make use of delta-function property, we will have to put the in-

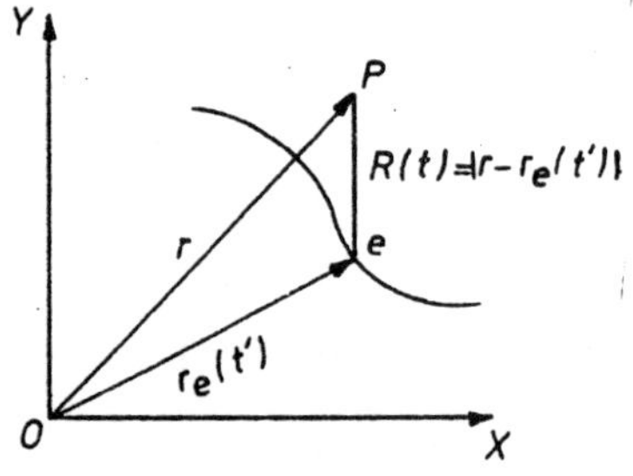

Figure 11.8.

tegral in the form $\int_{-\infty}^{\infty} f(x)\delta(x-x')$. One can see that the integral is equal to $f(x')$.

Let us introduce a new variable t'' such that

$$t'' \;=\; t' - t + \frac{|\mathbf{r}-\mathbf{r}_e(t')|}{c} \tag{11.77}$$

$$\therefore \qquad dt'' \;=\; dt' + \frac{1}{c}\frac{d}{dt'}|\mathbf{r}-\mathbf{r}_e(t')|dt' \tag{11.78}$$

In obtaining (11.78), we have taken $dt = 0$, because the observations is made at a fixed time. Let us choose the coordinates of the fixed point P be x_1, x_2, x_3 and those of the electron $x_{e_1}(t'), x_{e_2}(t'), x_{e_3}(t')$.

Now, we have

$$|\mathbf{r}-\mathbf{r}_e(t')| \;=\; \sqrt{\sum_i [x_{e_i}(t')]^2} \tag{11.79}$$

$$\therefore \qquad \frac{1}{c}\frac{d}{dt'}|\mathbf{r}-\mathbf{r}(t')| \;=\; \frac{1}{c}\sum_i \frac{\partial}{\partial x_{e_i}}|\mathbf{r}-\mathbf{r}_e(t')|\frac{dx_{e_i}}{dt'} \tag{11.80}$$

But $\frac{\partial}{\partial x_{e_i}}|\mathbf{r}-\mathbf{r}_e(t')|$ are the components of the gradient of $|\mathbf{r}-\mathbf{r}_e(t')|$ and $\frac{dx_e}{dt'}$ are the components of $\frac{dr_e}{dt}$, one obtains

$$\frac{1}{c}\frac{d}{dt'}|\mathbf{r}-\mathbf{r}_e(t')| = \frac{1}{c}\,\text{grad}_{r_e}|\mathbf{r}-\mathbf{r}_e(t')|\frac{dr_e}{dt'} \tag{11.81}$$

One can determine the gradient as

$$\text{grad}_{r_e}|\mathbf{r}-\mathbf{r}_e(t')| = -\frac{\mathbf{r}-\mathbf{r}_e(t')}{|\mathbf{r}-\mathbf{r}_e(t')|} = \frac{\mathbf{R}}{|\mathbf{R}|} \tag{11.82}$$

Here $\mathbf{R} = \mathbf{r}-\mathbf{r}_e(t)$. Also we have $\mathbf{u} = \frac{d\mathbf{r}_e}{dt}$, i.e. the velocity of the electron. Now, we can write

$$\frac{1}{c}\frac{d}{dt'}|\mathbf{r}-\mathbf{r}_e(t')| = -\frac{1}{c}\frac{\mathbf{R}}{|\mathbf{R}|}\cdot\mathbf{u} = -\frac{\boldsymbol{\beta}\cdot\mathbf{R}}{|\mathbf{R}|} \tag{11.83}$$

where $\beta = \frac{u}{c}$. Therefore

$$dt'' = dt'\left(1-\frac{\boldsymbol{\beta}\cdot\mathbf{R}}{|\mathbf{R}|}\right)$$

and

$$dt' = \frac{|\mathbf{R}|}{|\mathbf{R}|-\boldsymbol{\beta}\cdot\mathbf{R}}dt'' \tag{11.84}$$

Using (11.84), (11.76) takes the form as

$$\phi(\mathbf{r},t) = \frac{e}{4\pi\varepsilon_0}\int_{-\infty}^{\infty}\frac{\delta(t'')}{\mathbf{R}(t')}\frac{|\mathbf{R}(t')|}{|\mathbf{R}(t')|-\boldsymbol{\beta}(t)\cdot\mathbf{R}(t')} \tag{11.85}$$

$$= \frac{e}{4\varepsilon_0}\left[\frac{1}{|\mathbf{R}(t')|-\boldsymbol{\beta}(t')\cdot\mathbf{R}(t')}\right]_{t''=0} \tag{11.86}$$

We have $t''=0$ indicate $t' = t - \frac{R(t')}{c}$

$$\therefore \quad \phi(\mathbf{r},t) = \frac{e}{4\pi\varepsilon_0}\frac{1}{|\mathbf{R}-\boldsymbol{\beta}\cdot\mathbf{R}|}_{t'=t-\frac{R(t')}{C}} \tag{11.87}$$

Similarly, one finds the vector potential as

$$\mathbf{A}(\mathbf{r},t) = \frac{\mu_0 e}{4\pi}\left(\frac{\mathbf{u}}{R-\boldsymbol{\beta}\cdot\mathbf{R}}\right)_{t'=t-R\frac{(t')}{C}} \tag{11.88}$$

These potentials are velocity dependent but independent of the extent of the charge, i.e. any detailed electronic model. These were first given by A-Lienard and E. Wiechert for a moving point charge and after their names they are known as Lienard-Wiechert potentials.

11.9 THE ELECTROMAGNETIC FIELDS FROM LIENARD-WIECHART POTENTIALS OF A MOVING POINT CHARGE

So far we have obtained the electromagnetic potentials **A** and ϕ for a charge particle undergoing orbitrary motion. Let us now find the electromagnetic fields **E** and **B** of a moving charge. These can be found from the relations

$$\mathbf{E} = -\nabla\phi - \frac{\partial\mathbf{A}}{\partial t}$$

and $$\mathbf{B} = \nabla \times \mathbf{A}$$

Let us start with Lienard-Wiechert potentials to determine the fields of a charge moving with uniform velocity in a straight line.

Let the observer is in the rest frame of the charge, i.e. if observer is moving with the charge, the situation as observed by him is electrostatic. The scalar potential according to the observer is given by

$$\phi(r) = \frac{e}{4\pi\varepsilon_0 r}$$

Now, the question arises, what will be the potential if the observer is at rest and the charge is moving? One can obtain the desired expression, using relativistic transformations (see chapter 12). However, the relativity has its foundation in electrodynamics. Here we shall show that Maxwell's equations lead to what is known as Lorentz transformations.

Let us imagine that a charge e is moving along the x-axis with a velocity $u\hat{a}_x$. We are interested in determining the potential at the point $P(x, y, z)$ at time t. At time t the charge will be at the position (2) represented by $x = ut$. We are interested in knowing its position (1) at the retarded time $t' = t - \dfrac{R(t')}{C}$, where $R(t') = BP$, i.e. the distance of the point P from the position of the charge at time t'. From Fig. 11.9, we have

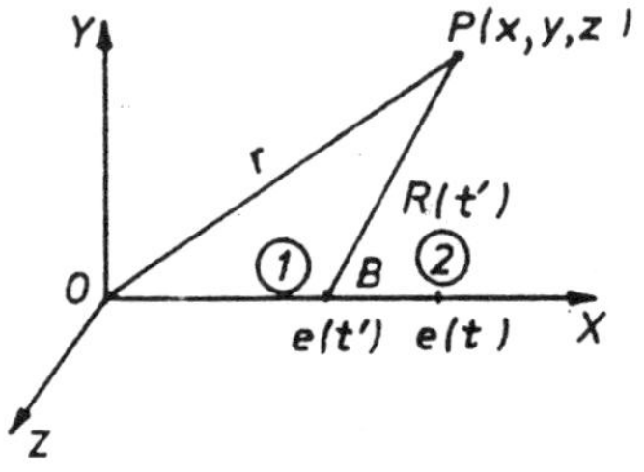

Figure 11.9.

$$R(t') = |\mathbf{r} - \hat{a}_x ut'|$$

where $\mathbf{r}$ is the position vector of the point P as shown in Fig. 11.9. We have

$$R(t') = [(x - ut')^2 + y^2 + z^2]^{\frac{1}{2}} \qquad \left(\because t' = t - \frac{Rt'}{c}\right)$$

$$\therefore \qquad R(t') = c^2(t - t')^2 = (x - ut')^2 + y^2 + z^2$$

or $$(c^2-u^2)t'^2-2(c^2t-xu)t'-r^2+c^2t^2=0$$

$$\therefore \quad t'=\frac{(c^2t-xu)-[(c^2t-xu)^2+(c^2-u^2)(r^2-c^2t^2)]^{1/2}}{(c^2-u^2)}$$

or $$\left(1-\frac{u^2}{c^2}t'\right) = \left(t-\frac{xu}{c^2}\right)-\frac{1}{c}\left[(x-ut^2)+\left(1-\frac{u^2}{c^2}\right)(y^2+z^2)\right]^{1/2}$$

or $$c(t-t') = \frac{u}{c}(x-ut')+\left[(x-ut^2)+\left(1-\frac{u^2}{c^2}\right)(y^2+z^2)\right]^{1/2} \tag{11.89}$$

We have

$$\beta\cdot\mathbf{R} = \frac{u}{c}R\cos\theta=\frac{u}{c}(x-ut')$$

$$\therefore \quad R-\beta\cdot\mathbf{R} = c(t-t')-\frac{u}{c}(x-ut')$$

$$= \left[(x-ut)^2+\left(1-\frac{u^2}{c^2}\right)(y^2+z^2)\right]^{\frac{1}{2}} \tag{11.90}$$

Now, we can express Lienard-Wiechert potential as

$$A(\mathbf{r},t) = \frac{\mu_0}{4\pi}\frac{u}{\left(1-\frac{u^2}{c^2}\right)^{1/2}\left[\frac{(x-ut)^2}{1-\frac{u^2}{c^2}}+y^2+z^2\right]^{\frac{1}{2}}} \tag{11.91}$$

and $$\phi(\mathbf{r},t) = \frac{e}{4\pi\varepsilon_0}\frac{1}{\left[(x-ut^2)+\left(1-\frac{u^2}{c^2}\right)(y^2+z^2)\right]^{\frac{1}{2}}} \tag{11.92}$$

on putting $u=0$, one obtains electrostatic formula for potential. We can conclude that in moving frame of reference, the transformation of coordinate should take place according to

$$x \rightarrow \frac{(x-ut)}{\sqrt{1-\frac{u^2}{c^2}}}$$

$$y \rightarrow y$$

$$z \rightarrow z \tag{11.93}$$

These transformations are known as *Lorentz transformations*.

11.10 ELECTRIC AND MAGNETIC FIELDS OF AN ACCELERATED CHARGE

In the case of an accelerated source charge, it is not possible to express the potential in terms, of coordinates of the 'present position' alone. Let the point of observation p be at $x_\alpha = x_1, x_2, x_3$ (Fig. 11.10) and $x'_\alpha(t') = x'_1(t'), x'_2(t'), x'_3(t')$ be the coordinates of the charge (source) at time t' at which a signal propagated with velocity c and arrive at x_α at time t. Thus the variables of source point and field point are related by

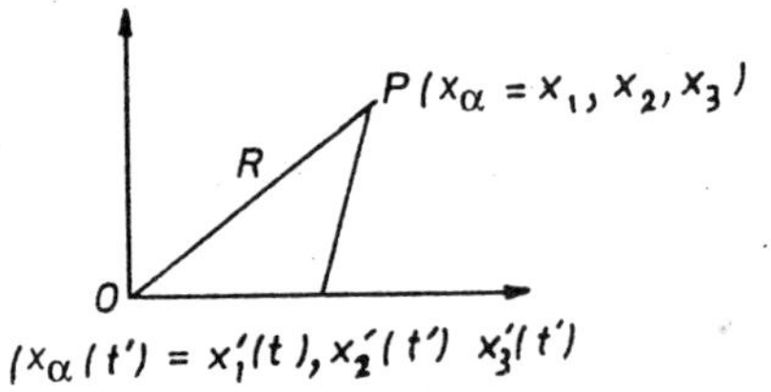

Figure 11.10.

$$R = \sum (x_\alpha - x'_\alpha)^{\frac{1}{2}} = c(t - t') \tag{11.94}$$

The electric and magnetic fields can be found using the basic equations

$$\mathbf{E} = -\nabla\phi - \frac{\partial \mathbf{A}}{\partial t} \tag{11.95}$$

and

$$\mathbf{B} = \nabla \times \mathbf{A} \tag{11.96}$$

The vector operator ∇ are partial space derivatives at constant time t (and not t') and the partial time derivatives are with respect to t (and not t') refer to a constant field point x. Since the coordinates and charge descriptions of the source point is known, it is easy to find the electric and magnetic fields at the point of observation. In effect, it is essential to transform $(\partial/\partial t)|_{x_\alpha=\text{Constant}}$ and $\nabla|_{x_\alpha=\text{Constant}}$ in terms of $(\partial/\partial t')|_{x_\alpha=\text{Constant}}$. This transformation is necessary because in the case of an accelerated charge it is not possible, in general to express the potentials in terms of the 'present position' alone. The potentials are given by

$$\mathbf{A} = \frac{\mu_0}{4\pi} e \left[\frac{\mathbf{u}}{R - \boldsymbol{\beta}\cdot\mathbf{R}}\right] = \frac{e}{4\pi\varepsilon_0} \frac{\mathbf{u}}{c^2 S} \tag{11.97}$$

and

$$\phi = \frac{e}{4\pi\varepsilon_0} \frac{1}{[R - \boldsymbol{\beta}\cdot\mathbf{R}]} = \frac{e}{4\pi\varepsilon_0} \frac{1}{S} \tag{11.98}$$

where

$$S = R - \beta \cdot \mathbf{R} \tag{11.99}$$

Now, we write

$$R[x_\alpha, x'_\alpha(t')] = [\Sigma(x_\alpha - x'_\alpha)^2]^{\frac{1}{2}} = c(t - t') \tag{11.100}$$

Since x'_α is a function t', we have R as a function of x_α and t'. We can express (11.100) as

$$R[x_\alpha, x'_\alpha(t')] = f(x_\alpha, t') = c(t - t') \tag{11.101}$$

$$\therefore \quad \frac{\partial R}{\partial t} = c\left(1 - \frac{\partial t'}{\partial t}\right) \tag{11.102}$$

Also

$$\frac{\partial R}{\partial t} = \frac{\partial R}{\partial t'} \cdot \frac{\partial t'}{\partial t} = -\frac{\mathbf{R} \cdot \mathbf{u}}{R} \cdot \frac{\partial t'}{\partial t} \tag{11.103}$$

From (11.102) and (11.103), we obtain

$$\frac{\partial t'}{\partial t} = \frac{1}{1 - \frac{\mathbf{R} \cdot \mathbf{u}}{Rc}} = \frac{R}{R - \frac{\mathbf{R} \cdot \mathbf{u}}{c}} = \frac{R}{S} \tag{11.104}$$

For brevity, we denote $\left(R - \dfrac{\mathbf{R} \cdot \mathbf{u}}{c}\right) = S$. This gives the desired transformation for time derivatives

$$\frac{\partial}{\partial t} = \frac{R}{S}\frac{\partial}{\partial t'} \tag{11.105}$$

Let us now proceed to evaluate the transformation of the differential operator (∇) from the coordinates of the field point to those of the source point. Because R is a function of x_α and t', one can write

$$\nabla R = \nabla_1 R + \frac{\partial R}{\partial t'}\nabla t' = \frac{\mathbf{R}}{R} - \frac{\mathbf{R} \cdot \mathbf{u}}{R}\nabla t' \tag{11.106}$$

where ∇' means differentiation with respect to the coordinates of the field point at constant retarded time t'. From (11.106), one obtains

$$\nabla R = -c\nabla t'$$

$$\therefore \quad -c\nabla t' = \frac{\mathbf{R}}{R} - \frac{\mathbf{R} \cdot \mathbf{u}}{R}\nabla t'$$

or

$$\nabla t' = -\frac{\mathbf{R}}{Sc} \tag{11.107}$$

with the help of (11.106) and (11.107), one obtains a general relation for transformation of source and field points as

$$\nabla = \nabla_1 - \frac{\mathbf{R}}{Sc}\frac{\partial}{\partial t} \tag{11.108}$$

Using these transformation equations, one can easily obtain expressions for electric and magnetic fields. We have

$$\begin{aligned}
\mathbf{E} &= -\frac{e}{4\pi\varepsilon_0}\nabla\left(\frac{1}{S}\right) - \frac{e}{4\pi\varepsilon_0}\frac{\partial}{\partial t}\left(\frac{\mathbf{u}}{Sc^2}\right) \\
&= \frac{e}{4\pi\varepsilon_0}\left[\frac{1}{S^2}\nabla_1 S - \frac{\mathbf{R}}{S^2 c}\frac{\partial S}{\partial t'} - \frac{R}{S^2c^2}\mathbf{u} + \frac{R\mathbf{u}}{c^2S^3}\frac{\partial}{\partial t'}\right] \\
&= \frac{e}{4\pi\varepsilon_0}\left[\frac{\mathbf{R}}{S^2R} - \frac{\mathbf{u}}{cS^2} + \frac{\mathbf{R}}{S^3c}\left(\frac{\mathbf{R}\cdot\mathbf{u}}{R}\right) - \frac{R}{S^3c}\frac{u^2}{c} + \frac{R}{S^3c}\left(\frac{\mathbf{R}\cdot\mathbf{u}}{c}\right)\right] \\
&\quad - \frac{Ru}{S^2c^2} - \frac{R}{S^3c^2}\mathbf{u}\left(\frac{\mathbf{R}\cdot\mathbf{u}}{c}\right) + \frac{R}{S^3c^3}\mathbf{u}u^2 \\
&\quad - \frac{R}{c^2S^3}\mathbf{u}\left(\frac{\mathbf{R}\cdot\mathbf{u}}{c}\right) \qquad \left(\because \nabla_1 S = \frac{\mathbf{R}}{R} - \frac{u}{c}\right)
\end{aligned} \tag{11.109}$$

After rearranging and combining the appropriate terms, one obtains

$$\begin{aligned}
\mathbf{E} &= \frac{e}{4\pi\varepsilon_0}\left[\frac{1}{S^3}\left(\mathbf{R} - \frac{R\mathbf{u}}{c}\right)\left(1 - \frac{u^2}{c^2}\right)\right. \\
&\quad \left. + \frac{1}{c^2S^3}\left\{\mathbf{R}\times\left(\left(\mathbf{R} - \frac{R\mathbf{u}}{c}\right)\times\dot{\mathbf{u}}\right)\right\}\right]
\end{aligned} \tag{11.110}$$

Similarly, one obtains for magnetic field as

$$\begin{aligned}
\mathbf{B} &= \frac{e}{4\pi\varepsilon_0 c^2}\left[\frac{\mathbf{u}\times\mathbf{R}}{S^3}\left(1 - \frac{u^2}{c^2}\right) + \frac{1}{cS^3}\frac{\mathbf{R}}{R}\right. \\
&\quad \left. \times\left\{\mathbf{R}\times\left(\left(\mathbf{R} - \frac{R\mathbf{u}}{c}\right)\times\dot{\mathbf{u}}\right)\right\}\right]
\end{aligned} \tag{11.111}$$

Comparison of (11.110) and (11.111) lead to the result

$$\mathbf{B} = \frac{\mathbf{R}\times\mathbf{E}}{Rc} \tag{11.112}$$

Obviously, the magnetic field emitted from a moving charge is always perpendicular to **E** and the retarded radius vector **R**.

Examining (11.110), we note that the electric field consists of two distinct parts. The first component given by the first term is a function of velocity **u**, where as the second is a function of acceleration. One can write

$$\mathbf{E} = \mathbf{E}_v + \mathbf{E}_a \tag{11.113}$$

where $\mathbf{E}_v$ is the velocity field and E_a the acceleration field. We also see that $E_v \propto \frac{1}{R^2}$ and $E_a \propto \frac{1}{R}$. Now we compute the Poynting vector for these fields. One obtains that the contribution to this vector due to the two components is

$$N_v \propto \frac{1}{R^4}$$

and

$$N_a \propto \frac{1}{R^2} \tag{11.114}$$

For the determination of energy radiated by the charged particle, one will have to integrate the normal component of N over the surface of the sphere of radius R. Because the element of surface area involves R^2, the integral involving N_v varies as $\frac{1}{R^2}$ while that containing N_a remains finite. Obviously, for large R, the contribution due to N_v tends to zero and that due to N_a is finite. This means, a particle moving with a uniform velocity cannot radiate energy. This shows that energy can be radiated only by accelerated charges. We must note that deceleration also produce radiation, e.g. when a electrons is projected into a block of material, the electrons are stopped and radiation is produced. This radiation is *"braking radiation"* and called *Bremsstrahlüng*.

11.11 RADIATION FROM AN ACCELERATED CHARGE AT LOW VELOCITY—LARMOR'S FORMULA

We have seen that accelerated charges radiate electromagnetic energy in space. Now, if the velocity of the charge particle is so small that $\frac{u}{c}$ can be neglected, then $S \simeq R$ and fields as obtained from (11.110) and (11.111) becomes

$$\mathbf{E}_a = \frac{e}{4\pi\varepsilon_0 R^3}[\mathbf{R} \times (\mathbf{R} \times \dot{\mathbf{u}})] \tag{11.115}$$

and

$$\mathbf{B}_a = \frac{e}{4\pi\varepsilon_0 c^3 R^3}(\dot{\mathbf{u}} \times \mathbf{R}) \tag{11.116}$$

One obtains the Poynting vector which contributes to the radiation as

$$\mathbf{N}_a = \mathbf{E}_a \times \mathbf{H}_a = \mathbf{E}_a \times \frac{\mathbf{B}_a}{\mu_0} = \mathbf{E}_a \times \frac{1}{\mu_0 c}\left(\frac{\mathbf{R} \times \mathbf{E}_a}{R}\right)$$

$\mathbf{E}_a$ is perpendicular to $\mathbf{R}$ and hence we have

$$\mathbf{N}_a = \frac{1}{\mu_0 c} E_a^2 \hat{a}_n = \sqrt{\frac{\varepsilon_0}{\mu_0}} E_a^2 \hat{a}_n \tag{11.117}$$

But

$$
\begin{aligned}
\mathbf{E}_a &= \frac{e}{4\pi\varepsilon_0 c^2 R^3}[\mathbf{R}\times(\mathbf{R}\times\dot{\mathbf{u}})] \\
&= \frac{e}{4\pi\varepsilon_0 c^2 R^3}[(\mathbf{R}\cdot\mathbf{u})R-(\mathbf{R}\cdot\mathbf{R})\dot{\mathbf{u}}] \\
&= \frac{e}{4\pi\varepsilon_0 c^2 R^3}\left[R\dot{u}\cos\theta\mathbf{R}-R^2\dot{\mathbf{u}}\right]
\end{aligned}
$$

Here θ is the angle between $\mathbf{R}$ and $\dot{\mathbf{u}}$. One obtains finally.

$$
\begin{aligned}
\mathbf{N}_a &= \frac{1}{\mu_0 c}\left(\frac{e^2}{16\pi^2\varepsilon_0^2 c^4 R^6}\right)[R\dot{u}\cos\theta\mathbf{R}-R^2\dot{\mathbf{u}}]^2\hat{a}_n \\
&= \left(\frac{e^2}{16\pi^2\varepsilon_0 c^3 R^6}\right)[R^4(\dot{u})^2\cos^2\theta+R^4(\dot{u})^2-2R^4(\dot{u})^2\cos^2\theta]\hat{a}_n \\
&= \frac{(\dot{u})^2 e^2}{16\pi^2\varepsilon_0 c^3 R^2}(1-\cos\theta)\hat{a}_n = \frac{(\dot{u})^2 e^2}{16\pi^2\varepsilon_0 c^3 R^2}\sin^2\theta\hat{a}_n \qquad (11.118)
\end{aligned}
$$

Relation (11.118) gives us the energy flow per unit area per unit time. One can obtain the power radiated per unit solid angle by multiplying by R^2 which is the area per unit solid angle. This can be written as

$$
\frac{dW}{d\Omega} = \frac{e^2(\dot{u})^2}{16\pi^2\varepsilon_0 c^3}\sin^2\theta \qquad (11.119)
$$

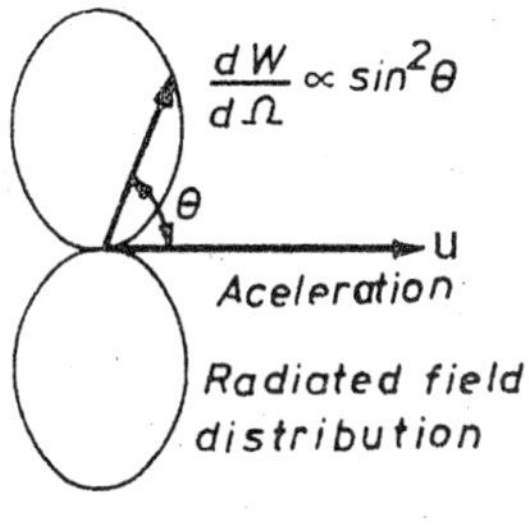

Figure 11.11.

This exhibits the characteristic $\sin^2\theta$ angular dependence, which is well-known result (Fig. 11.11). Integerating over the whole sphere, one obtain the total radiated power as

$$
\begin{aligned}
W &= \frac{e^2(\dot{u})^2}{16\pi^2\varepsilon_0 c^3 R^2}\int_0^{2\pi}\int_0^{\pi}(1-\cos^2\theta)R^2\sin\theta\,d\theta\,d\phi \\
&= 2\pi\frac{e^2(\dot{u})^2}{16\pi^2\varepsilon_0 c^3}\cdot\frac{4}{3} = \frac{e^2(\dot{u})^2}{6\pi\varepsilon_0 c^3} \qquad (11.120)
\end{aligned}
$$

This is the familiar *Larmor result* for a nonrelativistic accelarated charge.

Special Cases

(i) When the velocity and Acceleration of the particles are parallel (colinear)

One can express the radiation fields for this case as

$$\mathbf{E}_a = \frac{e}{4\pi\varepsilon_0 c^2 S^2}[\mathbf{R}\times(\mathbf{R}\times\dot{\mathbf{u}})] \quad \text{(i)}$$

and
$$\mathbf{B}_a = \frac{e\mathbf{R}}{4\pi\varepsilon_0 c^3 S^3}(\dot{\mathbf{u}}\times\mathbf{R}) \quad \text{(ii)}$$

$$\therefore \quad E_a^2 = \frac{e^2 R^4(\dot{u})^2}{16\pi^2\varepsilon_0^2 c^4 S^6}\sin^2\theta \quad \text{(iii)}$$

and
$$\mathbf{N}_a = \frac{1}{\mu_0 c}E_a^2\,\hat{a}_n = \frac{e^2 R^4(\dot{u})^2}{16\pi^2\varepsilon_0 c^3 S^6}\sin^2\theta\,\hat{a}_n \quad \text{(iv)}$$

Now, we obtain the energy radiated into a unit solid angle at θ and measured during the interval dt is

$$-dW(\theta) = \frac{e^2 R^6(\dot{u})^2}{16\pi^2\varepsilon_0 c^3 S^6}\sin^2\theta dt \quad \text{(v)}$$

The negative sign on the left hand side shows that this is the energy lost by the charged particle (here electron) in a time interval dt' during the emission of the signal. We have the energy lost per unit time per unit solid angle as

$$\frac{dW}{d\Omega} = -\frac{dW(\theta)}{dt'} = \frac{e^2 R^6(\dot{u}^2)}{16\pi^2\varepsilon_0 c^3 S^6}\sin^2\theta\frac{dt}{dt'}$$
$$= \frac{e^2 R^6(\dot{u})^2}{16\pi^2\varepsilon_0 c^3 S^6}\sin^2\theta\frac{S}{R} \quad \text{(vi)}$$

in obtaining (vi) we have used (11.104).

$$\therefore \quad \frac{dW}{d\Omega} = \frac{e^2 R^5(\dot{u}^2)}{16\pi^2\varepsilon_0 c^3 S^6}\sin^2\theta$$
$$= \frac{e^2 R^5(\dot{u})^2}{16\pi^2\varepsilon_0 c^3 R^5\left(1-\frac{\beta\cdot\mathbf{R}}{R}\right)^5}\sin^2\theta$$
$$= \frac{e^2(\dot{u})^2\sin^2\theta}{16\pi^2\varepsilon_0 c^3(1-\beta\cos\theta)^5} \quad \text{(vii)}$$

Relation (vii) gives the angular distribution of radiated energy.

For $\beta \ll 1$, i.e. $u \ll c$, one obtains (11.119) for low velocities. However, if $u \to c$, i.e. $\beta \to 1$, obviously the radiated energy increases in the forward direction (Fig. 11.12), but not exactly in the forward direction (i.e. for $\theta = 0$).

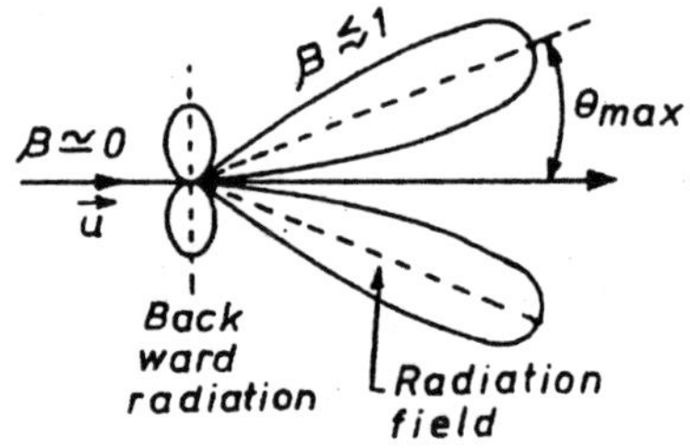

Figure 11.12. Radiation patterns of electrons for $u \rightsquigarrow c$.

(ii) Charged particle moving in a circular orbit

Now we consider the case of a charge particle moving in a circle of radius ρ with an angular frequency ω. We have

$$u = \rho\omega \quad \text{and} \quad \dot{u} = \rho\omega^2$$

This provides another example of angular distribution of radiation for a charge in instantaneously circular motion with its acceleration $\dot{\mathbf{u}}$ perpendicular to its velocity $\mathbf{u}$. Fig. 11.13 shows the orbit of the particle in the yz plane. We choose a coordinate system such that instantaneously the acceleration $\dot{\mathbf{u}}$ is directed towards the centre along the z direction and velocity $\mathbf{u}$ along the y-direction.

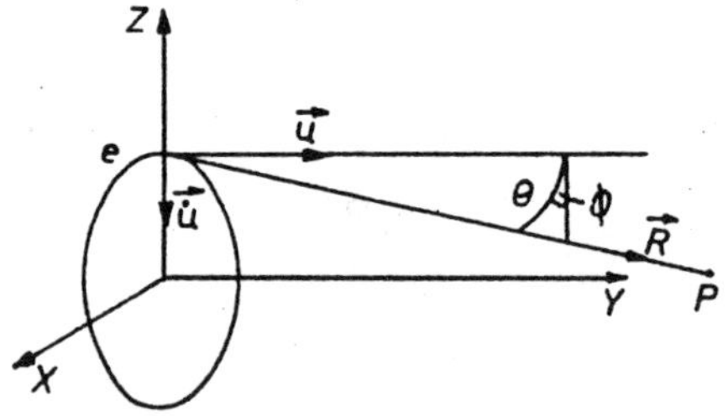

Figure 11.13.

Let θ be the angle between $\mathbf{u}$ and $\mathbf{R}$ and ϕ be the azimuthal angle. Now, we have

$$\mathbf{u} \cdot \mathbf{R} = uR\cos\theta \qquad \text{(viii)}$$

or

$$\dot{\mathbf{u}} \cdot \mathbf{R} = \dot{u} R \sin\theta \cos\phi \tag{ix}$$

The radiation field is given by

$$\mathbf{E}_a = \frac{e}{4\pi\varepsilon_0 c^2 S^3}\left[\mathbf{R} \times \left(\mathbf{R} - R\frac{\mathbf{u}}{c}\right) \times \dot{\mathbf{u}}\right]$$

and Poynting vector is given by

$$\mathbf{N}_a = \frac{1}{\mu_0 c} E_a^2 \hat{a}_n$$

Now, we have

$$\begin{aligned}
&\left[\mathbf{R} \times \left(\mathbf{R} - \frac{R\mathbf{u}}{c}\right) \times \dot{\mathbf{u}}\right]^2 \\
&= \left[(\mathbf{R}\cdot\mathbf{u})\left(\mathbf{R} - \frac{R\dot{\mathbf{u}}}{c}\right) - \left\{\mathbf{R}\cdot\left(\mathbf{R}\cdot\frac{R\mathbf{u}}{c}\right)\right\}\dot{u}\right]^2 \\
&= (\mathbf{R}\cdot\mathbf{u})^2\left(\mathbf{R} - \frac{R\mathbf{u}}{c}\right)^2 + \left\{R^2\dot{\mathbf{u}} - \frac{\mathbf{R}}{c}(\mathbf{R}\cdot\mathbf{u})\dot{\mathbf{u}}\right\}^2 \\
&\quad -2(\mathbf{R}\cdot\dot{\mathbf{u}})\left(\mathbf{R} - \frac{R\mathbf{u}}{c}\right)\cdot\left\{R^2\dot{\mathbf{u}} - \frac{\mathbf{R}}{c}(\mathbf{R}\cdot\mathbf{u})\dot{\mathbf{u}}\right\} \\
&= -R^2\left(1 - \frac{u^2}{c^2}\right)(\mathbf{u}\cdot\mathbf{R})^2 + R^4\left(1 - \frac{\mathbf{R}\cdot\mathbf{u}}{Rc}\right)^2(\dot{u})^2R^4 \\
&= -(\dot{u})^2R^4 - \frac{2\dot{u}^2R^4u\cos\theta}{c} + \frac{(\dot{u})^2R^4u^2\cos^2\theta}{c^2} \\
&= \left[\left(1 - \frac{u}{c}\cos\theta\right)^2 - \left(1 - \frac{u^2}{c^2}\right)\sin^2\theta\cos^2\phi\right](\dot{u})^2R^4
\end{aligned} \tag{x}$$

In obtaining (x) we have used (viii) and (ix). Now, we obtain

$$N_a = \frac{e^2(\dot{u})^2R^4}{16\pi^2\varepsilon_0^2\mu_0 c^5 S^6}\left[\left(1 - \frac{u}{c}\cos\theta\right)^2 - \left(1 - \frac{u^2}{c^2}\right)\sin^2\theta\cos^2\phi\right]\hat{a}_n \tag{xi}$$

One obtains the angular distribution of the radiated energy as

$$-\frac{dw(\theta)}{dt'}d\Omega = (\mathbf{N}_a\cdot\hat{a}_n)R^2\frac{dt}{dt'}d\Omega = \frac{e^2(\dot{u})^2}{16\pi^2\varepsilon_0 c^3}\frac{\left[\left(1 - \frac{u}{c}\cos\theta\right)^2 - \left(1 - \frac{u^2}{c^2}\right)\sin^2\theta\cos^2\phi\right]}{\left(1 - \frac{u}{c}\cos\theta\right)^5}d\Omega \tag{xii}$$

In obtaining (xii) we have used (11.104) and substituted for S and in (vii). The radiation pattern in the plane $\phi = 0$ for $\beta = 0.5$ is shown in Fig. 11.14

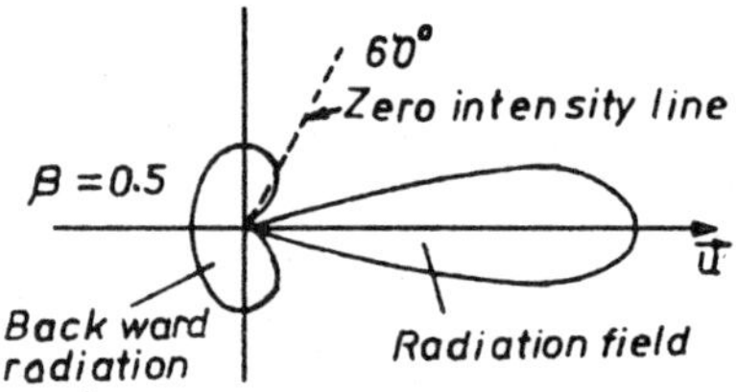

Figure 11.14. Radiation pattern in the plane $\phi = 0$ for $\beta = 0.5$.

11.12 RELATIVISTIC GENERALISATION OF LARMOR'S FORMULA

Larmoor's formula (11.120) can be generalised for arbitrary velocities of charge by using the fact that radiated electromagnetic energy behaves under Lorentz transformation like the fourth component of a four vector. We must note that if a pulse of electromagnetic radiation of finite spatial extent exists in charge and current free space, then total electromagnetic momentum and energy transform like a four vector. Since $W = \dfrac{dE_{\text{rad}}}{dt}$, i.e. $dE_{\text{rad}} = W\,dt$ and time t also behaves under Lorentz transformations like the fourth component of *space-time* four vector. This means that the power W is a Lorentz invariant quantity. Obviously, if one can find a Lorentz invariant reduces to non-relativistic Lormor's relation for $\beta = \dfrac{u}{c} \ll 1$, then one obtains the required relativistic generalization of Larmor's relation. However, from the general form of electromagnetic fields for a moving charge, it is evident that the general result must involve only β and $\dot{\beta}$. With this restriction on the order of derivatives which may appear, one finds unique result.

To obtain the appropriate generalisation of Larmor's formula let us use the familiar fact that $\mathbf{a} = \dfrac{\mathbf{F}}{m} = \left(\dfrac{d\mathbf{p}/dt}{m}\right)$. We can express Larmor's non-relativistic formula (11.120) in the form

$$W = \frac{1}{4\pi\varepsilon_0}\frac{2e^2(\mathbf{a}\cdot\mathbf{a})}{3c^3} = \frac{1}{4\pi\varepsilon_0}\frac{2}{3}\frac{e^2}{m^2c^2}\left(\frac{d\mathbf{p}}{dt}\cdot\frac{d\mathbf{p}}{dt}\right) \qquad (11.121)$$

where m is the mass of the charge and $\mathbf{p}$ its momentum. Eq. (11.121) reveals that its Lorentz invariant generalization must be

$$W = \frac{1}{4\pi\varepsilon_0}\frac{2}{3}\frac{e^2}{m^2c^2}\left(\frac{dp_\mu}{d\tau}\frac{dp_\mu}{d\tau}\right) \tag{11.122}$$

where p_μ is the charge particle's momentum energy four vector given by

$$p_\mu = \left(p, \frac{iE}{c}\right) \tag{11.123}$$

and $d\tau$ is proper time interval given by

$$\left.\begin{aligned} dt &= \frac{d\tau}{\sqrt{1-\frac{u^2}{c^2}}} = \frac{d\tau}{\sqrt{1-\beta^2}} = \gamma d\tau \\ \text{with}\quad \gamma &= \frac{1}{\sqrt{1-\beta^2}} \end{aligned}\right] \tag{11.124}$$

One can easily check that Eq. (11.122) reduces properly to Eq. (11.121) as $\beta \to 0$. For this one will have to evaluate the four-vector scalar product. We have

$$\begin{aligned} \frac{dp_\mu}{d\tau}\frac{dp_\mu}{d\tau} &= \left(\frac{dp_1}{d\tau}\right)^2 + \left(\frac{dp^2}{d\tau}\right)^2 + \left(\frac{dp_3}{d\tau}\right)^2 + \left(\frac{dp_4}{d\tau}\right)^2 \\ &= \left(\frac{dp}{d\tau}\right)^2 + \left\{\frac{d}{d\tau}\left(\frac{iE}{c}\right)\right\}^2 = \left(\frac{dp}{d\tau}\right)^2 - \frac{1}{c^2}\left(\frac{dE}{d\tau}\right)^2 \end{aligned}$$

But

$$d\tau = \frac{dt}{\gamma}$$

$$\therefore \quad \frac{dp_\mu}{d\tau}\frac{dp_\mu}{d\tau} = \gamma^2\left(\frac{dp}{dt}\right)^2$$

Using the above, (11.122) becomes

$$W = \frac{1}{4\pi\varepsilon_0}\frac{2}{3}\frac{e^2\gamma^2}{m^2c^2}\left[\left(\frac{dp}{dt}\right)^2 - \frac{1}{c^2}\left(\frac{dE}{dt}\right)^2\right] \tag{11.125}$$

To express the above equation in terms of velocity, acceleration, we make use of relativistic expressions

$$E = \frac{mc^2}{\sqrt{1-\beta^2}} \quad \text{and} \quad \mathbf{p} = \frac{m\mathbf{u}}{\sqrt{1-\beta^2}} \tag{11.126}$$

where m is the rest mass of the charge. From (11.126), one obtains

$$\frac{dE}{dt} = \frac{d}{dt}\left\{mc^2(1-\beta^3)^{-1/2}\right\} = \frac{mc^2\boldsymbol{\beta}\cdot\dot{\boldsymbol{\beta}}}{(1-\beta^2)^{3/2}} = mc^2\gamma^3\boldsymbol{\beta}\cdot\dot{\boldsymbol{\beta}} \qquad (11.127)$$

and

$$\frac{d\mathbf{p}}{dt} = \frac{d}{dt}\left\{\frac{m\mathbf{u}}{\sqrt{(1-\beta^2)}}\right\} = \frac{d}{dt}\left\{\frac{mc\boldsymbol{\beta}}{\sqrt{1-\beta^2}}\right\}\frac{mc[\dot{\boldsymbol{\beta}}(1-\beta^2)+\dot{\boldsymbol{\beta}}(\boldsymbol{\beta}-\dot{\boldsymbol{\beta}})]}{(1-\beta^2)^{3/2}}$$

or
$$\frac{d\mathbf{p}}{dt} = mc\gamma^3[\boldsymbol{\beta}\cdot(1-\beta^2)+\boldsymbol{\beta}(\boldsymbol{\beta}\cdot\dot{\boldsymbol{\beta}})] \qquad (11.128)$$

Using (11.127) and (11.128), Eq. (11.126) becomes

$$W = \frac{1}{4\pi\varepsilon_0}\frac{2}{3}\frac{e^2\gamma^3}{m^3c^3}\left[m^2c^2\gamma^6\left\{\dot{\boldsymbol{\beta}}(1-\beta^2)+\boldsymbol{\beta}(\boldsymbol{\beta}\cdot\dot{\boldsymbol{\beta}})\right\}^2 - m^2c^2\gamma^6(\boldsymbol{\beta}\cdot\dot{\boldsymbol{\beta}})^2\right]$$

On simplification, one obtains

$$W = \frac{1}{4\pi\varepsilon_0}\frac{2}{3}\frac{e^2}{c}\gamma^6\left[(\dot{\boldsymbol{\beta}})^2 - \beta^2\dot{\beta}^2 + (\boldsymbol{\beta}\cdot\dot{\boldsymbol{\beta}})^2\right]$$

But $(\boldsymbol{\beta}\times\dot{\boldsymbol{\beta}})^2 = (\boldsymbol{\beta}\cdot\dot{\boldsymbol{\beta}})^2 - \beta^2\dot{\beta}^2$

$$\therefore \quad W = \frac{1}{4\pi\varepsilon_0}\frac{2}{3}\frac{e^2\gamma^6}{c}[\dot{\beta}^2 - (\boldsymbol{\beta}\times\dot{\boldsymbol{\beta}})^2] \qquad (11.129)$$

Eq. (11.129) is the relativistic generalisation of Larmor's formula.

Special cases

(i) When the motion of the particle is non-relativistic

We have $\beta \to 0$ and $\gamma = \dfrac{1}{\sqrt{1-\beta^2}} \to 1$. Eq. (11.129) takes the form

$$W = \frac{1}{4\pi\varepsilon_0}\frac{2}{3}\frac{e^2}{c}(\dot{\boldsymbol{\beta}})^2$$

$$= \frac{1}{4\pi\varepsilon_0}\frac{2}{3}\frac{e^2}{c^3}\dot{\mathbf{u}}^2 \qquad (11.130)$$

This is non-relativistic Larmor's formula.

(ii) When the Velocity and acceleration of the particles are collinear (parallel)

We have $|\boldsymbol{\beta}\times\dot{\boldsymbol{\beta}}| = \beta\dot{\beta}\sin 0 = 0$. From Eq. (11.129), we have

$$W = \frac{1}{4\pi\varepsilon_0}\frac{2}{3}\frac{e^2\gamma^6\dot{\beta}^2}{c} = \frac{1}{4\pi\varepsilon_0}\frac{2}{3}\frac{e^2\gamma^6\dot{\mathbf{u}}^2}{3c^2} \qquad (11.131)$$

This type of radiations is *Bremsstrahlung radiations* which are produced from accelerated electrons with the assumption that the direction of motion of electrons does not change.

(iii) When the motion of the charged particle is such that its velocity and acceleration are perpendicular

We have $|\boldsymbol{\beta} \times \dot{\boldsymbol{\beta}}| = \beta\dot{\beta}\sin 90^\circ = \beta\dot{\beta}$

From Eq. (11.129), one obtains

$$\begin{aligned} W &= \frac{1}{4\pi\varepsilon_0}\frac{2}{3}\frac{q^2r^6}{3c}(\dot{\beta}^2 - \beta^2\dot{\beta}^2) \\ &= \frac{1}{4\pi\varepsilon_0}\frac{2}{3}\frac{q^2r^6}{3c}\dot{\beta}^2(1-\beta^2) \\ &= \frac{1}{4\pi\varepsilon_0}\frac{2}{3}\frac{q^2\gamma^4\dot{\beta}^2}{c} \end{aligned} \quad (11.132)$$

Such type of radiations occurs in circular accelerators like cylotron or synchrotron, where energy loss is drastic.

11.13 CERENKOV RADIATION

We have seen that the uniformly moving charge do not give rise to electromagnetic radiation. The Poynting vector corresponding to the induction field and other magnetic fields felt off rather rapidly with increasing distances and it is observed that it does not contribute to surface integral at large distances. However, this situation changes in the case of material media for which the dielectric constant $\varepsilon_r > 1$. The electromagnetic radiations from charge and current sources, in general, are obtained from the solution of inhomogeneous wave equations. For charges moving in a material medium and constituting a current source, one can express the inhomogeneous wave equation in terms of vector potential as

$$\nabla^2\mathbf{A} - \frac{M^2}{e^2}\frac{\partial^2 A}{\partial t^2} = \mu_0\mathbf{J} \quad (11.133)$$

The solution of (11.133) confirms to electromagnetic radiation moving with the phase velocity $v_p = \dfrac{c}{M}$, where $M = \sqrt{\varepsilon_r}$ is the index of refraction. One obtains the retarded time corresponding to this solution as $t' = \left(t - \dfrac{r}{v_p}\right) = \left(t - \dfrac{r\sqrt{\varepsilon_r}}{c}\right)$. When the intervening medium in a certain frequency range behaves like a medium of high refractive index, the situation may arise where the velocity of the charged particle u becomes greater than the phase velocity (v_p) of the electromagnetic radiations. We can see that in such circumstances the retardation denominator may vanish, i.e. $\left(1-\frac{u\sqrt{\varepsilon_r}}{c}\cos\theta\right)\to 0$. For $\cos\theta = \frac{c}{\sqrt{\varepsilon_r}}u = \frac{v_p}{u}$, the charged particle velocity is greater than the phase velocity of the emitted waves in the given medium and the electromagnetic radiation becomes large is confined to cone of angles around the direction of motion of the charged particles, i.e. the electromagnetic waves propagate along conical surfaces, making an angle α with the direction of propagation given by

$$\sin\alpha = \frac{u}{v_p}. \tag{11.134}$$

These waves are called *Cerenkov radiation* first observed by Cerenkov in 1934.

Cerenkov radiation is Similar to the *Mach sonic shock wave* developed when bodies travel at velocities greater than the phase velocities of the elastic waves in the given medium.

From (11.134) it is obvious that the effective direction of propagation of wave front is related to the velocity of the charged particle, it may be used to measure it. Devices used for this purpose are called *Cerenkov detectors*, which are widely used in experiments with fundamental particles because they provide direct information about the velocity of the particles.

EXAMPLE 11.1: A dipole with moment **p** vibrates with a frequency ω. A molecule with a polarizability α is placed at a point with a radius vector **r** (**r** is perpendicular to **p**). If, where $r \ll \lambda$, where λ is the radiation wavelength, show that the power radiated by the system is given by

$$W = \frac{\omega^4 p^2}{12\pi\varepsilon_0 c^3}\left(1-\frac{\alpha}{4\pi r^3}\right)^2.$$

Solution: We have the electric field vector of the dipole at a point with the radius vector **r** (**r** perpendicular to **p**) is

$$\mathbf{E}_1 = \frac{3(\mathbf{p}\cdot\mathbf{r})\mathbf{r}}{4\pi\varepsilon_0 r^5} - \frac{\mathbf{P}}{4\pi\varepsilon_0 r^3} = -\frac{\mathbf{P}}{4\pi\varepsilon_0 r^3} \tag{i}$$

The induced moment $\mathbf{P}_1$ in the molecule due to the field is given by

$$\mathbf{P}_1 = \varepsilon_0 \alpha \mathbf{E}_1 = -\frac{\varepsilon_0 \alpha \mathbf{p}}{4\pi\varepsilon_0 r^3} \tag{ii}$$

The resultant dipole moment $\mathbf{P}' = \mathbf{P} + \mathbf{P}_1$. This creates the net field. The power generated in the radiation zone is given by

$$W \doteq \frac{cP^2}{12\pi\varepsilon_0}\left(\frac{2\pi}{\lambda}\right)^4 = \frac{\omega^4 p^2}{12\pi\varepsilon_0 c^3}\left(1 - \frac{\alpha}{4\pi r^3}\right)^2$$

QUESTIONS

1. What are retarded potentials? Explain the concept of retardation and origin of potential functions ϕ and **A**.
2. What are Lienard-Wiechert potentials? How do you get them from retarded potentials? Discuss the transformations by using δ-function (Dirac).
3. Write down the Lienard Wiechert potentials produced by a moving point charge. Obtain the corresponding electric and magnetic fields and identify their respective induction and radiation parts. Show that a charge must be accelerated in order to radiate.
4. Starting from the expression for retarded potentials, obtain the electromagnetic field of a point charge in arbitrary motion.
5. Derive an expression for the rate of radiation of energy from an accelerated charge at low velocity.
6. Derive expressions for the field radiated by an accelerated charge at low velocity. Explain, what are Bremsstrahung and Cherenkov radiations.
7. Obtain the expression for the power radiated by an accelerated charged particle in the non-relativistic limit and discuss its angular distribution.
8. Obtain the electric and magnetic fields from an oscillating electric dipole and calculate the time averaged power radiated per unit solid angle and the total power radiated.
9. Obtain expressions for Poynting vector and power radiated by electric quadrupole.

10. Obtain an expression for the radiation resistance of centre fed half wave antenna. Explain, why it is efficient radiator in comparison to dipole antenna?
11. What is an antenna array? Explain its directional properties analytically.
12. Discuss the principle of working of an antenna array. Derive an expression for the resultant field in an direction making an angle ϕ with the line of an array consisting of N equally spaced antenna carrying currents of equal magnitude and having successful phase difference.
13. Write Short notes on (i) Lineard-Wiechart potentials (ii) Bremsstralilung and (iii) Cerenkov radiations.

PROBLEMS

1. Show that when two particles with identical charge and mass collide, there is no dipole radiation.

   [**Hint:** Let m_1, and m_2 be masses of two particles and e_1 and e_2 their charges and r is the separation between them. For identical particles, we have $m_1 = m_2 = m$ and $e_1 = e_2 = e$. In the centre of mass system, the dipole moment of two particles is

   $$\mathbf{P} = e_1\mathbf{r}_1 + e_2\mathbf{r}_2 = \frac{m_1 m_2}{m_1 + m_2}\left(\frac{e_1}{m_1} - \frac{e_2}{m_2}\right)\mathbf{r}$$

   Here $\mathbf{r}_1 = \mathbf{r}_2 = \mathbf{r}$, $m_1 = m_2 = m$ and $e_1 = e_2 = e$ $\therefore$ $p = 0$. Obviously, dipole moment is zero].

2. Prove in the case of uniformly moving charge that the ratio $\frac{E_{90^\circ}}{E_{0^\circ}} = \gamma^{-3}$, where $\gamma = \frac{1}{\sqrt{1-\beta^2}}$, E_{90° is the field at right angles to the direction of charge motion, while E_0 is in the direction of charge motion. Interpret numerically the ratio given by (i) for $\beta \to 0$ and $\beta = 0.3$.

3. Show that sine square distribution law is followed by power radiated by a low velocity accelerated charged particle. Also show that the total power radiated in such a case is given by Larmor formula.

4. Taking the total radiated power by an arbitrary accelerated charge given by Lienard radiation formula

$$W = \frac{1}{4\pi\varepsilon_0}\frac{2e^2}{3c}\gamma^6[\dot{\beta}^2 - |\beta \times \dot{\beta}|^2]$$

where $\beta = \frac{u}{c}$ and $\dot{\beta} = \frac{\partial \beta}{\partial t}$ with u as velocity of the charge, discuss (i) non-relativistic case (ii) collinear velocity and acceleration and (iii) perpendicular velocity and acceleration.

5. A radiator approximates to an electric dipole of length 250 m at a frequency of 60 k Hz. Assuming that the current is maintained over the length, show that the radiation resistance of the radiator is 1.97 Ω.

6. An antenna of length L carries alternating current of angular frequency ω. Treating it as an oscillating dipole, determine the total power radiated. Show that the equivalent resistance to produce the same power loss (the radiation resistance) is

$$R = \frac{2\pi}{3}\sqrt{\frac{\mu_0}{\varepsilon_0}}\left(\frac{L\omega}{2\pi c}\right)^2.$$

7. Show that the resultant field in a direction making an angle 0 with the line of an array consisting of N aerials equally spaced d apart and carrying equal aerial currents and phase is given by

$$E = E_0\frac{\sin(N\phi/2)}{N\sin\phi/2}$$

where E_0 is the value of maximum field and $\phi = \frac{2\pi}{\lambda}(d\cos\theta)$.

8. A practicle short dipole is a centre fed antenna with a length much smaller than the wavelength. The current distribution may be assumed to be linear, the current being maximum at the centre and zero at the ends. Show that the radiation resistance is $20\pi^2\left(\frac{L}{\lambda}\right)^2$ Ω, where L is the length of the antenna.

9. The charges $-e$, $2e$ and $-e$ form a linear quadrupole oscillator. The positive charge $2e$ is stationary at the origin. One of the negative charges is at $z_1 = a\cos\omega t/2$ and the other is at $z_2 = -a\cos\omega t$. Find the fields at large distances.

SHORT ANSWER QUESTIONS

1. For many calculations electrons are viewed as charge of finite size. What it is?

   [**Ans.** The *classical radius* of electron is given $r_0 = \dfrac{\mu_0 e^2}{4\pi m_0}$. The rest mass (m_0) of the charge changes relativistically as it moves.]

2. What is the relation between magnetic and electric fields of a uniformly moving charge moving with velocity u?

   [**Ans.** $\mathbf{B} = \dfrac{1}{c^2}(\mathbf{u} \times \mathbf{E})$]

3. What is non-radiating or induction field?

   [*Ans.* While studying the electric and magnetic fields of an accelerated electron or charge, we have seen that electric field can be expressed as, $E = E_v + E_a$, where E_v is the velocity field and E_a is a function of acceleration. E_v around the charge is conserved and does not show any loss of electrical energy. This is called as non-radiating or induction field.]

4. What is the nature of radiant energy from an accelerated charge?

   [*Ans.* The radiation fields show a loss of energy from the accelerated charge and contribute to the energy flow in the surrounding space. The radiant energy has a angular distribution in space. The angular distribution of radiated energy has an angular dependence of $\sin^2\theta$. The maximum field is radiated at $\theta = 90°$ and minimum field is radiated along $\theta = 0°$, i.e. along the direction of the accelerated charge.]

5. What are distant sources of electromagnetic sources in space?

   [**Ans.** Radio stars.]

6. What are the two distinct processes in which the electromagnetic radiation in the presence of a magnetic field is generally divided?

   [*Ans.* (i) *Cylotron Process*: The electromagnetic radiation produced by low energy charged particles gyrating around a static magnetic field and (ii) *Synchrotron Process*: The electromagnetic radiation produced by relativistic charged particles gyrating around static magnetic field.]

OBJECTIVE QUESTIONS

1. Radiation resistance of a half wave (linear) dipole antenna is
   (a) 377Ω, (b) 73Ω, (c) 37Ω, (d) 552Ω. **[b]**

2. Radiation resistance in a given direction is the
   (a) power radiated per square metre.
   (b) energy radiated per square metre.
   (c) power radiated per unit solid angle in that direction.
   (d) none of the above. **[c]**

3. In the case of dipole antenna, for large values of r, the ratio of only non zero field components E_θ and H_ϕ is equal to
   (a) $\dfrac{E_\theta}{H_\phi} = \sqrt{\dfrac{\mu_0}{\varepsilon_0}}$, (b) $\dfrac{E_\theta}{H_\phi} = \dfrac{\mu_0}{\varepsilon_0}$,
   (c) $\dfrac{E_\theta}{H_\phi} = 1$, (d) $\dfrac{E_\theta}{H_\phi} = \left(\dfrac{\mu_0}{\varepsilon_0}\right)^2$ **[a]**

4. In the case of dipole antenna for $l \ll \lambda$, the radiation resistance is given by
   (a) $R_{\text{rad}} = pi^2 \left(\dfrac{l}{\lambda}\right)^2 \Omega$, (b) $R_{\text{rad}} = 80\pi^2 \left(\dfrac{l}{\lambda}\right)^2 \Omega$,
   (c) $R_{\text{rad}} = (\pi)^2 \dfrac{l}{\lambda^2} \Omega$, (d) $R_{\text{rad}} = 80\pi^2 \Omega$. **[b]**

5. The average total power radiated by a half-wave antenna is given by
   (a) $< W >= 36.57 I_0^2$ Watt, (b) $< W >= 73 I_0^2$ Watt,
   (c) $< W >= 377 I_0^2$ Watt, (d) $< W >= 730 I_0^2$ Watt.

6. The radiated power in a travelling wave antenna in comparison to the standing wave antenna for the same length and frequency is
   (a) more, (b) less, (c) equal, (d) infinite. **[a]**

7. The radiation resistance for a small planar circular loop antenna of radius a is
   (a) $R_r = \dfrac{(\omega a)^4}{c^4 \pi^2} \Omega$, (b) $R_r = \dfrac{(\omega a)^4}{20\pi^2 c^4} \Omega$,
   (c) $R_r = 20\pi^2 \dfrac{(\omega a)^4}{c^4} \Omega$, (d) $R_r = 20\pi^2 \left(\dfrac{\omega}{ac}\right)^4$. **[b]**

8. If we neglect phase delay term along the line in case of half wave antenna, then the radiation resistance will be
   (a) 73.14 Ω, (b) 37 Ω, (c) 370 Ω, (d) 80 Ω. **[d]**

9. Power gain of an antenna in a given direction is the ratio of
   (a) radiation intensity in that direction to the total input radiated power.
   (b) radiation intensity in that direction to the average radiated power.
   (c) average power to maximum power.
   (d) none of the above. **[b]**

10. Radiation resistance of a current element of length dl is
   (a) $80\left(\frac{\lambda dl}{\pi}\right)^2 \Omega$, (b) $80\left(\frac{\lambda dl}{2\pi}\right)^2 \Omega$,
   (c) $800\left(\frac{\pi dl}{\lambda}\right)^2 \Omega$, (d) $80\left(\frac{\pi dl}{\lambda}\right)^2 \Omega$ **[d]**

11. In a broadside array, the maximum radiation occurs
   (a) at 45° to the array, (b) along the array,
   (c) at 60° to the array, (d) perpendicular to the array. **[d]**

12. Electrical length of an antenna is
   (a) small than physical length.
   (b) greater than physical length.
   (c) equal to its physical length
   (d) none of the above. **[b]**

Chapter 12

Propagation of Electromagnetic Waves Through Ionized Media (Plasma Physics)

12.1 INTRODUCTION

Plasma is a state of matter characterized by a high, or even complete, ionization of its particles. Sometimes described as fourth state of matter, plasma, exhibits various basic features which are also common to conductors and semi-conductors. Generally, plasma can be assumed to be a mixture of three components: free electrons, positive ions and neutral atoms (or molecules). Depending upon the degree of ionization α the ratio of the concentration of ionized particles to the total concentration, a plasma is classified as: *weakly ionized* (α is a fraction of one percent); moderately ionized (α equals several percent) and fully ionized (α close to 100%). In nature, weakly ionized plasma is found in the ionosphere. The Sun, hot stars, and certain interstellar clouds are examples of fully ionized plasma, formed at very high temperatures (high temperature plasma). Artificially, plasma is produced in gas discharges and gas discharge tubes.

Because the particles in a plasma are charged, its behaviour differs in some respects from that of a gas. The control of its motion is the basis for the use of plasma as the working medium in various engines and for the direct conversion of its internal energy into electrical energy (magnetohydrodynamics, or MHD, generators, plasma sources of electric energy; etc.).

The concentration of the current carriers in a plasma is very high. Therefore a plasma is an excellent conductor. The mobility of the electrons is about three orders of magnitude greater than that of the ions. Hence, the current in a plasma is mainly set up by its electrons.

The propagation of electromagnetic waves in plasma, i.e. ionized medium has gained overwhelming importance because of various wave-wave and wave-particle interactions which have been used as diagnostic techniques for plasma fields and other plasma parameters. The studies in plasma phenomena, therefore, are of great interest and one has to understand the mechanism of various processes occuring in a plasma.

12.2 CONDITIONS FOR PLASMA EXISTENCE

In order to produce a plasma it is necessary to free electrons that are normally bound into atoms. The ionization energies for many elements run from several electron volts (eV) to a few tens of eV, e.g. for hydrogen it is 13.6 eV and for nitrogen it is 15.8 eV. There are various means where by these energies may be added to the atomic system to produce ionization. A convenient way is to impart the energy by collision events of one sort or another.

On the addition of the ionization energy to a fraction of the atoms of a neutral gas, an *ionized gas* is formed. If sufficient total energy can be added, the gas may be completely ionized with no neutral particles remaining.

One of the most powerful influences upon plasma behaviour is the electromagnetic interaction of the charged particles. Even a partially ionized gas can be affected by externally applied electric and magnetic fields and can conduct electric currents. The ions and electrons serve as charge carriers in somewhat the same way as they do in an electrolyte. Most importantly, the Coulomb electrostatic fields of the charged particles in the ionized gas will produce the interesting consequences of the collective effects. The presence of collective effects constitutes the primary plasma criterion. Plasma differs from a simple accumulation of charged particles in that it must have a minimum density. This is determined from the condition that $L \gg D$, where L is any length characterizing the plasma and D is the *Debye Screening radius* (also called the *Debye shielding distance* and *Debye length*). Now, we shall obtain expression for Debye length.

12.3 QUASI-NEUTRALITY OF A PLASMA: DEBYE SHIELDING

The criterion for the existence of a plasma can be written as

$$n_i \approx n_e \tag{12.1}$$

where n_i is ion density and n_e is electron density in a sample of plasma contained in a sphere of 1 cm radius. According to Schottky, one of the definitive properties of plasma is its quasineutrality expressed by (12.1). In an ensemble of ionized gases having approximately an equal number of electrons and ions, the phenomenon of charge particle shielding occurs. The negative and positive plasma particles are surrounded by particles of opposite charge. The electric forces which bind the oppositely charged particles in plasma provide for its quasineutrality, i.e. its tendency to balance negative and positive space charges in each macroscopic volume element. Since the velocities of electrons and ions are different ($v_e > v_i$, i.e. velocity of electrons, v_e being more than that of positive ions, v_i), it is obvious that more electrons leave the plasma boundary (e.g. discharge tube) than positive ions. Any separation of charges when electrons shift with respect to ions, gives rise to electric fields. These electric fields act in the direction of restoring neutrality. If plasma is not cold and has finite temperature, then some of the charged particles from the edge of the surround cloud escape out. This violates quasineutrality in a small volume of plasma. However, the space charge densities adjust themselves so that the major part of the plasma is shielded from the field. For a given temperature and concentration, the Debye shielding distance (δ) is determined by the balance of the potential and kinetic energies of the total number of charged particles taking part in the shielding process. Within a volume with the linear size x, if $x \gg \delta$, then the concentration of the opposite charges in this volume are approximately equal and neutrality is maintained. However, if $x \ll \delta$, one finds that the separation of charges has no significance on the motion of particles and the neutrality is isolated. We must note that the electric field is screened off at a distance equal to Debye length (δ). Now we will estimate δ.

Suppose that the charges, within the plasma are completely separated in a volume element with linear size δ. If T be the plasma temperature, then the potential energy of a charged particle in a given volume of plasma will be of the order of thermal motion of the particles, i.e. kT, where k is Boltzmann's constant. If we consider the electric field to be a continuous function of distance, then the electric field in a volume satisfies the Poisson's equation

$$\nabla \cdot \mathbf{E} = \frac{\rho}{\varepsilon_0} \tag{12.2}$$

where ρ is the charge density. We must remember that the condition of quasi-neutrality, and non-zero value of electric field ($\nabla \cdot \mathbf{E} \neq 0$), are known as plasma approximations. If ϕ is the potential, then

$$\mathbf{E} = \nabla \phi \tag{12.3}$$

The charge particle velocity distribution function in a plasma obeys the Maxwellian velocity distribution law, If n_i and n_e are the number of ions and electrons per unit volume respectively, then

$$\rho = (n_i e - n_e e) \tag{12.4}$$

where e is electronic charge. In the light of (12.3) and (12.4), Eq. (12.1) becomes

$$\nabla \cdot (-\nabla \phi) = \frac{n_i e - n_e e}{\varepsilon_0}$$

or

$$\nabla^2 \phi = \frac{-(n_i - n_e)e}{\varepsilon_0} \tag{12.5}$$

If n_0 is the mean density of charged particles in the *quasineutral plasma*, then one obtains from Boltzmann distribution law,

$$n_i = n_0 \, e^{-e\phi/kT} \quad \text{and} \quad n_e = n_0 \, e^{e\phi/kT} \tag{12.6}$$

where T is the absolute temperature of the plasma. Using (12.6), Eq. (12.5) takes the form

$$\begin{aligned} \nabla^2 \phi &= \frac{n_0 e}{\varepsilon_0} \left[\exp\left(-\frac{e\phi}{kT} \right) - \exp\left(\frac{e\phi}{kT} \right) \right] \\ &= \frac{2en_0}{\varepsilon_0} \sin h \left(\frac{e\phi}{kT} \right) \end{aligned}$$

In the region $\frac{e\phi}{kT} \ll 1$, one can take $\sin h \left(\frac{e\phi}{kT} \right) \approx \frac{e\phi}{kT}$. Thus we have

$$\nabla^2 \phi = \frac{2en_0}{\varepsilon_0} \frac{e\phi}{kT} = \frac{2n_0 \, e^2 \phi}{\varepsilon_0 kT} \tag{12.7}$$

or

$$\nabla^2 \phi = \frac{\phi}{\delta^2} \tag{12.8}$$

where

$$\delta = \left(\frac{\varepsilon_0 kT}{2n_0 e^2}\right)^2 \tag{12.9}$$

If charge distribution is spherically symmetric, it is independent of angular coordinates, i.e. and θ and ϕ (this ϕ is different from ϕ we have used earlier). Thus, one can express $\nabla^2\phi$ as

$$\nabla^2\phi = \frac{1}{r^2}\frac{d}{dr}\left(r^2\frac{d\phi}{dr}\right) \tag{12.10}$$

Using (12.10), Eq. (12.8) takes the form

$$\frac{1}{r^2}\frac{d}{dr}\left(r^2\frac{d\phi}{dr}\right) = \frac{2N_0 e^2\phi}{\varepsilon_0 kT} = \frac{\phi}{\delta^2} \tag{12.11}$$

Differential equation (12.11) has finite number of solutions. We are interested in one solution which satisfies the boundary conditions,

$$\phi = \frac{1}{4\pi\varepsilon_0}\frac{Q}{r} \text{ at } r \to 0 = \phi_0 \quad \text{(say) and} \quad \phi \to \text{ at}$$

$r \to \infty$. One can express the solution as

$$\phi = \frac{1}{4\pi\varepsilon_0}\frac{Q}{r}\exp\left(-\frac{r}{\delta}\right) = \phi_0\exp\left(-\frac{r}{\delta}\right) \tag{12.12}$$

$\delta = \left(\frac{kT\varepsilon_0}{2n_0e^2}\right)^{1/2}$ is known as *Debye length.* This indicates the shortest distance at which electrons moving at random in the plasma, screen the Coulomb field of the probe particle. At distance of δ (also called as *Debye Screening radius* or *Debye shielding distance*), the Coulomb field of an arbitrary charge of the plasma is shielded because charge is predominantly surrounded by oppositely charged particles. The Debye shielding distance δ depends directly on electron temperature and inversely on the electron density. Eq (12.12) further reveals that the effect of the plasma is to reduce the potential due to a charge object exponentially. A useful relation often used is

$$\lambda_D\left(=\frac{\delta^2}{2}\right) = 6.9\times10^{-2}\left(\frac{T_e}{n_0}\right)^{1/2} m \tag{12.13}$$

where T_e is in Kelvin, $\lambda_D = 7.4\left(\frac{kT}{n_0}\right)^{1/2}$ m when kT_e is in eV.

Langmuir defined plasma as an ionized gas. For quasi neutrality of plasma $n_e = n_i$ to better than one part in 10^6; it is necessary that the dimension of plasma $L \gg \delta$, i.e. the Debye length is small compared to linear dimension occupied by the gas. The plasma under such condition is *quasi-neutral* but not exactly neutral so that all the interesting electromagnetic forces also vanish.

12.4 PLASMA PARAMETER

In a plasma of density n_0 the average distance between particles is $n_0^{-1/3}$. The average potential energy is

$$|V| \approx \frac{q^2}{r} \approx n_0^{1/3} e^2 \tag{12.14}$$

The kinetic energy of a particle is given by

$$\frac{1}{2} m_s \langle v \rangle^2 = \frac{3}{2} kT = \frac{3}{2} m_s v_s^2 \tag{12.15}$$

For a plasma to exist, we require that the kinetic energy be much greater than the potential energy, i.e.

$$n_0^{1/3} e^2 \ll kT \tag{12.16}$$

Rearranging terms gives

$$n_0^{1/3} \frac{e^2}{kT} \ll 1 \tag{12.17}$$

or

$$\frac{kT}{n_0^{1/3} e^2} = \varepsilon_0 n_0^{2/3} \left(\frac{kT}{\varepsilon_0 n_0 e^2} \right) \gg 1 \tag{12.18}$$

We define

$$\wedge_s = 2\sqrt{2} n_0 \left(\frac{\varepsilon_0 kT}{2 n_0 e^2} \right)^{3/2} = 2\sqrt{2}\, n_0 \delta^3 \gg 1 \tag{12.19}$$

where $\wedge_s$ is the plasma parameter, as a necessary condition for a plasma to exist.

12.5 PLASMA OSCILLATIONS

Since a plasma has a tendency to be macroscopically neutral, it will tend to return to its neutral equalibrium state after a local perturbation is introduced in the form of an excess of positive or negative charge. If, for example, in Fig. 12.1 the striped region has an excess of electrons, this excess produces an electric field urging these electrons in the direction of the arrows. In this motion, the electrons acquire a certain kinetic energy, and at the end of a given time they will be able to pass through the equilibrium position. If too many electrons leave the striped region, there will be a deficiency of electrons here, and the electric field will tend to pull them back. One can thus picture electron oscillations in a plasma. We have assumed in these oscillations that the ions stay fixed.

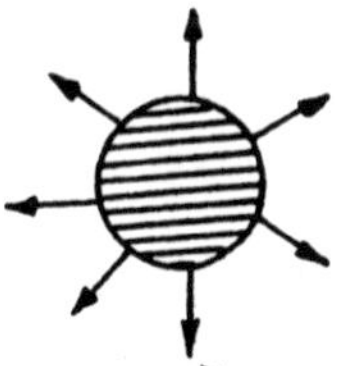

Figure 12.1. Electrical field produced by a region having an excess of electrons.

Figure 12.2. Scheme for calculating the frequency of oscillations.

The frequency of these oscillations can be calculated by means of a simple scheme of Fig. 12.2. In this figure a plane slice is considered containing a block of immobile ions, and the electrons which were associated with them are assumed to move out en masse. An electric field is created given by

$$E = \frac{\sigma}{\varepsilon_0} \qquad (12.20)$$

where σ is the surface density in the thin seath of positive charge on the left hand side of the figure or of negative charge on the right

$$\sigma = n_0 e x \qquad (12.21)$$

and

$$E = \frac{n_0 e x}{\varepsilon} \qquad (12.22)$$

Each electron of the cloud is then subject to a restoring force

$$f = -\frac{n_0 e^2 x}{\varepsilon_0} \qquad (12.23)$$

Oscillations will take place, the frequency being given by

$$\omega_p^2 = \frac{n_0 e^2}{m\varepsilon_0} \qquad (12.24)$$

i.e.

$$\omega_p = \left(\frac{n_0 e^2}{m\varepsilon_0}\right)^{1/2}. \qquad (12.24a)$$

while obtaining Eq. (12.24) by solving general equations of electron space charge oscillation, we have made following approximations: (i) the ions are assumed fixed and their density uniform (ii) thermal agitation is negligible (iii) collisions are negligible, (iv) the oscillations are of small amplitude and (v) there is no externally imposed magnetic field.

Rewritting 12.24(a) in the following form, we have

$$\nu_p = \frac{\omega}{2\pi} = \frac{1}{2\pi}\left(\frac{n_0 e^2}{m\varepsilon_0}\right)^{1/2} \tag{12.25}$$

The frequency ν_p is called the *plasma frequency* one obtains

$$\nu_p = 9\times 10^3 n_0^{1/2} \text{ cycles/s} \tag{12.26}$$

The hydrodynamic equations for the electrons can be written as

$$\frac{\partial n_e}{\partial t} + \nabla\cdot n_e\mathbf{v}_e = 0 \tag{12.27}$$

$$n_e m_e\left(\frac{\partial}{\partial t} + \mathbf{v}_e\cdot\nabla\right)\mathbf{v}_e = n_e q_e \mathbf{E} \tag{12.28}$$

In Eq. (12.27), the kinetic stress term, the collision term, and the magnetic field resulting from the oscillation have been neglected. Using Boltzmann formula, $n_e = n_0 \exp\left(\frac{e\phi}{kT}\right)$, (12.26) and (12.27) can be simplified in the form

$$n_e(\mathbf{r},t) = n_0 + a(\mathbf{r},t) \tag{12.29}$$

where $a(\mathbf{r},t)$ is a small perturbation. Eqs. (12.26) and (12.27) can therefore be linearized by neglecting the infinitesimals of higher order. If quantities vary at frequency ω, one can finally express (12.18) and (12.19) as

$$i\omega a + n_0\nabla\cdot\mathbf{v}_e = 0 \tag{12.30}$$

$$n_e m_e i\omega\mathbf{v}_e = n_e q_e \mathbf{E} \tag{12.31}$$

But on the other hand, we have

$$\nabla\cdot\mathbf{E} = \frac{(n_i q_i + n_e q_e)}{\varepsilon_0} = \frac{q_e a}{\varepsilon_0} \tag{12.32}$$

Eliminating $\mathbf{v}_e$ and a between these equations successively, one obtains

$$\mathbf{v}_e = q_e(i\omega m_e)\mathbf{E} \tag{12.33}$$

$$a = \left(\frac{n_0 q_e}{\omega^2 m_e}\right) \nabla \cdot \mathbf{E} \tag{12.34}$$

From which by (12.30),

$$\nabla \cdot \mathbf{E}\left(1 - \frac{\omega_p^2}{\omega^2}\right) = 0 \tag{12.35}$$

which requires that $\omega = \omega_p$.

The plasma oscillations represent stationary internal oscillations. If, as supposed, the electrons and ions do not have thermal motion, there is no wave propagation. The study of plasma oscillations in a situation involving thermal agitation has shown that they are able to propagate as a consequence of this agitation with a speed which is of the order of the thermal agitation speed of electrons and that, furthermore, their frequency is slightly modified and is expressed as

$$\omega^2 = \omega_p^2 + \left(\frac{3kT_e}{m}\right) k_1^2 \tag{12.36}$$

with

$$k_1 = \frac{2\pi}{\lambda} \tag{12.37}$$

We have considered electron osicllations in a plasma. Actually, ion oscillations in a plasma could be defined in a similar fashion, but their existence is subject to doubt. In reasoning about the electron oscillations in a plasma, it is validly assumed that the ions are fixed, as a consequence of their great inertia. The corresponding assumption made in the case of ion oscillations about a cloud of electrons seems a little unrealistic.

A more general investigation shows that three types of waves can exist in an ionized gas: (i) *Electromagnetic waves*, which in the simplest cases are only transverse. These waves are analogus to the usual electromagnetic waves propagating in a vacuum. Their speed is of the order of the speed of light in a vacuum. (ii) *The plasma waves* that are longitudinal and which only propagate due to thermal agitation with a low speed of the order of the speed of thermal agitation. (iii) *Magnetohydrodynamic waves*, which are somewhat similar to acoustic waves. They cause motion of electrons and ions with electromagnetic coupling. Their propagation speed is of the order of speeds of thermal agitation.

12.6 OCCURRENCE OF PLASMA

The various systems and situations in which plasmas occur fall into three broad categories: various *plasma devices* such as "neon" lights that are in common use, *laboratory plasmas* that are used to investigate plasma properties and to develop devices such as thermonuclear fusion reactors, and *cosmic or space plasma* that occur throughout most of the extra terrestrial universe.

For a plasma to be in a stationary state, processes are needed, that replenish the stocks of ions diminishing as a result of recombination. In high-temperature plasma, this is achieved as a result of thermal ionization, in gas-discharge plasma, as a result of collision ionization by electrons accelerated by an electric field. The ionosphere (one of the layers of the atmosphere) is a special variety of plasma. The high degree of ionization of molecules ($\sim 1\%$) is maintained in the ionosphere by photoionization due to the sun's short wave radiation.

12.7 PLASMA BEHAVIOUR IN ELECTRIC AND MAGNETIC FIELDS

We have read that charged particles are always present in plasma and the plasma is very frequently exposed to the effect magnetic and electric fields. Obviously, it would be worth while to study the behaviour of charge in these fields. When a charged particle is subjected to constant electric field then it leads to the formation of a thin sheath of space charge, which shields the major part of the plasma. This is not of any particular interest. Now, we consider following cases

(i) Charged particles in Homogeneous Magnetic Fields

Let us suppose that a particle of mass m, charge q be shot into a magnetic field of induction **B** with velocity **v** in a direction perpendicular to the lines of force as shown in Fig. 12.3. The magnetic force F_m on charge particle, i.e. Lorentz force acting at right angles to **v** as well as to B is

$$\mathbf{F}_m = q(\mathbf{v} \times \mathbf{B}) \qquad (12.38)$$

The magnitude of this force is $qvB\sin\theta$, where θ is the angle between **v** and **B** and the direction of force is normal to the plane containing **v** and **B**. The Lorentz force causes the particle to trace a circular path (as $\theta = 90°$) in the plane of the pater (we must remember that

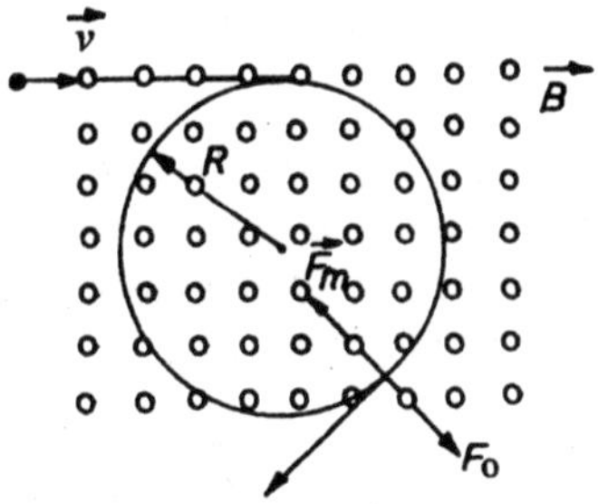

Figure 12.3. Charge particle in a homogeneous magnetic field.

when the particle enters parallel to the magnetic field, i.e. $\theta = 0$, then $F_m = qvB\sin 0 = 0$, i.e. the path of the particle remains undeflected in the magnetic field).

The magnetic force $\mathbf{F}_m$ must in every instant be in equalibrium with the necessary centripetal force for circular path, i.e.

$$qvB = \frac{mv^2}{R} \tag{12.39}$$

where R is the radius of the orbit. We have

$$R = \frac{mv}{qB} \tag{12.40}$$

This radius is known as the *Larmor radius* of the orbit. From (12.38), one obtains the angular frequency

$$\omega_c = \frac{v}{R} = \frac{qB}{m} = 2\pi v_c \tag{12.41}$$

This is known as *Cyclotron frequency*. We can see that ω_c is independent of the speed $\mathbf{v}$ of the particle. This provides the physical mechanism for the design of low energy cyclotron. $v_c = \dfrac{\omega_c}{2\pi}$ is known as linear frequency.

Let us consider the case when we shoot the charge particle into the magnetic field at an angle other than a right angle and $\theta \neq 0$, then one can resolve velocity v in two components, i.e. $\mathbf{v}_{11}$ parallel to the magnetic Field B and $v_\perp$ in the plane perpendicular to magnetic field B. Since $\mathbf{v}_{11} = v\cos\theta$ is unaffected by the field and path of the particle due to this is the straight line, one can write (38) as

$$q\mathbf{v}_\perp B = \frac{mv_\perp^2}{R}$$

or
$$R = \frac{mv_\perp}{qB} \tag{12.42}$$

The resulting path will be the superposition of the straight line and circular paths i.e. spiral or helix as shown in Fig. 12.4. Obviously, the particle is gyrating about the line of magnetic force. In other words the particle is "locked" to the field line.

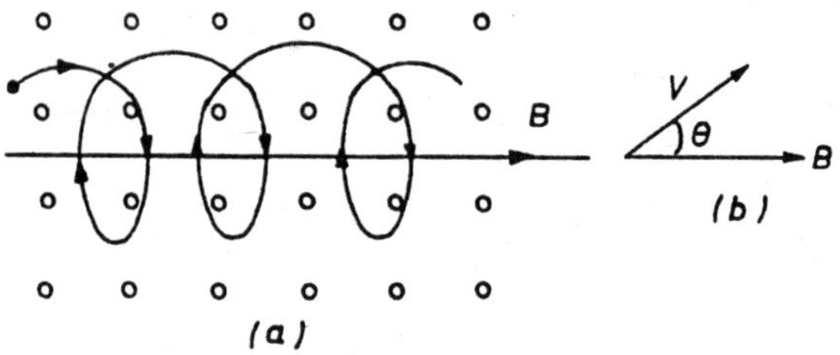

Figure 12.4. Helix or spiral path of a charge particle in a magnetic field.

(ii) Charged Particle in Simultaneous Electric and Magnetic Fields

When a charged particle of charge q and mass m enters simultaneously electric and magnetic fields **E** and **B** respectively, then the total force on the particle is

$$\mathbf{F} = \mathbf{F}_e + \mathbf{F}_m = q\mathbf{E} + q(\mathbf{v} \times \mathbf{B}) = q[\mathbf{E} + \mathbf{v} \times \mathbf{B}] \tag{12.43}$$

v being the velocity of the particle.

Let electric and magnetic fields are mutually perpendicular and the velocity **v** of the particle is perpendicular to both **E** and **B**. The force experienced by the particle due to the electric field $\mathbf{F}_e$ is along **E** and that exerted by magnetic field, **B**, $\mathbf{F}_m$ on the particle is perpendicular to **B**, i.e. parallel or antiparallel to **E**. If the magnitudes and directions of **E** and **B** fields, i.e. $\mathbf{F}_e$ and $\mathbf{F}_m$ are so adjusted that $\mathbf{F}_e$ and $\mathbf{F}_m$ are equal and opposite, the path of the charge particle will remain undeflected. Let us consider the case when **E** is along y-axis and **B** along z-axis with **v** along x-axis, then $\mathbf{F}_e = q\mathbf{E} = qE\hat{j}$ and $\mathbf{F}_m = q(\mathbf{v} \times \mathbf{B}) = q(v\hat{i} + B\hat{k}) = -q\mathbf{v}B\hat{j}$. Obviously $\mathbf{F}_m = \mathbf{F}_e$. The net result is zero.

If the magnitudes of $\mathbf{F}_e$ and $\mathbf{F}_m$ are equal, then $qE = q\mathbf{v}B$, i.e. $\mathbf{v} = \frac{E}{B}$. The velocity of the particle perpendicular to both **E** and **B** is called *drift velocity* and is denoted by v_D, i.e. $v_D = \frac{E}{B}$.

When the initial velocity of the particle is not equal to $\mathbf{v}_D$ or is not perpendicular to **B**, one obtains more complicated results. However, one can describe all these cases by particle gyration about the direction of **B** at an angular frequency $\omega_c = \frac{qB}{m}$, superimposed upon a linear drift motion of velocity $\mathbf{v}_0 = \dfrac{E}{B}$. As an example, let us consider that a positively charged particle of charge q moves with velocity $\mathbf{v} = \mathbf{v}_D + \mathbf{v}'$, then the force

$$\begin{aligned} \mathbf{F} &= q(\mathbf{E} + \mathbf{v} \times \mathbf{B}) = q(\mathbf{E} + (\mathbf{v}_D + \mathbf{v}') \times \mathbf{B}] \\ &= q(\mathbf{E} + \mathbf{v}_D \times \mathbf{B} + \mathbf{v}' \times \mathbf{B}) \end{aligned} \tag{12.44}$$

If we choose

$$\mathbf{v}_D = \frac{\mathbf{E} \times \mathbf{B}}{B^2} \tag{12.45}$$

then the first two terms on right hand side of (12.44) cancel each other and the force is given by

$$\mathbf{F} = q(\mathbf{v}' \times \mathbf{B})$$

which is same as studied earlier.

Fig. 12.5 shows the trajectory of charged particle. It is obvious from the figure that at point A the magnetic and electric forces act in the same sense and so thè radius of curvature is smaller, at the point P, the forces partly balance each after and therefore the radius of curvature is larger.

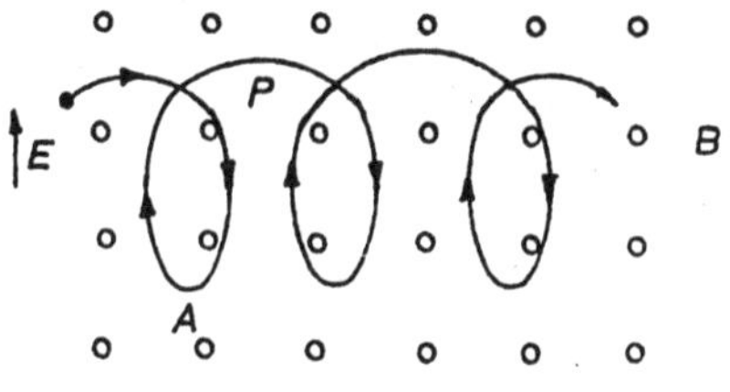

Figure 12.5.

(iii) Charge Particle in non-homogeneous Field

Let us consider that a charge particle of charge $+q$ enter a non-homogeneous magnetic field expressed by the density of lines of force in Fig. 12.6. One can easily see from figure that where the field is large, the lines of forces are more densely spaced and the curvature of the trajectory is more marked and vice versa. The result of this variation of radius of curvature during particle motion is also to induce a particle drift as can be seen in Fig. 12.6.

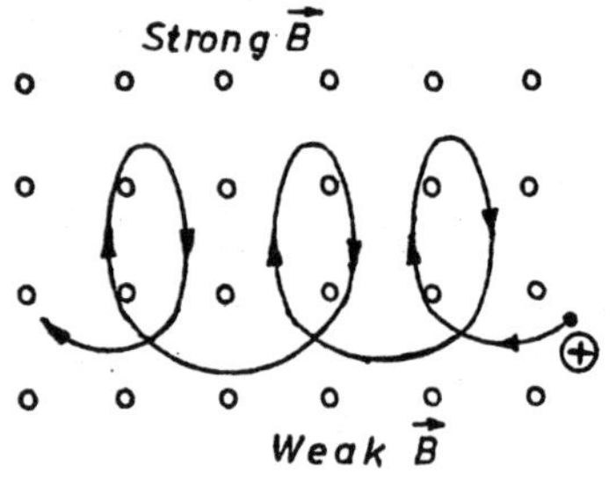

Figure 12.6. Charge particle in a non-homogeneous magnetic field.

A gyrating charged particle gives rise to a magnetic moment μ which is defined as

$$\mu = IA = \text{current due to particle gyration} \times \text{Area enclosed by that current}$$

$$= \frac{qv_\perp}{2\pi R}\pi R^2 = \frac{qv_\perp R}{2} = \pi R^2 B\left(\frac{q^2}{2\pi m}\right) = \frac{q^2}{2\pi m}\phi$$

$$\because \quad B = \frac{mv_\perp}{qR} \text{ and } \phi = \pi R^2 B) \tag{12.44a}$$

Obviously, the magnetic moment is proportional to the flux ϕ.

One can easily show that for slow variation of B either in space or time, μ is essentially constant.

The induced e.m.s. ε around a closed path in accordance with Faraday's law is given by

$$\varepsilon = -\frac{d\phi}{dt} = -\pi R^2\frac{dB}{dt} = \pi R^2\frac{dB}{dt} \text{ (numerically)} \tag{12.45}$$

When the charge q in the field is small, the work done upon a particle in one orbit is

$$W = q\varepsilon \tag{12.46}$$

But this work done must be equal to $\frac{dW}{dt} \times t$, where t is the time taken by the particle to go through one orbit, i.e. $t = \frac{2\pi R}{\mathbf{v}}$. We have

$$q\varepsilon = \frac{2\pi R}{\mathbf{v}}\frac{dW}{dt} = \frac{2\pi R}{\mathbf{v}}\frac{dW}{dB}\frac{dB}{dt} = q\pi R^2\frac{dB}{dt} \tag{12.47}$$

$$\therefore \quad \frac{dW}{dB} = \frac{qvR}{2} = \mu B \tag{12.48}$$

We also have

$$W = \frac{1}{2}mv^2 = \mu B \tag{12.49}$$

$$\therefore \quad \frac{dW}{dB} = B\frac{d\mu}{dB} + \mu \tag{12.50}$$

Comparing (11.49) with (11.47), one finds $\frac{d\mu}{dB} = 0$, i.e. μ is independent of B.

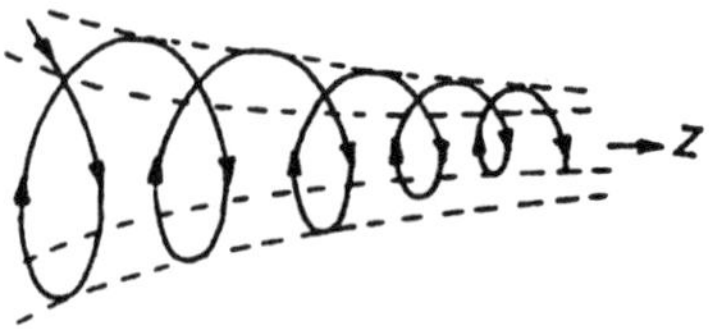

Figure 12.7.

The effect of such magnetic field variations can be treated by knowing the consequences of requiring the constancy of magnetic moment μ. For μ to be constant, Eq. 12.44(a) implies that the flux ϕ within particle orbit must remain constant. Obviously, in a magnetic field that varies slowly with position, as the magnetic field lines converge, the particle orbit is compressed in such a way as to maintain ϕ constant. This means that the particle is seen to move along the surface of a flux tube as shown in Fig. 12.7.

12.8 MAGNETOHYDRODYNAMICS

Now, we study the waves developing in a high-conductivity plasma which travels in a magnetic field. The magnetic lines of force in this case are "frozen" into the plasma so that if the plasma current is shifted somewhat so as to displace the magnetic lines of force, the magnetic field acts so as to shift the plasma current back. This gives rise to waves known as magnetohydrodynamic waves. In this section we shall treat a conducting fluid in particular plasma subject simultaneously to the laws of electromagnetic and hydrodynamics. The subject is usually known as Magnetohydrodynamics or simply hydromagnetics.

Let us consider an electrically neutral, conducting non-permeable fluid in electromagnetic fields. One can characterized it by a matter density $\rho(\mathbf{r},t)$, a velocity $\mathbf{v}(\mathbf{r},t)$, a pressure $p(\mathbf{r},t)$ taken to be a scalar and a real conductivity σ. One can express hydrodynamic equations as:

We have the continuity equation

$$\frac{d\rho}{dt} + \nabla \cdot (ev) = 0 \tag{12.51}$$

The force equation is

$$\rho \frac{d\mathbf{v}}{dt} = -\nabla \mathbf{p} + (\mathbf{J} \times \mathbf{B}) + \mathbf{F}_{\mathbf{v}} + \rho g \tag{12.52}$$

where $-\nabla \mathbf{p}$ and $\mathbf{J} \times \mathbf{B}$ are the pressure and magnetic force terms respectively, $F_{\mathbf{v}}$ is the vicous force and ρg is the gravitational force. We have the viscous force for an incompressible fluid as

$$F_v = \eta \nabla^2 \mathbf{v} \tag{12.53}$$

where η is coefficient of viscosity of the plasma. The time derivative of velocity is the *connective derivative* expressed as

$$\frac{d}{dt} = \frac{\partial}{\partial t} + \mathbf{v} \cdot \nabla \tag{12.54}$$

One obtains from (12.54) the total rate of change of a quantity moving instantaneously with velocity $\mathbf{v}$.

One can write the electromagnetic fields in the fluid after neglecting the displacement current as

$$\nabla \times \mathbf{E} = -\frac{\partial \mathbf{B}}{\partial t} \tag{12.55}$$

$$\nabla \times \mathbf{B} = \mu_0 \mathbf{J} \tag{12.56}$$

The condition $\nabla \cdot \mathbf{J} = 0$ implies the neglect of displacement current follows from (12.55) and from Faraday's law, we have $\frac{\partial}{\partial t} \nabla \cdot \mathbf{B} = 0$ and the requirement $\nabla \cdot \mathbf{B} = 0$ can be imposed as an initial condition. When displacement current is neglected, one may ignore Coulomb's law. The reason behind this is that the electric field is completely determined by curl equations and Ohm's law. If we retain displacement current in Ampere's law and $\nabla \cdot \mathbf{E} = \frac{\rho}{\varepsilon_0}$, where ρ being electric charge density, is take into account, we find that the corrections of only of the order of $\frac{\mathbf{v}^2}{c^2}$ result. However, these are completely negligible for normal magneto-hydrodynamic-problems.

To justify the specification of dynamical equations it is essential to specify the relation between current density $\mathbf{J}$ and $\mathbf{E}$ and $\mathbf{B}$. For a simple conducting medium having conductivity σ, Ohm's law is applicable and the current density is expressed as

$$\mathbf{J}' = \sigma \mathbf{E}' \tag{12.57}$$

where $\mathbf{J}'$ and $\mathbf{E}'$ are measured in the rest frame of reference. For a medium moving to laboratory frame the non-relativistic transformation of field **E** and **J** are as follows:

$$\mathbf{E} = \mathbf{E}' + \mathbf{v} \times \mathbf{B}' \tag{12.58}$$

and

$$\mathbf{J} = \mathbf{J}' + \rho \mathbf{v} \tag{12.59}$$

we have $\rho = 0$ for a one-component conducting fluid. Thus Ohm's law can be expressed as

$$\mathbf{J} = \sigma(\mathbf{E} + \mathbf{v} \times \mathbf{B}) \tag{12.60}$$

When the conductivity of the fluid is effectively infinite, then one finds that the condition is satisfied.

$$\mathbf{E} + (\mathbf{v} \times \mathbf{B}) = 0 \tag{12.61}$$

Eqs. (12.51), (12.52), (12.53), (12.56) and (12.60) supplemented by an equation of state for the fluid are called equations of magnetohydrodynamics.

Behaviour of Fluid in Electromagnetic Fields

The behaviour of a fluid in the electromagnetic fields is mainly governed by electrical conductivity of fluid. One finds that the effects are both electromagnetic and mechanical.

We have from (12.55), the time dependence of magnetic field as

$$\frac{\partial \mathbf{B}}{\partial t} = -\nabla \times \mathbf{E} \tag{12.62}$$

From (12.60), we have

$$\mathbf{E} = \frac{\mathbf{J}}{\sigma} - \mathbf{v} \times \mathbf{B} \tag{12.63}$$

Using (12.63), (12.62) becomes

$$\begin{aligned} \frac{\partial \mathbf{B}}{\partial t} &= -\nabla \times \left(\frac{\mathbf{J}}{\sigma} - \mathbf{v} \times \mathbf{B}\right) \\ &= \nabla \times (\mathbf{v} \times \mathbf{B}) - \nabla \times \left(\frac{\mathbf{J}}{\sigma}\right) \end{aligned}$$

Assuming σ to be constant in space and making use of (12.56), one obtains

$$\frac{\partial \mathbf{B}}{\partial t} = \nabla \times (\mathbf{v} \times \mathbf{B}) - \frac{1}{\mu_0 \sigma} \nabla(\nabla \times \mathbf{B})$$

$$= \nabla \times (\mathbf{v} \times \mathbf{B}) - \frac{1}{\mu_0 \sigma} [\nabla(\nabla \cdot \mathbf{B}) - \nabla^2 \mathbf{B}]$$

Since $\nabla \cdot \mathbf{B} = 0$, one obtains

$$\frac{\partial \mathbf{B}}{\partial t} = \nabla \times (\mathbf{v} \times \mathbf{B}) + \frac{1}{\mu_0 \sigma} \nabla^2 \mathbf{B} \tag{12.64}$$

If fluid is at rest, then $\mathbf{v} = 0$. Eq. (12.64) reduces to

$$\frac{\partial \mathbf{B}}{\partial t} = \frac{1}{\mu_0 \sigma} \nabla^2 \mathbf{B} \tag{12.65}$$

This is *diffusion equation.*

When the conductivity is large enough, one can neglect second term on R.H.S. in (12.64). In this case, one obtains the temporal behaviour of the magnetic field from

$$\frac{\partial \mathbf{B}}{\partial t} = \nabla \times (\mathbf{v} \times \mathbf{B}) \tag{12.66}$$

Eq. (12.66) indicates that magnetic lines of force also move with velocity $\mathbf{v}$ with the fluid. This means that the magnetic flux through any circuit moving with local fluid velocity is constant in time. We have assumed that conductivity is sufficiently large and therefore Eq. (12.61) is applicable.

With the help of vector analysis, the component of $\mathbf{v}$ perpendicular to $\mathbf{B}$ and symbolized by $\mathbf{w}$ is obtained as

$$\mathbf{w} = \frac{(\mathbf{B} \times \mathbf{v}) \times \mathbf{B}}{B^2} \tag{12.67}$$

Using (12.61), (12.67) reduces to

$$\mathbf{w} = \frac{(\mathbf{E} \times \mathbf{B})}{B^2} \tag{12.68}$$

One can understand the so called "$\mathbf{E} \times \mathbf{B}$" drift of both fluid and lines of force in terms of individual particle orbits of electrons and ions in the crossed electric and magnetic fields.

One can study the mechanical behaviour of the system with the form of Eq. (12.52).

$$\rho\frac{d\mathbf{v}}{dt} = \mathbf{F} - (\mathbf{J}\times\mathbf{B}) \tag{12.69}$$

where $\mathbf{F} = -\nabla p + \mathbf{F}_v + \rho g$ = sum of all the non-electromagnetic forces. Using sum of all the non-electromagnetic forces. Using (12.60), one obtains

$$\begin{aligned}\rho\frac{d\mathbf{v}}{dt} &= \mathbf{F} - \sigma\{\mathbf{E} + (\mathbf{v}\times\mathbf{B})\}\times\mathbf{B} \\ &= \mathbf{F} - \sigma B^2(\mathbf{v}_\perp - \mathbf{w})\end{aligned} \tag{12.70}$$

where $\mathbf{w} = \dfrac{(\mathbf{B}\times\mathbf{v})\times\mathbf{B}}{B^2}$ and $\mathbf{v}_\perp = \dfrac{(\mathbf{E}\times\mathbf{B})}{B^2}$ We have for fluids of infinite conductivity

$$\mathbf{v}_\perp = \mathbf{w}, \quad \text{i.e.} \quad \rho\frac{d\mathbf{v}}{dt} = \mathbf{F} \tag{12.71}$$

This shows that flow parallel to **B** is governed by non-electromagnetic forces only.

In the limit of large conductivity, one can conveniently relate **J** in force equation to **B** via Ampere's law and use infinite conductivity condition (12.61) to eliminate **E** from Faraday's law to yield (12.66).

Using $\nabla\times\mathbf{B} = \mu_0\mathbf{J}$, one can write the magnetic force term in Eq. (12.52) as

$$\mathbf{J}\times\mathbf{B} = \frac{1}{\mu_0}(\nabla\times\mathbf{B})\times\mathbf{B} = -\frac{1}{\mu_0}\mathbf{B}\times(\nabla\mathbf{B}) \tag{12.72}$$

Making use of vector identity

$$\nabla(\mathbf{B}\cdot\mathbf{B}) = 2(\mathbf{B}\cdot\nabla)\mathbf{B} + 2\mathbf{B}\times(\nabla\times\mathbf{B})$$

i.e.

$$\mathbf{B}\times(\nabla\times\mathbf{B}) = \frac{1}{2}\nabla(\mathbf{B}\cdot\mathbf{B}) - (\mathbf{B}\cdot\nabla)\mathbf{B}$$

one obtains (12.72) as

$$\begin{aligned}\mathbf{J}\times\mathbf{B} &= -\frac{1}{\mu_0}\left[\frac{1}{2}\nabla(\mathbf{B}\cdot\mathbf{B}) - (\mathbf{B}\cdot\nabla)\mathbf{B}\right] \\ &= \left[-\nabla\left(\frac{B^2}{2\mu_0}\right) + \frac{1}{\mu_0}(\mathbf{B}\cdot\nabla)\mathbf{B}\right]\end{aligned} \tag{12.73}$$

From (12.73), one finds that the *magnetic force* is equivalent to a *mag-*

netic hydrostatic pressure $p_m = \dfrac{B^2}{2\mu_0}$ plus a term which may be taken as an additional tension along magnetic field direction.

If one neglects the viscous effects and assume that the gravitational force is derivable from a potential, say ψ such that $\mathbf{g} = -\nabla\psi$, then one finds that (12.52) takes the form

$$\rho\frac{d\mathbf{v}}{dt} = -\nabla(p + p_m + \rho\psi) + (\mathbf{B}\cdot\nabla)\frac{\mathbf{B}}{\mu_0} \tag{12.74}$$

When one is dealing with simple geometrical situations, e.g. **B** having only one component, the last term in (12.74) representing additional tension, vanishes. One obtains

$$\rho\frac{d\mathbf{v}}{dt} = -\nabla(p + p_m + \rho\psi) \tag{12.75}$$

One finds that the static properties of the fluid can be described by

$$p + p_m + \rho\psi = \text{constant} \tag{12.76}$$

This obviously reveals that *apart from gravitational effects, any change in mechanical pressure is balanced by an opposite change in magnetic pressure*. If one wants that the fluid is to be confined within a certain region so that the mechanical pressure p falls rapidly to zero outside the region, then the magnetic pressure will have to rise equally rapidly.

12.9 PLASMA CONFINEMENT

In a high temperature plasma, the nuclei and electrons move in all directions at average speeds of several kilometers per second. Hence, unless confined in some way, all the particles would soon strikes the walls of a containing vessel, thereby imparting some of their energy to the walls. The particles would then return to the plasma with less energy, that is, at a lower temperature. Thus, the plasma would be cooled, and the fusion temperature could not be realized or maintained. Two quite different methods are currently being investigated (i) Magnetic confinement and (ii) Inertial confinement. We shall restrict ourself to the study of confinement by self magnetic fields plasma.

Magnetic confinement: Pinch Effect

The use of various magnetic field arrangements for confining high temperature plasmas to achieve controlled thermonuclear fusion have been made.

If one neglects the viscous and gravitational effects, the condition for equilibrium is $\mathbf{F} = 0$, i.e.

$$\nabla p - \mathbf{J} \times \mathbf{B} = 0 \quad \text{or} \quad \nabla p = \mathbf{J} \times \mathbf{B} \tag{12.77}$$

The scalar product with **B** and **J** yields

$$\mathbf{B} \cdot \nabla p = \mathbf{B} \cdot (\mathbf{J} \times \mathbf{B}) = 0 \tag{12.78}$$

and

$$\mathbf{J} \cdot \nabla p = \mathbf{J} \cdot (\mathbf{J} \times \mathbf{B}) = 0 \tag{12.79}$$

Eqs. (12.78) and (12.79) reveals that both **B** and **J** lie on the surfaces of constant pressure. If these surfaces are closed (say), then in accordance with (12.78) and (12.79), no **B**-lines or **J** lines can cross them. Obviously, one can view them as made up from a winding of **B** lines and of **J** lines (Fig. 12.8).

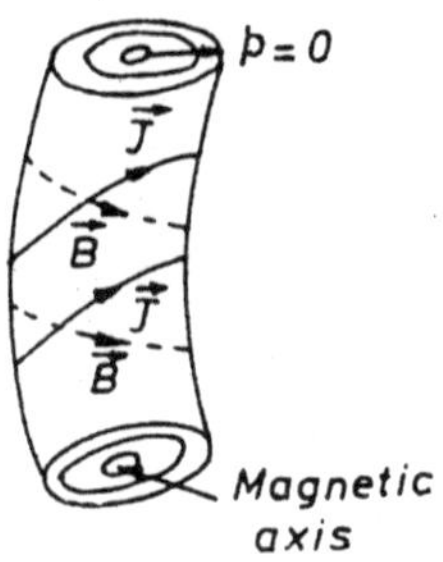

Figure 12.8.

One can see that the pressure increases from outside towards the axis and the force $\mathbf{J} \times \mathbf{B}$ also points towards the axis in these isobaric surfaces. This shows that the plasma is contained by the force $\mathbf{J} \times \mathbf{B}$ which is electromagnetic in nature. This is the basic principle of magnetic confinement of plasma.

The phenomena of confinement of a plasma by self magnetic fields is termed as "Pinch effect". To understand the pinch effect, let us consider an infinite cylindrical plasma or conducting fluid with an axial current density $J_z = J(r)$ and a resulting azimuthal magnetic induction $B_\phi = B(r)$. For simplicity, we have assumed the dependence of magnetic field, the current density, pressure etc. on distance r from the cylindrical axis. Moreover, we will neglect the effects of viscous and gravitation. Let us first consider whether a steady state condition can exist in which the plasma is mainly confined within a certain radius $r = R$ by the action of self magnetic field. We have $\mathbf{v} = 0$ for a steady state and hence one can write the equation of motion for the fluid

$$\rho \frac{d\mathbf{v}}{ds} = -\nabla(p + p_m + \rho\psi) + \frac{1}{\mu_0}(\mathbf{B} \times \nabla)\mathbf{B} \text{ in the form}$$

$$0 = -\frac{dp}{dr} - \frac{d}{dr}\left(\frac{B^2}{2\mu_0}\right) + \frac{B^2}{\mu_0 r} \tag{12.80}$$

we have Ampere's law in integral form relating to $B(r)$ to the current enclosed as

$$B(r) = \frac{\mu_0}{r}\int_0^r rJ(r)dr \tag{12.81}$$

when the fluid lies almost entirely inside $r = R$, then Eq. (12.81) for the magnetic induction inside the fluid can be expressed as

$$B(r) = \frac{\mu_0 I}{2\pi r} \tag{12.82}$$

where I is the total current flowing in the cylinder and given by

$$I = \int_0^R 2\pi r J(r) \tag{12.83}$$

One can write (12.80) as

$$\frac{dp}{dr} = -\frac{1}{2\mu_0 r^2}\frac{d}{dr}(r^2B^2) \tag{12.84}$$

Integration of (12.84) yields

$$p(r) = p_0 - \frac{1}{2\mu_0}\int_0^r \frac{1}{r^2}\frac{d}{dr}(r^2B^2)dr \tag{12.85}$$

where p_0 is the pressure of fluid at $r = 0$. If we take that the matter is confined to $r \leq R$, the pressure drops to zero at $r = R$. One obtains axial pressure as

$$p_0 = \frac{1}{2\mu_0}\int_0^R \frac{1}{r^2}\frac{d}{dr}(r^2B^2)dr \tag{12.86}$$

We can see from (12.82) that (12.86) vanishes for $r \geq R$. This means that one can replace the upper limit of the integration by infinity. Using (12.86), Eq. (12.85) reduces to

$$p(r) = \frac{1}{2\mu_0}\int_0^R \frac{1}{r^2}\frac{d}{dr}(r^2R^2)dr \tag{12.87}$$

One can express the overage pressure inside the cylinder as

$$\langle p\rangle = \frac{2\pi\int_0^R rp(r)dr}{\pi R^2} \tag{12.88}$$

Integrating and making use of (12.82) and (12.87), one obtains,

$$\langle p\rangle = \frac{2I^2}{\mu_0 R^2} \tag{12.89}$$

Obviously, we have obtained a relation between average pressure, total current I and the radius of cylindrical plasma confined by its self magnetic field. We must note that the magnetic pressure, $\frac{B^2}{2\mu_0}$, is equal to the average pressure of matter at the surface of the cylinder.

We must note that the temperature of hot plasma used for the production of controlled thermonuclear fusion is of the order of $10^8 K (\approx 80\,\text{keV})$ and density $\approx 10^{21}$ particles/m^3. These parameters correspond to pressure of $\approx 1.4 \times 10^6$ N/m^2 (= 14 atmosphere). Calculations reveal that a magnetic induction $B = 1.9$T at the surface, corresponding to a current $\approx 9 \times 10^2$R amperes is required for the confinement of this high temperature plasma. Obviously, very high currents are required to confine plasma at about 10^8 K temperature.

12.10 INSTABILITIES

Studies reveals that there is no stable equilibrium pinch, i.e. there is a slight deviation from the equilibrium state which tends to increase and finally results into the disintegration of the form of plasma. There is a large class of instability phenomena that is important to the dynamics of plasma. Most instabilities involve the collective motions of the plasma particles. Instabilities are important in a wide range of physical situations in which plasmas play a part.

That two important instabilities are: (i) magnetohydrodynamic instabilities, which can be explained in terms of magnetic pressure and stresses, and (ii) instabilities associated with particle motions.

Let us consider what is called a *kink instability* which develops when a linear pinch is bent due to some perturbation (Fig. 12.9). We can see that the magnetic lines of force attenuate, i.e. the field is weakened on the outside of the curve, and they grow denser, i.e. the field is intensified, on its inside.

Figure 12.9. Kink instability.

One finds that the stronger field causes stronger magnetic pressure in the bend which is conductive to further bending. This results ultimately into the breaking up of the linear pinch.

The second type of distortion observerd in plasma is *sausage instability* which occurs if the plasma column is pinched by chance (Fig. 12.10). In the neighbourhood of constriction the azimuthal induction increase (as $B \propto \frac{1}{r}$) causing a greater inward pressure at the neck, i.e. squeezed region than elsewhere. This causes further pinching ultimately disrupting the column.

In addition to the above mentioned instabilities, there are several other kinetic instabilities exist which are caused either by chance or by artificial deviation of the distribution of particles from Maxwellian distribution.

Attention has been given to inhibiting the sausage and kink instabilities by providing a "frozen in" axial magnetic field along the pinch. This axial field tends to slow the growth of kink instabilities by requiring that they expend energy in stretching this field. Sausage instabilities also tend to be inhibited because of the compression of the axial "*frozen in*" field required by their development. Nevertheless, it has not been possible in practice to prevent either form of pinch instability by use of such fields.

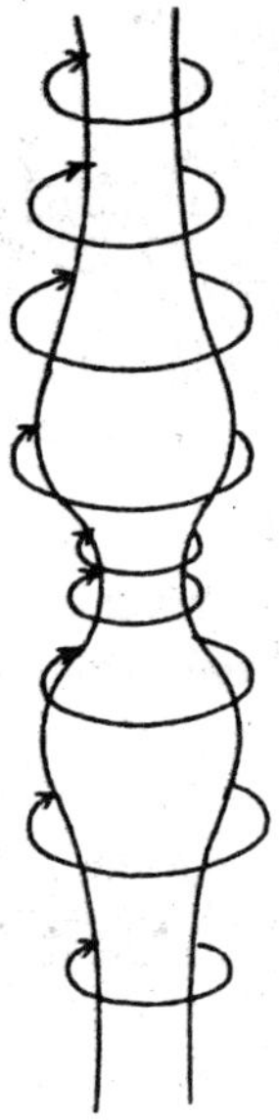

Figure 12.10. Sausage instability.

Various schemes for containing plamsa and overcoming the difficulty of instability are proposed. One such most promising system is '*tokamak*', -an axially symmetric toroidal system. In this system the plasma is contained by the magnetic field of the current which flows along the axis. This makes it possible to achieve values of $N\tau$ less than 10^{19} s/m^3 and ion temperature less than 10^7K. A very powerful magnetic field is applied parallel to the current which suppresses the plasma instability.

12.11 PLASMA WAVES

Although plasma, as a consequence of the interaction between particles, is a gas, oscillations and noises play a much greater role than in ordinary gas. First, in a plasma that is located in external fields and is not homogeneous, a wide variety of oscillations can occur because of the long-range interaction between particles. Secondly, these oscillations vary frequently and become amplified to a relatively high energy.

In this case, the plasma oscillations determine its parameters and development. We shall analyse briefly two cases

(i) Plasma-electron oscillations

When electrons are displaced with respect to ions in a system, high energy electron oscillations result. A plasma in which electrons and ions have been slightly but collectively displace relative to each other is shown in Fig. 12.11. The displacement distance δ is small compared to plasma thickness L. The displacement of electrons in a neutral plasma disturbs it. The displaced charges produce electric field in the inerior of the plasma. This field is given by

$$E = \frac{Ne\delta}{\varepsilon_0} \tag{12.90}$$

where N represents the density of charges in the plasma. In the inerior of the plasma, the force per unit area acting on the electrons is given by

$$NeEL = \frac{N^2e^2\,\delta L}{\varepsilon_0} \tag{12.91}$$

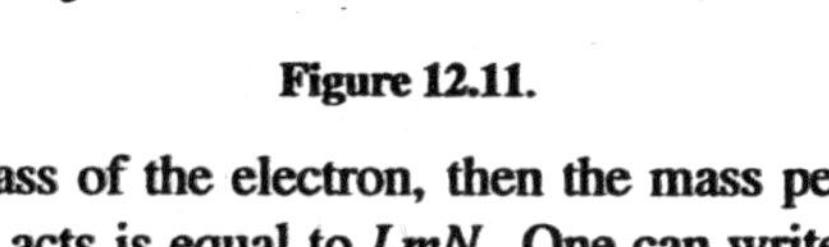

Figure 12.11.

If m is the mass of the electron, then the mass per unit area upon which this force acts is equal to LmN. One can write the equation of motion for each electron in the plasma as

$$LmN\frac{d^2\delta}{dt^2} + \frac{N^2e^2L\delta}{\varepsilon_0} = 0 \tag{12.92}$$

or

$$\frac{d^2\delta}{dt^2} + \frac{N^2e^2}{m\varepsilon_0}\delta = 0 \tag{12.93}$$

The solution of (12.93) is

$$\delta = F\exp[i\omega_p t] \tag{12.94}$$

where F is constant and

$$\omega_p = \left(\frac{Ne^2}{m\varepsilon_0}\right)^{1/2} = 2\pi\nu_p \tag{12.95}$$

Thus

$$\nu_p = \frac{1}{2\pi}\left(\frac{Ne^2}{m\varepsilon_0}\right)^{1/2} \tag{12.96}$$

is the oscillation frequency. ν_p is generally known as *plasma frequency*.

We must note that plasma oscillations are longitudinal in contrast to electromagnetic oscillations. Hence, the electric field due to plasma waves is directed along the wave vector.

One can also consider the oscillations that are due to the motion of the ions in a uniform isotropic plasma. The pecularities of these oscillations is determined by the large mass of the ions. Owing to its small mass, an electron can follow the motion of an ion so that the plasma remains quasineutral in any large volume containing a large number of charged particles: $n_i = n_e$. One obtains the following dispersion relation between the frequency (ω) and the wave number (k)

$$\omega = \left(\frac{T_e}{M}\right)^{1/2} k \tag{12.97}$$

where M is the ion mass.

The oscillations caused by the ion motion are known as the *ion sound*. Similar to the plasma oscillations, the ion sound is a longitudinal wave, i.e., the wave vector **k** is parallel to the oscillation vector of the electric field, **E**. The dispersion relation for the ion sound is similar to that for ordinary sound. This is due to the fact that both types of oscillations are characterized by a short-range interaction. In the case of the ion-sound, the interaction is short ranged because the electric field of propagating wave is screened by the plasma. This screening is effective if the wavelength of the ion sound is considerably larger than the *Debye-Huckel radius* (r_D) for the plasma where the sound propagation occurs,

$$kr_D \ll 1 \tag{12.98}$$

The dispersion relation (12.97) is valid for the ion sound if (12.98) is satisfied.

(ii) Alfven waves or Hydromagnetic waves

When ions are considered to be massive as compared to electrons, then it is justified to ignore their effect on the high frequency wave propagation. However, for low frequency electromagnetic wave propagation, the gyrofrequency of ions becomes comparable to the propagating wave frequency and the effect of ions play an important role.

Let us study the waves developing in a high conductivity plasma which travels in a magnetic field. The magnetic lines of force in this case are "frozen" into the plasma so that if the plasma current is "shifted" somewhat so as to displace the magnetic lines of force, the magnetic field acts so as to shift the plasma current back. This gives rise to waves known as magnetohydrodynamic waves.

For wave periods that are short compared to the diffusion times the magnetic fields of the wave are frozen into the fluid. They thereby force the fluid to move along the flux lines. Since the fluid is thus tied to the wave fields, the wave velocity is slowed by the inertial effects of the fluid. This situation is analogous to the propagation of waves along a string where the wave velocity is given in terms of tension T and the linear mass density ρ of the string as

$$V = \left(\frac{T}{\rho}\right)^{1/2} \tag{12.99}$$

Making the analogy to hydrodynamic plasma waves, we replace the string tension T by the magnetic tension along the field lines, $\frac{B^2}{\mu_0}$, and ρ becomes the plasma mass density. Thus for these hydrodynamic waves we might expect the velocity to be

$$V_A = \frac{B}{(\mu_0\rho)^{1/2}} \tag{12.100}$$

The velocity V_A is known as the *Alfven velocity*, and the hydrodynamic waves are known as the *Alfven waves*, after Hannes Alfven, who first recognized that such hydrodynamic waves are possible.

12.12 IONOSPHERE (REFLECTION FROM A PLASMA)

Aside from lighting and the aurora, perhaps the longest-studied aspect of the plasma environment of the earth is the ionosphere. As early as 1902 the presence of an ionized layer in the upper atmosphere was postulated by Heaviside and Kennelly to account for the anomalously long distances radio waves could be propagated.

Present day-techniques for study of the ionosphere include the reflection and transmission of radio signals by the ionized material. The reflection technique, which has been used since 1925, involves the use of ground-based radio transmitters and receivers that measure the transit time required for signals transmitted from the earth to be reflected from various levels of ionosphere. The reflection occurs from that region at which the local plasma frequency equals that of the electromagnetic signal. Now, we shall obtain the conditions for reflection from an ionized layer.

We have the following relation in the case of gases

$$\chi = \alpha N \tag{12.101}$$

where α is the polarizability of the atom and χ is the electrical susceptibility. Now,

$$\varepsilon_{r-1} = \alpha N \quad \text{or} \quad n^2 = 1 + \alpha N \tag{12.102}$$

The equation of a charged particle having charge q, in the presence of an electric field **E** is as

$$qE = m(\ddot{x} + l\dot{x} + \omega_0^2 x) \tag{12.103}$$

where $m\omega_0^2 x$ is the restoring force and $ml\dot{x}$ is the damping force. The solution of (12.103) is

$$x = \frac{qE/m}{-\omega^2 + il\omega + \omega_0^2} \tag{12.104}$$

we obtain the induced dipole moment as

$$\mathbf{p} = q\mathbf{x} = \frac{q^2\mathbf{E}/m}{-\omega^2 + il\omega + \omega_0^2} \tag{12.105}$$

But

$$\mathbf{p} = \varepsilon_0 \alpha \mathbf{E} \tag{12.106}$$

$$\therefore \quad \alpha = \frac{q^2/m\varepsilon_0}{-\omega^2 + il\omega + \omega_0^2} \tag{12.107}$$

We obtain

$$n^2 = 1 + \frac{Nq^2/m\varepsilon_0}{-\omega^2 + il\omega + \omega_0^2} \tag{12.108}$$

One can take $\omega_0 = 0$, as the free charges in the plasma are not tightly bound to ions. obviously, the restoring force is almost negligible. Eq. (12.108) reduces to

$$\begin{aligned} n^2 &= 1 + \frac{Nq^2/m\varepsilon_0}{-\omega^2 + il\omega} = 1 + \frac{(Nq^2/m\varepsilon_0)}{\omega^4 + l^2\omega^2}(-\omega^2 - il\omega) \\ &= 1 - \frac{Nq^2}{m\varepsilon_0}\left(\frac{\omega^2}{\omega^4 + l^2\omega^2}\right) - i\frac{Nq^2(l\omega)}{m\varepsilon_0(\omega^4 + l^2\omega^2)} \end{aligned} \tag{12.109}$$

One can express the complex refractive index in terms of its real and imaginary parts, i.e.

$$n = n_r - in_i \tag{12.110}$$

Squaring (12.110) and comparing with (12.109), one obtains

$$n_r^2 - n_i^2 = 1 - \frac{Nq^2}{m\varepsilon_0}\left(\frac{1}{\omega^2 + l^2}\right) \tag{12.111}$$

and

$$2n_r n_i = \frac{Nq^2}{m\varepsilon_0}\left(\frac{1}{\omega^3 + l\omega}\right)$$

or

$$n_i = \frac{Nq^2}{2m\varepsilon_0 n_r}\left(\frac{1}{\omega^2 + l^2\omega}\right) \tag{12.112}$$

Using (12.112), (12.111) becomes

$$n_r^4 - \left(1 - \frac{Nq^2}{m\varepsilon_0(\omega^2 + l^2)}\right)n_r^2 - \left(\frac{Nq^2 l}{2m\varepsilon_0(\omega^3 + l^2\omega)}\right)^2 = 0 \tag{12.113}$$

Let

$$a_1 = 1 - \frac{Nq^2}{m\varepsilon_0(\omega^2 + l^2)} \quad \text{and} \quad a_2 = \frac{Nq^2 l}{2m\varepsilon_0(\omega^3 + l^2\omega)}$$

One obtains

$$n_r^4 - a_1 n_r^4 - a_2^2 = 0$$

$$\therefore \quad n_r^2 = \frac{a_1 \pm \sqrt{a_1^2 + 4a_2^2}}{2}$$

We can see that for $\omega \ll l$, the term a_2 dominates the calculation of n_r. Obviously, we can take

$$n_r \approx \sqrt{a_2} = \left[\frac{Nq^2 l}{2m\omega\varepsilon_0(\omega^2 + l^2)}\right]^{1/2} \approx \left(\frac{Nq^2}{2m\varepsilon_0\omega l}\right)^{1/2} \tag{12.114}$$

Relation (12.114) is the expression for real part of the refractive index (n_r) at low frequencies. When $\omega \gg l$, i.e. at high frequencies, we can see that the term a_1 dominates the calculation of n_r we have

$$n_r \simeq \sqrt{a_1} = \left(1 - \frac{Nq^2}{m\varepsilon_0(\omega^2 + l^2)}\right)^{1/2}$$

$$\simeq \left(1 - \frac{Nq^2}{m\varepsilon_0\omega^2}\right)^{1/2} \qquad (12.115)$$

This means that at high frequencies the real part of the refractive index will be less than 1.

We must note that there is no sharp boundary to the ionosphere. Experiments have shown that the degree of ionization and, hence N varies in the ionospheric layer. Obviously, the value of n_r changes with the position. It is found that at each point the local value of n_r is related to the angle of incidence (θ_i) and angle of refraction (θ_T) through the following relation

$$n_r = \frac{\sin\theta_i}{\sin\theta_T} \qquad (12.116)$$

We can see that as the wave enters the ionospheric layer, n_r decreases because N increases, i.e. θ_T increases. We can further see that for $\theta_T = 90°$, the wave is just reflected. It is interesting to note that the reflection in the ionosphere corresponds to a gradual turning of the incident wave until it is out of the ionospheric layer as shown in Fig. 12.12.

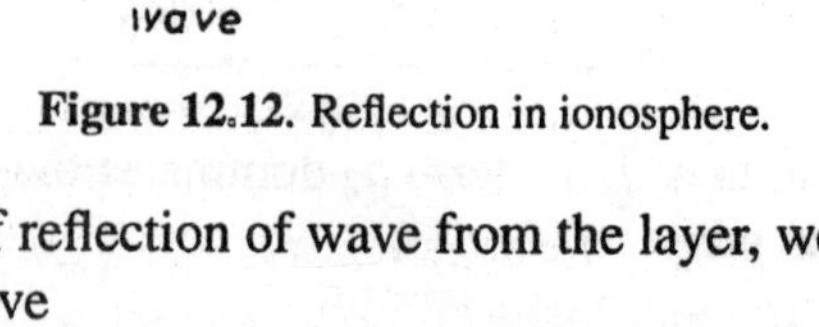

Figure 12.12. Reflection in ionosphere.

At the point of reflection of wave from the layer, we have $\theta_T = 90°$. Therefore, we have

$$n_r = \left(1 - \frac{Nq^2}{m\varepsilon_0\omega^2}\right)^{1/2} = \sin\theta_i$$

$$\therefore \quad \frac{Nq^2}{m\varepsilon_0\omega^2} = \cos^2\theta_i$$

or

$$\omega = \frac{1}{\cos\theta_i}\left(\frac{Nq^2}{m\varepsilon_0}\right)^{1/2} \tag{12.117}$$

When the wave is making normal incidence we have

$$\omega = \left(\frac{Nq^2}{m\varepsilon_0}\right)^{1/2} \tag{12.118}$$

or

$$2\pi\nu_c = \left(\frac{Nq^2}{m\varepsilon_0}\right)^{1/2}$$

or

$$\nu_c = \frac{1}{2\pi}\left(\frac{Nq^2}{m\varepsilon_0}\right)^{1/2} \tag{12.119}$$

where ν_c is the *critical or maximum* frequency for which reflection will occur at normal incidence. From (12.119) it is evident that $\nu_c \propto \sqrt{N}$, i.e. ν_c depends on the degree of ionization. We must note that ionization mainly depends on the radiation from the sun. This means ionization varies from place to place around the earth and from time to time. The time behaviour of the ionization regions is interesting and some what complex. Associates with solar flares are sudden ionospheric disturbances. During these disturbances the D region ionization is markedly increased. In the polar region the increased ionization may be so extensive as to prevent almost all radio noise from sources beyond the solar system from penetrating the ionosphere.

The day-to-night variation in the E region ionization is very great, amounting to something like two orders of magnitude in electron density. The F-region is divided in daytime, but the division disappears at night. The variability of this region is much greater than the others.

EXAMPLE 12.1: Consider a plasma consisting of two homogeneous streams of electrons with equal densities and equal but opposite velocities of magnitude v'. Show that waves with wave number $k < k_c = \sqrt{2}\,\frac{\omega_p}{v'}$, where ω_p is the plasma frequency, are unstable.

Solution: One can solve this problem easily in a frame of reference where one stream of electrons has velocity zero and the other has velocity $\mathbf{v}_0 = 2\mathbf{v}'$. We may then linearize about equilibrium with the assumptions

$$n = n_0 + n_1 \tag{1}$$

and

$$\mathbf{v} = \mathbf{v}_0 + \mathbf{v}_1 \tag{2}$$

where n_0 and $\mathbf{v}_0$ do not vary in time or space. The equation of continuity for the stream with velocity $2\mathbf{v}'$ is

$$\frac{\partial}{\partial t}(n_{01}+n_{11})+\nabla\cdot[(n_{01}+n_{11})(\mathbf{v}_{01}+\mathbf{v}_{11})]=0 \tag{3}$$

which reduces to

$$\frac{\partial}{\partial t}n_{11}+\mathbf{v}_{01}\nabla n_{11}+n_{01}\nabla\cdot\mathbf{v}_{11}=0 \tag{4}$$

Similarly, the equation of motion is

$$m_e(n_{01}+n_{11})\left[\frac{\partial}{\partial t}(\mathbf{v}_{01}+\mathbf{v}_{11})\ +\{(\mathbf{v}_{01}+\mathbf{v}_{11})\cdot\nabla\}(\mathbf{v}_{01}+\mathbf{v}_{11})\right] \tag{5}$$
$$(n_{01}+n_{11})qE_1$$

or

$$m_e n_{01}\frac{\partial}{\partial t}\mathbf{v}_{11}+m_1 n_{01}\mathbf{v}_{01}\nabla\cdot\mathbf{v}_{11}=n_{01}qE_1 \tag{6}$$

where we have assumed $E_0=0$. By comparison, the equation of continuity for the stream of electrons at rest is

$$\frac{\partial}{\partial t}n_{12}+n_{02}\nabla\cdot\mathbf{v}_{12}=0 \tag{7}$$

and the equation of motion is

$$m_e n_{02}\frac{\partial}{\partial t}\mathbf{v}_{12}=n_{02}qE_1 \tag{8}$$

Since we are looking for plane wave solutions, we let $\nabla\to ik$, and $\frac{\partial}{\partial t}\to -i\omega$. Eqs. (7) and (8) reduces to

$$-i\omega m_e n_{01}\mathbf{v}_{11}+ikm_e n_{01}\mathbf{v}_{01}\mathbf{v}_{11} = n_{01}qE_1 \tag{9}$$

and

$$-i\omega m_e n_{02}\mathbf{v}_{12} = n_{02}qE_1 \tag{10}$$

Using the fact that $\mathbf{v}_{01}=2\mathbf{v}'$ and $q=-e$, one obtains

$$\mathbf{v}_{12} = \frac{eE_1}{i\omega m_e} \tag{11}$$

and

$$\mathbf{v}_{22} = \frac{eE_1}{-i\omega m_e+ik2m_e\mathbf{v}'} \tag{12}$$

Eqs. (9) and (10) reduces to

$$-i\omega n_{11}+ikv_{01}n_{11}+ikn_{01}v_{11} = 0 \tag{13}$$

and

$$-i\omega n_{12}+ikn_{02}v_{12} = 0 \tag{14}$$

Using the values for $\mathbf{v}_{11}$ and $\mathbf{v}_{12}$ given in (11) and (12), one finds

$$n_{11} = n_{01}\left(\frac{-ikeE_1}{m_e(\omega-2v'k)^2}\right) \tag{15}$$

and

$$n_{12} = n_{02}\left(\frac{-ikeE_1}{m_e\omega^2}\right) \tag{16}$$

Substituting these expressions into Poisson's equation

$$ikE_1 = \nabla\cdot\mathbf{E} = \frac{\rho}{\varepsilon_0} = -\frac{e}{\varepsilon_0}(n_{11}+n_{12}) \tag{17}$$

yields

$$\begin{aligned} 1 &= \frac{e^2 n_0}{m_e\varepsilon_0}\left[\frac{1}{\omega^2}+\frac{1}{(\omega-2v'k)^2}\right] \\ &= \omega_{p,e}^2\left[\frac{1}{(\omega+v'k)^2}+\frac{1}{(\omega-v'k)^2}\right] \end{aligned} \tag{18}$$

This in turn reduces to

$$(\omega+v'k)^2(\omega-v'k)^2 = \omega_{p,e}^2\left[(\omega+v'k)^2+(\omega-v'k)^2\right] \tag{19}$$

or

$$\omega^4-\omega^2\left(2\omega_{p,e}^2\right)+\left(v'^4k^4-2v'^2k^2\omega_{p,e}^2\right)=0 \tag{20}$$

Solving for ω, one obtains

$$\omega^2 = \omega_{p,e}^2 \pm \sqrt{\omega_{p,e}^4+2\omega_{p,e}^2 v'^2k^2-v'^4k^4} \tag{21}$$

It is seen that if $k<\sqrt{2}\frac{\omega_{p,e}}{v'}$, then $\omega^2<0$, ω is imaginary, and the wave is unstable.

EXAMPLE 12.2: Consider a plasma in a uniform static magnetic field B_0. The plasma consists of singly charged ions of mass m_i and electrons of mass m_e with equilibrium plasma density n_0. Assuming that the ions are very cold in comparison to the electrons, $T_i \sim 0$, and that the electrons have a finite temperature T_e. Derive the dispersion relation for small amplitude electrostatic ion-cyclotron waves propagating nearly perpendicular to B_0. Assume that the electrons move along the magnetic field and are in Boltzmann equilibrium.

Solution: When we linearize about equilibrium, we have

$$n_e \to n_{0e}+n_{1e} \tag{1}$$

$$n_i \to n_{0i}+n_{1i} \tag{2}$$

$$\mathbf{v}_e \rightarrow \mathbf{v}_{0e} + \mathbf{v}_{1e} \tag{3}$$

$$\mathbf{v}_i \rightarrow \mathbf{v}_{1i} \tag{4}$$

The equations for continuity for the electrons and ions are

$$\frac{\partial}{\partial t} n_{e1} + n_{e0} \nabla \cdot \mathbf{v}_{el} = 0 \tag{5}$$

and

$$\frac{\partial}{\partial t} n_{i1} + n_{i0} \nabla \cdot \mathbf{v}_{i1} = 0 \tag{6}$$

respectively. The fact that the electrons are in Boltzmann equilibrium implies that

$$n_e = n_{e0} \exp\left(\frac{-eV}{kT_e}\right) \quad \text{or} \quad n_{e1} = n_{e0}\left(\frac{e\phi_1}{kT_e}\right) \tag{7}$$

where $\mathbf{E}_1 = -\nabla\phi_1$. Similarly because we are looking for plane wave solutions, Eq. (6) implies that

$$n_{i1} = n_{i0} \frac{k}{\omega} \mathbf{v}_{i1} \tag{8}$$

From the problem definition we may assume that both n_0 and B_0 are constant and uniform, and that $\mathbf{v}_0 = E_0 = 0$. The equations of motion for electrons and ions are

$$m_e n_{0e} \frac{\partial}{\partial t} \mathbf{v}_{el} = -e n_{e0}[\mathbf{E}_1 + \mathbf{v}_{e1} \times \mathbf{B}_0] + \mathbf{v} \times \mathbf{P}_e \tag{9}$$

$$m_i n_{0i} \frac{\partial}{\partial t} \mathbf{v}_{i1} = +e n_{i0}[\mathbf{E}_1 + \mathbf{v}_{i1} \times \mathbf{B}_0] \tag{10}$$

respectively. Because $m_e \ll m_i$, we assume $m_i = 0$. With the proper choice of geometry, B_0 in the z-direction with k, E in the $x-y$ plane, the vector equation of motion for the electrons reduces to

$$e n_{e0}[E_{1,x} + \mathbf{v}_{e1,y} B_0] - ik(k_B)T_e n_{e1} = 0 \tag{11}$$

The vector equation of motion for the ions gives

$$-i\omega m_i n_{0i} \mathbf{v}_{i1,x} = -e n_{i0}[-ik\phi_1 + v_{i1,y} B_0] \tag{12}$$

and

$$-i\omega m_i n_{0i} \mathbf{v}_{i1,y} = -e n_{i0}[v_{i1,x} B_0] \tag{13}$$

Solving these last two equations for $\mathbf{v}_{i1,x}$ gives

$$\mathbf{v}_{i1,x} = \frac{ek\phi_1}{\omega m_i}\left[1 - \frac{e^2 B_0^2}{\omega^2 m_i^2}\right]^{-1} \tag{14}$$

By definition $\Omega_i = \dfrac{e\,B_0}{m_i}$ is the ion cyclotron frequency, so

$$\mathbf{v}_{il,x} = \frac{ek\phi_1}{\omega m_i}\left(1 - \frac{\Omega_i^2}{\omega^2}\right)^{-1} \tag{15}$$

From Eq. (6),

$$\phi_1 = \frac{n_{e1}}{n_{e0}}\left(\frac{kT_e}{e}\right) \tag{16}$$

$$\mathbf{v}_{i1,x} = \left(1 - \frac{\Omega_i^2}{\omega^2}\right) = \frac{ek}{\omega m_i}\,\frac{n_{e1}}{\omega m_i}\left(\frac{k_B T_e}{e}\right) \tag{17}$$

In the plasma approximation, $n_e \sim n_i$, so

$$\frac{n_{e1}}{n_{e0}} \approx \frac{n_{i1}}{n_{01}} = \frac{k}{\omega}v_{i1,x} \tag{18}$$

Using this in the preceding equation gives

$$\mathbf{v}_{i1,x}\left[1 - \frac{\Omega_i^2}{\omega^2}\right] = \frac{ek}{\omega m_i}\,\frac{k_B T_e}{e}\,\frac{k}{\omega}\mathbf{v}_{i1,x} \tag{19}$$

or

$$\left(1 - \frac{\Omega_i^2}{\omega^2}\right) = \frac{k^2}{\omega^2}\,\frac{k_B\,T_e}{m_i} \tag{20}$$

Rearranging terms, gives

$$\omega^2 = \Omega_i^2 + k^2\left(\frac{k_B\,T_e}{m_i}\right) = \Omega_i^2 + k^2\mathbf{v}_s^2 \tag{21}$$

which is the dispersion relation for electrostatic ion cyclotron waves.

QUESTIONS

1. What do you understand by a plasma. Mention and explain the conditions for the existence of a plasma.
2. Derive the magneto-hydrodynamic equations and show that the magnetic force is equivalent to magnetic hydrostatic pressure $p_m = B^2/\mu_0$ plus an additional tension along the magnetic lines of force.
3. What is Pinch effect in plasma? Give a brief account of dynamical model of Pinch effect.

4. Explain kink, sausage and flute instabilities in a plasma. How can the plasma confinement can be overcome due to these instabilities.
5. Explain, what are hydromagnetic waves? Show that the phase velocity for plasma waves approximates to Alfven Velocity.
6. Write Short notes on
   (i) Pinch effect in Plasma.
   (ii) Kink, sausage and flute instabilities in plasma.
   (iii) Alfven waves.
   (v) Reflection from a plasma (Ionosphere).

PROBLEMS

1. A cylindrical column of gas of radius R carries a uniform axial current intensity $\mathbf{J}$. The column is surrounded by vacuum. Show that the pressure distribution necessary for equilibrium is distributed parabolically with radius, reaching maximum at $r = 0$, i.e. on the centre-line and can be represented as $P = \frac{\mu_0 J^2}{4}(R^2 - r^2)$.
2. The sun's corona consists of ionized hydrogen with approximately 10^{12} particles per cubic meter. If the field in the corona, $B \approx 1000$ gauss, show that the phase velocity of the Alfven waves is 8.9×10^3 m/s.
3. A current flows axially in a cylindrical tube of plasma. It is observed that the plasma pressure vanishes at $r = r_0$ and $P = N(r)kT$. If $N = \int_0^{r_0} n(r) 2\pi r dr$, then show that $NkT = \frac{\pi}{4\mu_0}(rB)^2_{r=r_0}$.

SHORT ANSWER QUESTIONS

1. What do you understand by a plasma state of matter?

   [**Ans.** It is a state of matter characterized by a high, or even complete, ionization of its particles.]

2. Depending upon the degree of ionizations, in how many categories a plasma can be classified?

   [**Ans.** (a) weakly ionized, (b) moderately ionized, and (c) fully ionized.]

3. What is the basis for the use of plasma as the working medium in various engines and for MHD, generators, plasma sources of electric energy, etc.

   [**Ans.** The control of Plasma motion.]

4. What is the property of plasma which brings it near to those of conductors? [**Ans.** High electric conductivity.]

5. How a plasma is regarded thermodynamically?

   [**Ans.** As an ideal gas.]

6. What is the condition for the production of a fully ionized plasma?

   [**Ans.** $kT = e\phi_i$, where ϕ_i is the ionization potential of the atoms of the gas. For hydrogen and deuterium, this corresponds to a temperature $T \simeq 160{,}000$ K.]

7. What kinds of radiation of plasma are in the optical range and in the far ultraviolet?

   [**Ans.** Breaking (bremsstrahlung) radiation of the electrons, occuring when they are retarted by the ions; recombination radiation, accompanying the recombination processes; and ordinary radiation produced by the excited particles-of the spectral lines. Moreover, in a magnetic field, betatron (synchrotron) radiation is also possible.]

OBJECTIVE QUESTIONS

1. The Characteristic linear parameter δ (Debye length) for a plasma is given by

   (a) $\delta \simeq \dfrac{\varepsilon_0 kT}{N_0 e^2}$ (b) $\delta \simeq \left(\dfrac{\varepsilon_0 kT}{N_0 e^2}\right)^{1/2}$

   (c) $\delta \simeq \left(\dfrac{\varepsilon_0 kT}{N_0 e^2}\right)^{2}$ (d) $\delta \simeq \left(\dfrac{\varepsilon_0 kT}{N_0 e^2}\right)^{3/2}$ **[b]**

2. In magnetic confinement-Pinch effect, the force which contain the plasma is given by

   (a) $\mathbf{J} \times \mathbf{B}$ (b) $\mathbf{E} \times \mathbf{B}$ (c) $\frac{B^2}{2\mu_0}$ (d) $\frac{\mathbf{E}}{\mathbf{B}}$ **[a]**

3. The condition for stable equilibrium of plasma as a conducting fluid-MHD is

   (a) $\frac{B^2}{2\mu_0} = \text{Constant}$ (b) $P + \frac{B^2}{2\mu_0} = 0$

   (c) $P + \frac{B^2}{2\mu_0} = \text{Constant}$ (d) $P = 0$ **[c]**

4. Plasma frequency is given by

   (a) $\nu_p = \frac{1}{2\pi}\left(\frac{Ne^2}{m\varepsilon_0}\right)$ (b) $\nu_p = \frac{1}{2\pi}\left(\frac{Ne^2}{m\varepsilon_0}\right)^{3/2}$

   (c) $\nu_p = \frac{1}{2\pi}\left(\frac{Ne^2}{m\varepsilon_0}\right)^2$ (d) $\nu_p = \frac{1}{2\pi}\left(\frac{Ne^2}{m\varepsilon_0}\right)^{1/2}$ **[d]**

5. The Alfven velocity is given by

   (a) $\mathbf{v}_A = \frac{B}{\mu_0 e}$ (b) $\mathbf{v}_A = \frac{B}{(\mu_0 e)^{1/2}}$

   (c) $\mathbf{v}_A = \sqrt{\frac{B}{\mu_0 \rho}}$ (d) $\mathbf{v}_A = \frac{B^2}{\mu_0}$ **[b]**

Chapter 13

Relativistic Electrodynamics

13.1 INTRODUCTION

We all are aware that the theory of special relativity was developed to prove the invariance of Maxwell's field equations and the unify the electrical and magnetic phenomena. We have seen that Maxwell studied the consistency of existing experimental laws, such as Gauss, Faraday and Ampere's and unified these laws by adding a new term in Ampere's Law. The special theory of relativity was propounded by Einstein in 1905 with a view to reconcile with certain results of Maxwell's field equations, i.e. Maxwellian electrodynamics. In the present chapter we will see that the laws of Ampere and Faraday and their generalization to Maxwellian electrodynamics are consistent with the implications and predictions of Einstein's special theory of relativity.

13.2 GALILEAN RELATIVITY

According to the '*principle of relativity*' all laws of nature are identical in all inertial systems of references. Obviously, the equations expressing the laws of nature retain their form in different inertial systems; i.e. they are invariant with respect to transformation of coordinates and time from one system to another. Large number of carefully performed experiments has shown that the principle is valid.

Let us consider two inertial frames S and S' with parallel axes as shown in Fig. 13.1. Suppose S' is moving along x-direction with a uniform velocity $\mathbf{v}$ relative to frame S. Any event occuring at a point will be specified by an observer in S by the coordinates (x, y, z, t). The

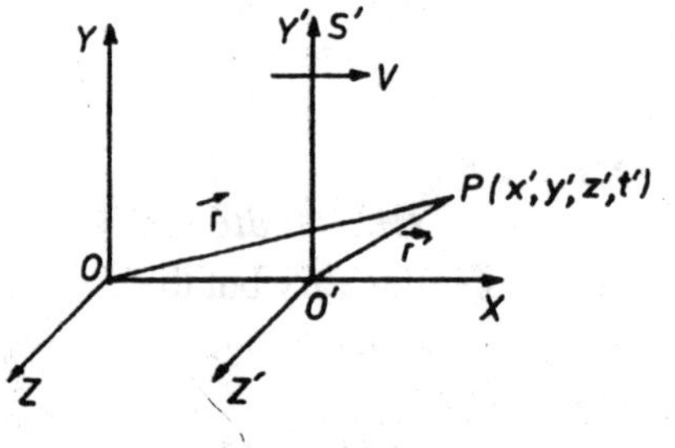

Figure 13.1.

same event will be described by an observer in S' by (x', y', z', t'). One can write the transformation of coordinates of a point from one system to another is clearly

$$x' = x - \mathbf{v}t, \quad y' = y, \quad z' = z \quad \text{and} \quad t' = t \qquad (13.1)$$

The last relation in (13.1) is a consequence of the implicit assumption of classical physics that time is *absolute*, i.e. independent of a frame of reference. These simple transformations are known as *Galilean transformations*. The form of Eqs. (13.1) depends, of course, on the relative motion of two frames of reference, but it also depends upon certain assumptions regarding the nature of space and time. As stated earlier that it is assumed that *time t is independent* of any particular frame of reference, i.e., if t and t' be the times recorded by the observer o and o' of an event occuring at P, then $t' = t$. The other assumption, regarding the nature of space, is that the distance between two points or particles is quite independent of any particular frame of reference, i.e. according to the Galilean transformation the space-interval measurements are absolute, i.e. they are same for all inertial observers. From (13.1) we have

$$\frac{dx'}{dt'} = \frac{dx'}{dt} = \frac{dx}{dt} - \mathbf{v} \quad (\because \quad t' = t)$$

$$\frac{dy'}{dt'} = \frac{dy}{dt}$$

$$\frac{dz'}{dt'} = \frac{dz}{dt}$$

and, hence,

$$\frac{d^2x'}{dt'^2} = \frac{d^2x}{dt^2}$$

$$\frac{d^2y'}{dt'^2} = \frac{d^2y}{dt^2}$$

$$\frac{d^2z'}{dt^2} = \frac{d^2z}{dt^2}$$

This shows that though velocities vary but the acceleration of a particle relative to s and s' frames are identical. The Newton's equation of motion in the two systems is

$$\mathbf{F}_i = m\ddot{\mathbf{x}}_i = m\ddot{\mathbf{x}}'_i = \mathbf{F}'_i, \quad \text{where} \quad i = 1,2,3. \tag{13.2}$$

Obviously, the laws of classical mechanics invariant under the Galilean transformation. The real difficulty was experienced when these transformations were applied to electric and magnetic fields predicted by Maxwell's field equations. Now we examine the behaviour of laws of electromagnetism under Galilean transformations. We know that the following scalar equation results from Maxwell's theory of electromagnetism

$$\nabla^2\psi - \frac{1}{c^2}\frac{d^2\psi}{dt^2} = 0 \tag{13.3}$$

Galilean transformation gives

$$\frac{\partial}{\partial x} = \frac{\partial}{\partial x'} + \frac{1}{\mathbf{v}}\frac{\partial}{\partial t'}$$

$$\therefore \quad \frac{\partial^2}{\partial x^2} = \frac{\partial}{\partial x'^2} + \frac{1}{\mathbf{v}^2}\frac{\partial^2}{\partial t'^2} + \frac{2}{\mathbf{v}}\frac{\partial^2}{\partial t'\partial x'}$$

$$\text{and} \quad \frac{\partial}{\partial y^2} = \frac{\partial^2}{\partial y'^2}, \frac{\partial^2}{\partial z^2} = \frac{\partial^2}{\partial z'^2} \quad \text{and} \quad \frac{\partial^2}{\partial t^2} = \frac{\partial^2}{\partial t'^2} \tag{13.4}$$

Using (13.4), Eq. (13.3) gives

$$\nabla'^2\psi - \frac{1}{c^2}\frac{\partial^2\psi}{\partial t'^2} + \frac{1}{\mathbf{v}^2}\frac{\partial^2\psi}{\partial t'^2} + \frac{2}{\mathbf{v}}\frac{\partial^2\psi}{\partial t'\partial x'} = 0 \tag{13.5}$$

We see that (13.4) and (13.5) are not the same wave equations, i.e. the form of the wave equation is not preserved by the substitution (13.1). One can easily see that Maxwell's equations are variant with the Galilean transformations. We obtain the electric field in the moving frame as

$$\mathbf{E}' = \mathbf{E} + \mathbf{v} \times \mathbf{B} \tag{13.6}$$

and $$\mathbf{B}' = \mathbf{B} - \frac{1}{c^2}(\mathbf{v} \times \mathbf{B}) \tag{13.7}$$

where $$\mathbf{v} = \hat{x}\mathbf{v}_x, \quad \mathbf{v}_y = \mathbf{v}_z = 0$$

We can easily see that these are inconsistent and reveals that the invariance of physical laws with Galilean transformation (13.1) is not valid. This can also be seen if we recall that the electromagnetic waves, as found from Maxwell's electromagnetic theory propagate with a constant velocity c, while the velocity measured by the observer in frame S' moving with velocity $\mathbf{v}$ relative to the frame S will be $c + \mathbf{v}$ or $c - \mathbf{v}$ depending on the direction of relative motion. Obviously, the velocity of electromagnetic waves is not invariant under the Galilean transformation, i.e. Maxwell's equations will also change their form on transformation from one system to another as evident from (13.6) and (13.7). In order to resolve this conflict, the bold new concept of the special theory of relativity was introduced by *A*. Einstein in 1905.

13.3 BASIC POSULATES OF SPECIAL THEORY OF RELATIVITY

The two postulates are:

(i) *All physical phenomena* (mechanical, electromagnetic etc.) *can be described by the same physical laws by two observers moving with constant linear velocity with respect to each other* or in other words, it is impossible to ascertain, by means of any experiments, whatsoever, conducted in a closed system of bodies, whether the system is at rest or is travelling at uniform velocity in a straight line with respect to some inertial frame of reference.

The laws of physics may be expressed in equations having the same from in all frames of reference moving at constant velocity with respect to one another.

This postulate expresses the absence of a universal frame of reference.

(ii) *The free space speed of light is same for all observers regardless of their motion.*

or

The speed of light in vacuum is independent of the speed of the source of this light and is same in all directions.

According to the first postulate, this speed is the same in all inertial frames of reference, being thereby a universal constant. This postulate follows directly form the results of the Michelson-Morley experiment and many others. Recently, it has been decisively verified. For example the speed of radiation emitted in the decay of π-meson as measured Farley et al. was $(2.9979 \pm 0.0003) \times 10^8$ m/s although the mesons were moving at a speed greater than 0.99975C.

Of the two postulates, the second represents an experimental fact, whereas the first is a generalization from a wide range of physical experience. The first postulate is in no way self evident; it is a hypothesis to be tested by experiment. Einstein decided that the conflict between them must be due to an imperfection of the classical ideas about *simultaneity.*

13.4 THE THEORY OF SIMULTANEITY

Time in classical relativity (in contrast to the coordinates) is regarded as being independent of the frame of reference. Two events which are simultaneous in one inertial frame are considered to be simultaneous in any other inertial frame of reference. This contradicts the special theory of relativity which leads to relativity of simultaneity (events). According to the principle of special relativity, time elapses different frames of reference, and the specification of an interval of time between two events has a meaning only if the frame of reference is indicated i.e., the *concept of simultaneity is only a relative concept.*

13.5 LORENTZ TRANSFORMATION

A concept of inertial frame was developed. Inertial frames are those frames in which laws of inertia hold good. In an inertial frame of reference all bodies move with a constant velocity unless they are subjected to external force. All inertial frames move with a constant velocity relative to one another. Now we obtain the equation of transformation between two inertial systems which would replace Galilean transformation and which would keep the speed of light constant. The new set of transformation equations we are going to obtain is known as *Lorentz transformation* equations. These were developed to relate space and time coordinates in different inertial frames of reference. We consider two inertial frames of reference S and S' having a common origin and coincident axes system at $t = 0$, and S' moving along x-axis with a constant relative $\mathbf{v}$ as shown in Fig. 13.2.

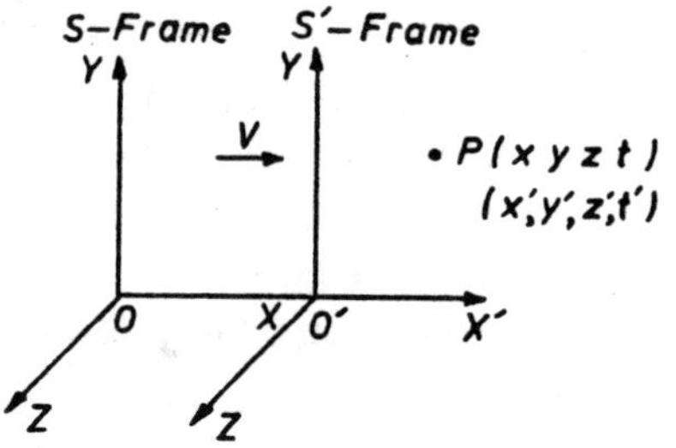

Figure 13.2.

Now consider a light signal emitted in free space at the instant $t = 0$ from the coincident origins of the systems S and S'. When the light signal reaches at the point P, let the position and times measured by the observers O and O' be (x, y, z, t) and (x', y', z', t') respectively. If the velocity of light signal is c, then the time required by the light signal in traversing the distance OP in frame S is

$$t = \frac{OP}{c} = \frac{\sqrt{x^2 + y^2 + z^2}}{c}$$

or

$$x^2 + y^2 + z^2 - c^2t^2 = 0 \tag{13.8}$$

The time required by the same light signal in travelling the distance $O'P$ in frame S', as seen by the observer at rest at the origin of S' is

$$t' = \frac{O'P}{c} = \frac{\sqrt{x'^2 + y'^2 + z'^2}}{c}$$

or

$$x'^2 + y'^2 + z'^2 - c^2t'^2 = 0 \tag{13.9}$$

The velocity of light c is same in both the frames of references (according to second postulate of Einstein's special relativity).

The Galilean transformation connects measurements in the two frames according to the transformation equations (13.1). Substituting (13.1) in (13.9), one obtains

$$x^2 - 2xvt + v^2t^2 + y^2 + z^2 = c^2t^2$$

and this is not in agreement with (13.8) in any way. This clearly indicates that the Galilean transformation fails. Therefore, we want new transformation equations which transforms Eq. (13.9) to Eq. (13.8). The transformation must be linear in x and t, because we want to get a sphere which expand at a uniform rate. Therefore, we try the relations

$$\left.\begin{aligned} x' &= k(x - vt) \\ y' &= y \\ z' &= z \\ t' &= k'(t - bx) \end{aligned}\right\} \tag{13.10}$$

where k, k' and b are constants. Making these substitutions in (13.9), one obtains

$$k^2(x^2 - 2xvt + v^2t^2) + y^2 + z^2 = c^2k'^2(t^2 - 2bxt + b^2x^2)$$

or

$$(k^2 - k'^2c^2b^2)x^2 - 2(2k^2v + k'^2c^2b)xt + y^2 + z^2 = (k'^2 - k^2v^2/c^2)c^2t^2$$

This must be identical to (13.8). Equating these two, one obtains

$$\begin{aligned} k^2 - k'^2b^2c^2 &= 1 \\ k^2v^2 - k'^2bc^2 &= 0 \\ k'^2 - k^2v^2/c^2 &= 1 \end{aligned}$$

Solving these equations for k, k' and b, one obtains

$$k = k' = \left(1 - \frac{v^2}{c^2}\right)^{-1/2}$$

and $$b = \frac{\mathbf{v}}{c^2}$$

The new transformation equations which are essentially based on the invariance of the velocity of light, is then

$$\left.\begin{aligned} x' &= k(x - vt) = \frac{x - vt}{\sqrt{1 - v^2/c^2}} \\ y' &= y \\ z' &= z \\ t' &= k(t - bx) = \frac{t - \frac{vx}{c^2}}{\sqrt{1 - v^2/c^2}} \end{aligned}\right\} \tag{13.11}$$

This set of transformation equations is known as the Lorentz transformation. The Lorentz transformations are symmetrical and retain their form in the transformation from S' to S except that the sign of V is changed. Thus the corresponding inverse transformation equations are

$$\left.\begin{aligned} x &= \frac{x' + vt'}{\sqrt{1 - v^2/c^2}}, \quad y = y', \; z = z' \\ \text{and} \quad t &= \frac{\left(t' + \frac{vx'}{c^2}\right)}{\sqrt{1 - v^2/c^2}} \end{aligned}\right\} \tag{13.12}$$

The Lorentz transformations are linear and at low velocities ($v/c \ll 1$) they become the Galilean transformations. In most of the cases, which we encounter on earth, $v \ll c$ so that the results of Lorent transformations do not differ to any great extent from those of the Galilean transformations.

From these equations, we can easily confirm the statement that events which happen at the same place at different times, as viewed from one frame, may be seen from another frame to happen at different place as well. Similarly, a difference in spatial position w.r.t. One frame may correspond to a difference in both space and time w.r.t. another. Thus a space difference can be converted partly into a time difference, or vice versa, merely by changing the frame of reference that is being used. Obviously, Lorentz transformations represent a most profound conceptual change specially with regard to *space and time*.

It is convenient to make use of some standard notations. Abbreviating the quantities v/c and $\left(1 - \frac{v^2}{c^2}\right)^{1/2}$ as $\beta = \frac{v}{c}$ and $\gamma = \frac{1}{\sqrt{1 - v^2/c^2}}$, Lorentz transformations can be expressed as

$$\left.\begin{aligned} x' &= \frac{x - vt}{\sqrt{1 - \beta^2}} = \gamma(x - vt) \\ y' &= y \\ z' &= z \\ \text{and} \quad t' &= \frac{\left(t - \frac{vx}{c^2}\right)}{\sqrt{1 - \beta^2}} = \gamma\left(t - \frac{vx}{c^2}\right) \end{aligned}\right\} \tag{13.13}$$

The transformation equations for the velocity of a moving object can be obtained by taking the derivatives of (13.13) with respect to t and t'.

13.6 CONSEQUENCES OF LORENTZ TRANSFORMATIONS

Two features of Lorentz transformation equations are particularly important in evaluating the electric and magnetic fields: (i) time dilatation (ii) length contraction.

(i) Length Contraction

One consequence of the Lorentz transformation is the contraction of moving rod standard in the direction of its motion.

Let a rod in S be carried along the x-axis and let it be at rest there. If x_1 and x_2 are the abscissal of the ends of the rod, then the length l_0 in the frame S is given by

$$l_0 = x_2 - x_1$$

This is the *proper length* of the rod. Proper length is the linear dimension l_0 of a body in the system of coordinates in which it is at rest. Now

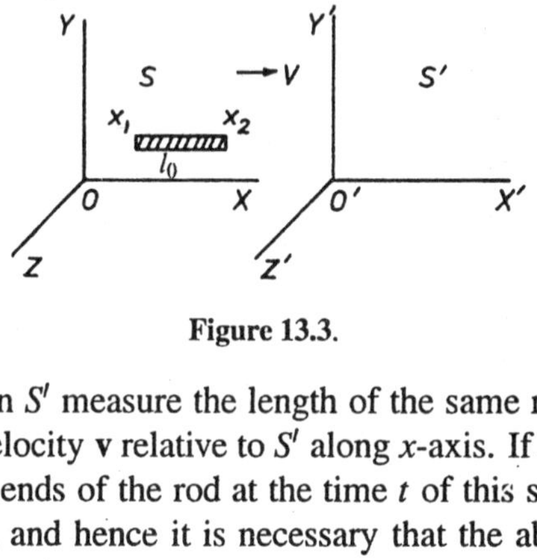

Figure 13.3.

let an observer in S' measure the length of the same rod when S' is in motion with a velocity **v** relative to S' along x-axis. If x_1' and x_2' are the abscissae of the ends of the rod at the time t of this system (since the rod is in motion and hence it is necessary that the abscissae must be noted simultaneously, i.e. at the same moment by this observe); then the length L of the rod for him is given by

$$l = x_2' - x_1'$$

Now from Lorentz transformation equations, one obtains

$$x_1 = \frac{(x_1' + vt')}{\sqrt{1 - v^2/c^2}} \tag{13.14}$$

and

$$x_2 = \frac{(x_2' + vt')}{\sqrt{1 - v^2/c^2}} \tag{13.15}$$

Subtracting (14) from (15), one obtains

$$x_2 - x_1 = \frac{x_2' - x_1'}{\sqrt{1 - v^2/c^2}}$$

or

$$l_0 = \frac{l}{\sqrt{1 - v^2/c^2}}$$

or

$$l = l_0\sqrt{1 - v^2/c^2} \tag{13.16}$$

Clearly, a moving rod appears shorter or contracted by a factor $\sqrt{1-\beta^2}$ in its direction of motion, the transverse section remains unaffected, as $\Delta y = \Delta y'$ and $\Delta z' = \Delta z$. This is called *Fitz-gerald Contraction.* It is important to note that the effect is *mutually reciprocal between different observers.*

The phenomenon of length contraction can hardly be noticed in everyday life because the motions ordinarily encountered are negligibly slow in comparison with that of light ($v \ll c$).

(ii) Time Dilatation

The time interval measured in two inertial frames in motion are not same. Suppose that a clock at a fixed point x' in the frame measures a time interval $\Delta t' = (t_2' - t_1')$ between two events that occurred at x' at times t_1' and t_2'. To an observer in the frame S, the two events appear to occur at times t_1 and t_2, given by

$$t_1 = \frac{t_1' + \left(\frac{vx'}{c^2}\right)}{\sqrt{1 - (v^2/c^2)}} \tag{13.17}$$

and

$$t_2 = \frac{t_2' + (vx'/c^2)}{\sqrt{1 - (v^2/c^2)}} \tag{13.18}$$

or

$$t_2 - t_1 = \Delta t = \frac{t_2' - t_1'}{\sqrt{1 - (v^2/c^2)}} = \frac{\Delta t'}{\sqrt{1-\beta^2}} \tag{13.19}$$

Obviously, the time interval in the S-frame is larger than the time measured by the clock in the S'-frame. Time interval measured on a clock that travels together with a system is called its *proper time.* This means that the proper time noted by a moving observer is always less than the corresponding interval of time in a stationary frame. A *stationary observer finds that a moving clock runs slower than a stationary one.*

The effect due to time dilatation has to be considered while considering Doppler shift in radiation.

Time dilatation has been best verified in experiments on a nuclear particle, called meson. There are subatomic particles which can be created in the laboratory by high-energy particle accelerators. Rossi and Hall in 1941 observed that the apparent life-time of fast moving mesons in the cosmic rays was considerably larger than the life-time of slow moving mesons in the laboratory. [The mean proper life-time of a positive μ-meson is 2.20×10^{-6}. The average path of such a particle, travelling at a velocity $0.98c$, should be 660m in air before decaying. Actually, the average path of μ-meson in air is much longer because its mean life time, measured in a stationary frame fixed in the earth (in the air) is substantially larger than 2.20×10^{-6} s]. They attributed this result to the time dilatation effect, according to which the meson clock, being in fast motion with respect to the terrestrial clocks, would seem to go relatively slow; consequently, a cosmic ray meson which lives its normal span of life according to its own proper time would appear to have lived longer according to the terrestrial clocks. More recently, the dilatation caused by the thermal vibration of the nuclei in certain crystals has also been verified.

We have seen that the lengths and time intervals are changed by Lorentz transformations. However, observers in the two inertial frames S and S' find that the light travels with a constant velocity c. When an electromagnetic signal starts from origin of the reference frame at time $t = 0$ and propagates in space as a spherical wave, then an observer in the S frame describes the locus of spherical wave front at time by the equation

$$x^2 + y^2 + z^2 - c^2t^2 = 0 \tag{13.20}$$

Similarly, as seen by an observer in the S' frame, the locus of the spherical wavefront at time t' is given by

$$x'^2 + y'^2 + z'^2 - c^2t'^2 = 0 \tag{13.21}$$

One can easily see that (13.20) and (13.21) are consistent with the Lorentz transformation, i.e.

$$x'^2 + y'^2 + z'^2 - c^2t'^2 = x^2 + y^2 + z^2 - c^2t^2 \tag{13.22}$$

A quantity $(x^2 + y^2 + z^2 - c^2t^2)$ is said to be *Lorentz invariant*, i.e. this has always the same value for observers in S and S' frames.

13.7 TRANSFORMATION PROPERTIES OF MOMENTUM, ENERGY, MASS AND FORCE

One of the important consequences of special relativity is that mass increases with velocity. The mass of a particle moving with relativistic speed v in S frame increases and the increased mass is given by

$$m = \frac{m_0}{\sqrt{1-\beta^2}} \tag{13.23}$$

where m_0 is the rest mass, $\beta = \frac{v}{c}$ and v is the constant relative velocity of the moving particle in the same frame. The total energy of the particle moving with relativistic velocity v consists of two parts: (i) the rest energy m_0c^2 and (ii) the kinetic energy of the particle. In terms of the relativistic mass, one can express the kinetic energy of the particle as

$$\text{K.E.} = mc^2 - m_0c^2 \tag{13.24}$$

Therefore, the total energy of the particle is

$$U = mc^2 - m_0c^2 = m_cc^2 + \text{K.E.} \tag{13.25}$$

We can define the momentum of a particle in the S-frame by $\mathbf{P} = mv$, where m is the relativistic mass given by (11.23). One can write the components of momentum and total energy of the particle moving with a velocity $\mathbf{v}$ in the S-frame are

$$p_x = \frac{m_0v_x}{\sqrt{1-\beta^2}}, \quad p_y = \frac{m_0v_y}{\sqrt{1-\beta^2}}$$

$$p_z = \frac{m_0v_z}{\sqrt{1-\beta^2}}, \quad \text{and} \quad U = \frac{m_0c^2}{\sqrt{1-\beta^2}} \quad (\beta = v/c) \tag{13.26}$$

If v' be the velocity of the particle in S' frame, the components of momentum and total energy of the particle are

$$p'_x = \frac{m_0\mathbf{v}'_x}{\sqrt{1-\beta'^2}}, \quad p'_y = \frac{m_0v'_y}{\sqrt{1-\beta'^2}}$$

$$p'_z = \frac{m_0\mathbf{v}'_z}{\sqrt{1-\beta'^2}}, \quad \text{and } U' = \frac{m_0c^2}{\sqrt{1-\beta'^2}} \quad (\beta' = \mathbf{v}'/c) \tag{13.27}$$

We can write the Lorentz transformation for momentum of particles in S frame and S' frame moving with a relative velocity u along the x-axis as

$$\left.\begin{aligned} p_x &= \frac{1}{\sqrt{1-(u^2/c^2)}}\left(p'_x+\frac{U'u_x}{c^2}\right) \\ p_y &= p'_y, \quad p_z = p'_z \\ \text{and} \quad U &= \frac{1}{\sqrt{1-u^2/c^2}}(U'+u_x p'_x) \end{aligned}\right\} \tag{13.28}$$

Changing u to $-u$ and interchanging primed and unprimed quanties, one obtains the inverse transformations as

$$\left.\begin{aligned} p'_x &= \frac{1}{\sqrt{1-(u^2/c^2)}}\left(p_x-\frac{Uu_x}{c^2}\right) \\ p'_y &= p_y; \quad p'_z = p_z \\ \text{and} \quad U' &= \frac{1}{\sqrt{1-(u^2/c^2)}}(U-u_x p_x) \end{aligned}\right\} \tag{13.29}$$

Obviously, the momentum has four-dimensional form whose time component is total energy. If total energy and momentum both are conserved in a frame of reference, then the energy and momentum will be conserved in any frame of reference. In case if only momentum is conserved, energy must also be conserved.

One can write the force components experienced by a particle moving with velocity in a reference frames as

$$F_x = \frac{d}{dt}(mv_x), \quad F_y = \frac{d}{dt}(mv_y) \quad \text{and} \quad F_z = \frac{d}{dt}(mv_z) \tag{13.30}$$

The force components in the S' frame are

$$F'_x = \frac{d}{dt}(m'v'_x), F'_y = \frac{d}{dt}(m'v'_y) \quad \text{and} \quad F'_z = \frac{d}{dt'}(m'v'_z) \tag{13.31}$$

The S' frame is moving with a velocity u with respect to frame S. Making use of Lorentz transformations, one can show that the force components in the two frames S and S' are related as

$$F_x = \frac{F'_x+(u/c^2)(\mathbf{v}\cdot\mathbf{F})}{1+(v'_x u/c^2)},$$

$$F_y = \frac{F'_y\sqrt{1-u^2/c^2}}{1+(v'_x u/c^2)}$$

and
$$F_z = \frac{F'_z\sqrt{1-u^2/c^2}}{1+(v'_x u/c^2)} \tag{13.32}$$

From (13.32) it is obvious that the force in S frame is related with the power generated by the force in S' frame $\mathbf{v}\cdot\mathbf{F}$. Moreover, one can also see that if a particle is at rest in the S' frame where it is subjected to a force with components F'_x and F'_y and F'_z and $v' = 0$ in this frame, then the force transformation equations reduce to

$$F_x = F'_x,\ F_y = F'_y\sqrt{1-(u^2/c^2)} \text{ and } F_z = F'_z\sqrt{1-(u^2/c^2)} \tag{13.33}$$

Eqs (13.33) are widely used in studying the motion of a test particle in electromagnetic fields.

13.8 RELATIVITY OF CHARGES

We have seen that *electric charge is absolutely conserved.* In an atom, the electrons and protons are moving about, yet the atom is electrically neutral, i.e. positive and negative charges exactly cancel. obviously the charge on each elementary particle is $\pm e$ whether or not the particle is moving relative to the observer. This means that the *charge on an elementary particle* and, hence, *the net charge carried by any macroscopic system is Lorentz invariant.* Now, the question arises. Whether the *charge density* is also invariant? To answer this question, let us consider a series of positive charges equally spaced along a line, each with charge $+e$. Let the charges are at rest in frame S and span a distance of unit length. The linear charge density as seen by an observer in S frame is Ne Coulomb/m, where N is the number of charges in unit length. In the light of Lorentz transformations, an observer is S' frame moving with a speed u parallel to the charges finds that the charges are squashed up into a distance $\sqrt{1-(u^2/c^2)}$. Obviously, the linear charge density according to observer in S' frame is $\dfrac{Ne}{\sqrt{1-(u^2/c^2)}}$ Coulomb/m.

13.9 RELATIVITY OF MAGNETIC AND ELECTRIC FIELDS

We here consider the forces exerted by electric and magnetic fields, as viewed by observers in S and S' frames. Let us consider a long straight conducting wire in the frame S in which free electrons move with a drift velocity $\mathbf{v}$ in the x-direction. Let n be the number of electrons per unit length of the conductor. Obviously, this is also the number of positive ions per unit length. Let us consider that a charge q is moving parallel to the conductor at a speed $\mathbf{v}$. The magnetic force experienced by this charge, $\mathbf{F} = q(\mathbf{v} \times \mathbf{B})$ in the S frame, where $\mathbf{B}$ is the magnetic field produced by the current in the conductor.

Now, we consider an observer in S', moving with velocity $\mathbf{v}$ in the x-direction. For this observer the charge q is at rest and hence, the magnetic force $\mathbf{F} = 0$. However, all inertial frames are equivalent, i.e. observer in S' must agree with the observer in S. This means that there is a force on charge q though it is not a magnetic force. Now, the question arises, how we should interpret this force? In the frame S the current is constituted by conduction electrons of linear charge density (λ^-) moving with velocity $\mathbf{v}$, where $\lambda^- = -ne$. The positive charge, i.e. the immobile background charge density due to positive ions is $\lambda^+ = ne$. The net linear charge density

$$\lambda = \lambda^- + \lambda^+ = 0 \tag{13.34}$$

Now, we obtain the current density in the wire as

$$\begin{aligned} \mathbf{J}_x^- &= ne\mathbf{v} = \lambda^- \mathbf{v}, \quad \mathbf{J}_x^+ = 0 \\ \therefore \quad \mathbf{J}_x &= \mathbf{J}_x^+ + \mathbf{J}_x^- = \lambda^- \mathbf{v} \end{aligned} \tag{13.35}$$

The magnetic field B produced by the current is given by

$$B = -\mu_0 \frac{nev}{2\pi r}$$

Obviously, B causes a force F to act on the charge. We have

$$|\mathbf{F}| = |q(\mathbf{v} \times \mathbf{B})| = -\mu_0 \frac{nev^2 q}{2\pi r} \tag{13.36}$$

Now, we are interested in the observation of the observer in S' about this force. The positive charges which were at rest to the observer in S are moving at a speed $\mathbf{v}$ in the opposite direction to the observer in S' frame. Now, applying Lorentz transformation, the unit length in S frame is equivalent to the length $\sqrt{1-\beta^2}$ in S' frame and the linear charge density due to the positive ions $\lambda^+ = \dfrac{ne}{\sqrt{1-\beta^2}}$. The electrons are at rest in this frame and they span a distance $\dfrac{1}{\sqrt{1-\beta^2}}$ in S' frame. (We must remember that the distance is of unit length in S frame). Obviously, the linear charge density is $\lambda' = -ne\sqrt{1-\beta^2}$. Therefore, the net charge density in S' is given by

$$\lambda' = \lambda'^{-} + \lambda'^{+} = -ne\sqrt{1-\beta^2} + \frac{ne}{\sqrt{1-\beta^2}}$$

$$= ne\frac{\beta^2}{\sqrt{1-\beta^2}} \qquad (13.37)$$

we can see that this is positive and not zero as found by the observer in S frame. Thus the net charge in the moving wire is positive. Obviously, the observer in S' frame, therefore, does not think that the wire is electrically neutral. According to the observer in S' frame, there is an electrostatic field around the wire given by

$$E' = \frac{\lambda'}{2\pi\varepsilon_0 r} = \frac{ne\beta^2}{2\pi\varepsilon_0 r\sqrt{1-\beta^2}} \qquad (13.38)$$

The magnitude of the electric force in S' frame can be easily obtained

$$|\mathbf{F}| = |q\mathbf{E}'| = \frac{qne\beta^2}{2\pi\varepsilon_0 r\cdot\sqrt{1-\beta^2}}$$

$$= \frac{\mu_0 nev^2 q}{2\pi r\sqrt{1-\beta^2}} = \frac{|\mathbf{F}|}{\sqrt{1-\beta^2}} \qquad (13.39)$$

$$\left(\text{we have} \quad c = \frac{1}{\sqrt{\varepsilon_0\mu_0}}\right)$$

We have obtained a very striking result. The force that was purely magnetic in S frame appears to be purely electrostatic in S'. This shows that the magnetic and electric effects are just two ways of looking at the same thing. Obviously, the special theory of relativity provides the *unification* of electric and magnetic effects.

The charges moving with velocity v in the S frame constitute a current. We have

$$J = \lambda^{-}v = \frac{\lambda_0 v}{\sqrt{1-v^2/c^2}} \qquad (13.40)$$

one can write the components of the current density in S frame as

$$J_x = \frac{v_x}{\sqrt{1-(v^2/c^2)}}, \quad J_y = \frac{v_y}{\sqrt{1-(v^2/c^2)}} \quad \text{and} \quad J_z = \frac{v_z}{\sqrt{1-(v^2/c^2)}} \qquad (13.41)$$

Let us find the relation between current density and the charge density. We can see that this has a close similarity with the relation between momentum and energy. Consistent with these transformations, one can write three invariant equations as

$$x^2+y^2+z^2-c^2t^2=c^2t^2 \;=\; \text{constant} \tag{13.42}$$

$$p_x^2+p_y^2+p_z^2-c^2m^2=c^2m_0^2 \;=\; \text{constant} \tag{13.43}$$

and

$$J_x^2+J_y^2+J_z^2=c^2m^2=c^2\rho_0^2 \;=\; \text{constant} \tag{13.44}$$

Eq. $\lambda=\dfrac{\lambda_0}{\sqrt{1-(v/c)^2}}$ together with (13.40) yields

$$\lambda=\frac{\lambda_0}{m_0}m \quad \text{and} \quad \mathbf{J}=\frac{\lambda_0}{m_0}\mathbf{p} \tag{13.45}$$

Obviously, the charge density λ and current density **J** transform exactly as m and momentum **p**-respectively. The transformation of current density in inertial frames S and S' moving with a constant velocity u along the x-axis can be expressed as

$$J_x \;=\; \frac{J_x+\lambda_u'}{\sqrt{1-(u/c)^2}}; \quad J_y=J_y' \quad \text{and} \quad J_z=J_z' \tag{13.46}$$

and

$$\lambda \;=\; \frac{\lambda'+(uJ_x'/c^2)}{\sqrt{1-(u/c)^2}} \tag{13.47}$$

This means that in the S frame, the net charge density is zero and there is no observed electric field. The motion of charges gives rise to a net current density giving rise to a magnetic field around the conducting wire. We have seen that in the S frame there is only the **B** field and the charge outside the conductor experiences a magnetic force directed towards to the conductor. On the other hand, in the S' frame, there is a positive net charge density, as well as, a current density which gives rise to E' and B' fields. However, the net charge in the S' frame is not zero and the charge outside the conducting wire experiences an electrical force directed towards the wire. This leads one to conclude that whether an electromagnetic field is purely magnetic field or purely electric field or electromagnetic field depends on the inertial frame of reference in which sources are observed. This is an important result of special theory of relativity.

13.10 THE LORENTZ TRANSFORMATION AS AN ORTHOGONAL TRANSFORMATION

Let us consider that a frame of reference is rotated through an angle θ about z-axis. Obviously, the coordinates of point G in the rotated frame will not be the same as those in the original frame of reference.

One can express the coordinates in the rotated frame as (Fig. 13.4)

$$\begin{aligned} x' &= x\cos\theta + y\sin\theta \\ y' &= -x\sin\theta + y\cos\theta \\ z' &= z \end{aligned} \tag{13.48}$$

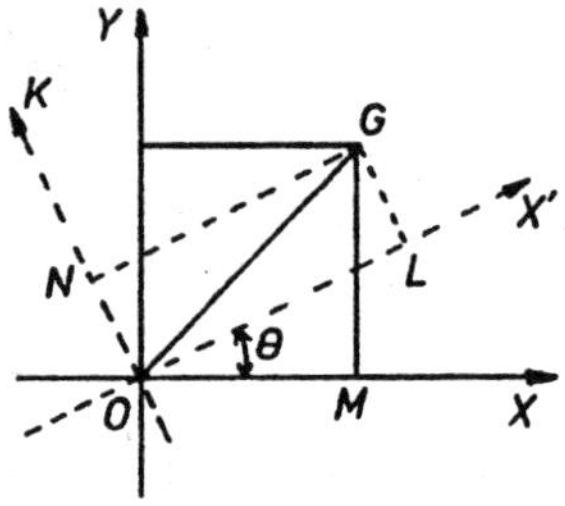

Figure 13.4.

We can see that (13.48) is a linear combinations of original coordinates. Let the point G is specified by a position vector $\mathbf{r}$, then its length in the original frame is given by

$$|\mathbf{r}|^2 = x^2 + y^2 + z^2$$

One can see from (13.48) that

$$|\mathbf{r}|^2 = x^2 + y^2 + z^2 = x'^2 + y'^2 + z'^2 \qquad 13.48(a)$$

This means that (13.48(a)) leaves the length of the vector unchanged. Such transformations are called *orthogonal transformations*.

When the new coordinates can be expressed as a linear combination of the old coordinates, then the coordinate transformation is said to be *linear*. Thus we have

$$\begin{aligned} x'_1 &= a_{11}x_1 + a_{12}x_2 + a_{13}x_3 \\ x'_2 &= a_{21}x_1 + a_{22}x_2 + a_{23}x_3 \\ x'_3 &= a_{31}x_1 + a_{32}x_2 + a_{33}x_3 \\ - &\quad ---------- \\ - &\quad ---------- \end{aligned}$$

i.e.,

$$x'_i = \sum_j a_{ij}x_j \tag{13.49}$$

is a linear transformation. One can express it in the matrix form as

$$\begin{bmatrix} x'_1 \\ x'_2 \\ x'_3 \end{bmatrix} = \begin{bmatrix} a_{11} & a_{12} & a_{13} \\ a_{21} & a_{22} & a_{23} \\ a_{31} & a_{32} & a_{33} \end{bmatrix} \begin{bmatrix} x_1 \\ x_2 \\ x_3 \end{bmatrix} \tag{13.50}$$

For (13.50) to be an orthogonal transformation, we have

$$\sum_i (x_i')^2 = \sum_k x_k^2$$

i.e.,

$$\sum_{j,k,l}^{i} a_{ij} a_{ik} x_j x_k = \sum_k x_k^2 \tag{13.51}$$

(13.51) is possible if

$$\sum_i a_{ij} a_{ik} = \begin{matrix} 0 \text{ if } j \neq k \\ 1 \text{ if } j = k \end{matrix}$$

In other words, we can express it in terms of delta function as

$$\sum_i a_{ij} a_{ik} = \delta_{jk} \tag{13.52}$$

We can easily see that for the transformation to be orthogonal the elements of the transformation matrix

$$A = \begin{bmatrix} a_{11} & a_{12} & a_{13} \\ a_{21} & a_{22} & a_{23} \\ a_{31} & a_{32} & a_{33} \end{bmatrix}$$

must satisfy the condition (13.52).

We have seen that the quantity $x^2 + y^2 + z^2 - c^2t^2$ is Lorentz invariant, i.e. it has the same value in S and S' frames. Now we find that the Lorentz transformation, like the rotational transformation, is a linear relation between different coordinate systems, but it mixes up time with the space coordinates. One can specify an event occuring at a particular time and place in a frame S by four coordinates (x, y, z, t).

As compared to three dimensional scalar product r^2, the $S^2 (= x^2 + y^2 + z^2 - c^2t^2)$ has the dimension of length. Obviously, one can interpret S^2 as the square of the "length" of the "four vector" S, i.e. one can visualize S^2 as a scalar product of two vectors. The four "coordinates" are $x_\mu = x_1, x_2, x_3$ and x_4, where

$$x_1 = x, \quad x_2 = y, \quad x_3 = z \quad \text{and} \quad x_4 = ict \tag{13.53}$$

Obviously, a new time variable namely '*ict*' in place of time has been choosen. One can write the four-vector scalar product as

$$S^2 = x_\mu x_\mu = x_1^2 + x_2^2 + x_3^2 + x_4^2 = x^2 + y^2 + z^2 - c^2t^2 \tag{13.54}$$

This non-Euclidean four-dimensional space is known as *Minkowaski space*. The four-vector of an event is generally represented by x_μ. The line element is called a *tensor of rank zero* and is expressed as

$$dS^2 = g_{ij} dx^i dx^j \tag{13.55}$$

where dx^i and dx^j are contravariant components of basic displacement vector dx. In writing (13.54), we have used the summation convention. We must remember that whenever a Greek suffix (e.g. μ) appears twice in a single term, the term is assumed to be summed from 1 to 4.

Now, one can express the Lorentz transformation equations in new notations as

$$\begin{aligned} x_1' &= \frac{x_1 + i\beta x_4}{\sqrt{1-\beta^2}}, \\ x_2' &= x_2, \\ x_3' &= x_3 \end{aligned}$$

and

$$x_4' = \frac{x_4 - i\beta x_1}{\sqrt{1-\beta^2}} \tag{13.56}$$

One finds the inverse transformation equations as

$$\begin{aligned} x_1 &= \frac{x_1' - i\beta x_4'}{\sqrt{1-\beta^2}}, \\ x_2 &= x_2', \\ x_3 &= x_3' \end{aligned}$$

and

$$x_4 = \frac{x_4' + i\beta x_1'}{\sqrt{1-\beta^2}} \tag{13.57}$$

One can write the matrix of Lorentz transformation as

$$A = \begin{bmatrix} \dfrac{1}{\sqrt{1-\beta^2}} & 0 & 0 & \dfrac{i\beta}{\sqrt{1-\beta^2}} \\ 0 & 1 & 0 & 0 \\ 0 & 0 & 1 & 0 \\ -\dfrac{-i\beta}{\sqrt{1-\beta^2}} & 0 & 0 & \dfrac{1}{\sqrt{1-\beta^2}} \end{bmatrix} \tag{13.58}$$

One can easily see that (13.58) satisfies (13.52). This shows that *Lorentz transformation is an orthogonal transformation.*

The momentum and energy of the particles in an inertial frame forms a four-vector set and one can express their scalar product as

$$p^2 = p_\mu p_\mu = p_x^2 + p_y^2 + p_z^2 - \frac{E^2}{c^2} = -m_0^2 c^2 \tag{13.59}$$

One can see that this is also a *Lorentz invariant*.

13.11 COVARIANT FORMULATION OF ELECTRODYNAMICS

In order to investigate the relativistic behaviour of Maxwell's equations it is necessary to write the equations in four dimensional form and to evaluate the transformation properties of spatial and temporal differential operators. Applying the rules of partial differentiation, one can write

$$\begin{aligned}\frac{\partial}{\partial x_1'} &= \frac{\partial x_1}{\partial x_1'}\frac{\partial}{\partial x_1} + \frac{\partial x_2}{\partial x_1'}\frac{\partial}{\partial x_2} + \frac{\partial x_3}{\partial x_1'}\frac{\partial}{\partial x_3} + \frac{\partial x_4}{\partial x_1'}\frac{\partial}{\partial x_4} \\ &= \frac{1}{\sqrt{1-\beta^2}}\left(\frac{\partial}{\partial x_1} + i\beta\frac{\partial}{\partial x_4}\right)\end{aligned} \tag{13.60}$$

where we have used (13.57). This transformation is similar to that shown for x_1' in (13.56). Similar transformation equations can be obtained for $\frac{\partial}{\partial x_2'}$, $\frac{\partial}{\partial x_3'}$ and $\frac{\partial}{\partial x_4'}$. This shows that they all transform according to the Lorentz transformation and the differential operators also constitute a set of four-vectors and is represented by $\frac{\partial}{\partial x_\mu}$. One can write the scalar product of this four-vector differential operator as

$$\begin{aligned}\frac{\partial}{\partial x_\mu}\frac{\partial}{\partial x_\mu} &= \frac{\partial^2}{\partial x_1^2} + \frac{\partial^2}{\partial x_2^2} + \frac{\partial^2}{\partial x_3^2} + \frac{\partial^2}{\partial x_4^2} \\ &= \frac{\partial^2}{\partial x^2} + \frac{\partial^2}{\partial y^2} + \frac{\partial^2}{\partial z^2} - \frac{1}{c^2}\frac{\partial^2}{\partial t^2} = \Box^2\end{aligned} \tag{13.61}$$

where $\Box^2$ is known as *d'Alembertion operator*. This operator is also a Lorentz invariant and is widely used in wave equation, e.g.

$$\nabla^2\psi - \frac{1}{c^2}\frac{\partial^2\psi}{\partial t^2} = \Box^2\psi = 0 \tag{13.62}$$

Maxwell's Equations in Four-Vector Form

We have already seen that the electric and magnetic fields transform into one another in different moving inertial frames. We have achieved this with the help of Lorentz transformation equation and the invariance of total charge in the moving inertial frame. The mathematical statement of charge conservation is contained in the equation of continuity

$$\nabla \cdot \mathbf{J} + \frac{\partial \rho}{\partial t} = 0 \tag{13.63}$$

This equation must, therefore, have the same form in all inertial frames. Let us see how this can be expressed in an invariant form. We must remember that in the theory of relativity, the current density and charge density are not distinct entities since the charge distribution that is static in one reference frame will appear to be a current distribution in a moving reference frame. Let us introduce four vector $J_\mu = (J, ic\rho)$, where the current density J is the space-like part of J_μ and the charge density is used to denote the fourth component $J_4 = ic\rho$. The four components of J_μ being J_1, J_2, J_3 and $ic\rho$. In terms of four-vectors, the continuity equation can be expressed as

$$\nabla \cdot \mathbf{J} + \frac{\partial \rho}{\partial t} = \frac{\partial J_1}{\partial x_1} + \frac{\partial J_2}{\partial x_2} + \frac{\partial J_3}{\partial x_3} + \frac{\partial (ic\rho)}{\partial (ict)} = 0 \tag{13.64}$$

or

$$\frac{\partial J_\mu}{\partial x_\mu} = 0 \tag{13.65}$$

obviously, the current continuity equation expressed in four-vector form is also *Lorentz invariant.*

Now, representing the four dimensional divergance operator by 'Div', one may express the equation of continuity in the four-dimensional for as

$$\text{Div } \mathbf{J} = 0 \tag{13.66}$$

we have seen that in free space the potential **A** and ϕ can be written in the form

$$\nabla^2 \mathbf{A} - \frac{1}{c^2}\frac{\partial \mathbf{A}}{\partial t^2} = -\mu_0 \mathbf{J} \tag{13.67}$$

$$\nabla^2 \phi - \frac{1}{c^2}\frac{\partial^2 \phi}{\partial t^2} = -\frac{\rho}{\varepsilon_0} \tag{13.68}$$

We can write these equations in terms of four-dimensional D' Alembertion operator as

$$\Box \mathbf{A} = \frac{\partial^2 \mathbf{A}}{\partial x_\mu^2} = -\mu_0 \mathbf{J} \tag{13.69}$$

and

$$\Box \phi = \frac{\partial^2 \phi}{\partial x_\mu^2} = -\frac{\rho}{\varepsilon_0} \tag{13.70}$$

The operator $\Box$ operates on the scalar and vector potentials. We find that the vector and scalar potential form a set of another four vector $\left(A, \frac{i\phi}{c}\right)$. The four vector potential A_ν can be defined as

$$A_\nu = (A_1, A_2, A_3, A_4) \equiv \left(A_1, A_2, A_3, \frac{i\phi}{c}\right) = \left(A, \frac{i\phi}{c}\right) \tag{13.71}$$

where ν takes values from 1 to 4 to denote four-vectors. The Lorentz condition is

$$\text{div}\mathbf{A} + \frac{1}{c^2}\frac{\partial \phi}{\partial t} = 0$$

Now

$$\begin{aligned} \text{Div}\mathbf{A}_\nu &= \frac{\partial A_1}{\partial x_1} + \frac{\partial A_2}{\partial x_2} + \frac{\partial A_3}{\partial x_3} + \frac{\partial A_4}{\partial x_4} \\ &= \text{div}\mathbf{A} + \frac{\partial (i\phi/c)}{\partial (ict)} = \text{div}\mathbf{A} + \frac{1}{c^2}\frac{\partial \phi}{\partial t} = 0 \end{aligned} \tag{13.72}$$

Eqs. (13.69) and (13.70) can now be expressed in a single equations

$$\frac{\partial^2 \mathbf{A}}{\partial x_\mu^2} = -\mu_0 \mathbf{J} \tag{13.73}$$

where the fourth component of (13.73) is found by making substitution $A_4 \to \dfrac{i\phi}{c}$ and $J_4 \to i\rho c$. From these equations, one finds that the Lorentz condition used for making the potentials unique is also *Lorentz invariant*. One can also write in terms of four vectors

$$\nabla \cdot \mathbf{A} - \frac{1}{c^2}\frac{\partial \phi}{\partial t} = \frac{\partial A_\mu}{\partial x_\mu} = 0 \tag{13.74}$$

We find that Maxwell's equations are consistent with the basic axioms of theory of special relativity. We have seen that the Maxwell's field equations, the continuity equation and the conditions of uniqueness of potentials and fields are all *Lorentz invariant*. As the electric and magnetic fields are *Lorentz invariant*, the Lorentz force acting on the charged particles is also *Lorentz invariant.* This shows that observers in different reference frames will have a consistent description of the motion of charged particles in electromagnetic fields.

13.12 ELECTROMAGNETIC FIELD TENSOR

Now, we examine whether the Lorentz force is invariant. For this purpose, one has to investigate the transformation of **E** and **B**. We know that **E** and **B** are not four vectors, but we can first express them in terms of potentials and then we can have transformations. Let us see how the six components E_1, E_2, E_3, B_1, B_2 and B_3 can be used to define a tensor of rank two called the electromagnetic field tensor. One can easily see that in n-dimensional space, a tensor of the mth rank is a set of n quantities and this transform as

$$T'_{abcd} = \sum_{i,j,k,l,\ldots} \lambda_{ai}\lambda_{bj}\lambda_{ck}\lambda_{dl}\ldots T_{ijkl}$$

Since we are dealing with a 4-dimensional space and hence we need to consider only the tensors of rank two which transform as

$$T'_{ij} = \sum_{k,l} \lambda_{ik}\lambda_{jl}T_{kl} \tag{13.75}$$

We have

$$E_x = -\frac{\partial\phi}{\partial x} - \frac{\partial A_x}{\partial t} \text{ etc.} \tag{13.76}$$

Since $A_\mu = \left(A_1, A_2, A_3, A_4 = \frac{i\phi}{c}\right)$ and $x_4 = ict$, one can write (13.76) as

$$E_1 = ic\left(\frac{\partial A_4}{\partial x_1} - \frac{\partial A_1}{\partial x_4}\right)$$

$$E_2 = ic\left(\frac{\partial A_4}{\partial x_2} - \frac{\partial A_2}{\partial x_4}\right)$$

and

$$E_3 = ic\left(\frac{\partial A_4}{\partial x_3} - \frac{\partial A_3}{\partial x_4}\right) \tag{13.77}$$

Similarly, one obtains

$$B_1 = \frac{\partial A_3}{\partial x_2} - \frac{\partial A_2}{\partial x_3};$$

$$B_2 = \frac{\partial A_1}{\partial x_3} - = \frac{\partial A_1}{\partial x_3} - \frac{\partial A_3}{\partial x_1}$$

and
$$B_3 = \frac{\partial A_2}{\partial x_1} - \frac{\partial A_1}{\partial x_2} \tag{13.78}$$

Let us define a set of quantities

$$F_{\mu\nu} = \frac{\partial A_\nu}{\partial x_\mu} - \frac{\partial A_\mu}{\partial x_\nu} \tag{13.79}$$

Thus one obtains

$$E_1 = icF_{14}; \quad E_2 = icF_{24} \quad \text{and} \quad E_3 = icF_{34} \tag{13.80}$$

and
$$B_1 = F_{23}; \quad B_2 = F_{31} \quad \text{and} \quad B_3 = F_{12} \tag{13.81}$$

Since $F_{\mu\nu} = -F_{\nu\mu}$ i.e. tensor is antisymmetric and $F_{\mu\mu} = 0$, one can express the tensor $\{F\}$ in terms of the elements $F_{\mu\nu}$, as

$$\begin{aligned}\{F\} &= \begin{Bmatrix} F_{11} & F_{12} & F_{13} & F_{14} \\ F_{21} & F_{22} & F_{23} & F_{24} \\ F_{31} & F_{32} & F_{33} & F_{34} \\ F_{41} & F_{42} & F_{43} & F_{44} \end{Bmatrix} \\ &= \begin{Bmatrix} 0 & B_3 & -B_2 & -iE_1/c \\ -B_3 & 0 & B_1 & -iE_2/c \\ B_2 & -B_1 & 0 & -iE_3/c \\ iE_1/c & iE_2/c & iE_3/c & 0 \end{Bmatrix}\end{aligned} \tag{13.82}$$

This tensor is usually called as the *electromagnetic* antisymmetric tensor of rank two. Now we shall see how this tensor can be used to express Maxwell's equations. Let us consider the equation

$$\frac{\partial F_{\lambda\mu}}{\partial x_\nu} + \frac{\partial F_{\mu\nu}}{\partial x_\lambda} + \frac{\partial F_{\nu\lambda}}{\partial x_\mu} = 0 \tag{13.83}$$

Now assigning λ, μ, ν values 1, 2, 3 in this or in any other order, one finds that (13.83) reduces to

$$\frac{\partial F_{12}}{\partial x_3} + \frac{\partial F_{23}}{\partial x_1} + \frac{\partial F_{31}}{\partial x_2} = 0$$

which means

$$\frac{\partial B_3}{\partial x_3} + \frac{\partial B_1}{\partial x_1} + \frac{\partial B_2}{\partial x_2} = 0$$

or

$$\nabla \cdot \mathbf{B} = 0$$

Obviously, we have obtained one of the Maxwell's equations.

Let us now assign to one of the indices the value 4, say $\lambda = 2, \mu = 3$ and $\nu = 4$, one obtains from (13.83)

$$\frac{\partial F_{23}}{\partial x_4} + \frac{\partial F_{34}}{\partial x_2} + \frac{\partial F_{42}}{\partial x_3} = 0$$

or
$$\frac{\partial B_1}{\partial (ict)} + \frac{1}{ic}\frac{\partial E_3}{\partial x_2} - \frac{1}{ic}\frac{\partial E_2}{\partial x_3} = 0$$

or
$$\frac{\partial B_1}{\partial t} + \frac{\partial E_3}{\partial x_2} - \frac{\partial E_2}{\partial x_3} = 0$$

or
$$(\nabla \times E)_x + \frac{\partial B_x}{\partial t} = 0 \quad (x = 1) \tag{13.84}$$

Obviously, (13.84) is the x-component of $\nabla \times \mathbf{E} + \frac{\partial \mathbf{B}}{\partial t} = 0$. This shows that (13.83) also represents the other homogeneous Maxwell's equations. One can easily obtain the other two inhomogeneous Maxwell's equations from (13.83) as follow. We have

$$\frac{\partial F_{\mu\nu}}{\partial x_\nu} = \mu_0 J_\mu \tag{13.85}$$

Let $\mu = 1$, Eq. (13.85) takes the form

$$\frac{\partial F_{11}}{\partial x_1} + \frac{\partial F_{12}}{\partial x_2} + \frac{\partial F_{13}}{\partial x_3} + \frac{\partial F_{14}}{\partial x_4} = \mu_0 J_1$$

or
$$0 + \frac{\partial B_3}{\partial x_2} - \frac{\partial B_2}{\partial x_3} + \frac{1}{ic}\frac{\partial E_1}{\partial (ict)} = \mu_0 J_1$$

or
$$(\nabla \times B)_x - \frac{1}{c^2}\frac{\partial E_x}{\partial t} = \mu_0 J_x \quad (x = 1)$$

which is the x component of

$$\nabla \times \mathbf{B} - \frac{1}{c^2}\frac{\partial \mathbf{E}}{\partial t} = \mu_0 \mathbf{J} \tag{13.86}$$

Now taking $\mu = 4$, one can express (13.85) as

$$\frac{\partial F_{41}}{\partial x_1} + \frac{\partial F_{42}}{\partial x_2} + \frac{\partial F_{43}}{\partial x_3} + \frac{\partial F_{44}}{\partial x_4} = \mu_0 J_4$$

or
$$-\frac{1}{ic}\frac{\partial E_1}{\partial x_1} - \frac{1}{ic}\frac{\partial E_2}{\partial x_2} - \frac{1}{ic}\frac{\partial E_3}{\partial x_3} + 0 = ic\mu_0\rho$$

or
$$\frac{\partial E_1}{\partial x_1}+\frac{\partial E_2}{\partial x_2}+\frac{\partial E_3}{\partial x_3}=\nabla\cdot\mathbf{E} = \frac{\rho}{\varepsilon_0}$$

Obviously, the four Maxwell's electromagnetic equations are represented by only two tensor equations involving operations on the components of the field tensor.

13.13 LORETZ TRANSFORMATION OF ELECTRIC AND MAGNETIC FIELDS

Let us now obtain the transformation of electric and magnetic fields. For this we first evaluate E'_x in the S' frame in terms of **E** and **B** in S frame. We have

$$E'_x = -\frac{\partial\phi'}{\partial x'}-\frac{\partial A'_x}{\partial t'}$$

Since $A_\mu = \left(A_1, A_2, A_3, \frac{i\phi}{c}\right)$ and $\frac{\partial}{\partial x_\mu} = \left(\frac{\partial}{\partial x_1}, \frac{\partial}{\partial x_2}, \frac{\partial}{\partial x_3}, \frac{\partial}{\partial(ict)}\right)$,

we have
$$E'_x = icF'_{14} = ic\left\{\frac{\partial A'_4}{\partial x'_1}-\frac{\partial A'_1}{\partial x'_4}\right\}$$

We know that A_μ, $\frac{\partial}{\partial x_\mu}$ are Lorentz covariant, we have

$$A'_4 = \frac{1}{\sqrt{1-\beta^2}}(A_4 - i\beta A_1) = \gamma(A_4 - i\beta A_1) \quad \left(\text{Here } \gamma = \frac{1}{\sqrt{1-\beta^2}}\right)$$

$$\frac{\partial}{\partial x'_4} = \gamma\left(\frac{\partial}{\partial x_4} - i\beta\frac{\partial}{\partial x_1}\right)$$

$$A'_1 = \gamma(A_1 + i\beta A_4); \quad \frac{\partial}{\partial x'_1} = \gamma\left(\frac{\partial}{\partial x_1} + i\beta\frac{\partial}{\partial x_4}\right)$$

Therefore, one obtains

$$\begin{aligned}E'_x &= ic\left[\gamma\left(\frac{\partial}{\partial x_1}+i\beta\frac{\partial}{\partial x_4}\right)\{r(A_4 - i\beta A_1)\}\right.\\ &\left.\quad -\gamma\left(\frac{\partial}{\partial x_4}-i\beta\frac{\partial}{\partial x_1}\right)\{\gamma(A_1 + i\beta A_4)\}\right]\\ &= ic\gamma^2\left[\frac{\partial A_4}{\partial x_1}+i\beta\frac{\partial A_4}{\partial x_4}-i\beta\frac{\partial A_1}{\partial x_1}\right.\\ &\left.\quad +\beta^2\frac{\partial A_1}{\partial x_4}-\frac{\partial A_1}{\partial x_4}+i\beta\frac{\partial A_1}{\partial x_1}-i\beta\frac{\partial A_4}{\partial x_4}-\beta^2\frac{\partial A_4}{\partial x_4}\right]\end{aligned}$$

$$= ic\gamma^2(1-\beta^2)\left(\frac{\partial A_4}{\partial x_1}-\frac{\partial A_1}{\partial x_4}\right)$$

$$= ic\left(\frac{\partial A_4}{\partial x_1}-\frac{\partial A_1}{\partial x_4}\right) = E_x$$

Similarly, one can obtain the other components. Finally, one obtains

$$\begin{aligned} E'_x &= E_x \quad &; \quad B'_x &= B_x \\ E'_y &= \gamma[E_y - c\beta B_z] \quad &; \quad B'_y &= \gamma[B_y + \frac{\beta}{c}E_z] \\ \text{and} \quad E'_z &= \gamma[E_z + \beta c B_y] \quad &; \quad B'_z &= \gamma[B_z - \frac{\beta}{c}E_y] \end{aligned} \tag{13.87}$$

We can see that the components of **E** and **B** in the direction of motion are unaffected but in the transverse direction are modified.

13.14 ELECTROMAGNETIC FIELD DUE TO A POINT CHARGE IN UNIFORM MOTION

Let us consider that a charge is stationary in the frame S. Obviously, $\mathbf{B} = 0$ in this frame and **E** is given by

$$\mathbf{E} = \frac{q}{4\pi\varepsilon_0}\frac{\mathbf{r}}{|\mathbf{r}|^3} \tag{13.88}$$

Now, we are interested in finding the fields in frame S' which is moving with relative velocity **v** relative to S along positive direction of common x-axis. Let us assume that the charge to be at rest at the origin of S and the field is çalculated at the instant $t = 0$, i.e. at $t = 0$ two origins of S and S' coincide. From (13.87), we have

$$E'_1 = E_1; \quad E'_2 = \gamma E_2 \quad \text{and} \quad E'_3 = \gamma E_3$$
$$(\because \quad B_2 = B_3 = 0 \text{ in the rest frame } S) \tag{13.89}$$

We have at $t = 0$

$$r_1 = \gamma x'_1; \quad x_2 = x'_2 \quad \text{and} \quad x_3 = x'_3$$

If r be the distance of the observation point P from the origin, then

$$r = \sqrt{x_\mu x_\mu} = \sqrt{\gamma^2 x_1'^2 + x_2'^2 + x_3'^2} \tag{13.90}$$

Now, one obtains the components of the electric field in frame S' as

$$E'_1 = E_1 = \frac{q}{4\pi\varepsilon_0}\frac{x_1}{r^3} = \frac{q}{4\pi\varepsilon_0}\frac{\gamma x'_1}{(\gamma^2 x_1'^2 + x_2'^2 + x_3'^2)^{3/2}}$$

$$E_2' = \gamma E_2 = \frac{q}{4\pi\varepsilon_0}\frac{\gamma x_2}{r^3} = \frac{q}{4\pi\varepsilon_0}\frac{\gamma x_2}{(\gamma^2 x_1'^2 + x_2'^2 + x_3'^2)^{3/2}}$$

$$E_3' = \gamma E_3 = \frac{q}{4\pi\varepsilon_0}\frac{\gamma x_3}{r^3} = \frac{q}{4\pi\varepsilon_0}\frac{\gamma x_3'}{(\gamma^2 x_1'^2 + x_2'^2 + x_3'^2)^{3/2}}$$

Thus, we have

$$\mathbf{E}' = \frac{q\gamma}{4\pi\varepsilon_0}\frac{\mathbf{r}'}{\left(\gamma^2 x_1'^2 + x_2'^2 + x_3'^2\right)} \tag{13.91}$$

We have $x_1' = r'\cos\theta$, where θ is the angle that r' makes with the x-axis and

$$x_1'^2 + x_2'^2 + x_3'^2 = r'^2$$

This leads to

$$x_2'^2 + x_3'^2 = r'^2 \sin^2\theta$$

Now, we have

$$\begin{aligned} \gamma^2 x_1'^2 + x_2'^2 + x_3'^2 &= \gamma^2 r'^2 \cos^2\theta + r'^2 \sin^2\theta \\ &= r'^2\gamma^2\left(\cos^2\theta + \frac{\sin^2\theta}{\gamma}\right) \\ &= \gamma^2 r'^2 (1 - \beta^2 \sin^2\theta) \end{aligned} \tag{13.92}$$

or

$$\mathbf{E}_1 = \frac{q}{4\pi\varepsilon_0}\frac{\mathbf{r}'(1-\beta^2)}{r'^3(1-\beta^2\sin^2\theta)^{3/2}} \tag{13.93}$$

Expression (13.93) once again illustrates that Maxwell's field equations are relativistically correct.

From (13.93) one can see that the electric field is no longer spherically symmetric. Along the line of motion ($\theta = 0$), one obtains the field as

$$\mathbf{E} = \frac{q}{4\pi\varepsilon_0}\frac{\mathbf{r}}{r^3}(1-\beta^2) \tag{13.94}$$

This shows that Coulomb field is reduced by a factor $(1-\beta^2)$. The field perpendicular to the line of motion, i.e. $\theta = \frac{\pi}{2}$, we have

$$\mathbf{E} = \frac{q}{4\pi\varepsilon_0}\frac{\mathbf{r}}{r^3}\frac{1}{\sqrt{1-\beta^2}} \tag{13.95}$$

This shows that the electric field perpendicular to the line of motion is enhanced by a factor $\dfrac{1}{\sqrt{1-\beta^2}}$.

EXAMPLE 13.1: Show that he three dimensional volume element $dxdydz$ is not invariant, but four dimensional volume element is invariant under Lorentz transformation.

Solution: According to Lorentz-Fitzgerald length contraction, one finds

$$dx' = dx\sqrt{1-(v/c)^2}, \quad dy' = dy \quad \text{and} \quad dz' = dz$$

∴ The three dimensional volume element in frame S' is equal to

$$\begin{aligned} = dx'dy'dz' &= dx\sqrt{1-\beta^2}\,dydz \\ &= dxdydz\sqrt{1-\beta^2} \\ &\neq dxdydz \end{aligned}$$

Obviously, three dimensional volume element is not invariant under Lorentz transformations. We have

$$dt' = \frac{dt}{\sqrt{1-\beta^2}} \quad \text{where} \quad \beta = \frac{v}{c}$$

∴ Four dimensional volume element in frame S' is

$$\begin{aligned} = dx'\,dy'\,dz'\,dt' &= dx\sqrt{1-\beta^2}\,dydzdt/\sqrt{1-\beta^2} \\ &= dxdydzdt \end{aligned}$$

Obviously, four dimensional volume element is invariant under Lorentz transformations.

EXAMPLE 13.2: S' frame is moving relative to frame S with velocity $\mathbf{v} = (4\hat{\imath} + 3\hat{\jmath}) \times 10^7$ m/s. What will be the electric field in S' frame when the field due to a stationary charge in S frame is $\mathbf{E} = (400\hat{\imath} + 500\hat{\jmath})$ V/m.?

Solution: We have the unit vector along velocity **v** is

$$\hat{\mathbf{v}} = \frac{\hat{\mathbf{v}}}{|\mathbf{v}|} = \frac{(4\hat{\imath}+3\hat{\jmath})\times 10^7}{5\times 10^7} = \frac{4\hat{\imath}+3\hat{\jmath}}{5}$$

Now, the component of **E** parallel to **v** is given by

$$E_{11} = \mathbf{E}\cdot\hat{\mathbf{v}} = (400\hat{\imath}+500\hat{\jmath})\cdot\frac{(4\hat{\imath}+3\hat{\jmath})}{5} = 620\ \text{V/m}$$

Representing E_{11} vectorially, we have

$$\mathbf{E}_{\|} = 620\left(\frac{4\hat{\imath}+3\hat{\jmath}}{5}\right) = (496\hat{\imath}+372\hat{\jmath})\ \text{V/m}$$

$$\therefore \quad \mathbf{E}_{\perp} = \mathbf{E} - \mathbf{E}_{\parallel} = (420\hat{i} + 500\hat{j}) - (496\hat{i} + 372\hat{j})$$
$$= (-96\hat{i} + 128\hat{j}) \text{ V/m}.$$

Now, making use of Lorentz transformation equations, one obtains the component of $\mathbf{E}'$ (field in the frame S') as

$$\mathbf{E}'_{\perp} = \frac{\mathbf{E}_{\perp}}{\sqrt{1 - \frac{v^2}{c^2}}} = \frac{(-96\hat{i} + 128\hat{j})}{\left[1 - \left|\frac{(4\hat{i}+3\hat{j})\times 10^7}{3\times 10^8}\right|^2\right]} = \frac{9 \times (-96\hat{i} + 128\hat{j})}{8.75}$$
$$= 1.028(-96\hat{i} + 128\hat{j})$$
$$= (-98.74\hat{i} + 131.6\hat{j}) \text{ V/m}.$$

Thus the electric field in S' frame will be

$$\mathbf{E}' = \mathbf{E}'_{\parallel} + \mathbf{E}'_{\perp} = (496\hat{i} + 372\hat{j}) + (-98.74\hat{i} + 131.6\hat{j})$$
$$= (397.26\hat{j} + 523.6\hat{j}) \text{ V/m}$$

QUESTIONS

1. Show that the Maxwell's electromagnetic equations are invariant under the Lorentz transformation.
2. Express Maxwell's equations in covariant form and derive the transformation laws for the electric and magnetic fields.
3. What is an inertial frame? Show that under Lorentz transformations all inertial frames are equivalent for the description of electromagnetic phenomenon.
4. Show that the charge is not altered by the motion relative an observer.
5. Prove the Lorentz covariance of the first pair of Maxwell's equations of the electromagnetic field: $\nabla \cdot \mathbf{B} = 0$ and $\nabla \times \mathbf{E} = \frac{\partial \mathbf{B}}{\partial t}$.
6. Using the tensor expression for the electromagnetic field components or otherwise find the transformation equations for the electric and magnetic fields.
7. Write down the equation for field tensor and express Maxwell's equations in 4-tensor form. Using this form show that these are invariant under Lorentz transformations.

8. Show that $C^2B^2 - E^2$ and $\mathbf{E} \cdot \mathbf{B}$ are invariant under Lorentz transformations.
9. Derive the Lorentz transformations for electromagnetic potential four vector and current density four vector.
10. Write Maxwell's equations in the presence of a point charge in motion. Show that the total charge, energy and momentum of the system are conserved.
11. Write the Maxwell's equations in a covariant form ad show that the electric and magnetic induction are elements of a second rank antisymmetric field strength tensors.

PROBLEMS

1. If two reference frames are related to each other by the Lorentz transformations

$$x' = \frac{(x - \mathrm{v}t)}{\sqrt{1-(\mathrm{v}^2/c^2)}},\ y' = y,\ z' = z \text{ and } t' = \frac{t-(\mathrm{v}x/c^2)}{\sqrt{1-(\mathrm{v}^2/c^2)}}$$

how are the electromagnetic potential (ϕ, A) and electromagnetic field strengths (**E**, **B**) in the two systems related ?

2. Let frames S' and S'' move with velocities $\mathbf{v}_1$ and $\mathbf{v}_2$ relative to a frame of reference along z-axis. Write the Lorentz transformation equations between S' and S and between S'' and S. Find the law of addition for two velocities along z-axis.

[**Ans.** $u = \dfrac{(v_1 - v_2)}{[1 - \frac{v_1 v_2}{c^2}]}$]

3. If p is momentum of particles and **A** and ϕ are vector and scalar potentials, what will be the total energy of the charged particle in an electromagnetic field?

[**Ans.** $W = \left(\left[\left(p - \dfrac{eA}{c}\right)^2 c^2 + m_0^2 c^2\right]^{1/2} + e\phi\right)$]

4. The frequency of vibration of an oscillator is ω_0. When this oscillator is placed in a magnetic field, show that the frequency will be

$$\omega = \sqrt{\omega_0^2 + \left(\frac{eB}{2\mu_0 mc}\right)^2} \pm \frac{eB}{2\mu_0 mc}.$$

Also show that for weak field $\omega \approx \omega_0 \pm \dfrac{eB}{2\mu_0 mc}$.

5. Show explicitly that two successive Lorentz transformations in the same direction are equivalent to a single Lorentz transformation with a velocity $v = \dfrac{v_1 + v_2}{\left[1 + \frac{v_1 v_2}{c^2}\right]}$.

6. Show that if the medium has a velocity $\mathbf{v} = c\boldsymbol{\beta}$ with respect to some inertial frame that the 3-vector current in the frame is

$$\mathbf{J} = \gamma\sigma[\mathbf{E} + \boldsymbol{\beta} \times \mathbf{B} - \boldsymbol{\beta}(\boldsymbol{\beta} \cdot \mathbf{E})] + \rho\mathbf{v}$$

where ρ is the charge density observed in the frame.

7. Show that for liquid flow at a speed **v** parallel or antiparallel to the path of the light the speed of light, as observed in the laboratory, is given to first order in v by

$$u = \frac{c}{n(\omega)} \pm \mathbf{v}\left(1 - \frac{1}{n^2} + \frac{\omega}{n}\frac{dn(\omega)}{d\omega}\right)$$

where ω is the frequency of the light in the laboratory (in the liquid and outside it) and $n(\omega)$ is the index of refraction of the liquid.

8. Show that the relativistic form of the 3-dimensional acceleration vector for a charged particle in an electromagnetic field is as follows:

$$\mathbf{a} = \frac{d\mathbf{u}}{dt} = \frac{e}{m\gamma}\left[\mathbf{E} + \frac{1}{c}(\mathbf{u} \times \mathbf{B}) - \frac{1}{c^2}(\mathbf{E} \cdot \mathbf{u})\mathbf{u}\right]$$

SHORT ANSWER QUESTIONS

1. Is electromagnetic theory invariant under Newtonian mechanics? **[Ans.** No]

2. Is all inertial frames are equivalent? **[Ans.** Yes]

3. What do you understand by the postulate of the constancy of the speed of light?

   [Ans. The speed of light has the same value in all inertial frames.]

4. What do you understand by Fitzgerald contraction?
   [**Ans.** $L = L_0\sqrt{1-\beta^2}$. The rod appears contracted by a factor $\sqrt{1-\beta^2}$, i.e. every body appears to be longest when at rest relatively to the observer.]

5. What is time dilatation?
   [**Ans.** $\Delta t = \dfrac{\Delta t'}{\sqrt{1-\beta^2}}$ i.e. moving clocks appear to run *slow*.]

6. A force which is purely magnetic in S frame appears to be purely electrostatic in S' frame moving relative to S frame with velocity **v** in x-direction. What this signifies?

   [**Ans.** The theory of relativity provides the unification of electric and magnetic fields, i.e. the magnetic and electric effects are just two ways of looking at the same thing. This also leads to the conclusion that whether an electromagnetic field is purely magnetic or purely electric or electromagnetic depends on the inertial reference frame in which sources are observed.]

7. What the transformation equations of E signifies?

   [**Ans.** The form of transformation equations of **E** in general asserts that the component of E parallel to the relative velocity of the two frames is unchanged, whereas the components of E perpendicular to the relativity velocity transform to a mixed electric and magnetic fields and depend on the magnitude and direction of the relative velocity.]

8. What is Lorentz ionization?

   [**Ans.** The microscopic object of atomic size can be easily accelerated to a velocity of the order of 5×10^6 m/s. Such an atom moving in a magnetic field of 1 tesla in the S' frame (at rest with respect to the atom) experiences an electric field $\sim 5 \times 10^6$ V/m. The neutral atoms with captured electrons moving at this speed are most likely to be in the excited state. Such atoms subjected to electric fields of 5×10^6 V/m are quickly ionized and form an ion beam. This is termed as Lorentz ionization. This process seems quite feasible on the laboratory scale and can provide a possible means of injecting ions into fusion reactor plasma.]

9. What is the difference between a four vector and an ordinary vector?

   [**Ans.** A four vector differs from an ordinary vector in one important respect. The length of a four vector can be zero and yet the four vector can have non-zero components. The length of a photon's four-momentum, i.e. its mass is zero, and yet the photon can have momentum and energy. Thus a four vector can be called as time like, or null, depending on whether the square of its magnitude is negative, positive or zero respectively.]

10. What is eigen time or proper time?

    [**Ans.** This is the time measured in inertial frame in which the body is instantaneously at rest.]

OBJECTIVE QUESTIONS

1. The net charge density (λ) in frame S is zero. In frame S' moving with a velocity **v** relative to S along the x-axis, the charge density (λ') will be (n number of electrons per unit length of the conducting wire)

   (a) $\lambda' = \text{zero}$, (b) $\lambda' = ne\beta^2$,

   (c) $\lambda' = \dfrac{ne\beta^2}{\sqrt{1-\beta^2}}$, (d) $\lambda' = \dfrac{ne\beta^2}{2}$ [c]

2. In problem 1 the electrostatic field around the conducting wire according to the observer in S' is

   (a) $E' = 0$, (b) $E' = \dfrac{ne\beta^2}{2\pi\varepsilon_0 r}$,

   (c) $E' = \dfrac{ne\beta^2}{2\pi\varepsilon_0 r\sqrt{1-\beta^2}}$, (d) $E' = \dfrac{ne\beta^2}{4\pi\varepsilon_0 r}$ [c]

3. The square of the 'length' of the 'four vector' S is

   (a) $x^2+y^2+z^2-c^2t^2$, (b) $x^2+y^2+z^2$,

   (c) $(x^2+y^2+z^2-c^2t^2)^{1/2}$, (d) zero. [a]

4. The d'Alembertian operator ($\square^2$) is

(a) $\frac{\partial^2}{\partial x^2}+\frac{\partial^2}{\partial y^2}+\frac{\partial^2}{\partial z^2}$, (b) $\frac{\partial^2}{\partial x^2}+\frac{\partial^2}{\partial y^2}+\frac{\partial^2}{\partial z^2}-\frac{1}{c^2}\frac{\partial^2}{\partial t^2}$,

(c) $\frac{\partial}{\partial x}+\frac{\partial}{\partial y}+\frac{\partial}{\partial z}$, (d) $\left(\frac{\partial}{\partial x}+\frac{\partial}{\partial y}+\frac{\partial}{\partial z}-\frac{1}{c}\frac{\partial}{\partial t}\right)^2$ [b]

5. Electromagnetic field tensor is a tensor of rank

(a) Four, (b) Three, (c) Two, (d) Six [c]

6. When four-vector scalar product is written as $S^2 = x_\mu x_\mu = x_1^2 + x_2^2 + x_3^2 + x_4^2$, then the line element is called a tensor of rank

(a) four, (b) two, (c) zero, (d) three [c]

7. In terms of four-vectors, the continuity equation is written as

(a) $\frac{\partial J_\mu}{\partial x_\mu} = 0$, (b) $\frac{\partial J_\mu}{\partial x_\mu} = \rho$,

(c) $\frac{\partial J_\mu}{\partial x_\mu} = \mathbf{E}$, (d) $\frac{\partial J_\mu}{\partial x_\mu} = \frac{1}{2}$ [a]